BIOSYNTHESIS OF NATURAL PRODUCTS

TO ERSILIA AND OUR CHILDREN

BIOSYNTHESIS OF NATURAL PRODUCTS

PAOLO MANITTO
Professor of Organic Chemistry
University of Milan

Translation Editor:
P. G. SAMMES
Professor of Organic Chemistry
University of Leeds

ELLIS HORWOOD LIMITED
Publishers · Chichester

Halsted Press: a division of
JOHN WILEY & SONS
New York · Chichester · Brisbane · Toronto

First published in 1981 by
ELLIS HORWOOD LIMITED
Market Cross House, Cooper Street, Chichester, West Sussex, PO19 1EB, England

The publisher's colophon is reproduced from James Gillison's drawing of the ancient Market Cross, Chichester.

Distributors:

Australia, New Zealand, South-east Asia:
Jacaranda-Wiley Ltd., Jacaranda Press,
JOHN WILEY & SONS INC.,
G.P.O. Box 859, Brisbane, Queensland 40001, Australia

Canada:
JOHN WILEY & SONS CANADA LIMITED
22 Worcester Road, Rexdale, Ontario, Canada.

Europe, Africa:
JOHN WILEY & SONS LIMITED
Baffins Lane, Chichester, West Sussex, England.

North and South America and the rest of the world:
Halsted Press: a division of
JOHN WILEY & SONS
605 Third Avenue, New York, N.Y. 10016, U.S.A.

© 1981 P. Manitto/Ellis Horwood Ltd., Publishers
British Library Cataloguing in Publication Data
Manitto, Paolo
  Biosynthesis of natural products.
  1. Biosynthesis
  2. Natural products
  I. Title
  574.1'929   HQ345   80–41739
ISBN 0–85312–062–5 (Ellis Horwood Limited, Publishers)
ISBN 0–470–27100–0 (Halsted Press)

Typeset in Press Roman by Ellis Horwood Limited.
Printed in the U.S.A. by Eastern Graphics Inc., Old Saybrook, Connecticut

Table of Contents

Author's Preface

Natural products have always attracted chemists and biologists. Many of them challenge the organic chemist's analytical, synthetic and speculative capacities. The structures of terpenes, alkaloids, polyketides, plant pigments, etc., are extremely varied and often very complex. They show a wide reactivity range and sometimes physiological properties. The organic chemist considers these secondary metabolites as the *par excellence* natural substances.

Since the beginning of this century research chemists have increasingly been pointing out the peculiarity of natural organic substances, that is their formation by living organisms. This 'biosynthesis' can be regarded as their 'fourth dimension'!

The technological development of the last two decades has made the detection of isotopic nuclides (radioactive or not) easier and easier; in addition, isotopically labelled compounds are increasingly more available. As a consequence, biosynthetic studies based on tracers have expanded enormously. The large amount of experimental data so far collected makes it possible to form a picture of the synthetic capabilities of plant and animal organisms. Such a picture, although incomplete and with many speculative points, clearly indicates a biochemical evolution. The secondary metabolites, which reflect the differentiation of the species at a chemical level, appear to originate from general biosynthetic schemes common to all or nearly all species populating our planet.

In my opinion a university course in Chemistry of Natural Products should largely deal with biosynthesis. In fact, most *in vivo* chemical reactions have extremely interesting mechanisms and stereochemistry. The student should be able to recognise the origin of natural compounds from their structures: he should therefore know the fundamental biogenetic principles, the reaction mechanisms most frequently involved in enzyme-catalysed processes, the key roles played by coenzymes, and the most important pathways to secondary metabolites.

It has been my intention to write a textbook covering basic biosynthetic topics and to be used by: (a) chemistry students interested in biological-organic trends, (b) graduate chemists wishing to widen their vision of current biosynthetic

Author's Preface

problems and methodologies, and (c) biologists and pharmacologists, with a limited knowledge of organic chemistry, yet interested in chemical aspects of secondary metabolism.

Thus, from my teaching experience and taking into account the different background knowledge of potential readers, I have emphasised some biological concepts in Chapters 1–3 for chemists lacking the essential rudiments of biology and biochemistry, and I have included Appendices 1–4 containing basic notions of reaction mechanisms and stereochemistry to assist biologists.

The text is largely based on my lecture notes used at Milan University since 1970 in the course of Chemistry of Organic Natural Substances for chemistry and biology students. Many paragraphs could certainly have been more detailed, but obvious reasons of brevity have limited them. For the same reasons alkaloid biosynthesis has been omitted. The references given at the end of each Chapter should offer useful sources for the reader wishing to go beyond this book. In order to achieve this aim, the most important texts, reviews and articles of the last fifteen years have been chosen.

I wish to thank Dr. Guido Serra-Errante, who translated the Italian manuscript; Prof. P. G. Sammes, who critically read the English version; Prof. A. Guerritore for his useful biological suggestions; Prof. L. Canonica for stimulating my interest in the field of natural product biosynthesis; and my wife, who helped me in preparing this book with her patience and constant encouragement.

Milan,
December, 1980   Paolo Manitto

Chapter 1

Primary and Secondary Metabolism

1.1 INTRODUCTION

Even before the end of the Eighteenth Century, organic materials had been isolated from living organisms and their products. The German chemist, Karl Wilhelm Scheele (1742-1786) was particularly distinguished in this art: he was able to extract some simple compounds, including glycerol and oxalic, lactic, tartaric, and citric acids from various organic sources, both vegetable and animal. Friederich W. Sertürner (1783-1841) obtained morphine from opium in 1806 and Pelletier and Caventou isolated strychnine, brucine, quinine, cinchonine, and caffeine in the next fifteen years.

Because of their chemical complexity and important physiological properties, these alkaloids should be considered as the first of the typical, natural organic substances isolated by man as pure compounds. The isolation of many thousands of other compounds from natural sources rapidly followed and continues to this day.

Organic chemistry developed in parallel with the isolation and study of natural substances. Because of their large structural variety, the molecules produced by living organisms are valuable to the organic chemist: he can work on them to widen and deepen his knowledge of organic reactions and, in particular, they can be used to verify certain hypotheses, such as on mechanisms. Some established examples are, steroids − conformational analysis; cyclic terpenes − molecular rearrangements; tropolones − ring chemistry; pigments − electronic absorption spectroscopy, etc.

Besides giving useful substrates for mechanistic and stereochemical studies, living organisms also stimulate the organic chemist in two other ways. On one hand is the challenge to elucidate certain extraordinarily complicated structures (such as that of vitamin B_{12}) and to synthesize them *in vitro*; on the other hand is the unravelling of the secrets of enzymatic reactions (biocatalysis) and the 'building up' processes by which the cells make organic molecules (biosynthesis or biogenesis)†. The information available from such investigations, together with various theories and hypotheses, form a well-defined and largely autonomous branch of chemistry called the biogenesis of natural products.

The biogenesis of natural products, although connected with traditional organic chemistry and biochemistry, differs from both by its different aims. Organic chemistry mainly deals with structural studies, the physical and chemical properties of compounds, and the synthesis of compounds, natural or otherwise, using *in vitro* methods, but tends to ignore the characteristic peculiarity of natural compounds, that is to say, their mode of formation and their biological role. Biochemistry, on the other hand, tries to answer the most general questions about life on our planet, preferentially dealing with primary metabolism (see later), and neglecting secondary processes such as the formation of alkaloids, terpenes, etc.

In Table 1.1 some fundamental dates in the history of natural products are listed. It illustrates how the greatest progress in biogenetic research has taken place in the last twenty five years. This is principally due to the more extensive use of compounds labelled with either stable or radioactive isotopes. Since 1950 substances containing ^2H, ^3H, ^{13}C, ^{14}C, ^{15}N, and ^{18}O have been commercially available and sophisticated instruments for detecting such isotopes have been developed: examples of this are scintillation counters, gas chromatographs coupled to mass spectrometers, nuclear magnetic resonance (n.m.r.) spectrometers with Fourier transform facilities, and so on.

1.2 PRIMARY AND SECONDARY METABOLISM

Polysaccharides, proteins, fats, and nucleic acids are the fundamental building blocks of living matter and are thus considered to be **primary metabolites**. The whole range of processes by which organisms synthesise and demolish these substances, in order to survive, constitute the *primary metabolism processes.* The primary metabolism of all organisms, even those which are genetically very distant, is similar.

Other chemical processes take place only in certain species or else give different products according to the type of species. Such reactions do not appear to be essential for the existence of the organism and hence are called *secondary metabolic processes*. Products from secondary metabolism tend to coincide with the traditional natural products of the organic chemist, such as terpenes, alkaloids, pigments. Although not essential for the existence of the

---

† The terms **biosynthesis** and **biogenesis** both mean the formation of natural substances by living organisms. Although they are often used synonymously, there is a slight difference in their meanings; the first one refers to the acquisition of experimental data, whilst the second one emphasises the speculative aspect of certain facts. Current practice further restricts the word biosynthesis to the elaboration of natural molecules from *less* complex structures by endoergonic reactions. Such reactions are typical of anabolic processes in secondary metabolism.

individual, secondary metabolites often play a key role in the survival of the species over others. Included here are the defence chemicals, sex attractants, and the pheromones. The reason why many secondary products are produced remains a mystery. Some authors believe they are detoxification products of poisonous or overabundant metabolites, which cannot be otherwise eliminated by the organism. This explanation would account for the greater production of secondary metabolites in plants rather than in animals. Whilst animals have sophisticated processes for removing secondary products through their liver and kidneys, plants are obliged to transform their unwanted materials into compounds which can be stored in the cellular vacuoles, in cell walls, etc. Some scientists consider such metabolites to be a store of energy and food in plants, which can be utilised in times of necessity. Whatever the purpose of secondary metabolites, a lot is known about their chemistry and how they are formed.

Table 1.1 Some important dates in the study of natural products

Isolation of some important natural products

1806	morphine	Sertürner
1818	strychnine	Pelletier and Caventou
1845	camphor	Bouchardat
1859	cocaine	Niemann
1864	bilirubin	Städeler
1909	cholesterol	Windaus
1934	progesterone	Butenandt
1938	penicillin	Florey and Chain
1948	aureomycin	Lederle Laboratories

Structure elucidation and synthesis

1932	cholesterol	Wieland
1946	strychnine	Robinson
1954	strychnine synthesis	Woodward
1955	vitamin B_{12}	Hodgkin
1959	colchicine synthesis	Eschenmoser
1960	chlorophyll synthesis	Woodward
1968	β-amyrin synthesis	Barton
1971	erythronolide B	Corey

Biogenetic researches

1953	acetate hypothesis	Birch
1955	shikimic acid pathway	Davis
1956	mevalonic acid discovered	Merck Laboratories
1959	isoprene rule	Ruzicka
1964	oxidative coupling	Barton

Intermediate metabolism usually indicates certain reactions which allow exchange of material between different metabolic pools and producing energy required by both the single cells and the whole organism. Intermediate metabolism corresponds, approximately, to the central area of Fig. 1.1 and usually involves small molecules (acetic acid, etc). These molecules are partly burnt to CO_2 and H_2O with production of energy. Because of its vital importance the intermediate metabolism is generally discussed with primary metabolism.

Chapters 4, 5, 6, 7, and 8 deal with chemical aspects of secondary metabolism: Chapter 2 surveys enzymic aspects of metabolism, whilst Chapter 3 describes some of the more important primary metabolic processes. A good knowledge of primary metabolism is essential for an understanding of the biogenesis of natural products, since a close connection exists between the various chemical processes occurring inside a cell. In fact, the distinction between primary and secondary metabolism is merely artificial; the concept is a useful simplification of the complex phenomena which occur *in vivo* (Fig. 1.1).

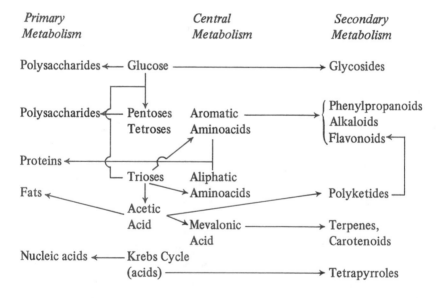

Fig. 1.1 Relationships of primary and secondary metabolism.

1.3 THE BIOGENETIC PATH

The *biogenetic path* is the sequence of reactions through which an organism synthesises a substance X starting with a molecule A. The products B, C, D . . . , into which A is transformed during this progression towards X are called *natural intermediates* and usually increase in chemical complexity during the biosynthetic process (Fig. 1.2(a)). Substance X can, in turn, be changed to another compound, Y, thus becoming an intermediate of the latter product.

A *natural product* is not always the result of one unique biogenetic path. It can well occur at the end of several pathways and, under these circumstance, one of the convergent paths usually predominates over others (Fig. 1.2(b)). The natural intermediates of a unique or overwhelmingly predominant pathway are called *obligatory*; blocking the synthesis of such intermediates causes an almost total interruption of formation of the final compound.

A *biogenetic pathway* can be ramified at one or several points (Fig. 1.2(c)). The combination of convergent and ramified paths leads to a *metabolic grid* (Fig. 1.2(d)). Quite frequently, a substance is formed and used at about the same rate and thus its concentration in the organism is virtually constant. A stationary state then exists (an equilibrium between anabolic and catabolic phases): the amount of substance either formed or transformed per unit of time is called the *turnover rate*.

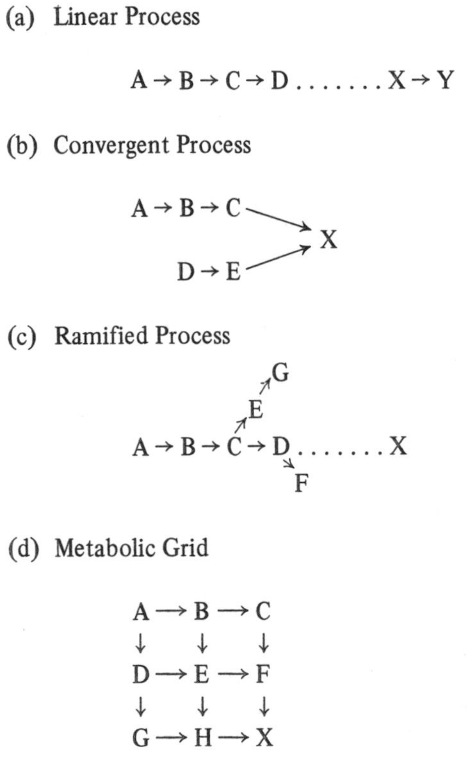

(a) Linear Process

$$A \to B \to C \to D \ldots \ldots X \to Y$$

(b) Convergent Process

$$A \to B \to C \searrow$$
$$ X$$
$$D \to E \nearrow$$

(c) Ramified Process

$$ G$$
$$ E$$
$$A \to B \to C \to D \ldots \ldots X$$
$$ F$$

(d) Metabolic Grid

$$A \longrightarrow B \longrightarrow C$$
$$\downarrow \quad \downarrow \quad \downarrow$$
$$D \longrightarrow E \longrightarrow F$$
$$\downarrow \quad \downarrow \quad \downarrow$$
$$G \longrightarrow H \longrightarrow X$$

Fig. 1.2 Various types of metabolic pathways.

1.4 PLANT AND ANIMAL ORGANISMS

The chemist studying natural products depends heavily on the type of living organism chosen for his biogenetic research. It is important in such studies to understand the essential biological and taxonomic features of the selected organism. It is also important to collaborate with experts from other scientific disciplines such as biology, pharmacology, microbiology and so on, in order to maximise the value of such studies.

Living organisms are traditionally divided into the two large *animal* and *plant* kingdoms. The universal unit of organisation is the *cell*. On the basis of fundamental structural, cellular features living organisms are divided into:

(i) *Procaryotes,* in which the genetic material (DNA) is *not* enclosed in a well-defined nucleus;

(ii) *Eucaryotes,* which contain a membrane-surrounded nucleus and various cytoplasmic organelles, each delineated by membranes (mitochondria, chloroplasts, etc).

Despite their ability to reproduce, *viruses* are not included by most biologists amongst living organisms. These nucleoprotein particles are devoid of any metabolic processes of their own and can only multiply within living cells.

Animals are typical encaryotic organisms and they utilise the oxidation of organic substances as the only source of chemical and physical energy. The animal kingdom has ten major *phyla*: Protozoa (protozoans), Porifera (sponges), Coelenterata (coelenterates), Aschelminthes (aschelminthes), Platyhelminths (flatworms), Echinodermata (echinoderms), Mollusca (molluscs), Anellida (anellid worms), Arthropoda (arthropods), and Chordata (chordates). The plant kingdom, which includes all other forms of life is more heterogeneous and is divided into *divisions.* The four main groups in the plant kingdom are:

1. **The Schizophytes** (procaryotes), which are further subdivided into the Cyanophytes (blue algae) and Bacteriophytes (bacteria).
2. **The Algae** (other than the Cyanophytes). These are eucaryotes and are divided into the Phytoflagellates, Chlorophytes (green algae), Phaeophytes (brown algae), and Rhodophytes (red algae).
3. **Fungi and Lichens** (eucaryotes).
4. **Euphytes** (eucaryotes), which includes all the common green plants.

These four divisions need to be discussed in more detail, as follows.

Schizophytes (from the Greek words $\Sigma\chi\iota\zeta\omega$ = ro divide, and $\phi\upsilon\tau o\nu$ = plant) are inferior vegetables, usually unicellular, which reproduces by cell division. In some cases the formation of *spores* or *cysts* is noted. These are durable cells with a thick membrane and dense protoplasm. The pigmentation of the Schizophytes is closely connected with their metabolic activity. Colourless organisms, and the nearly yellow and red pigmented cells, for example bacteria, are *etherotrophic* (they use organic matter for energy and food). They can be classified as follows:

1. *Symbiotes*, when they exist with other organisms to which they usually bring obvious advantages (e.g. *Rhizobium leguminosarum*, which fixes atmospheric nitrogen, and helps the formation of root tubules in the Leguminosae).
2. *Parasites* (all pathogenic bacteria).
3. *Commensal organisms*, when they live in live and healthy organisms without causing any advantages or disadvantages (e.g. *Escherichia coli* in the human intestine).
4. *Saprophites*, when they live and reproduce in dead organic tissues.

Autotrophic organisms are those which have the capacity to live independently from other organisms (alive or dead) and without using any organic substances derived from the latter. Such organisms are generally green, blue-green or blue-grey in colour due to photosynthetic pigments. The pigments (chlorophyll and phycoxanthins) act as *photoreceptors* and aid both the conversion of light energy into chemical energy and the assimilation of carbon dioxide. The Cyanophytes fall into this category.

The **Cyanophytes** represent the most ancient group of organisms still in existence on our planet: traces of them have been found in Precambrian strata, of more than five hundred million years age. Of the Bacteriophytes, the Streptomycetaceae family (order Actinomycetales) are very important. These microorganisms, are very similar to moulds in appearance, and produce secondary metabolites (such as antibiotics and vitamins) of great importance to man.

The **Algae, Fungi, and Lichens** are considered to be members of the somewhat artifical sub-kingdom of Thallophytes. Although this group varies widely in dimension and shape and can be uni- or polycellular, they do not show a well-defined structure having roots, stems and leaves. The separation of the Thallophytes (eucaryotes) from the Schizophytes (procaryotes) is justified both by differences in cellular structure and by the presence of sexual reproduction phases in the former and by their absence from the latter group.

The Algae are *autotrophic* and they all contain photosynthetic pigments. These pigments can vary with the type of algae. Brown algae are particularly rich in xanthophylls, i.e. cartenoid substances which mask the green colour of chlorophyll; the red algae contain chromoproteins, e.g. phycoerythrin, which is red.

The Fungi (or *mycetes*: μυκησ = fungus) are typically *etherotropic* organisms. Some authors treat the fungi as a separate kingdom from those of the animal and plant kingdoms. The appearance of the vegetative body of the fungi, i.e. the thallus, varies widely. In some cases it is formed by a *mycelium*, which is a body of intertwined microscopic filaments called *hyphae*. The hyphae are characterised by a cellular wall which encloses the cytoplasm and many nuclei. They are sometimes divided by a porous, transverse *septa*, through which the cytoplasm and, in some cases, nuclei can circulate.

Mycelium development occurs during the aggressive phase of fungal growth, when the hyphae proliferate. Nutritive material is absorbed by the whole mycelium surface and this facilitates the efficient growth of fungi in synthetic media in which all essential nutrients are added. The controlled growth of fungi in an artificial ambience also allows the isolation of metabolites expelled into this surrounding media.

Two types of mycelium can be distinguished, the vegatative one, which proliferates within the supporting media and which assimilates the nutritative material; and the aerial part, which specialises in producing spores. The spores mature when ambient conditions for the production of the mycelium become unfavourable and they are dispersed by carrying agents such as the wind and water. Fungi are classified according to the type of spores which are produced and to the nature of the mycelium. There are four classes, the Phycomycetes, Ascomycetes, Basidiomycetes, and the Fungi Imperfecti.

The most primitive class are the Phycomycetes (i.e. algae-fungi). The lower species are predominantly aquatic and unicellular; the mycelium is aseptate and reproduction can either be asexual or sexual. The Phycomycetes include some plant parasites, e.g. potato blight, *Phytophthora infestans* and the mildews, as well as some moulds, as seen on jam, bread, fruit, etc.

The Ascomycetes take their name from the *ascus*, an organ which is sometimes quite large and club-shaped and which contains the sexual spores (ascospores). The asci can be bunched in containers called *ascocarpi*. Besides sexual reproduction asexual spores (conidia) can form, produced by differentiation of the aerial hyphae into *conidiophores*. The mycelium is septate. Ascomycetes include very simple organisms such as the yeasts (Saccharomycetales), which usually reproduce by 'budding', and much more complex organisms, such as truffles. Many of the Ascomycetes are also plant parasites, such as *Claviceps purpurea*, which is responsible for the disease in man called ergotism, caused by eating infected rye. Formerly, this disease was called 'St. Anthony's fire' or 'the devil's curse'.

The Basidiomycetes are the most highly evolved class of fungi. The sexual spores (basidiospores) are carried on the external surface of club-shaped or cylindrical cells called *basidia*. The latter cells are associated in fruiting-bodies (basidiocarpi) characterised by the well-known shape of the mushrooms with their stem and hat. The mycelium is often perennial and develops either on the ground or on tree trunks. Most edible mushrooms are Basidiomycetes (e.g. *Boletus edulis*), and they are often symbiotic with trees and bushes. Some Basidiomycetes are also injurious to a wide variety of plants, developing rusts and smuts.

The Fungi Imperfecti do not have phases of sexual reproduction, as those described above do, and produce conidiophores. These have the characteristic appearance of moulds and include the important *Penicillium* and *Aspergillus* species. Thorough examination of certain species of the imperfect fungi has revealed them as conidic forms of Ascomycetes. For example, *Fusarium moniliforme* is the conidic form of *Gibberella fujikuroi*.

Lichens are symbiotic associations of an Ascomycete or Basidiomycete fungus with a photosynthetic, unicellular alga, either a Cyanophyte or a Chlorophyte. The fungal component (mycobiont) utilises the organic substances made by the alga (phycobiont) autotrophically. In some cases it has been possible to separate the fungal and algal components of some lichens in the laboratory and, consequently, it has been discovered that most of the secondary metabolites arise from the mycobiont.

The Euphytes (or sub-kingdom Embryophytes) consist of either plants in which a true *corm*, with roots, stalk, and leaves (Cormophytes) can be recognised, or plants in which no true corm is apparent but only a differentiation of tissues which preclude the existence of a corm (Protocormophytes, or Bryophytes). This second situation is common in the mosses and liverworts. Cormophytes include the Pteridophytes, Gymnosperms, and Angiosperms — this latter group including the mono- and dicotyledon forms. The Pteridophytes can be considered as a *trait d'union* between the Bryophytes and higher Cormophytes. As Bryophytes they lack a floral apparatus and reproduce by spores; whilst, as higher Cormophytes, they are vascular plants with a grassy, bushy, or arboreal stalk. In the Devonian and Carboniferous eras (200–300 million years ago) the Pteridophytes developed important tree-like forms, due to the simultaneous evolution of an important new metabolic process, the biosynthesis of lignin (see p. 388), which dramatically altered the evolution of plant forms.

The Pteridophytes currently in existence form the three large classes of the Lycopodinae, Equisetinae, Filicinae (e.g. club mosses, horsetails, and ferns respectively).

The Gymnosperms have floral organs and produce seeds for reproduction and propagation. The flowers of the Gymnosperms (e.g. pines, firs, and cypresses) are extremely simple and non-showy: the naked ovules are inserted in fruit-bearing leaves which remain open. By contrast the Angiosperms are characterised by the presence of an ovary, enclosing the ovules, which develop into fruit. The classical distinction between monocotyledon Angiosperms, such as the tulip, lily, and orchid, and the dicotyledon types, like the anemones, poppies, and daisies, is due to the presence of either one or two embryonic leaves, called cotyledons and is also justified by other morphological differences, such as the shape of their leaves and stems, etc. The phylogenetic relationships between various plants are schematised in Fig. 1.3.

The classification of living organisms, or *taxonomy* is a very complicated branch of biology, which is continuously being updated, and causes many disagreements amongst specialists. In this short account it is sufficient to remember that the two fundamental systematic categories by which an individual is classified are the *species*‡ and the *genus* (to which the species belongs). A binary nomen-

‡ The species is defined as 'a group of interrelated individuals, separate from other, similar groups by the fact that they can reproduce only within themselves'.

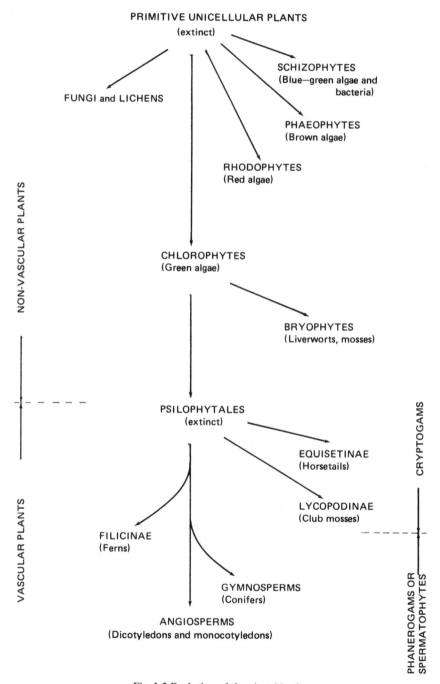

Fig. 1.3 Evolution of the plant kingdom.

clature for both the plant and animal kingdoms results, with the names of the genus and species written in italics, (the genus starting with a capital letter). For instance, the common basil is called *Ocimum basilicum* Linn. (Labiatae). This means that this plant is of the *basilicum* species, of the *Ocimum* genus. Linn. is a short form of C. Linnaeus (1707–1778), who was the botanist who first described that species. The *family* name, in which the genus is included, often follows and is written in brackets, in this case Labiatae. The taxa, which are superior to the family (name ending in -ae) are respectively the *order* (-ae or ales in the plant kingdom), the *sub-class,* the *class,* the *sub-division* and the *division* (-a; -phyta or -mycota in the plant kingdom). Labiatae, according to the scheme of A. Engler, belong to the Tubiflorae order, Sympetalae sub-class, Dicotyledoneae class, Angiosperm sub-division, Spermatophyta division. Appendix VI lists all the families of the Spermatophyta division classified by orders.

1.5 THE CELL

All plant and animal activities – assimilation, biosynthesis, secretion, excretion, movement, and reproduction – result from the coordinated activities of the various cells, each of which can be considered as a 'physiological unit'. In Fig. 1.4 two typical cells – a plant and an animal one – are schematized as they might appear from electron micrographs. Both show a continuous external envelope, the *plasma membrane* (or *plasmalemma*), which separates them from their surroundings. In the plant cell the plasma membrane is often reinforced by a thick cellulose wall.

The internal cell structure contains a *nucleus,* which itself contains one or more *nucleoli,* and the *cytoplasm,* which is the part of the cell between the plasma membrane and the nuclear membrane. The cytoplasm is where the majority of primary and secondary metabolic processes take place: it appears as a non-homogeneous fluid, or matrix, in which various cytoplasmic organelles (mitochondria, chloroplasts, etc.) are suspended. The fluid, called *cytosol* is crossed by a network of inner membranes known as the *endoplasmic reticulum.* Each organelle is separated from the cytosol by a membrane very similar to the plasmatic one. An analogous membrane defines the various vacuoles within the cell: they appear to be particularly numerous and of large dimensions in plant cells; they contain fluid which has a very different composition from that of the cytosol.

A knowledge of cellular membrane properties is quite fundamental to the study of biosynthetic processes; in most cases the metabolism of substances given to an organism takes place when they make contact with intracellular enzymes. For instance, the successful incorporation of labelled precursors in a biosynthetic experiment (see Section 1.6) largely depends on whether the plasma membrane is permeable to such a precursor.

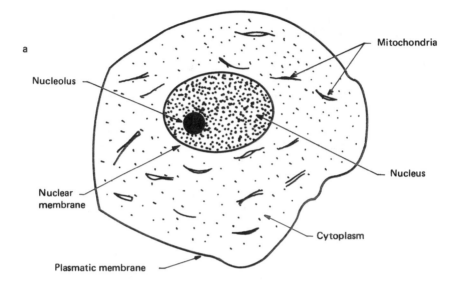

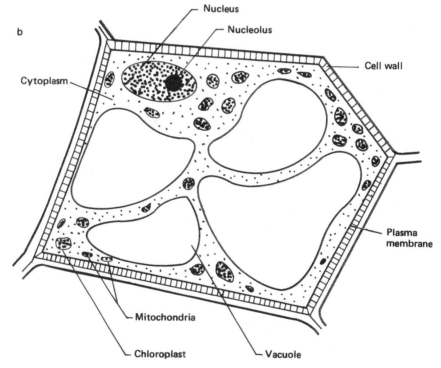

Fig. 1.4 The essential features of typical plant and animal cells. After *Enciclopedia Italiana delle Scienze. Invertebrati,* 1, De Agostini, Novara, 1972, p.3.

The 'ultrastructure' of a typical plasma membrane, deduced from its chemical composition and from investigations by electron microscopy, is schematised in Fig. 1.5. The membrane is 7.5–8.0 nm thick and is generally composed to two thin protein films, one facing the internal part of the cell, and the other arranged towards the external medium. Between these two protein phases there is a double layer of phospholipid molecules, oriented in such a way that their hydrophilic heads protrude towards the protein films, whilst the hydrophobic tails tend to align themselves in order to minimise their contact with water.

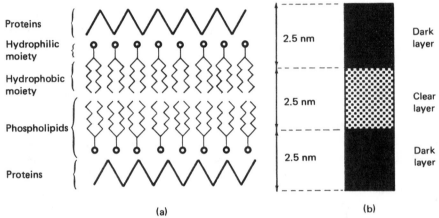

(a)   (b)

Fig. 1.5 Ultrastructure of a plasmotic membrane, (a) schematic representation; (b) appearance under the electron microscope. (After A. Berkaloff *et al*. 'Biologie et Physiologie Cellulaire', Hermann, Paris, 1967).

The lipids are mostly phopholipids but cholesterol is also involved. Besides peripheral proteins, loosely associated with the membrane and easily solubilised, there are other proteins, called *intrinsic* or *internal* proteins, which are embedded in the lipid bilayer. The membrane proteins play important biological roles such as receptors, transport carriers, enzymes, etc. After staining with osmium tetroxide, the membrane appears to be formed of three layers: two opaque ones at the margins, corresponding to the proteic regions, and to the hydrophilic parts of the lipid section, to which osmium tetroxide is strongly bound, and a transparent internal one, corresponding to the hydrophobic portions of the lipid molecules.

The ultrastructure of the other cell membranes — those, for example, of the endoplasmic reticulum of the Golgi apparatus, of the mitochondria and of the chloroplasts — is almost identical to the plasma membrane. For this reason we normally refer to these as a 'unitary' membrane. Very complex transfer mechanisms can operate during migration of substances between the cytosol and the extracellular medium, and between the cytosol and the cellular organelles. Reference should be made to more specialised books on this subject (see listing at the end of this chapter).

Usually the membranes show a high permeability towards lipid soluble compounds, and are rapidly destroyed by solvents of fatty materials. This is in agreement with the largely lipid nature of the membranes. The higher penetration into cells of ethers, ketones, esters, etc., in contrast to other more water-soluble compounds, arises from the tendency of the former substances to dissolve in the lipid phases of the plasma membrane. Nevertheless it has been found that some highly water-soluble substances, such as methanol, ethylene glycol and water itself, also penetrate easily into the cytosol. Because of the small molecular dimensions of these substances, they are thought to pass freely through 'porous' membrane zones, constituted of channels of fibrous proteins. The membrane is thus not a homogenous structure, but is rather a mosaic of zones, which differ in their compositions and functions. There are, in particular, some regions with special enzymic systems, called permeases, specialised in carrying hydrosoluble compounds with dimensions far larger than those of the aforementioned porous zones. It is probable that glucose passes through the membranes by binding to certain membrane components, which act as carriers.

In the cytosol there are many soluble enzymes, which are responsible for many primary and secondary metabolic reactions. Some other processes, such as photosynthesis, take place instead in the organelles immersed in the cytosol and which are specialised in performing fundamental processes. It is pertinent here to give a brief description of those organelles which are especially important in metabolic processes connected with the biosynthesis of natural substances. Examination, by electron microscopy, of the cytoplasm of cells, e.g. hepatocytes, reveals a whole series of intercommunicating sacs, vesicles and tubules, appearing as spongy material. Zones can be distinguished in which some sacs intermingle in a disordered way, whilst in others the sacs are regularly arranged in piles.

The first structures constitute the *endoplasmic reticulum*, whilst the second are called the *Golgi apparatus*. In both cases the cavities are separated from the cytosol by the usual three-layered type of cellular membrane. The endoplasmic reticulum cavities can be either flattened, with a thickness of 100–500 nm., or much wider, as in plant cell vacuoles. The reticulum forms a shell around the nucleus (nuclear membrane), and there are some rather regularly distributed pores (*ca.* 50 nm in diameter) which allow communication between the nucleoplasm and the cytosol. On the wall of the reticular membrane facing towards the cytosol one often finds granules, of about 15 nm diameter, which are opaque after fixing with osmium tetroxide. They are composed of RNA and proteins, and are called *ribosomes*. They can exist as isolated units on the reticulum (or even separate units in the cytosol) or, more frequently, they can be disposed in series, when they are tied to each other by a very fine filament. Such aggregates, called *polysomes*, synthesise the proteins and the enzymes, using the codes of the messenger – RNAs produced by the nucleus on the basis of information contained in the cellular DNA. The part of the endoplasmic reticulum which carries the ribosomes is called the *ergastoplasm* and the smooth part the *agranular reticulum*.

On homogenising cells the reticulum is fragmented, and the membrane fragments form closed vesicle structures, to the outside of which ribosomes can be attached. These vesicles, which can be isolated after centrifugation, are called *microsomes,* and can be further divided, after treating with sodium desoxycholate, into the ribosomal fraction and another of the pure membrane fragments. The content of the sacs, as well as the complete physiological roles of the endoplasmic reticulum, are yet to be explored.

The vacuoles of plant cells, often of remarkable dimensions, are filled with water, in which inorganic and organic salts are dissolved (sometimes crystalline precipitates result after saturation), as well as glycosides of flavonic and other organic pigments.

Many enzymic activities are localised on the membrane of the endoplasmic reticulum besides protein synthesis, e.g. cholesterol is converted into the various sterols, particularly into the steroidal hormones. Steroid syntheses (see Chapter 6) are particularly developed in the agranular reticulum of some specialised vertebrate cells, (interstitial cells of the testicle, corpus luteum cells, adrenal cortex cells). The cholesterol which is so utilised probably originates from the lipid layer of the reticulum membrane, and it is thus continuously renewed. Certain substances, like proteins, enzymes, and steroids, are concentrated in the endoplasmic reticulum, either being produced by the reticulum itself, or synthesised elsewhere. Such substances can also be stored as reserve materials, as for instance in aleuron granules in the vegetable cells of the developing seed.

The Golgi apparatus is formed by piles of very flattened sacs (10-20 nm thick, and with a 1000-3000 nm diameter), dispersed in the cytosol. The Golgi body of the plant cells often has a netted appearance and is therefore called a *dictyosome.* It is sided by vesicles formed by separation of the same peripheral zones of the sacs. The relationship between the endoplasmic reticulum and the Golgi apparatus has emerged from experiments with guinea pig pancreas cells, to which a labelled amino-acid (tritiated leucine) had been given. The radioactivity was initially detected in the endoplasmic reticulum where it was incorporated into enzymes. It was then passed into the Golgi apparatus where a concentration was effected followed by its incorporation into the vesicles. Finally, the labelled enzymes were carried out of the cell into ducts, whence they could be transported to carry out their digestive function.

Besides concentrating proteins, the Golgi bodies (dictyosomes) synthesise polysaccharides and mucopolysaccharides, and conjugate these species to proteins, giving the mucoproteins. Pectin, for instance, which is bonded to cellulose in the support wall of young plant cells, is produced in the Golgi apparatus: it is carried out of the cell by the vesicles, which are generated at the edges of the main sacs, migrate into the cytosol, break on contacting the plasma membrane and then pour out their contents.

Mitochondria are either spherical or rod-shaped corpuscles, with a diameter of 300-700 nm and a length of some 1000-2000 nm. Their number vary according to the cells: there are about 600-800 of them in a hepatocyte, and up to 50,000 in large cells, like amoeba. The ultrastructure of mitochondria shows the normal external membrane, 7.5 nm thick, and a second inner membrane, also approximately 7.5 nm thick, which forms numerous and deep folds on bending itself (*mitochondrial cristae*). There is a space of about 10 nm between the two membranes.

The internal membrane, enclosing the *mitochondrial matrix*, is spotted with knoblike particles of 8.5 nm diameter, immersed in the matrix. Primary metabolic processes take place in the mitochondria (see Chapter 3): in the matrix, fatty acids are oxidised and acetyl CoA is burnt to CO_2 (Krebs' cycle). The respiratory chain process takes place on the inner membrane, where the knob-like particles couple the redox reactions of the electron carriers (NAD, flavoproteins, cytochromes) to the ADP phosphorylation (see Fig. A1.1). ATP, produced by oxidative phosphorylation, and carbon dioxide (resulting from the combustion of acetyl-SCoA), passes into the cytosol; whilst the mitochondria remove ADP, oxygen, inorganic phosphorus, and oxidisable substrates (fatty acids, pyruvic acid, amino-acids) from this medium.

One peculiarity of plant cells is the presence of certain organelles, called *plastids*, which synthesise and accumulate substances: starch is accumulated in amyloplasts, the lipids in the elaioplasts, and the proteins in the proteoplasts. Besides these colourless organelles there are often other plastids rich in pigments (chromoplasts). Amongst them the *chloroplasts* have the greatest physiological importance. Photosynthesis takes place inside them, that is the conversion of light energy into chemical energy and the transformation of carbon dioxide into glucose (see Chapter 3). The chloroplasts appear as ellipsoid-organelles, of remarkable dimensions (3000-10,000 nm diameter, and 1000-2000 nm thick) and are coloured an intense green by the chlorophyll pigment which they contain. Sometimes their green colour is masked by other colours, as in the brown and red algae. In the inner part of the plastid, which is separated by a double membrane from the cytosol (as in mitochondria), there is an amorphous substance called *stroma*, in which are immersed lipid drops, starch granules, and a lamellar apparatus, which is opaque to electron microscopy. This lamellar structure is the most characteristic feature of the chloroplast. It consists of some very flattened vesicles (*lamellae*), arranged in different ways according to the class to which the vegetable belongs. In mosses and vascular plants they are almost parallel to each other, and the wider ones alternate with piles of much smaller lamellae, which appear as dark granules (the *grana*) in an optical microscope. A more accurate inspection by electron microscopy reveals a granular structure of the lamellae, Their membranes are composed of particles (*quantasomes*), about 5.5 nm thick, having a diameter of about 15 nm, in which chlorophyll, carotenes, and other molecules are arranged in an orderly array (Fig. 1.6). From polarised light studies the tetrapyrrolic rings of the chlorophyll molecules appear to be parallel with respect to the lamellar surface, whilst the fatty phytol chain is perpendicular.

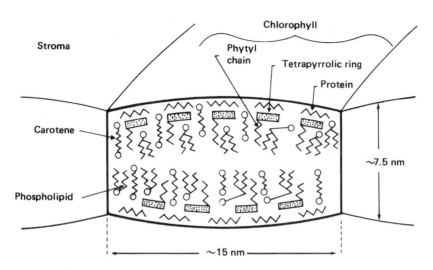

Fig. 1.6 Fine structure of a quantasome. (After A. Berkaloff *et al;* 'Biologie et physiologie cellulaire', Hermann, Paris, 1967).

The relative position of the chlorophyll molecules and that of the other pigments is extremely important for the efficiency of the phytophysical and photochemical processes (photoexcitation and the photolysis of water), which take place in the quantosome (see Chapter 3). The quantasome is therefore the fundamental biological unit for chlorophyll photosynthesis. In the Cyanophytes and photosynthetic bacteria (procaryotic organisms), where cytoplasmic membranes do not exist, the chlorophyll photosynthesis is not localised in well defined cellular organelles such as the chloroplasts, but takes place in lamellae distributed throughout the cytoplasm or in the *chromatophores* (spherical vesicles containing chlorophyll) fixed to the cell membrane.

From the previous discussion it is apparent that the cell is divided into a large number of compartments. The effect of such compartmental divisions can often be noticed during biosynthetic experiments based on feeding a living organism with one or more substances labelled with radioisotopes. In order to take part in a certain metabolic process these substances must get inside the cell, passing through the plasma membrane, as well as reaching the specific cell compartment in which the desired process takes place: they therefore have to overcome barriers of various kinds, and sometimes have to bind to certain carrier or activator molecules. These obstacles are not met by the same substances of endogenous origin, which are mostly synthesised in their 'active' form (e.g. acetyl-SCoA in the oxidation of fats) and in the compartments in which they undergo their further transformations. For instance, mevalonic acid, which is synthesised in the chloroplasts, is used *in situ* for the biosynthesis of terpenoids and carotenoids (Chapter 5).

Higher organisms, being multicellular with differentiated cells, can produce certain natural substances in specialised tissues or organs. The localisation of secondary metabolic processes in particular anatomic parts (for example the liver, the tegumental cells, or certain specific glands in some animals; the oil-glands and the resin ducts in plants) is another phenomenon to be accounted for in biosynthetic studies, since the access of exogenous substances into these compartments can be limited or even totally forbidden.

The technique of growing plant or animal cells *in vitro* (tissue cultures) is becoming increasingly important in biosynthetic studies, since they provide a more instant contact between exogenous substances and the metabolising cells.

As far as the aims of biosynthetic researches are concerned, such cell cultures can be treated in the same way as cultures of microorganisms. In some instances, however, significant metabolic differences have been noticed between the tissue culture cells and those from the intact natural organism.

Another phenomenon, very frequently associated with higher organisms and not to be forgotten in incorporation experiments, should finally be mentioned: it is the phenomenom of carrying and storing metabolites into accumulation sites, well separated from the site of synthesis. Nicotine, for instance, is synthesised in the younger portions of *Nicotiana* plant roots, and is accumulated in its leaves. Any labelling experiment involving potential alkaloid precursors which are carried out using only the leaves of *Nicotiana* prove to be unsuccessful.

1.6 RESEARCH METHODS AND TECHNIQUES

Two different methods can be used whilst studying natural phenomena: either (a) direct observation of the objects implicated in the phenomenon itself, causing the least amount of interference possible with their natural organisation; or (b) by provoking a strong interference, in order to understand how inter-dependent the variables are, even to the extent of purposely modifying the system.

The first method is usually preferred in biogenetic studies, because the isolation of a biosynthetic process from its metabolic grid reflects back on the process under investigation, often in a manner difficult to control.

In practice, some molecules of a particular metabolite are appropriately marked by substituting certain atoms with the corresponding stable or radio-active isotopes, and these are then injected into the metabolic pool of the system itself, at the same time trying to disturb the organism in question as little as possible. The labelled molecules (called 'tracers'), behave just like the natural ones, because living organisms cannot distinguish between isotopically different molecules, and as the treated molecules carry a 'label' on them it becomes easy to recognise the products arising from their biotransformations (provided the labelled atoms have not been lost). In this way, it is possible to follow the metabolic pathway of the marked molecules, and thus of the corresponding natural ones.

These experiments (*precursor incorporation expts*) presume that enzymes and cellular membranes are unable to distinguish between molecules only differing isotopically. This is the so called 'tracer principle', which is an empirical rule, valid within approximate limits, but by no means axiomatic. The limits of its applicability depends on the presence and importance of isotopic effects in biotransformations. Transformation involving hydrogen atoms isotopically substituted with deuterium or tritium show remarkable isotopic effects. An isotopic effect is more effective the larger is the ratio of the competing isotopic nuclides. The deuterated molecules, even more the tritiated ones, usually undergo slower transformations than the corresponding molecules containing protium.

The isotopic effects — the primary and secondary kinetic ones are of particular interest in bioorganic chemistry — and their origins are discussed more extensively in Appendix 4. However, three points are mentioned here, so that the reader should not under- or over-estimate the importance of isotropic effects in bioorganic studies:

(1) *Primary* kinetic isotope effects are the variations of reaction rates due to breaking or forming isotopically different bonds, e.g. the fission of a C–T bond instead of the analogous C–H one.
    Secondary kinetic effects are observed when the isotopically substituted atoms are near the reaction centres, but not involved in breaking or forming bonds. The secondary effects are usually quantitatively far smaller than the primary ones.

(2) An enzymatic reaction can be considered as resulting from two successive steps: an enzyme-substrate complex (ES) is formed in the first step, whilst the very reaction of the substrate within the enzyme active site is the second step. The isotopic substitution does not alter the shape and polarity of molecules and produces minimal variations in the equilibrium: E(enzyme) + S(substrate) $\rightleftharpoons$ ES, which essentially depends on the complementarity between the surfaces of the substrate and of the enzyme active site. These variations have sometimes been measured and depend on secondary kinetic effects, relative to the rates of association and dissociation reactions between enzyme and substrate. The yeast alcohol dehydrogenase (YADH) catalysing the oxidation-reduction of the couple, ethanol-acetaldehyde, shows an affinity for CH_3CD_2OH slightly larger than for CH_3CH_2OH in forming the enzyme-substrate complex. (H. R. Mahler and J. Douglas, *J. Amer. Chem. Soc.*, 1957, **79**, 1159). Similarly small isotopic effects can be presumed when organic molecules pass through cellular membranes. More significant kinetic isotope effects (especially primary ones) can be observed when the substrate reacts in the active site of the enzyme. Thus, natural ethanol is transformed into acetaldehyde by the above mentioned YADH at a 1.8-fold faster rate than CH_3CD_2OH (H. R. Mahler and J. Douglas, *loc. cit.*).

(3) The isotopic effects can be ignored in many experiments *in vivo*, being

unimportant for the conclusions. This takes place with compounds labelled with one nuclide only in one specific position of the molecule, or when, although molecules with multiple labels are employed, only secondary isotopic effects intervene and determine variations in the relative concentrations of differently labelled molecules within the limits of the experimental errors.

The second method quoted above is applied when an organism is fed with compounds which can specifically inhibit certain enzymes. The metabolic processes are blocked at the reactions catalysed by the inhibited enzymes, and thus information on the biogenetic relationships amongst the substances leading up to this point can be gained. For instance, if some ^{14}C-labelled mevalonic acid, and a specific enzymic inhibitor (2-diethylaminoethyl phosphate) are simultaneously given to tobacco plant tissues, radioactive 2,3-epoxysqualene accumulates. Incorporation of ^{14}C into the phytosterols, which normally derive from mevalonic acid and which can be isolated as labelled molecules in the absence of the inhibitor, is suppressed.

These facts indicate a close biosynthetic connection between 2,3-epoxysqualene and the phytosterols. The deduction is that the phytosterols derive from 2,3-epoxysqualene through subsequent successive transformations and one of these, presumably the one just after the synthesis of 2,3-epoxysqualene, is blocked by the given inhibitor (see Chapter 5.7.2).

Enzymatic blocks can be applied to microorganisms through 'almost' natural methods, such as the *mutagenic technique*. This method consists of modifying the genetic character of certain cells by mutagenic agents, for instance, ultraviolet radiation. The cells which lack an enzyme implied in a certain biosynthetic process are successively isolated from a culture §.

Consider that these cells lack the enzyme corresponding to the reaction C→D of the path illustrated in Fig. 1.2(a) (A→B→C→D→→→X→Y). If the final product of the interrupted biosynthetic sequence is essential for the growth of the microorganism, the mutant cells will not grow in the culture medium of minimum requirement established for the primitive cells. (This 'minimum' medium is that containing only those nutritive substances essential for developing and multiplying the original, primitive strain). Addition of the component X will enable growth to occur.

By trials it is then possible to determine the nature of the compound X, which is needed for growth of the microrganism. As soon as this compound is given, the microorganisms will grow in a regular manner, but they will accumulate the obligatory intermediate C, which is normally transformed in the primitive cells by the enzyme, which is missing in the mutant variety.

The mutant cells can also grow if we add to the 'minimum' culture medium an earlier intermediate found after the enzyme block, for example the intermediate

§The probability of simultaneously suppressing more than one enzyme of a single biosynthetic route within the same cell is extremely small.

D. By such experiments, a biosynthetic path with all its essential ramifications can be fully elucidated. For example, by use of auxotrophic mutants, Davis brilliantly discovered the important metabolic route to shikimic acid and its derivatives, such as the aromatic amino acids (Chapter 7). Davis has also introduced a very convenient method for isolating mutant cells from primitive ones. This method is based on the different action of antibiotics, particularly of penicillin, towards cells which are growing and the ones which are resting (in 'stasis'). The antibiotics destroy the growing cells, but do not kill the latter, resting ones. If a bacterial culture, in which some mutations have previously been induced, is placed in the minimum medium for the primitive cells (so that only these can develop and reproduce) and it is then treated with the antibiotic, the growing cells, genetically intact, are destroyed, whilst the latter, mutated ones, which are inactive, survive.

One of the most difficult experimental tasks is the feeding of higher plants or animals with substrates or inhibitors (labelled or otherwise). The most common tricks are described in paragraph 1.6.2. One other method, which avoids feeding difficulties, is known as *sequential analysis*. It is a rigorous application of the first of the two general research methods quoted above; it consists in the evaluation of the content of certain metabolites (by techniques such as thin layer chromatography or gas chromatography) in an organism at different moments of its development. If a metabolite C appears together with another one of a very similar structure (let us call it B), and C accumulates whilst B gradually disappears, it is then possible that the substances are related with B as precursor and C as product.

In *Conium maculatum*, for instance, the maximum content of γ-coniceine (Fig. 1.7(a)(1)) is reached one week before that of coniine (2), and the second alkaloid very likely originates from the first one. The success of this particular technique depends on the co-occurrence of several, relatively rare, circumstances including a slow growth of the organism, a slow metabolic rate, and a method for estimating the metabolites, etc.

The stages of biogenetic studies on a natural substance X generally progresses:

1. Determination of the origin of the carbon atoms which form the skeleton of substance X. This means being able to identify compounds which are involved in the biosynthetic region intermediate between the primary and secondary metabolism, (e.g. acetic acid, mevalonic acid; (see Fig. 1.1).

    This problem is usually solved by trying incorporation experiments (Section 1.6.1), in order to 'screen' for the possible precursors of X*.

2. Determination of the metabolic route, that is the sequence of the intermediates and reactions leading to the formation of substance X.

---

*Choice of suitable precursors are not randomly selected but are made by using certain working hypotheses. These are set up before experimentation and are based on either general biosynthetic patterns (deductive) or by comparison with similar, known situations (analogy).

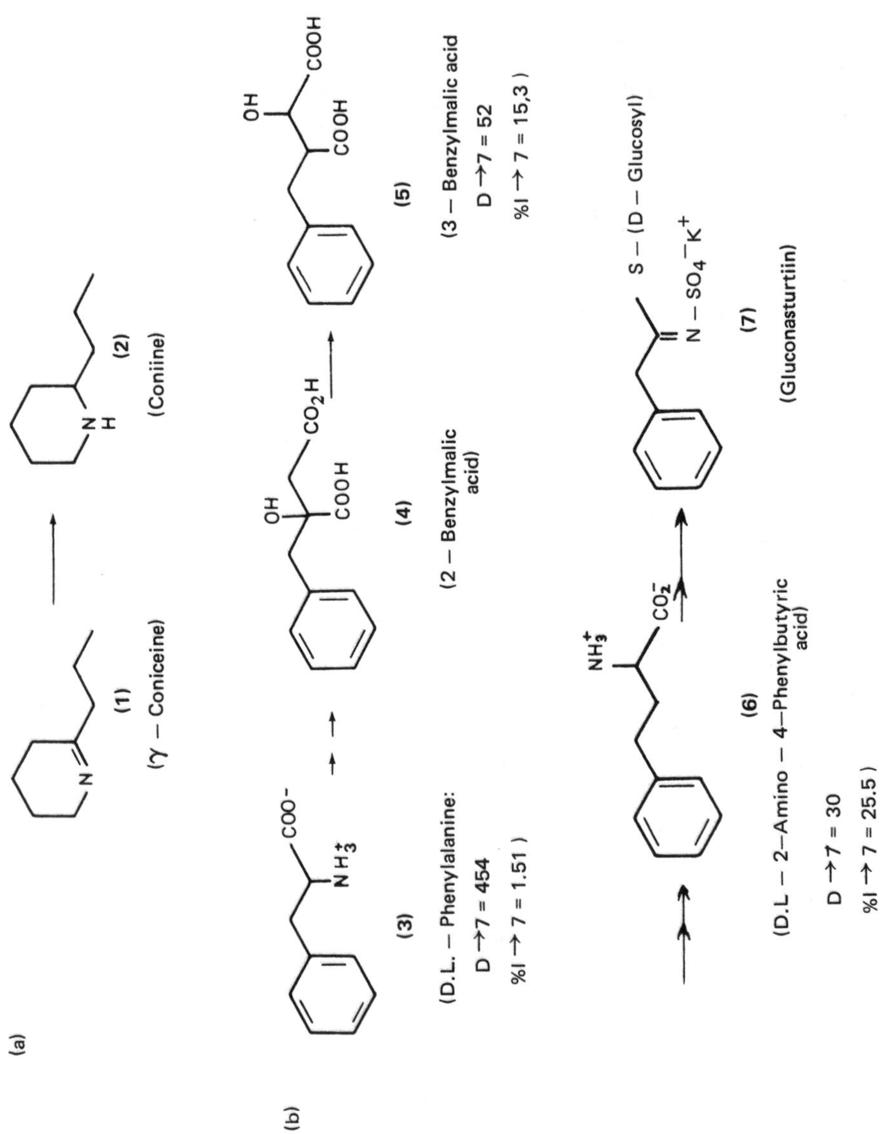
(a)
(1)
(γ – Coniceine)
(2)
(Coniine)
(b)
(3)
(D.L. – Phenylalanine:
D → 7 = 454
%I → 7 = 1.51)
(4)
(2 – Benzylmalic acid)
(5)
(3 – Benzylmalic acid
D → 7 = 52
%I → 7 = 15,3)
(6)
(D.L – 2–Amino – 4–Phenylbutyric acid)
D → 7 = 30
%I → 7 = 25.5)
(7)
(Gluconasturtiin)
S – (D – Glucosyl)
N – SO4⁻K⁺

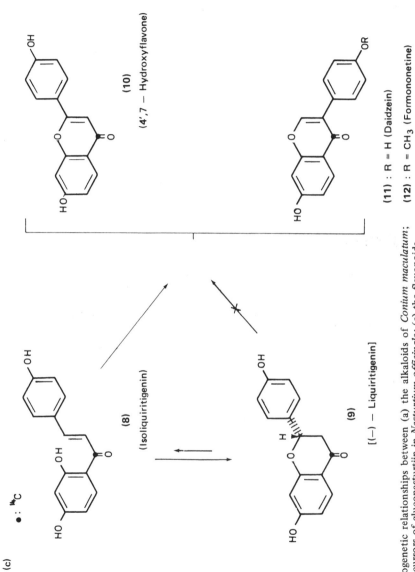

Fig. 1.7 Biogenetic relationships between (a) the alkaloids of *Conium maculatum*; (b) the precursors of gluconasturtiin in *Nasturtium officinale*; (c) the flavonoids of *Trifolium subterraneum*.

3. Determination of the source of oxygen and, if present, nitrogen atoms, etc. For oxygen there is generally the question as to whether it arises from water or atmospheric oxygen. In this case separate experiments employing either labelled oxygen, with $H_2^{18}O$ or with $^{18}O_2$, are used.

4. The nature of the enzymatic processes involved in each step of the formation of X. This biosynthetic aspect, which is certainly the most chemical one, is best studied *in vitro*, utilising isolated enzymes, prepared in a pure form. The isolation of enzymes from living organisms is usually quite difficult and cumbersome. Despite these problems verifying an enzymatic reaction *in vitro* remains the most convincing proof for a reaction step.

1.6.1 Isotopic tracer methods

It is first of all necessary to distinguish between radioactive and stable isotopes, as the techniques and the scope of use are very different for these two cases. Radioactive isotopes most frequently used in biogenetic studies are tritium (^3H, or T) and ^{14}C. They both decay by emitting β-rays, with a maximum energy, and a half-life time ($t_{\frac{1}{2}}$) of 0.0185 MeV and 12.35 years for tritium, and 0.156 MeV and 5730 years for ^{14}C.

The radioactivity of a milligram-atom of carrier free (100% isotopic abundance) tritium is 29.2 Ci; a milligram atom of carrier free ^{14}C is 62.4 mCi.*

These values also represent the specific molar radioactivities of molecules in which a hydrogen or carbon atom has been entirely substituted by its respective radioactive isotope, therefore they are the maximum molar radioactivities for compounds specifically and singly labelled with ^3H or ^{14}C. The specific activity of products commonly used in incorporation experiments corresponds to isotopic abundances which approach 100% for ^{14}C, and 0.1-1% for ^3H.

The instruments used for measuring radioactivity, for example a scintillation counter, gives the number of counts per unit of time. If we want to transform the number of counts per minute (cpm) into the actual disintegration number per miniute (dpm) we use the relationship:

$$dpm = \frac{cpm}{\zeta}$$

where ζ is the instrument efficiency for that particular isotope, and is generally obtainable after counting standard references (it is usually between 0.5 and 0.9). Even if the same sample is labelled with both ^3H and ^{14}C, it is possible to distinguish between them, by counting the radioactivity of each nuclide (even concurrently) on two different energy windows (channels) of the instrument. In practice, obtaining actual cpm and dpm figures for each nuclide is slightly more difficult for two nuclides than for a singly labelled molecule. The different

*1 Ci = 2.22×10^{12} disintegrations per minute (dpm).

efficiency of the two channels must be taken into consideration. Also, since, in the tritium channel, counts from both tritium and ^{14}C are collected, the number of the ^{14}C cpm must be subtracted from the total figure appearing on the tritium channel, in order to obtain a real value of cpm arising from the tritium atom only. The number to be subtracted is proportional to the ^{14}C cpm number, and the proportionality value can only be obtained by counting standard references.

The specific radioactivity is the radioactivity per mass unit (mg or millimole) of the product: for instance mCi/mmol, Ci/mg, dpm/mmol, etc.

The substances given to living organisms must be radiochemically pure, and as chemically pure as possible. The *radiochemical purity* is defined as the percentage of the total radioactivity contained in a unique and well-defined chemical compound, and, usually, in a specific site of such species. The *chemical purity* is the percentage weight of a certain chemical species, mixed with other chemically different species (called impurities). These two purities are completely independent from each other. A compound can be radiochemically pure and chemically very impure, such as [formyl-^3H] benzaldehyde containing benzonitrile; it can also be chemically very pure and radiochemically impure, such as (R)-[2-^3H] glycine mixed with small amounts of (S)-[2-^3H] glycine. What is said for the radiochemical purity can obviously be extended to any isotopic substitution.

The specific activity of compounds used for feeding experiments must be rather high, in order to avoid disturbing the normal metabolism of the plant or animal by having to introduce massive amounts of the exogenous materials.

A product can be labelled with a radioisotope, particularly with ^3H and ^{14}C, in three different ways:

1. **Generally labelled (G)**, when the average distribution* of the radioisotope in the various positions of the molecule is irregular. For instance, in D,L-[G-^3H] phenylalanine the protons in every position are partially substituted by tritium atoms in a random manner. Such labelling is statistical and labelled molecules differ in the number and positions of the tracer atoms.
2. **Uniformly labelled (U)**, when the average distribution of the radioisotope is about the same in all the possible positions of the molecule. For instance, if we isolate, as CO_2, the carbon atoms from the various positions of L-[U-^{14}C] phenylalanine, the same abundance of ^{14}C will be found in each CO_2 sample. Certain natural compounds may be purchased with uniform labelling. They are produced biosynthetically using photosynthetic microorganisms and $^{14}CO_2$ as the sole source of carbon.
3. **Specifically labelled**, when the radioisotopes are localised in certain, well-defined positions and well-defined relative abundances (% of the total specific abundance in ^3H or ^{14}C). The specific label can be single, double, or multiple according to the number of labelled positions. An example of single labelling is given by D,L-[1-^{14}C] phenylalanine, in which all of the ^{14}C radioactivity is concentrated in the carboxylic group (C-1).

*That is, calculated on a very large number of molecules.

Sometimes, if the same radioisotope is present in several positions, considered to be distinguishable, the total number of such positions is quoted on the right hand of the nuclide symbol; for instance in L-$[3-^3H_2]$ phenylalanine the tritium is distributed between the 3-*pro-R* and 3-*pro-S* positions (see Appendix II for definitions). A specific double, or multiple, label can be *intramolecular* or *statistical*. In the former case the label is identical for every labelled molecule or, at least, for the overwhelming majority of the labelled molecules. Statistically labelled compounds are almost always used in practice; they are obtained by mixing appropriate amounts of the same chemical species, each having a specific single label. If D,L-$[1-^{14}C]$ phenylalanine and D,L-$[3-^{14}C]$ phenylalanine are mixed in known amounts, a doubly labelled sample, D,L-$[1,3-^{14}C_2]$ phenylalanine, is obtained.

During feeding experiments, one often uses mixtures of samples labelled with either ^3H or ^{14}C. If L-$[1-^{14}C]$ phenylalanine and L-$[3-^3H_2]$ phenylalanine are mixed, we obtain a mixture characterised by a certain $^3H/^{14}C$ ratio (dpm ^3H/dpm ^{14}C), and containing the following molecular species, (assuming the natural molecules to be ^1H, ^{12}C and ^{14}N): natural molecules of phenylalanine, molecules in which carbon 1 is substituted by ^{14}C, and molecules in which a protium atom (either 3-*pro-R* or 3-*pro-S*) is substituted by tritium (monotritiated molecules). Compounds containing hydrogen atoms exchangeable in acidic or basic medium in two or more specific positions afford a mixture of non-tritiated (T_0), monotritiated (T_1), ditritiated (T_2) etc. molecules on exchange with tritiated water. The relative abundance of such species can be calculated on a kinetic or statistical basis. The intramolecularly labelled samples can only be prepared through more accurate and cumbersome synthetic procedures, usually very expensive. The labels are successively introduced in the desired positions so that the isotopic abundance approaches 100% in each position. Then the sample may be diluted with unlabelled carrier. Their use is limited to particular situations: e.g., when it is necessary to take precautions with respect to possible biological discriminations between molecules having identical structures but isotopic differences (isotopic effects). As has already been mentioned, if we want to study the biogenesis of substance X a certain number of labelled substrates (as few as possible) are utilised in incorporation experiments in order to indicate the best precursor of X. The precursors of X are those specifically-labelled compounds which, once given to the organism producing X, give rise to radioactively labelled X.

In order to interpret the incorporation data correctly it is necessary to check for the absence of *randomisation* of the label during the biotransformation of a precursor into the product.

The randomisation of radioactivity in a product X arising from a labelled compound can arise as follows: the labelled compound is degraded *in vivo* into smaller fragments (for instance acetic acid (C_2) and formic acid (C_1)), which are then utilised in the intermediate metabolic pools. Subsequently such fragments are successively utilised *in vivo* for synthesising X; as the fragments used are partially labelled, an extensive and irregular randomisation of the distribution of radioactivity in X can result. The localisation of the radioactivity in X generally requires complex chemical work (setting up special degradation methods, etc.), which is made more difficult by the small amounts generally available and by the low specific radioactivities of the isolated, labelled products. The mechanisms of the key steps in such processes should be completely known, especially on localising tritium. In case of doubts, control reactions are carried out with specially synthesised molecules deuterated at particular sites. The contribution of the organic chemist is thus vital both in experiments before the incorporation (in preparing labelled precursors) and in establishing the position of labelling in the natural product by the degradative work. Figure 4.6 is pertinent here: it illustrates an example of chemical degradations aimed at localising the radioactivity of a typical metabolite.

A precursor is very good, good, poor, or very poor according to the amount of radioactivity recovered in X (% incorporation), and according to the degree of radioactivity dispersed in the various positions of the molecule X. When the precursor is very good, the incorporation is very high, and the randomisation practically non-existent. In evaluating the percentage incorporation one must always bear in mind the type of experiment made, the sort of organism fed with the precursor, the method for administering the labelled compound, etc. A good precursor into bacterial or fungal cultures would be incoorporated in amounts greater than 50%, whilst incorporations of 0.1–1.0% can be considered more than satisfactory in higher plants*.

In a case where U, V, and W have been fed to an organism and V and W, but not U, have acted as precursors of X, the following questions arise:
1. Do the compounds V and W coincide with the corresponding, natural intermediates of X?
2. Must the compound U be necessarily excluded as an intermediate of X, if it gives an extremely poor % incorporation?
3. Is it possible, from the % incorporation of V and W, or from analogous data (see later), to obtain information about the position of such compounds along the metabolic pathway to X?

There are no universally valid criteria for answering these questions. In order to interpret the results of an incorporation experiment, the following generalisations, (1) and (2), should be considered:-

(1) 'V, a precursor of X, does not *necessarily* coincide with a natural and obligatory intermediate of X.'

---

*According to D. H. R. Barton, the lower significant limit can be around 0.001%: lower incorporation data are meaningless.

It could be that V is a substance foreign to the organism's normal metabolism, and could be transformed into X, for instance according to the scheme (b) of Fig. 1.8. This process often takes place in some microorganisms, which have remarkable capacities for metabolic adaptation. Many species of *Penicillium* incorporate diethyl malonate into polyketide molecules, since it penetrates into the cells, where it is transformed into the true, natural intermediate, that is, malonyl-CoA.

Although V may be present in the cell under normal conditions, it may not be an obligatory intermediate of X. It could be converted into X through ways which are normally of secondary importance, but which are made more efficient as soon as large amounts of exogenous W are introduced into the organism. Finally, V can be a natural compound in fast equilibrium with an actual bio-synthetic intermediate of X, for instance B in scheme (c) of Fig. 1.8.

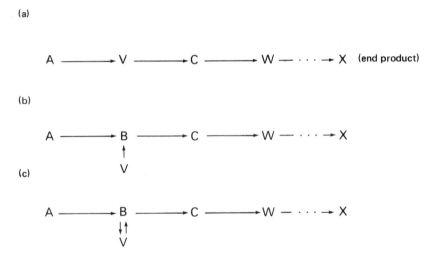

Fig. 1.8 Different metabolic routes possible from a substance V to a substance X.

In some cases it is possible to clarify the effective metabolic role of V — especially in simple biosynthetic situations, such as the use of cell homogenates or cultures of microorganisms — by using the *metabolic trap* method. Let us suppose V to be an intermediate of X, and A to be another intermediate as well; A is more distant than V along the common biogenetic path (Fig. 1.8(a)). If we feed the system synthesising X with A* (labelled) together with large amounts of the V compound (unlabelled), a dramatic decrease in the specific activity of X* will be observed, with respect to an analogous experiment performed without adding V; a certain amount of V* (labelled) will, moreover, be obtained, as a unique V pool is formed, into which V* produced by A*, and externally added V have arrived.

Such a metabolic trap is a *cold* one, since an unlabelled compound is added in order to trap a transient, labelled intermediate. A *hot* metabolic trap can also be used and more clear-cut results are often obtained. In such a case, if the system is fed with A* (labelled with an isotope, e.g. ^{14}C) and with V° (labelled with a different isotope, e.g. ^{3}H), if a direct path A→V→X exists, the compound $V*^{\circ}$ is recovered (labelled with both isotopes), together with $X*^{\circ}$, obviously doubly labelled.

If trapping experiments give results coherent with an (a) type metabolic path (Fig. 1.8), (b) type paths can be excluded, but not (c) type ones. These last ones are undoubtedly rare, but possible. The doubt between the two situation in which V is an intermediate between A and X (path (a)) or a side product in a fast equilibrium with an intermediate between A and X (path (c)) *cannot* be resolved with certainty through feeding experiments, that is, by purely chemical procedures. If two enzymes are proved to exist in the organism, respectively catalysing the formation of V from A, and the conversion of V into C (A and C being proved intermediates of X), V is then certainly a real intermediate in the synthesis of X, according to the sequence A→V→C as in Fig. 1.8(a).

Trapping results not coherent with type (a) or (c) biogenetic paths can be interpreted in various ways.

If the specific activity of X is low, but meaningful, and the recovered V is not radioactive, V *can* be *hypothesised* as a precursor of X, although not being between A and X (e.g. Fig. 1.8): this is a hypothesis which has to be demonstrated! If X and V are recovered inactive, the sequence A→V→X is still possible, if an enzymatic block before V takes place, owing to a 'negative feedback' from the large amount of V introduced into the metabolic system.

An example of the metabolic trap method was used in studies on the biosynthesis of erythromycin in *Streptomyces erythraeus*. Using $[1-^{14}C]$ propionic acid, and unlabelled erythronolide B recovered from high-yielding mutant strains of this species, it was shown that the latter was an intermediate in the formation of erythromycin, and not a secondary product (see Fig. 4.38).

(2) 'If a labelled compound U, hypothesised as an intermediate of X, is not incorporated into X, this is not sufficient evidence for discarding the original hypothesis.'

There are many factors which can inhibit an incorporation and they are not always avoidable. Amongst these can be quoted a slow speed of synthesis of X during the experiment, the impermeability of cellular membrane towards the exogenous material, an induced alteration or destruction of the compound U before reaching the synthetic sites for X, etc. Certain phosphoric esters, such as IPP, DMAPP, etc. are well recognised intermediates in many biosynthetic processes (to terpenes). Nevertheless, they are not used as exogenous precursors since they cannot pass through the cellular membranes of the plants.

If A_V and A_X are the specific activities (on a molar basis) respectively of labelled V, given to a certain organism, and of the labelled X, subsequently isolated from that organism and purified to constant activity, the *dilution value* ($D_{V \to X}$) is defined as the ratio A_V/A_X. The *percent incorporation value* ($\%I_{V \to X}$) is alternatively given by the expression, $\dfrac{A_X \cdot M_X}{A_V \cdot M_V} \cdot 100$, where M_X and M_V are the molar quantities of X and of V. Both dilution values and % incorporations are used for expressing quantitatively the results of incorporation experiments, and they both have reciprocal advantages and disadvantages.

The dilution value does *not* depend on the amount of X recovered, and it thus allows an accurate purification of X to constant activity A_X.

During these operations, however, it is not possible to dilute the isolated, labelled X with inactive X (carrier material), since the A_X value must not be altered: the procedure required for obtaining the exact dilution value reduces the possibility of carrying out complex chemical degradations on X, which is limited by the amount isolated and which is generally small. The major use of the dilution value derives from the supposition that, in experiments carried out under standard conditions (the organism, age, the season, temperature, the length of the experiment, etc.), the dilution values are bigger as the precursors get further from the product X. This is understandable, if we assume that a labelled compound is diluted at every stage of its metabolic evolution by the pool of the intermediate involved in that stage. If we consider the sequence in Fig. 1.8(a) it is very likely that, for two separate incorporation experiments of V and W,

$$\frac{A_V}{A_X} > \frac{A_W}{A_X}, \text{ that is } D_{V \to X} > D_{W \to X}$$

Figure 1.7(b) shows the biosynthetic path proposed for gluconasturtiin (7), a thioglucoside of *Nasturtium officinale*. The dilution values are obtained by feeding nasturtium plants with the various intermediates labelled with ^{14}C and measuring the specific activity of the gluconasturtiin produced: they are in good agreement with the metabolic sequence indicated [E. W. Underhill, *Can. J. Biochem.*, **46**, 401 (1968)].

The use of such criteria for unravelling a biogenetic route is susceptible to criticism as there are other factors, such as the permeability of cellular membranes, which can also alter the dilution values. Shikimic acid, for instance, sometimes gives dilution values in the product phenylpropanoids which are smaller than those from phenylalanine, even though phenylalanine is biogenetically much nearer to the phenylpropanoids than shikimic acid. Furthermore, comparing $D_{V \to X}$ to $D_{W \to X}$, obtained from two separate experiments carried on different (although genetically homogeneous) individuals, might lead to the wrong conclusions when the pools of X differ considerably in such individuals.

Differences of even 500% in the moles of X between individuals of the same age and kept in identical conditions are not exceptional. This latter source of error is avoided by giving simultaneously V^* and W° (labelled with different nuclides) and by measuring the dilution values into $X^{*\circ}$ with respect of each nuclide, i.e. $D_{V \to X}$ and $D_{W \to X}$.

The % incorporation value has the disadvantage that recovery of the labelled product X from the organism has to be achieved as accurately as possible. On the other hand one can dilute the material with even large amounts of inactive product X, and utilise this diluted product for subsequent chemical degradations.

In comparing the % incorporations, information can be gained on the metabolic sequences; higher % incorporations would be expected the closer, in metabolism, the given precursor is to X. Nevertheless, similar limitations, as mentioned for dilution values, apply to the biogenetic interpretation of % incorporations.

A technique often affording clear results is *isotopic competition.* This is based on the same principle as the already described 'cold trap'. An isotopic competion experiment may clarify the relative position of two intermediates, V and W, along the metabolic path to X . V^* is given, together with a large amount of inactive W to the biological system under investigation. The specific activity of X^* so isolated is compared to that of X^* obtained from a parallel experiment, where W^* and a large amount of inactive V were given. If V precedes W along the metabolic sequence (cf. Fig. 1.8(a)), the dilution value of V into X from the first experiment is likely to be much larger than that of W into X from the parallel experiment (and than that of V when only V^* is administered). The addition of inert W to the system increases in fact the weight of the metabolic pool of W itself: if V precedes W, V passing through W, will be more diluted the more the pool of W has been increased. For instance, in order to decide which compound 8 or 9, in slow equilibrium between each other, is the more immediate precursor of the flavonoids in *Trifolium terraneum* (Leguminosae) (Fig. 1.7(c)), the plant was fed, in parallel experiments, with [^{14}C] chalcone (8) together with the unlabelled flavanone (9), and with [^{14}C] flavanone (9) together with the unlabelled chalcone (8). The flavonoids isolated from the experiments with radioactive chalcone showed higher specific activities (that is lower dilution values) than those of the same flavonoids recovered from the parallel experiments with radioactive flavanone. Thus, it was inferred that the chalcone (8) is a more immediate precursor of the examined flavonoids [E. Wong, *Phytochemistry, 7,* 1751 (1968)].

Competition experiments, as well as cold traps, may be included in the second category mentioned at the beginning of this section as far as the methodology is concerned. They therefore suffer from the limitations typical of those experiments in that their conclusions may be criticised as being obtained by perturbing metabolic equilibria and stimulating unnatural enzymatic regulation mechanisms.

Often an organism is fed with precursors labelled doubly or more times, that is, marked in several positions with either the same isotopes or with different ones. Multiply-labelled precursors are usually tagged with radioactive ^{3}H or ^{14}C, or stable ^{13}C and ^{15}N. Such precursors are obtained mixing samples of the same chemical species marked with the same or different isotopes (and usually in a specific site). The important factor in using these precursors is the precise ratio between the isotopic abundances of the two different isotopes (or of the same isotope in two different positions) in the administered compound (labelled W), compared to the product (labelled X) and subsequently in the chemical degradation products from the latter.

The isotopic abundances are expressed as ^{14}C radioactivity, ^{3}H radioactivity, and as atom % excess for ^{15}N or ^{13}C.

Marking with different isotopes is widely used in biosynthetic experiments, and is quite useful since evaluation of the individual isotope abundances does not usually require any chemical degradation. The radioactivites of ^{14}C and ^{3}H are characteristic and can be simultaneously measured in the same sample, whilst the stable isotope ^{15}N is determined by mass spectrometry of the sample and ^{13}C by either mass spectrometry or nuclear magnetic resonance spectroscopy.

When a doubly labelled precursor W is incorporated into X, and a variation of the isotopic abundance ratios is observed it can be concluded that, during the bioconversion, the molecule W has broken certain bonds, and gains or loses fragments characterised by a different content of isotope label.

Consider, for instance, that W is specifically labelled with ^{3}H and ^{14}C, that χ_W is the ^{3}H/^{14}C ratio (dpm ^{3}H/dpm ^{14}C), and that χ_X is the analogous ratio of the product X, obtained after incorporating W. If $\chi_W > \chi_X$, it means that W, or one of its transformation products, breaks in such a way that the isotopically labelled atoms are separated in different fragments, Z_1 (only or mostly containing ^{3}H), and Z_2 (only or mostly containing ^{14}C), and that the product X is Z_2 or is formed from Z_2. It may correspond to the loss of the only tritium in W ($\chi_W \gg \chi_X$), or be in good agreement with a value corresponding to the loss of 1, 2 or more, tritium atoms (Fig. 1.9(a)). If this difference is not interpretable in the previous ways, it should be possible that X is formed as Z_1 and Z_2 recombine again, after they have been diluted in their respective pools (Fig. 1.9(b)).

If $\chi_W < \chi_X$ the above reasoning still holds, but that X derives from fragment Z_1 (Fig. 1.9(c)).

If $\chi_W \simeq \chi_X$ we can be sure that W is converted into X without separating the isotopes into different fragments (Fig. 1.9(d)); we cannot however exclude possible rearrangements of the isotope positions within the molecular skeleton, due to intramolecular processes (Fig. 1.9(e)). However, an unchanged ratio between the abundances of two tracers along a bioconversion (W→X) could also result from two phenomena which cannot be ruled out *a priori*, although highly improbable: (1) 'intermolecular' transposition of isotopically substituted atoms, i.e. exchange of a labelled atom between two W molecules when these

(a)

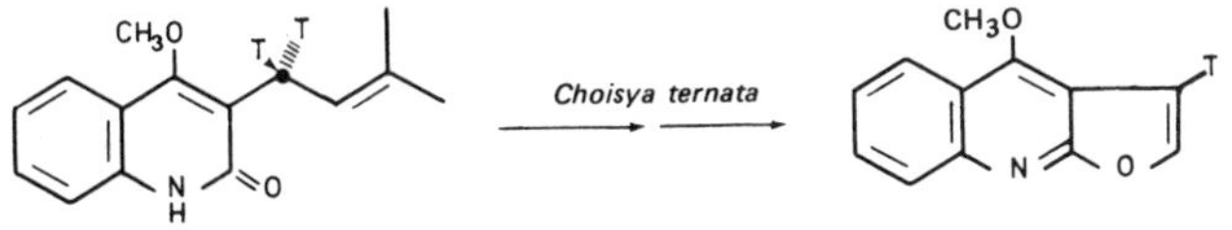
CH3O
T
Choisya ternata
CH3O
T
N
H
O
N
O

4 – Hydroxy – 3 – Prenyl – 2 – Quinolone

● = ^{14}C $^{3}H : ^{14}C = 2 : 1$

Skimmianine

$^{3}H : ^{14}C = 1.1 : 1$

[M.F. Grundon, D.M. Harrison and C.G. Spyropoulos, *Chem. Comm.* 51 (1974)]

(b)

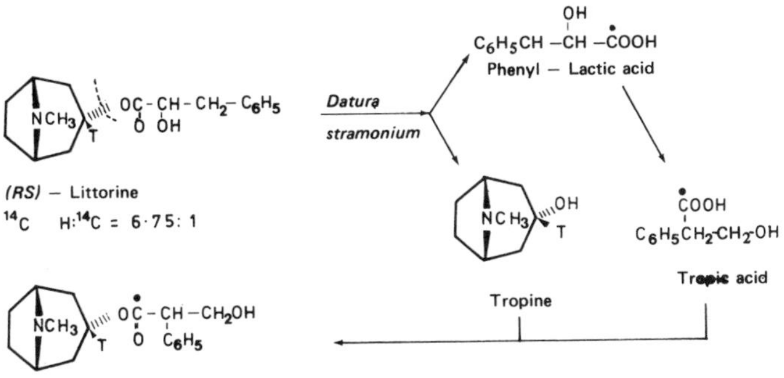
NCH3
T
OC-CH-CH2-C6H5
O OH
Datura
stramonium
OH
C6H5CH –CH –COOH
Phenyl – Lactic acid
NCH3
OH
T
COOH
C6H5CH2-CH2-OH
Tropine
Tropic acid
NCH3
T
OC-CH-CH2OH
O C6H5

(RS) – Littorine

● = ^{14}C $H:^{14}C = 6{\cdot}75 : 1$

(–) – Hyoscyamine

$^{3}H : ^{14}C = 33 : 1$

[E. Leete and E.P. Kirven, *Phytochemistry*, 1502, 13 (1974)]

(c)

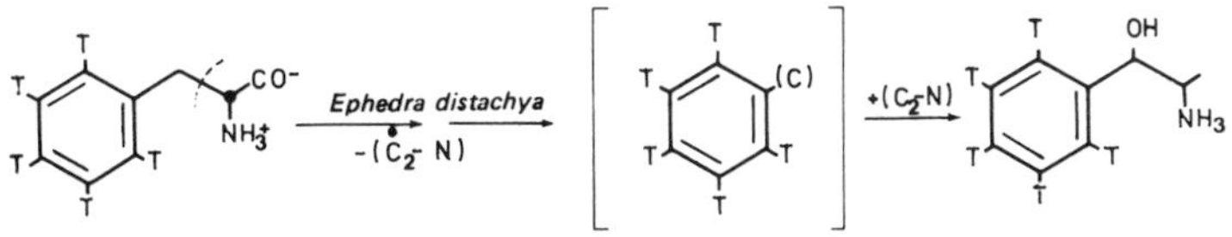
T
T
T
T
T
CO⁻
NH3+
Ephedra distachya
-(C2- N)
T
(C)
T
T
T
+(C2N)
T
OH
T
NH3
T
T
T

D,L – Phenylalanine

● = ^{14}C

(–) – Ephedrine

(^{14}C nil)

[K. Yamasaki, U. Sankawa and S. Shibata, *Tetrahedron Letters* 4099 (1969)]

(d)

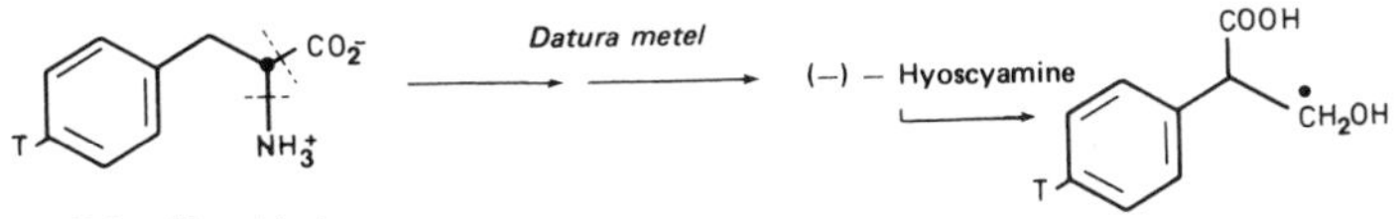
CO2-
T
NH3+
Datura metel
(–) – Hyoscyamine
COOH
CH2OH
T

D.L – Phenylalanine

● = ^{14}C $^{3}H : ^{14}C = 6.5 : 1$

(–) – Tropic acid

$^{3}H : ^{14}C = 6.9 : 1$

[H.W. Liebisch, G.C. Bhavsar and H.J. Schaller, in "Biochem. Physiol. Alkaloide, Int. Symp. 4th" K. Mothes, Ed., Akad.–Verlag, Berlin. 1972, p.233]

(e)

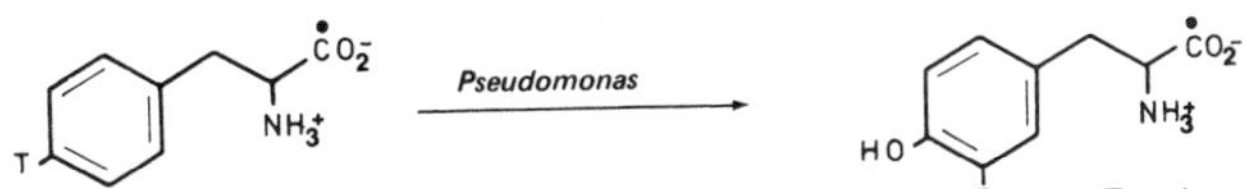
CO2-
NH3+
T
Pseudomonas
CO2-
NH3+
HO
T

Phenylalanine

● = ^{14}C $^{3}H : ^{14}C = 5.43$

Tyrosine.

$^{3}H : ^{14}C = 5.20$

[W.R. Bowman, W. Gretton and G.W. Kirby, *J. Chem. Soc, Perkin I*, 218 (1973)]

Fig. 1.9 Examples of incorporation experiments involving doubly-labelled precursors.

transform into two X molecules [cf. E. Leete, N. Kowanko, and R. A. Newmark, *J. Amer. Chem. Soc.,* **97**, 6826 (1975)]; (2) The isotopically labelled atoms are separated into fragments of W, but these are fortuitously diluted to an almost identical degree in their respective metabolic pools, before joining again to give X [D. J. Austin and M. B. Meyers, *Chem. Comm.,* 125 (1966)]. Reliable conclusions from incorporation experiments also depend on their number. A single experiment is usually not enough: the incorporations should be repeated on different biological samples, and the data statistically evaluated.

The use of stable isotopes, such as deuterium (^2H or D), ^{13}C, ^{15}N, ^{18}O, is limited to particular cases in biosynthetic studies, when the dilution value is not too high (not larger than 100). In practice, although the preparation of precursors with very high isotopic abundances is possible — for instance it is relatively easy to obtain samples containing almost exclusively specifically monodeuterated molecules ($D_1 = 100\%$) — the techniques used for revealing stable isotopes, such as mass spectrometry and nuclear magnetic resonance spectroscopy, do not permit measurements with sufficient accuracy for isotopic incorporations of less than 1%. This limit could well be lowered with the introduction of more sophisticated instruments in the future.

Stable isotopes are generally used as tracers in experiments with partially, or completely, purified enzymes, with homogenates and with microbial cultures or tissues; i.e. with systems characterised by high incorporations. The advantage of ^2H and ^{13}C is the absence of radioactivity, the absence of contamination dangers, and the possibility of localising the isotopically labelled atoms in the molecules by simple spectral interpretation, and without the need for cumbersome sample degradations.

1.6.2 Administration of labelled substrates to living organisms or cells

The administration of the labelled substrates is an experimental problem which should not be underestimated during a biogenetic enquiry. The success of any such experiments might be seriously jeopardised by low incorporations, etc.

The principal factors, which must be considered in order to optimise incorporations of a precursor, are the physical nature of the substrate, and the particular organism or cellular culture which is selected and, with living organisms or animals, the age of the individual(s) and the season. The season must always be considered, especially with plants, as these frequently synthesize certain secondary metabolites only, or mostly, in a given period of the year.

If the substrate is a gas, for instance $^{14}CO_2$ or $^{18}O_2$, the only serious difficulty consists in maintaining the individual or the culture in a well-closed vessel in the presence of the labelled substrate, which is almost always spontaneously absorbed by the cells. Compartmentation phenomena are not usually observed, as the small gas molecules easily pass through all cellular membranes. With solid or liquid substrates the techniques usually employed are:

1. *Administration through feeding.* This technique is often used with animals, for instance with insects.

2. *Injection* (intravenous, intraperitoneal, intramuscular, etc). This is advantageous when a systemic administration into animals is desired (rats, rabbits, frogs, etc). It can also be used for plants, the labelled substrate being dissolved or dispersed in water; a fine emulsion can be obtained using an emulsifier, for instance the polyoxyethylenesorbitan monooleate (TWEEN 80; 1%); the emulsion is then injected into the vascular system of the stalk.

3. *Perfusion.* This consists of maintaining an organ (isolated or not) in conditions of normal metabolic activity then circulating an appropriate liquid with the labelled substrate dissolved or suspended in it. Common examples are the perfusion of rat liver or heart, of human placenta, etc.

4. *Diffusion.* The substrate is dissolved or suspended in an aqueous medium, in which small slices of an organ, or the cells of a microbial culture or tissue are present.

5. *The cotton wick method.* This method is widely used for plants having a robust, but not too woody, stalk, and is a variation of the injection method. The end of a cotton wick (5-10 mm) is inserted, through a small longitudinal cut, into the plant stalk, whilst the other end is immersed into the solution (0.5-1.0 ml) to be incorporated, and which is usually contained in a small tube tied to the stalk itself.

   Because of the capillary action of the cotton wick and of the ascending flow of materials, the liquid rapidly diffuses into the aerial parts of the plant. This phenomenon can be observed experimentally by using a coloured solution, for instance eosin.

6. *Method of cutting the stalk.* The stalk of some plants can be cut near the base and the cut end is immediately immersed in water. About a further $\frac{1}{2}$ inch of the stalk is immediately cut underwater with some scissors; the small segment is discarded, whilst the stalk and its aerial parts (leaves, flowers) is quickly placed in a tube containing an aqueous solution of the precursor; the solution is rapidly sucked up by the plant because of transpiration through the leaves. This method is only suitable for studying rapid biosynthetic processes, since in most plants manipulated in this way the metabolic processes slow down and degenerate after a few hours. Cultivating plants for long periods in aqueous solutions containing the precursors is not a convenient method for incorporation experiments. This is due to the limited utilisation of the precursor, which is diluted in very large amounts of water, and to the poor permeability of organic molecules through the walls of the roots.

7. *The scarring method.* The stalk of the plant is scarred at various points with a blade, so that the more external veins are exposed. The liquid containing the labelled material is then added dropwise, with a microsyringe, to the scars.

8. *The brush method.* The labelled precursor is suspended in silicone oil, and the mixture is brushed onto the leaves of the plants, kept either in vases or in the ground. The silicone eases the penetration of the organic substances through the leaves. This method is convenient for long time range experiments.

SOURCE MATERIALS AND SUGGESTED READING

(a) *Chemistry of Natural Products (General)*

[1] L. Zechmeister, 1938-1980, *Fortschritte der Chemie Organischer Naturstoffe – (Progress in the Chemistry of Organic Natural Products),* 1-39, Springer Verlag.

[2] *The Chemistry of Natural Products, Special lectures presented at International Symposia,* **I** (Australia 1960), **II** (Prague 1962), **III** (Kyoto 1964), **IV** (Stockholm 1966), **V** (London 1968), **VI** (Mexico City 1969), **VII** (Riga 1970), **VIII** (New Delhi 1972), **IX** (Montreal 1974); Butterworths: **X** New Zealand 1976), Pergamon Press; *Pure and Applied Chemistry,* **51**, 681 (1979) (Bulgaria 1978).

[3] F. M. Dean, (1963), *Naturally Occurring Oxygen Ring Compounds,* Butterworths.

[4] B. Witkop, (1971), New Directions in the Chemistry of Natural Products: the Organic Chemist as a Pathfinder for Biochemistry and Medicine, *Experientia,* **27**, 1121.

[5] C. C. Chichester (Ed.), (1972), *The Chemistry of Plant Pigments,* Academic Press.

[6] T. K. Devon and A. I. Scott, *Handbook of Naturally Occurring Compounds,* (1975), **1**, (Acetogenins, Shikimates and Carbohydrates); (1972), **2**, (Terpenes).

[7] Rangaswami and Subba Rao, (1972), *Some Recent Developments in the Chemistry of Natural Products,* Prentice-Hall.

[8] J. M. Tedder, A. Nechvatal, A. W. Murray and J. Carnduff, (1972), *Basic Organic Chemistry,* Part 4, J. Wiley.

[9] J. B. Hendrickson, (1973), *The Molecules of Nature,* 3rd ed., Addinson-Wesley.

[10] K. Nakanishi, T. Goto, S. Ito, S. Natori, S. Nozce, *Natural Products Chemistry,* (1974), **1**, (1975), **2**, Academic Press.

[11] T. Robinson, (1975), *The Organic Constituents of Higher Plants,* Cordus Press.

[12] A. Ohsaka, K. Hayashi, and Y. Saway, (1976), *Animal Plant and Microbial Toxins,* 2 vols., Plenum Publishing Corporation.

[13] D. J. Faulkner, (1977), Interesting Aspects of Marine Natural Product Chemistry, *Tetrahedron,* **33**, 1421.

[14] IUPAC Commission on the Nomenclature of Organic Chemistry (CNOC), (1978), Nomenclature of Organic Chemistry. Section F: Natural Products and Related Compounds, *Eur. J. Biochem.*, **86**, 1.

[15] P. J. Scheuer (Ed.), (1978), *Marine Natural Products: Chemical and Biological Perspectives*, **1**, Academic Press.

[16] J. M. Brand, J. C. Young and R. M. Silverstein, (1979), *Insect Pheromones: A Critical Review of Recent Advances in Their Chemistry, Biology and Application*, in ref. 1, **37**.

(b) *Biosynthesis (Text-books)*
[17] J. D. Bu'Lock, (1965), *The Biosynthesis of Natural Products*, McGraw-Hill.

[18] T. A. Geissman and D. H. G. Crout, (1969), *Organic Chemistry of Secondary Plant Metabolism*, Freeman. Cooper and Co.

[19] M. Luckner, (1972), *Secondary Metabolism in Plants and Animals*, Chapman and Hall.

[20] J. Mann, (1978), *Secondary Metabolism*, Oxford University Press.

(c) *Biosynthesis (General)*
[21] R. Robinson, (1955), *The Structural Relations of Natural Products*, Clarendon Press.

[22] J. H. Richards and J. B. Hendrickson, (1964), *The Biosynthesis of Steroids, Terpenes, and Acetogenins*, W. A. Benjamin, Inc.

[23] J. B. Pridham and T. Swain (Eds.), (1965), *Biosynthetic Pathways in Higher Plants*, Academic Press.

[24] P. Bernfeld (Ed.), (1967), *Biogenesis of Natural Compounds, 2nd ed.*, Pergamon Press.

[25] J. D. Bu'Lock, *Essays in Biosynthesis and Microbial Development*, Wiley.

[26] D. M. Greenberg (Ed.), (1967–1972), *Metabolic Pathways, 3rd ed.*, **1–6**, Academic Press.

[27] H. Grisebach, (1967), *Biosynthetic Patterns in Micro-organisms and Higher Plants*, Wiley.

[28] S. Dagley and D. E. Nicholson, (1970), *An Introduction to Metabolic Pathways*, Blakwell Scientific Publications.

[29] E. Sondheimer and J. B. Simeone (Eds.), (1970), *Chemical Ecology*, Academic Press.

[30] *Biosynthesis: A Specialist Peridoical Report*, (T. A. Geissman Ed.), 1972–1975, **1–3**; (J. D. Bu'Lock, Ed.), 1976–1977, **4–5**.

[31] T. W. Goodwin and E. I. Mercer, (1972), *Introduction to Plant Biochemistry*, Pergamon Press.

[32] H. E. Street and W. Cockburn, (1972), *Plant Metabolism, 2nd ed.*, Pergamon Press.

[33] A. J. Birch, (1973), Biosynthetic Pathways in Chemical Phylogeny, *Pure and Applied Chemistry*, **33**, 17.

[34] B. V. Milborrow, (1973), *Biosynthesis and Its Control in Plants,* Academic Press.

[35] D. H. Northcote (Ed.), (1974), *Plant Biochemistry,* Butterworths.

[36] J. Bonner and J. E. Varner (Eds.), (1976), *Plant Biochemistry,* 3rd ed., Academic Press.

[37] H. Dalton, (1976), Microbial Metabolism, *Outline Studies in Biology Series,* Chapman and Hall.

[38] B. L. Goodwin, (1976), *Handbook of Intermediary Metabolism of Aromatic Compounds,* Chapman and Hall.

[39] T. W. Goodwin (Ed.), (1976), *Chemistry and Biochemistry of Plant Pigments,* 2 vols., 2nd ed., Academic Press.

[40] P. J. Lea and P. D. Norris, (1976), The Use of Amino-Acid Analogues in Studies of Plant Metabolism, *Phytochemistry,* **15**, 585.

[41] D. Ranganathan and S. Ranganathan, (1976), *Art in Biosynthesis,* Academic Press.

[42] A. R. Battersby, (1978), Ideas and Experiments in Biosynthesis in Further Perspectives in Organic Chemistry, (*Ciba Foundation Symposium 53*), Elsevier.

[43] A. J. Birch, (1978), Biosynthesis in Theory and Practice: Structure Determinations in Further Perspectives in Organic Chemistry (*Ciba Foundation Symposium 53*), Elsevier.

[44] U. Weiss and J. M. Edwards, (1978), *The Biosynthesis of Aromatic Compounds,* Wiley, New York.

(d) *Cell Biology, Microbiology, Physiology, Phytochemistry, Chemotaxonomy, Chemical Ecology.*

[45] R. Hegnauer, (1962-1973), *Chemotaxonomie der Pflanzen,* 7 vols., Birkhäuser Verlag, Basel.

[46] R. E. Alston and B. L. Turner, (1963), *Biochemical Systematics,* Prentice-Hall.

[47] P. H. Davis and V. H. Heywood, (1963), *Principles of Angiosperm Taxonomie,* Oliver and Boyd.

[48] T. Swain (Ed.), (1963), *Chemical Plant Taxonomy,* Academic Press.

[49] N. S. Cohn, (1964), *Elements of Cytology,* Harcourt, Brace and World.

[50] A. L. Lehninger, (1964), *The Mitochondrion,* Benjamin.

[51] M. R. J. Salton, (1964), *The Bacterial Cell Wall,* Elsevier.

[52] H. Linser, (1966), The Hormonal System of Plants, *Angew. Chem. Int. Ed. Engl.,* **5**, 776.

[53] T. Swain (Ed.), (1966), *Comparative Phytochemistry,* Academic Press.

[54] A. Berkaloff, J. Bourguet, P. Favard and M. Guinnebault, (1967), *Biologie et Physiologie Cellulaire,* Hermann, Paris.

[55] E. J. Du Praw, (1968), *Cell and Molecular Biology,* Academic Press.

[56] *Progress in Phytochemistry,* (L. Reinhold and Y. Liwschitz, Eds.), (1968-

1972), 1-3, Wiley-Interscience; (L. Reinhold, J. B. Harborne, T. Swain, Eds.), (1977), 4, Pergamon Press.

[57] Recent Advances in Phytochemistry, *Proceedings of the Annual Symposia of the Phytochemical Society of North America,* (T. J. Mabry, V. C. Runeckles, Eds.), (1968), 1; (M. K. Seikel and V. C. Runeckles, Eds.), (1969), 2; (C. Steelink and V. C. Runeckles, Eds.), (1970), 3; (V. C. Runeckles and J. E. Watkin, Eds.), (1971), 4, Appleton-Century-Crofts, (V. C. Runeckles and T. C. Tso, Eds.), 1972, 5; (V. C. Runeckles, and T. J. Mabry, Eds.), (1973), 6; (V. C. Runeckles, E. Sondheimer, and D. C. Walton, Eds.), (1974), 7; (V. C. Runeckles and E. E. Conn, Eds.), (1974), 8, Academic Press, (V. C. Runeckles, Ed.), (1975), 9; (J. W. Wallace and R. L. Mansell, Eds.), (1976), 10; (F. A. Loewus and V. C. Runeckles, Eds.), (1977), 11; (T. Swain, J. B. Harborne and C. F. Van Sumere, Eds.), (1979), 12; (T. Swain and G. Waller, Eds.), (1979), 13, Plenum Press.

[58] H. J. Rogers and H. R. Perkins, (1968), *Cell Walls and Membrane,* Chapman and Hall, London.

[59] H. W. Doelle, (1969), *Bacterial Metabolism,* Academic Press.

[60] J. B. Harborne and T. Swain (Eds.), (1969), *Perspectives in Phytochemistry,* Academic Press.

[61] W. Stiles and E. C. Cocking, (1969), *An Introduction to the Principles of Plant Physiology,* 3rd ed., Chapman and Hall.

[62] J. B. Harborne, (1970), *Phytochemical Phylogeny,* Academic Press.

[63] A. B. Novikoff and E. Holtzman, (1970), *Cells and Organelles,* Halt, Rinehart and Winston.

[64] E. Sondheimer and J. B. Simeone (Eds.), (1970), *Chemical Ecology,* Academic Press.

[65] R. Y. Stainer, M. Doudoroff and E. A. Adelberg, (1970), *The Microbial World,* 3rd ed., Prentice-Hall.

[66] J. T. O. Kirk, (1971), Chloroplast, Structure and Biogenesis, *Ann. Rev. Biochem.,* 40, 161.

[67] T. L. Lentz, (1971), *Cell Fine Structure,* Saunders.

[68] N. E. Tolbert, (1971), Microbodies − Peroxysomes and Glyoxysomes, *Ann. Rev. Plant Physiol.,* 22, 45.

[69] J. B. Harborne (Ed.), (1972), *Phytochemical Ecology,* Academic Press.

[70] G. Vanderkooi, (1972), Molecular Architecture of Biological Membranes, *Ann. N.Y. Acad. Sci.,* 196, 6.

[71] M. Davies, (1973), *Functions of Biological Membranes,* Chapman and Hall.

[72] J. B. Harborne, (1973), *Phytochemical Methods,* Chapman and Hall.

[73] W. D. James, (1973), *An Introduction to Plant Physiology,* 7th ed., Oxford University Press.

[74] L. P. Miller, (Ed.), (1973), *Phytochemistry,* 3 vols., Van Nostrand Reinhold.

[75] T. Swain, (Ed.), (1973), *Chemistry in Evolution and Systematics,* Butter-worth.

[76] J. Bendz and J. Santesson (Eds.), (1974), *Chemistry in Botanical Classification, Proceedings of the 25th Nobel Symposium held August 1973 at Sodergarn, Sweden,* Academic Press.

[77] M. Durand and P. Favard, (1974), *La Cellule – Structure et Anatomie Moleculaire,* Hermann.

[78] A. E. Radford, W. C. Dickinson, J. R. Massey and C. R. Bell, (1974), *Vascular Plant Systematics,* Harper and Row.

[79] C. D. Linden and C. F. Fox, (1975), Membrane – Physical State and Function, *Accounts Chem. Res.,* **8**, 321.

[80] D. S. Parson, (1975), *Biological Membranes,* Oxford University Press.

[81] J. E. Smith and D. R. Berry (Eds.), (1975), *The Filamentous Fungi,* 2 vols., E. Arnold.

[82] M. Barbier, (1976), *Introduction a l'Ecologie Chimique,* Masson.

[83] J. Haslam, (1976), *Membrane Assembly,* Chapman and Hall.

[84] J. B. Jones, (1976), Biochemical Concepts in Organic Chemistry in *Application of Biochemical Systems in Organic Chemistry,* ref. 98 of Chap. 2, **1**.

[85] D. Perlman, (1976), Procedures Useful in Studying Microbial Transformations in *Application of Biochemical Systems in Organic Chemistry,* ref. 98 of Chap. 2, **1**.

[86] H. Schildknecht, (1976), 'Chemical Ecology – A Chapter of Modern Natural Product Chemistry, *Angew. Chem. Int. Ed. Engl.,* **15**, 214.

[87] C. J. Sih and J. P. Rosazza, (1976), Microbial Transformations in Organic Synthesis in *Application of Biochemical Systems in Organic Chemistry,* ref. 98 of Chap. 2, **1**.

[88] P. M. Smith, (1976), *Chemotaxonomy of Plants,* Edward Arnold.

[89] E. Albone, Ecology of Mammals – A New Focus for Chemical Research, *Chem. in Britain,* **13**, 92.

[90] D. E. Fenton, (1977), Across the Living Barrier, *Chem. Soc. Rev.,* **6**, 325.

[91] K. H. Overton and D. J. Picken, (1977), Studies in Secondary Metabolism with Plant Tissue Cultures, in ref. 1 of Chap. 1, **34**.

[92] A. C. Wilson, S. S. Carlson and T. J. White, (1977), Biochemical Evolution, *Ann. Rev. Biochem.,* **46**, 573.

[93] B. Frank, (1979), Key Building Blocks of Natural Biosynthesis and Their Significance in Chemistry and Medicine, *Angew. Chem. Int. Ed. Engl.,* **18**, 429.

(e) *Isotope Methodology*

[94] S. Z. Roginskii and S. E. Shnol, (1965), *Isotopes in Biochemistry,* Davey and Co.

[95] C. H. Wang and D. L. Willis, (1965), *Radiotracer Methodology in Biological Science,* Prentice-Hall.

[96] D. Samuel and B. L. Silver, (1966), *Oxygen Exchange Reactions of Organic Chemistry,* (V. Gold Ed.), **3**, Academic Press.

[97] B. J. Wilson, (1966), *The Radiochemical Manual,* The Radiochemical Centre, Amersham, England.

[98] G. D. Chase and J. L. Rabinowitz, (1967), *Principles of Radioisotope Methodology,* Burgess.

[99] H. Simon and H. G. Floss, (1967), *Bestimmung de Isotopenverteilung in markierten Verbindungen,* Springer Verlag.

[100] V. F. Raaen, A. G. Ropp and H. P. Raaen, (1968), *Carbon-14,* McGraw-Hill.

[101] J. R. Catch, (1971), *Labelling Patterns: Their Determination and Signicance – Review 11,* The Radiochemical Centre, Amersham, England.

[102] A. F. Thomas, (1971), *Deuterium Labelling in Organic Chemistry,* Appleton-Century-Crofts.

[103] S. A. Brown, (1972), *Methodology* in ref. 30, 1.

[104] R. M. Caprioli, (1972), Use of Stable Isotopes in *Biochemical Applications of Mass Spectrometry,* (G. R. Waller, Ed.), Wiley-Interscience.

[105] R. A. Faires and B. H. Parks, (1973), *Radioisotope Laboratory Techniques,* 3rd ed., Butterworths.

[106] M. Tanabe, *Stable Isotopes in Biosynthetic Studies* in ref. 30, 1973; 2; (1975), 3; (1976), 4.

[107] I. M. Campbell, (1974), Incorporation and Dilution Values – Their Calculation in Mass Spectrally Assayed Stable Isotope Labelling Experiments, *Bioorganic Chem.,* 3, 386.

[108] A. Dyer, (1974), *An Introduction to Liquid Scintillation Counting,* Heyden and Son.

[109] E. A. Evans, (1974), *Tritium and Its Compounds,* 2nd ed., Butterworths.

[110] Y. Kobayaski and D. V. Maudsley, (1974), *Biological Application of Liquid Scintillation Counting,* Academic Press.

[111] K. D. Neame and C. A. Homewood, (1974), *Introduction to Liquid Scintillation Counting,* Butterworths.

[112] D. Staschewski, (1974), The Stable Isotopes of Oxygen in Research and Technical Applications, *Angew. Chem. Int. Ed. Engl.,* 13, 357.

[113] A. G. McInnes and J. L. C. Wright, (1975), Use of Carbon-13 Magnetic Resonance Spectroscopy for Biosynthetic Investigations, *Accounts Chem. Res.,* 8, 313.

[114] T. J. Simpson, (1975), Carbon-13 Nuclear Magnetic Resonance in Biosynthetic Studies, *Chem. Soc. Rev.,* 4, 497.

[115] I. M. Campbell, (1975), The Derivation and Use of Biosynthetic Incorporation and Dilution Values, *Phytochemistry,* 14, 683.

[116] E. Buncel and C. C. Lee (Eds.), *Isotopes in Organic Chemistry,* (1976); (Isotopes in Hydrogen Transfer Processes), (1977), 2; (Carbon-13 in Organic Chemistry), (1978), 3; (Tritium in Organic Chemistry), 4, Elsevier.

[117] I. M. Campbell, (1976), Determining the Structure of a Biosynthetic Pathway from Incorporation Measurements, *Phytochemistry,* 15, 1367.

[118] D. R. Hawkins, (1976), Tracing Metabolic Fates, *Chem. in Britain*, **12**, 379.

[119] H. W. Whitlock Jr., (1976), Enzymatic versus Chemical Synthesis of Molecules Labeled with Heavy Isotopes in *Application of Biochemical Systems in Organic Chemistry*, ref. 98 of Chap. 2, **2**.

[120] IUPAC Commission on the Nomenclature of Organic Chemistry (CNOC), (1978), Nomenclature of Organic Chemistry. Section H: Isotopically Modified Compounds, *Eur. J. Biochem.*, **86**, 9.

[121] M. J. Garson and J. Staunton, (1979), Some New N.M.R. Methods for Tracing the Fate of Hydrogen in Biosynthesis, *Chem. Soc. Rev.*, **8**, 539.

[122] F. W. Wehrli and T. Nishida, (1979), *The Use of Carbon-13 Nuclear Magnetic Resonance Spectroscopy in Natural Products Chemistry* in ref. 1, **36**.

Chapter 2

Enzyme Reactions

2.1 INTRODUCTION

Almost every reaction which takes place *in vivo* is catalysed by enzymes. If we think of a living organism as a highly specialised chemical laboratory, enzymes appear as extremely skilful operators, capable of performing the most sophisticated reactions at controlled rates and in high yields. The three fundamental properties of these biocatalysts are:

1. increase of reaction rate,
2. specificity towards reagents and products, and
3. kinetic control.

All known enzymes are proteins. A typical plant cell contains approximately $5{-}50 \times 10^8$ enzyme molecules, the diameter of which is usually between 20 to 100Å. They weigh from 10,000 up to several million Daltons, the number of amino-acids varying from 100 to 10,000 per molecule.

In order to understand how enzyme catalysis occurs some general comments on their structural features will be given.

2.2 PROTEIN STRUCTURE OF ENZYMES

An enzyme in its active state (the holoenzyme) is composed of a protein part (the apoenzyme) and, occasionally, by ions or molecules of a different nature (cofactors). The apoenzyme is always the largest part of the molecule, either by weight or volume, and its structure is most important for its catalytic effectiveness. The polypeptide character of enzymes is represented by the general formulae (1), typical of all apoenzymes, with R one of the side chains of the 20 most common amino acids, and with n larger than 75.

$$H_2N - \overset{\overset{\textstyle R^1}{|}}{CH} - CO.NH \left(\overset{\overset{\textstyle R^n}{|}}{CH} - CO.NH \right)_n \overset{\overset{\textstyle R^2}{|}}{CH} - CO_2H$$

(1)

The amino acids which form the enzymes are almost exclusively of the L-series, and their *sequence* determines the *primary structure* of the protein.

The peptide chain adopts a particular shape, the biologically active molecular conformation*, which stems from four factors:

(1) The partial double bond character of the –CO–N– bond, owing to delocalisation between the resonance formulae (2) and (3).

$$
\begin{array}{ccc}
\underset{\underset{\cdot\cdot}{-\,C\,-\,NH\,-}}{\overset{\overset{O}{\parallel}}{}} & \longleftrightarrow & \underset{\underset{+}{-\,C\,=\,NH\,-}}{\overset{\overset{O\,-}{|}}{}} \\[2em]
\textbf{(2)} & & \textbf{(3)}
\end{array}
$$

(2) A non-covalent interaction, called hydrogen bonding, between the amide nitrogen proton and the carbonyl oxygen atom of another, more or less distant, peptide bond. Each hydrogen bond (4) contributes about 0.8 kcal. mole^{-1} of stabilisation energy to the protein structure.

$$
\begin{array}{c}
\overset{\displaystyle R^3}{\diagdown} \\
R^1 \overset{\displaystyle}{N-H} \\
\diagdown \diagup \\
N-H \cdots\cdots O=C \\
\diagup \diagdown \\
O=C R^4 \\
\diagdown \\
R^2
\end{array}
$$

(4)

(3) The –S–S– bridges, formed by oxidative coupling of the thiol groups of cysteine; the cysteine units can be well apart from each other along the polypeptide chains.

(4) Other interactions such as the hydrogen bonds between hydroxy groups (e.g. from serine and tyrosine units) and carboxylic groups (e.g. glutamic and aspartic acid); dipole-electrostatic interactions; van der Waals' attractions and hydrophobic bonds. Hydrophobic bonds are the mutual attractions between the non-polar side chains of certain amino acids (e.g. leucine, isoleucine, valine, phenylalanine); they are due to the fact that these hydrocarbon groups prefer media with lower dielectric constants than that of water. The hydrophobic bond energies are of the same order as the hydrogen bond. The factors (1) and (2) are responsible for the *protein secondary structure* which can be in the α-helix, pleated, or completely random forms (Fig. 2.1).

*A protein is said to be 'native' if its atoms are spatially disposed as when it performs its functions *in vivo*.

The word 'denatured' is used for a protein which has, reversibly, changed its native structure, thanks to certain ambient conditions, e.g. experimental ones. If such a change is *irreversible*, the protein is said to be 'inactivated'.

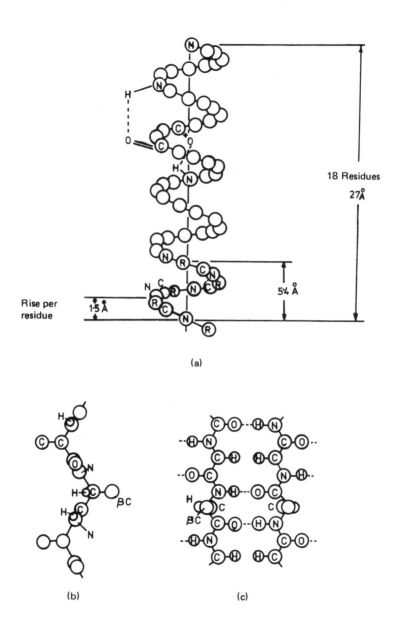

Fig. 2.1 Secondary structure of proteins: (a) α-helix; (b) pleated sheet, viewed parallel to the planes of the sheet; (c) pleated sheet, viewed perpendicularly to the planes of the sheet. (After E. H. Mercer, 'Kerotins and Kerotinization', Pergamon Press, New York, 1963, pp. 179, 182).

Disulphide bridges (3) and interactions of type (4) stabilise the *tertiary structure* of the polypeptide chain. The hydrophobic bonds, in particular, separate the amino acid side chains such that the non-polar ones are gathered within the internal part of the molecule, whilst the more polar and hydrophilic ones are on the surface, pointing towards the outside, in direct contact with the aqueous medium (Fig. 2.2).

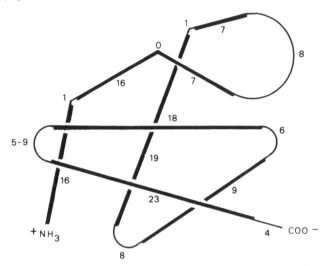

Fig. 2.2 Tertiary structure of myoglobin   a representation scheme of the poly-peptide chain; the thick lines indicate lengths of α=helical segments and the numbers the units of amino-acids along its length.

Some proteins can also exist as aggregates of identical or different sub-units. The haemoglobin molecule, for instance, is built up from four sub-units, in this case involving two pairs of identical sub-units. The sub-units are held together by hydrogen bonds and van der Waals' forces, which are responsible for the *protein quaternary structure*.

2.3 THE ACTIVE SITE

It is a universally accepted hypothesis, supported by many experimental facts, that biocatalysed reactions take place in well defined areas, or cavities, of the macromolecule forming the apoenzyme. These regions are called *active sites* and are of limited dimensions with respect to the whole protein surface. An enzyme can have one or many active sites, and a knowledge of the active site structure is essential for understanding the mechanism of enzyme catalysis.

The nature and the relative positions of the amino acids side chains contribute together, in a concerted manner, to make a site active. All the three (or four) levels of protein organisation thus determine the catalytic function. Because of their tertiary, three-dimensional structure, the functional groups of two amino acids, which are quite distant from each other along the polypeptide sequence, can be only a short distance apart in space and so, simultaneously or in immediate succession, interact with the substrate for the enzymic reactions. An example is the proteolytic enzyme chymotrypsin: in its active site the side chains of histidine (57th residue) and serine (195th residue) interact in the catalysis step: although they are 138 residues apart along the protein chain, they are only 2 or 3Å apart in space (Fig. 2.3(a)). (Note that the amino-acid residues are numbered conventionally, starting from the amine terminal of the polypeptide chain).

(a)   MECHANISM OF ACTION OF CHYMOTRYPSIN (*Endopeptidase*)

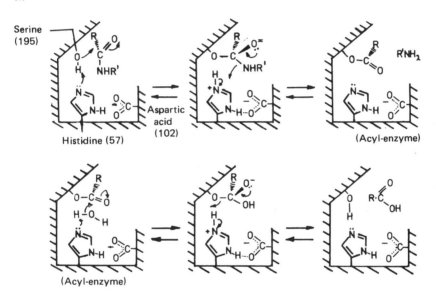

(b)   MECHANISM OF ACTION OF PHOSPHOGLUCOISOMERASE
       (D — Glucose — 6 — Phosphate ⇌ D — Frucrose — 6 — Phosphate)

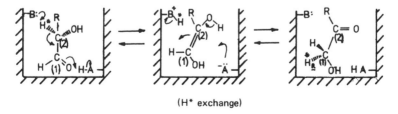

(H* exchange)

R = C(3) — C(6) of D — Glucose — 6 — Phosphate and D — Fructose — 6 — Phosphate

H* : exchangeable with medium (water) protons

B : base

AH : protic acid

Fig. 2.3 (contd. overleaf)

(c)   MECHANISM OF ACTION OF ALDOLASE   Fig. 2.3 (contd.)
       (Dihydroxyacetaone — Phosphate + D — Glyceraldehyde — 3 — Phosphate
       ⇌ D — Fructose — 1, 6 — Diphosphate)

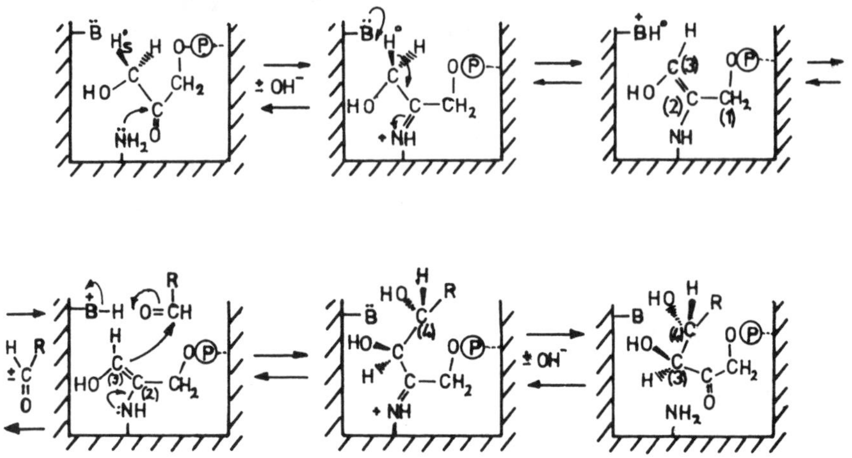

R = C(2) — C(3) of D — Glyceraldehyde — 3 — Phosphate or
    C(5) — C(6) of D — Fructose — 1,6 — Diphosphate

H● : exchangeable with protons from the medium (water)

Fig. 2.3 Mechanism of some enzymic reactions with schematic representations of
the active sites. (For definitions of H_R, H_S, re and si see Appendix 2).

In order for the reaction to take place, the reagent molecules must find
room in the active site of the apoenzyme. During the enzyme action several
components can be distinguished; the *substrate* (or substrates), that is the
compound (compounds) in which the chemical transformation occurs, the
cofactor molecules (see p. 61), and those of the *medium* (i.e. water). The last
two are not always necessary.

The specificity of enzymic reactions depends, first of all, on the structure
of the active site. In order to explain the specificity of enzymes the famous
German chemist Emil Fisher proposed a scheme according to which the active
site could be compared to a lock, and the substrate to the corresponding key.
This way of imagining the interaction between enzyme and substrate is called
the 'lock and key hypothesis' and still appears to be substantially valid. Fisher's
rigid model tends to be substituted with a more dynamic and elastic one, characterised by mutual interactions between enzyme and substrate (the 'induced-fit'
hypothesis). According to this last hypothesis the site should itself undergo
modification on interacting with the substrate, which in turn, would be altered
in its electronic distribution and in its molecular orbital structure during the
binding with the enzyme. An *enzyme-substrate* complex is so formed and the
precise electronic, steric, and entropic situation favours the ensuing chemical
reaction.

In order to describe the enzymic reaction mechanism, schematic drawings of the active site are often used, in which the functional groups, which interact with the reagent molecules and cause the catalysis, are stressed. Some enzymic mechanisms are shown in Fig. 2.3.

2.4 SPECIFICITY OF ENZYMES

The specificity of enzymes have various aspects. Before dealing with these it is necessary to clarify what certain words mean which are widely used in organic chemistry.

The reaction of a functional group, for instance of a double bond, which takes place without skeletal rearrangements and *exclusively* gives only one of two or more possible structural isomers, is called *regiospecific*. If one of the isomers prevails, but is not exclusive, the reaction is *regioselective*. An almost equimolar distribution of the isomers indicates a non-regiospecific reaction. The isomers in question are called *regioisomers*. The three cases are exemplified by reactions (a)–(c) of Fig. 2.4. The words regiospecific, regioselective, and non-regiospecific are also used with a wider meaning, to indicate the choice amongst two or more similar functional groups in the same molecule. For instance, glutamic acid in ethanol and hydrochloric acid is mostly esterified at the carboxylic group furthest from the amino group, as pointed out in reaction (e) of Fig. 2.4, and this would be called a regioselective process.

Regiospecific Reactions:

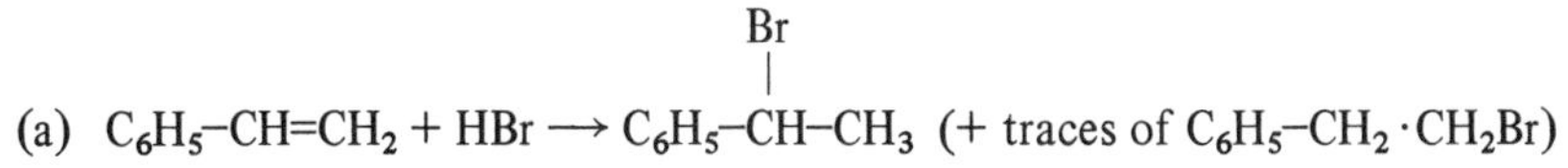

(a) $C_6H_5-CH=CH_2 + HBr \longrightarrow C_6H_5-\overset{\overset{\displaystyle Br}{|}}{C}H-CH_3$ (+ traces of $C_6H_5-CH_2 \cdot CH_2Br$)

(b) $CH_3 \cdot CH=CH_2 + HBr \xrightarrow{\text{PEROXIDE}} CH_3 \cdot CH_2 \cdot CH_2Br$ (sole product)

Regioselective Reaction:

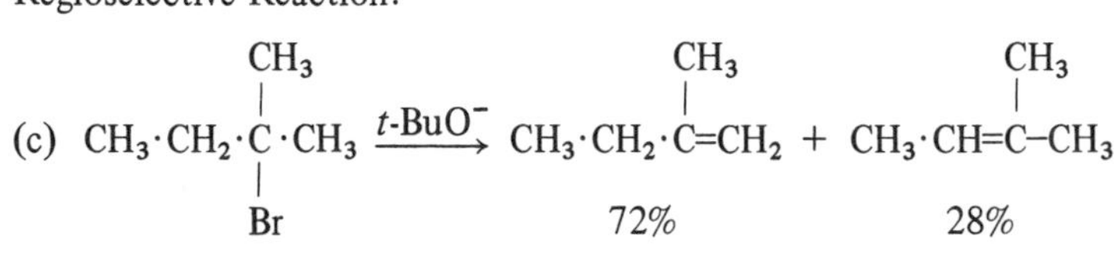

(c) $CH_3 \cdot CH_2 \cdot \overset{\overset{\displaystyle CH_3}{|}}{\underset{\underset{\displaystyle Br}{|}}{C}} \cdot CH_3 \xrightarrow{t\text{-BuO}^-} CH_3 \cdot CH_2 \cdot \overset{\overset{\displaystyle CH_3}{|}}{C}=CH_2 + CH_3 \cdot CH=\overset{\overset{\displaystyle CH_3}{|}}{C}-CH_3$

$\qquad\qquad\qquad\qquad\qquad\qquad\qquad\qquad 72\% \qquad\qquad\qquad 28\%$

Non-Regiospecific Reaction:

(d) $CH_3 \cdot \overset{\overset{\displaystyle CH_3}{|}}{C}H-\overset{\overset{\displaystyle CH_3}{|}}{C}H \cdot CH_3 \xrightarrow[\text{CCl}_4]{\text{Cl}_2,\ h\nu} Cl-CH_2 \cdot \overset{\overset{\displaystyle CH_3}{|}}{C}H-\overset{\overset{\displaystyle CH_3}{|}}{C}H \cdot CH_3 + CH_3 \cdot \overset{\overset{\displaystyle CH_3}{|}}{\underset{\underset{\displaystyle Cl}{|}}{C}}\!-\!-\!\overset{\overset{\displaystyle CH_3}{|}}{C}H \cdot CH_3$

$\qquad\qquad\qquad\qquad\qquad\qquad\qquad\qquad 55\% \qquad\qquad\qquad\qquad 45\%$

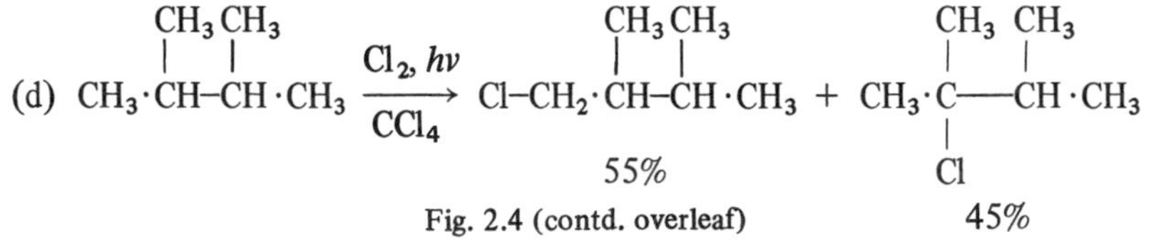

Fig. 2.4 (contd. overleaf)

Fig. 2.4 (contd.)

Regiospecificity also describes Preference in Reaction between Two or More Functions

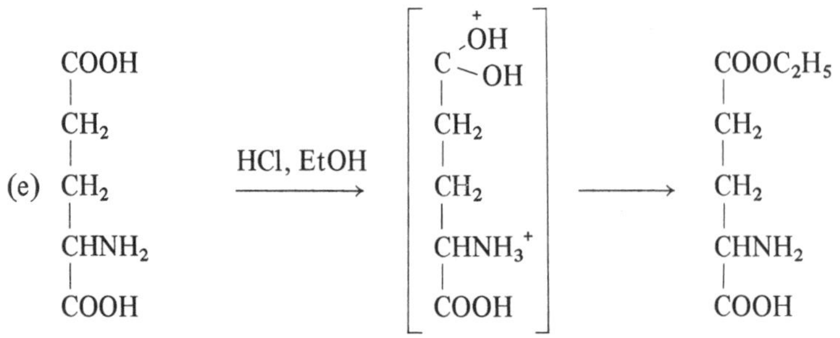

$$
\begin{array}{ccc}
\text{COOH} & \left[\begin{array}{c} \overset{+}{\text{C}}\!\!\overset{\text{OH}}{\underset{\text{OH}}{}} \end{array}\right. & \text{COOC}_2\text{H}_5 \\
| & | & | \\
\text{CH}_2 & \text{CH}_2 & \text{CH}_2 \\
| & \xrightarrow{\text{HCl, EtOH}} & | & | \\
\text{(e) CH}_2 & \text{CH}_2 & \longrightarrow & \text{CH}_2 \\
| & | & | \\
\text{CHNH}_2 & \text{CHNH}_3^+ & \text{CHNH}_2 \\
| & \left. | \right] & | \\
\text{COOH} & \text{COOH} & \text{COOH}
\end{array}
$$

Fig. 2.4 Regiospecificity and regioselectivity.

By analogy with the previous definitions it is advantageous to use the words *stereospecific, stereoselective,* and *non-stereospecific* in order to indicate the degree of steric specificity, that is the specificity with respect to the number of stereoisomers possible as reaction reagents or products (see Appendix 2). Use of these three terms is becoming more prevalent in bio-organic chemistry and some examples are given in (a)–(c) of Fig. 2.5.

STEREOSPECIFIC REACTION

(a)

94%   2%

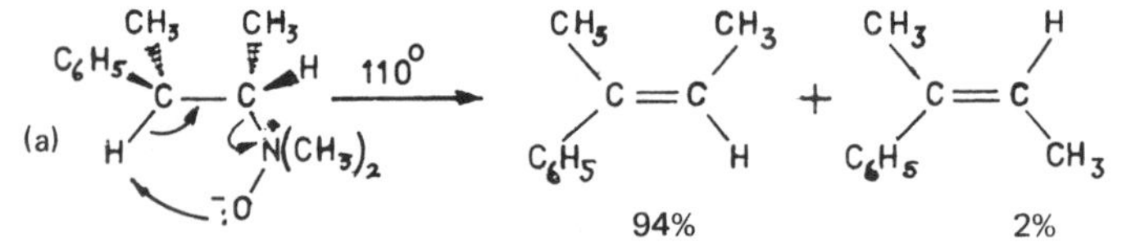

STEREOSELECTIVE REACTION

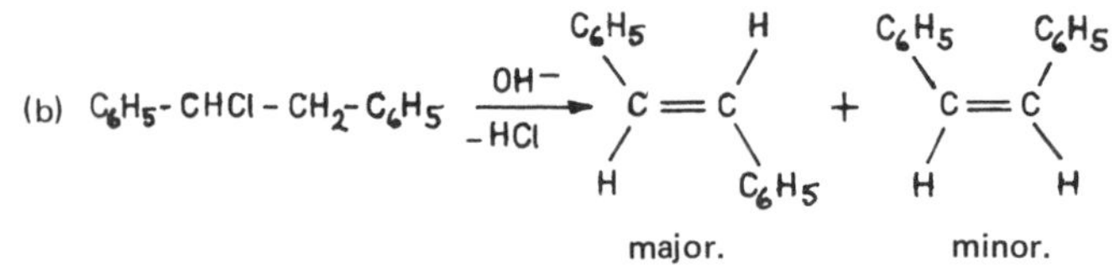

(b) $C_6H_5\text{-CHCl-CH}_2\text{-}C_6H_5 \xrightarrow[-\text{HCl}]{\text{OH}^-}$

major.   minor.

NON–STEREOSPECIFIC REACTION

(c)

50%   50%

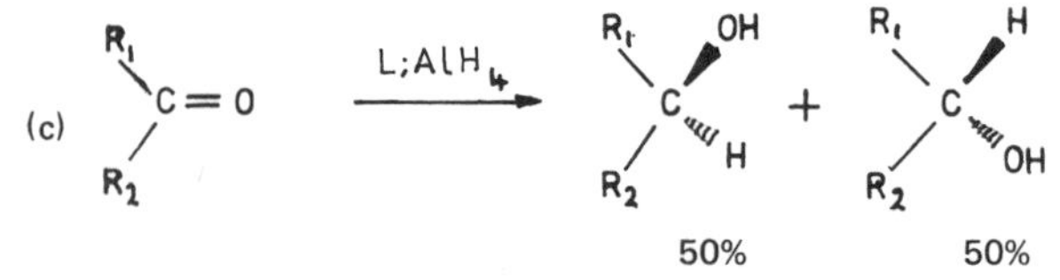

Fig. 2.5 Stereospecificity and stereoselectivity

The classical distinction between stereospecificity and stereoselectivity according to Eliel's criteria (see bibliography at the end of Appendix 2) is not very useful in dealing with enzyme reactions. Eliel defines as *stereospecific* transformations in which different stereoisomeric products are obtained from different stereoisomeric reagents: the formation of *meso*-2,3-dibromobutane by bromine addition to *trans*-2-butene, and of d,l-2,3-dibromobutane from *cis*-2-butene are thus stereospecific. *Stereoselective* reactions are instead characterised by a stereoisomeric preference as far as the products are concerned. So we should consider as stereoselevtive the high-temperature, free radical addition of hydrogen bromide to *trans*- (or *cis*-) 2-bromo-2-butene, which, in both cases gives 75% d,l-2,3-dibromobutane and 25% of the *meso*-isomer.

The peculiarity of enzymes is their specificity towards (a) the substrate, (b) the reaction product, and (c) substrate atoms or groups of atoms which are chemically identical (or very similar) but sterically different.

These various degrees of specificity hold, of course, for an enzyme reaction acting in either the forward or the reverse direction, which is because of *the principle of microscopic reversibility*: the molecule which acts as a substrate in the first direction is the product in the second one, and vice versa. The various types of enzymic specificity are illustrated in Fig. 2.6.

The specificity towards a *substrate* is *constitutional* when the enzyme catalyses the transformation of a compound, but it does not for other related compounds. For instance, phenylalanine-ammonia lyase (P.A.L.) converts L-phenylalanine into cinnamic acid, but does not convert L-tyrosine into *p*-hydroxy-cinnamic acid. If the *substrate* exists in different stereoisomeric forms, the term used is *stereospecificity* towards the substrate. P.A.L. acts on L-phenyl-alanine, but not on its D-enantiomer.

The specificity toward the reaction *product* can also be distinguished as both *constitutional* and *steric*. The first term concerns the enzyme capacity of catalysing a single reaction, and only that one, (for instance the ammonia elimination from the substrate L-phenylalanine in the P.A.L. case), or of forming only one regioisomer, for instance the Markownikow addition product (**5**), and not the anti-Markownikow (**6**) one. In this last case we could also refer to enzymic regiospecificity. The second type of specificity towards the product is defined by the exlusive formation of only one stereoisomer, for instance (*E*)-cinnamic acid* in the P.A.L. case, or L-malic acid, when oxaloacetic acid is reduced by oxaloacetic reductase.

Steric specificity towards the product results from two factors: the geometry of the active site of the enzyme, which determines the configuration of the new chiral centres, and basic chemical (stereoelectronic) requirements (see Appendix

*Double-bond stereoisomers, traditionally distinguished by the *cis* and *trans* prefixes, are indicated in this volume with the symbols Z and E respectively, according to the more recent nomenclature; (*E*)-cinnamic acid is the same as *trans*-cinnamic acid in the old nomenclature.

1, p. 437). The influence of the latter requirements are clear in the conversion of L-phenylalanine into (*E*)-cinnamic acid. The enzyme stabilises the most favourable transition state for the E2 type elimination of ammonia (corresponding to conformation A of Fig. 2.6): the steric structure of the reaction product stems unequivocally from this transition state geometry.

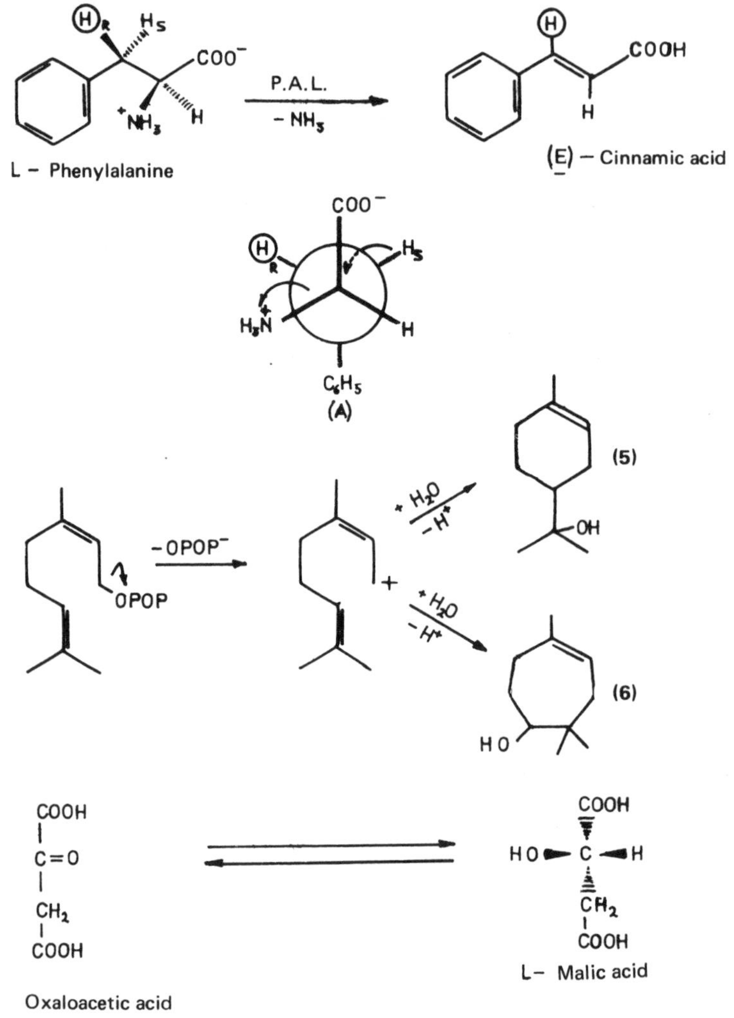

Fig. 2.6 Some examples of the specificity of enzymes.

Finally there is the most sophisticated type of enzymic specificity: it deals with the selection made by the enzyme, according to the aim of the reaction, for one atom or group of atoms, which are chemically identical (or almost so), but which are in the same molecule in *different* stereoisomeric surroundings. This type of stereospecificity is also found with P.A.L.: if we indicate the two benzylic hydrogens with the labels H_R and H_S,* only the H_R one is retained in (E)-cinnamic acid. The reason for such a choice can be found in the transition state (A), preferred for the elimination reactions. In many other cases, and always when *enantiotopic* atoms or atom groups are involved, the choice is related to the asymmetry of the active site of the enzyme.

2.5 COFACTORS, COENZYMES AND VITAMINS

There is a certain ambiguity in the biochemical literature about the definitions of the terms *substrate, cofactor, coenzyme, prosthetic group, carrier,* and *vitamin.*

The definition of substrate has already been given in the previous pages: an alcohol, oxidised to an aldehyde by a dehydrogenase, is the substrate of that particular enzyme.

The word *cofactor* should indicate all those inorganic ions (e.g. HPO_4^{--} or metallic ions, such as Zn^{++}, Mg^{++}) and organic substances (ATP, NAD(P), PLP, etc.) which co-operate with the enzyme in transforming the substrate. They are indispensable for the catalytic action of the enzyme. Nevertheless, whilst some cofactors are essential in making the active site of the enzymes receptive, and are found unchanged at the end of the reaction (metallic ions, for instance), other ones are transformed in a complementary manner to the substrate during the catalytic process. Such transformations are, however, reversible, and the cofactor in these cases is called a *coenzyme*; it is transformed into its initial form by other reactions. The conversion of an alcohol into the corresponding aldehyde requires the NAD coenzyme, which passes from its oxidised form (NAD^+) into its reduced form (NADH):

$$R-CH_2OH + NAD^+ \rightarrow R-CHO + NADH + H^+ \tag{1}$$

NADH can go back to NAD^+ by reduction of another substrate.

The co-enzymes, which are consumed and reformed in the complex play of metabolic processes, are called *carriers* when the reversible transformations which they undergo involve a temporary detachment from the apo-enzyme molecule: the pyridine nucleotides, for instance, bind themselves to a variety of enzymes, and carry a hydride ion from certain substrates to others (p. 82).

Prothetic groups are those coenzymes which are constantly bound to the enzymic part, although they transfer groups of atoms or electrons. An example is pyridoxal phosphate (PALD) (see p. 108).

---

*Such related atoms are called *diastereotopic*. For a deeper discussion of the various types of stereospecificity, its causes, and for stereochemical definitions the reader is referred to Appendix 2.

	Behaviour of the Coenzyme	Transferred Particles	Coenzyme Atoms Involved
	Carrier	H^- or $(2e^-, H^+)$	C – 4 of pyridine ring
	Carrier	H^- or $(2e^-, H^+)$	C – 4 of pyridine ring

(1)

R = H : NICOTINAMIDE ADENINE DINUCLEOTIDE OXIDISED FORM (NAD⁺)

REDUCED FORM (NADH)

(2)   Same as 1, R = PO₃H₂

NICOTINAMIDE ADENINE DINUCLEOTIDE PHOSPHATE
OXIDISED FORM (NADP⁺)
REDUCED FORM (NADPH)

(3)

$$COOH \quad CH_2SH$$
$$H_2N - CH_2 - CONH - CH - CONH - CH_2 - COOH$$

GLUTATHIONE
REDUCED FORM (GSH)

$$CH_2 - S - S - CH_2$$
$$- CH - \qquad - CH -$$

OXIDISED FORM (GSSG)

—S H

$2e^-$

Carrier

(4)

UBIQUINONE or
COENZYME Q (CoQ)
(see formula in figure 5.80)

Carbonyl groups

$2e^-$

Carrier

(5)

$$CH_2-CH-CH-CH-CH_2O-\overset{\overset{O}{\parallel}}{P}-OH$$
$$\quad\;\; OH \;\; OH \;\; OH \qquad\qquad OH$$

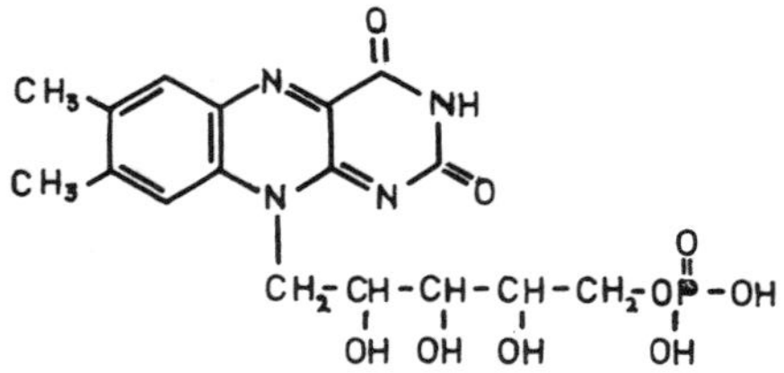

FLAVIN MONONUCLEOTIDE
OXIDISED FORM (FMN)

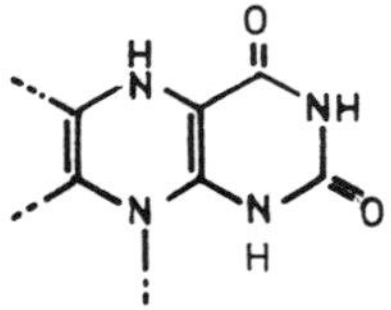

REDUCED FORM (FMNH$_2$)

$-N=C-C=N-$

$2e^-$

Prosthetic

Fig. 2.7 Principal coenzymes (continued next page)

Fig. 2.7 (*continued*)

	Behaviour of the Coenzyme	Transferred Particles	Coenzyme Atoms Involved
(6)	Prosthetic	$2e^-$	$- N = C - C = N -$
(7)	Carrier	Pi or PPi	Phosphate chain at $C - 5'$

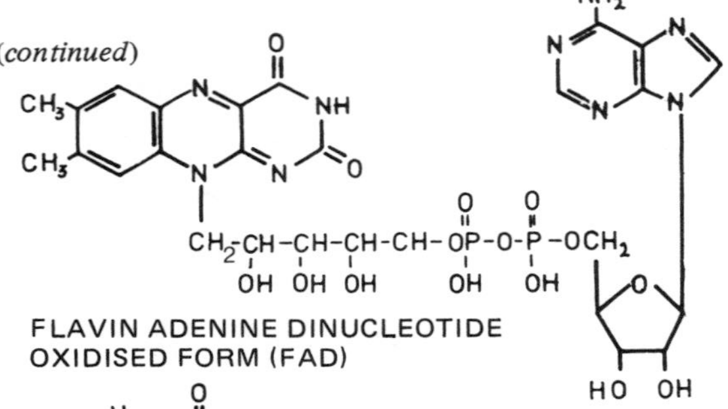

(6)

FLAVIN ADENINE DINUCLEOTIDE
OXIDISED FORM (FAD)

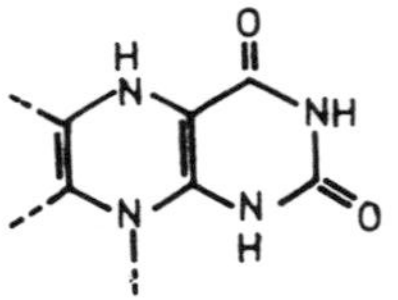

REDUCED FORM (FADH$_2$)

(7)

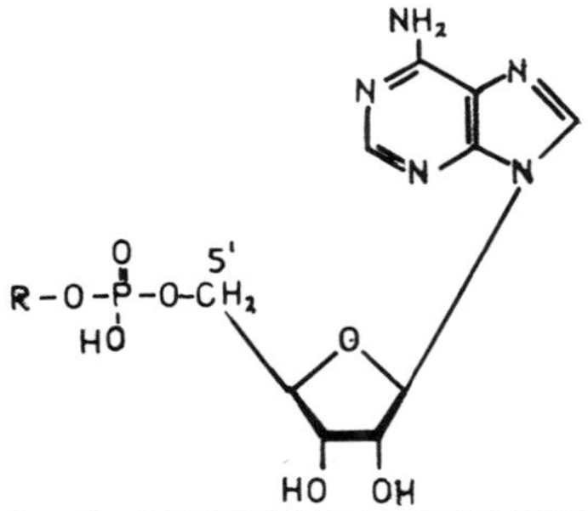

R = H : ADENOSINE MONOPHOSPHATE (AMP)

R = $- PO_3H_2$: ADENOSINE DIPHOSPHATE (ADP)

R = $- P(O)(OH) - OPO_3H_2$: ADENOSINE TRIPHOSPHATE (ATP)

(8)

by anology:

UMP, UDP, UTP (U = URIDINE)
CMP, CDP, CTP (C = CYTIDINE)
GMP, GDP, GTP (G = GUANOSINE)
IMP, IDP, ITP (I = INOSINE)

Carrier

Pi or PPi

Phosphate chain in C — 5^1

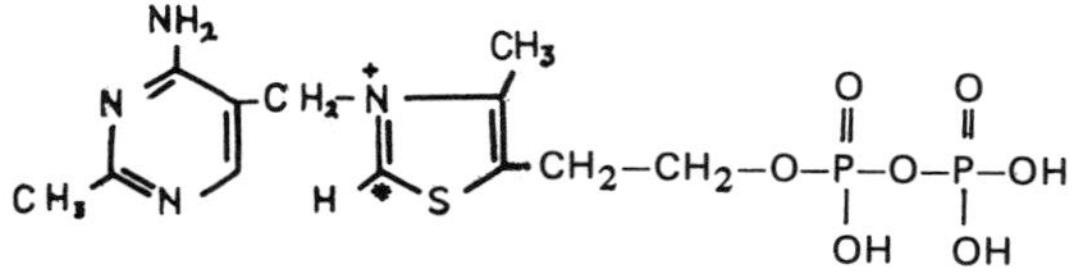

THIAMINE PYROPHOSPHATE (TPP o TPP$^+$)

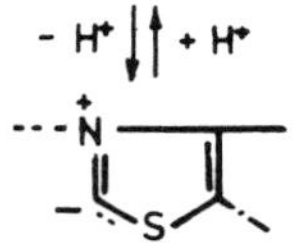

YLID FORM ($^-$TPP$^+$)

Prosthetic

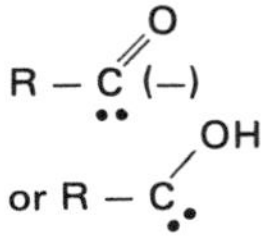

$$R-C(-)=O \quad \text{or} \quad R-C-OH$$

*C

(9)

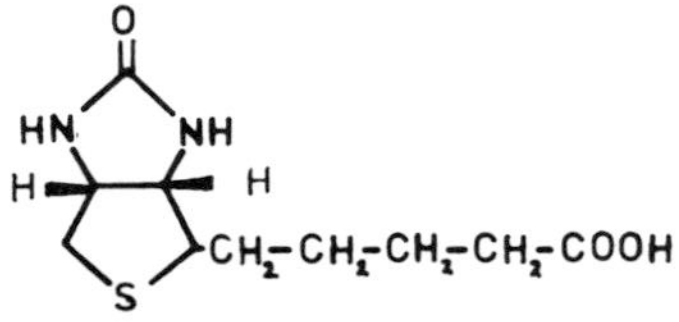

BIOTIN

Carrier

CO_2

not conclusively established

Fig. 2.7 Principal coenzymes (*continued next page*)

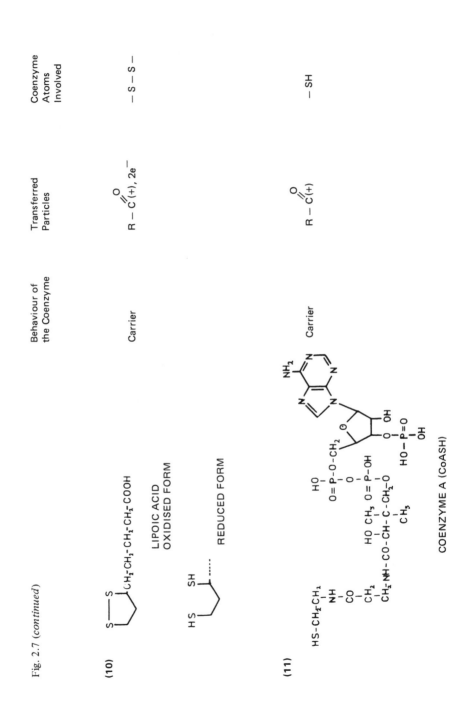

Fig. 2.7 (*continued*)

(12)	TETRAHYDROFOLIC ACID (FH$_4$; THF) (See Figure 7.8)	Carrier	one-carbon unit: −CH$_3$ / −CH$_2$OH / −CHO / −CH=NH (Fig 3.29)	N$_5$ / N$_5$ − N$_{10}$ / N$_5$ − N$_{10}$ / N$_5$, N$_5$ − N$_{10}$
(13)	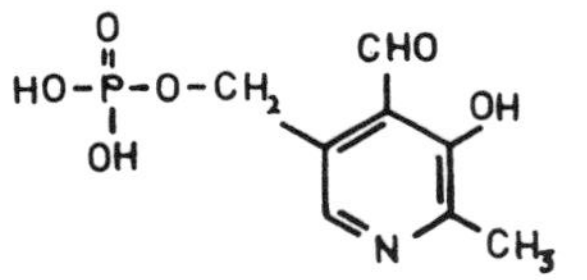 PYRIDOXAL PHOSPHATE (PLP)	Prosthetic	See Figure 2.37	−CHO
(14)	B$_{12}$ − COENZYMES (see Figure 2.41)	Carrier	−CH$_3$ (Figure 2.42) exchange a H atom with a ligand attached to an adjacent carbon atom (Figure 2.43)	Co^{+3}

Fig. 2.7 Principal coenzymes

In Fig. 2.7 are listed the most common coenzymes with their structural formulae; in each case there is indicated the type of bond with the apo-enzyme, the group which is transferred, and the molecular portion implicated in the 'transfer' reaction. The role of such coenzymes in biological reactions will be discussed further when dealing with the individual enzymic reations.

There is a close relation between coenzymes and *vitamins*. The latter are loosely defined as organic substances which are essential for the life and welfare of animals. They are usually either fundamental constituents of coenzymes (such as vitamin PP, or nicotinamide, contained in NAD or in NADP), or the coenzymes themselves (e.g. biotin).

2.6 ENZYME KINETICS

Kinetic and thermodynamic data on enzymic reactions generally indicate the following: (1) formation of a complex between the enzyme (active site) and reagents (substrate(s) and cofactor(s)); (2) reaction and subsequent formation of an enzyme-product complex; (3) dissociation of the enzyme-product complex, and regeneration of the active site in the appropriate state for receiving the reagents again.

The most simple case of the conversion of a substrate into a product is schematised in Fig. 2.8(b), where the transformation should be considered as reversible ones at the microscopic level. The enzymatic catalysis can be explained by comparing the reaction profile of the uncatalysed reaction with that of the catalysed one. In this last case there are two energy minima corresponding to the complexes enzyme-substrate (ES) and enzyme-product (EP). The rate of reaction of the enzymic process, determined by the slowest step, is a function of the energy difference between ES and the transition state $ES^{\ddagger}$; this energy difference is remarkably smaller than that observed for the corresponding uncatalysed reaction. The thermodynamic principle according to which a catalyst (the enzyme in this circumstance) does not alter the ratio of reagents and product at equilibrium still holds: the ratio depends on the *overall* free energy difference between reagents and products and not on the intermediate variations along the reaction path.

Many enzymic reactions can be described by the *Michaelis-Menten equation* (13), which holds within the following limitations:
(1) The concentration of only one substrate implied in the reaction can vary. Other substances can participate in the reaction only if their concentrations remain constant.
(2) The initial concentration of the substrate must be much larger than that of the enzyme.
(3) The whole enzyme catalysed process should fit the following scheme:

$$E + S \underset{k_{-1}}{\overset{k_1}{\rightleftharpoons}} ES \overset{k_2}{\rightarrow} E + P \tag{2}$$

There must therefore be a reversible reaction between the enzyme and substrate which allows the formation of the ES complex, and an irreversible step from this complex leading to the formation of the free enzyme and the product P. This is a further simplification of the enzymatic process showed in Fig. 2.8.

(a) UNCATALYSED REACTION

  $S \rightleftharpoons P$

  S = substrate,   P = product

(b) ENZYME – CATALYSED REACTION

  $S + E \rightleftharpoons ES \rightleftharpoons EP \rightleftharpoons P$

  E = enzyme

(c) SCHEMATIC STANDARD FREE ENERGY PROFILES FOR THE
   UNCATALYSED (- - - - - - -) AND ENZYME – CATALYSED (————)
   REACTION.

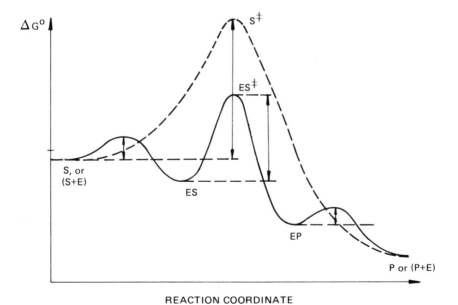

Fig. 2.8 Comparison between a simple uncatalysed and enzyme-catalysed reaction.

The following kinetic equations can be written for process (2):

$$-\frac{d[S]}{dt} = k_1[S].[E] - k_{-1}[ES] \tag{3}$$

$$\frac{d[ES]}{dt} = k_1[S].[E] - (k_{-1} + k_2).[ES] \tag{4}$$

$$\frac{d[E]}{dt} = (k_{-1} + k_2[ES] - k_1[S].[E] \tag{5}$$

$$\frac{d[P]}{dt} = k_2[ES] \tag{6}$$

Furthermore the conservation equation must be considered:

$$[E]_o = [E] + [ES] \tag{7}$$

Where $[E]_o$ is the initial concentration of the enzyme

Let us assume $k_2 \ll k_{-1}$:

This situation corresponds to a stationary state, in which the ES concentration is constant, and an equilibrium situation between such a complex and the free enzyme and substrate exists. That is,

$$K_S = \frac{k_{-1}}{k_1} = \frac{[E].[S]}{[ES]} \tag{8}$$

Where K_S is the dissociation constant of the enzyme-substrate complex.

If we combine equation (7) with equation (8) we have:

$$K_S = \frac{([E]_o - [ES])[S]}{[ES]} = \frac{[E]_o \cdot [S]}{[ES]} - [S]$$

which is

$$[ES] = \frac{[E]_o \cdot [S]}{K_S + [S]} \tag{9}$$

Substituting [ES] from equation (9) into equation (6), we obtain for the rate V:

$$V = \frac{d[P]}{dt} = \frac{k_2[E]_0 \cdot [S]}{K_S + [S]} \qquad (10)$$

At zero time, we shall have an expression for the initial rate of product formation V_0,

$$V_0 = \frac{k_2[E]_0 \cdot [S]_0}{K_S + [S]_0} \qquad (11)$$

that is $V_0 = f([S]_0)$ when $[E]_0$ is constant. If we put the initial rate V_0 measured experimentally as ordinate, and the initial concentrations $[S]_0$ (chosen for various experiments with a given $[E]_0$) as abscissa, we can draw a curve which, in agreement with equation (11), is a rectangular hyperbola (Fig. 2.9).

The asymptotic value of V_0 (which is V_{max}), obtained from the curve, mathematically corresponds to $K_2[E]_0$. This is understandable, if we consider that the product formation rate, expressed from equation (6), reaches its maximum value when the enzyme-substrate complex concentration is at a maximum, that is when $[ES] = [E]_0$.

Equation (11) then becomes:

$$V_0 = \frac{V_{max} \cdot [S]_0}{K_S + [S]_0} \qquad (12)$$

where V_{max} can be obtained graphically from Fig. 2.9.

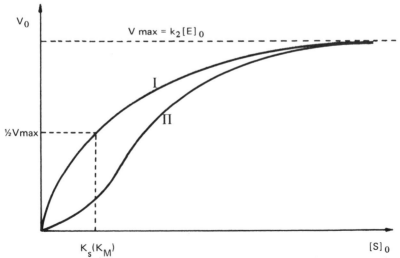

Fig. 2.9 A graphical representation of the Michaelis–Menten equation (curve I) and the kinetics of an allosteric enzyme reaction (curve II).

Finally, when $[S]_o = K_S$, we can write (since $K_2[E]_o = V_{max}$) according to equation (11):

$$V_o = \frac{V_{max} \cdot K_S}{2K_S} = \frac{V_{max}}{2}$$

which means that the K_S value (K_S is the dissociation constant) coincides with the substrate concentration such that the initial reaction rate is half of the maximum one. This particular concentration, obtained experimentally is indicated more frequently by the symbol K_M (Michaelis constant) and equation (12) can be written using the Michaelis-Menten formula

$$V_o = \frac{V_{max}[S]_o}{K_M + [S]_o} \tag{13}$$

The same equation, (13), can be obtained by another route, again accepting the condition for a stationary state, but not necessarily with $k_2 \ll k_{-1}$. In such a case the Michaelis constant, defined kinetically, will not equal K_S, but $\dfrac{k_{-1} + k_2}{k_1}$. Determining V_{max} and K_M through curves such as Fig. 2.9, will not generally be easy, as V_{max} tends to an asymptotic value as $S_o \to \infty$. Two linear equations, obtainable from (13) after some mathematical transformations are then used:

$$\frac{1}{V_o} = \frac{K_M}{V_{max}} \cdot \frac{1}{[S]_o} + \frac{1}{V_{max}} \tag{14}$$

$$V_o = V_{max} - K_M \cdot \frac{V_o}{[S]_o} \tag{15}$$

Expression (14) is the Lineweaver-Burk equation, and is graphically represented in Fig. 2.10; expression (15), the Eadie-Hofstee expression, is represented in Fig. 2.11.

A good agreement with the Michaelis-Menten equation (and so also with (14) and (15)) is found for the kinetics of enzymes such as hydrolases (E.C.3), many transferases (E.C.2) (only if a strong excess of acceptor is present), many isomerases (E.C.5) and lyases (E.C.4)*. In other cases much more complicated kinetic equations must be used: these reaction pathways are not expressed by sequence (2). An example is when a second complex (ES') between the enzyme and a substrate fragment is formed as an intermediate, according to the sequence:

$$E + S \underset{k_{-1}}{\overset{k_1}{\rightleftharpoons}} ES \overset{k_2}{\underset{+P_1}{\rightarrow}} ES' \overset{k_3}{\rightarrow} E + P_2 \tag{16}$$

*The classification of enzymes is described on p. 80 and in Appendix 5.

This happens for enzymes such as chymotrypsin, trypsin, cholinesterases, etc. which catalyse the hydrolysis of amide or ester groups. ES′ is called 'acyl-enzyme', as it results from the esterification of the substrate acid group with an OH group (from serine), or an-SH group (from cysteine), present in the enzyme active site (see Fig. 2.3(a)).

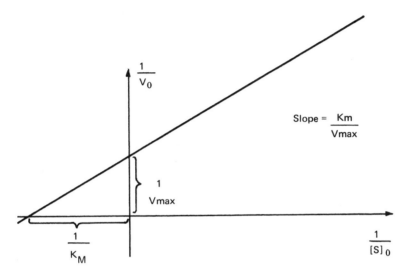

Fig. 2.10 Graphical representation of the Lineweaver–Burk equation .

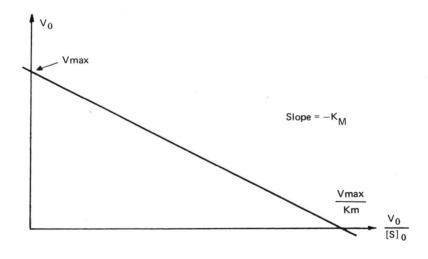

Fig. 2.11 Graphical representation of the Eadie–Hofstee equation.

There are many cases of enzymic processes in which two or more substrates are implied. An advantageous schematic representation of such processes has been introduced by Cleland, and is exemplified in Fig. 2.12 for two of the most frequent cases: the *ping-pong* mechanism and the *ordered* one. The ping-pong scheme is followed in transamination reactions, catalysed by transaminases with pyridoxal phosphate (PLP) as a prosthetic group (see p. 108), and dehydrogenation reactions catalysed by flavoproteins (see p. 85). The ordered mechanism is instead followed by most dehydrogenases, which are NAD and NADP dependent (see p. 84).

(a)   PING PONG MECHANISM

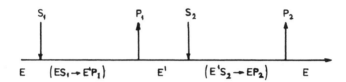

Sequence:

$$E + S_1 \rightleftharpoons ES_1 \rightleftharpoons E^1P_1 \rightleftharpoons E^1 + P_1$$
$$E^1 + S_2 \rightleftharpoons E^1S_2 \rightleftharpoons EP_2 \rightleftharpoons E + P_2$$

(b)   ORDERED MECHANISM

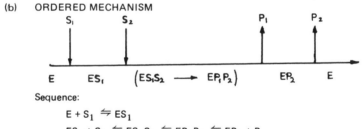

Sequence:

$$E + S_1 \rightleftharpoons ES_1$$
$$ES_1 + S_2 \rightleftharpoons ES_1S_2 \rightleftharpoons EP_1P_2 \rightleftharpoons EP_2 + P_1$$
$$EP_2 \rightleftharpoons E + P_2$$

Fig. 2.12 Principal enzyme mechanisms which implicate only two substrates (or one substrate and one coenzyme).

The four most important factors which can influence the rate of an enzymic process are: (1) the temperature, (2) the pH, (3) the presence of activators, and (4) the presence of inhibitors.

A temperature variation influences the rates of the single reaction stages, as it either modifies Arrhenius activation energies, or, according to the transition state theory, the equilibrium between the reagents and the corresponding activated states. Moreover, in enzymic reactions, it is possible to denature the enzymes, to a minor or major degree, on varying the temperature. Thus, conformational changes in the polypeptide chain can take place or the sub-units of enzyme complexes can change their degree and type of association, altering the active site, or the rates of processes occurring in it.

Almost all enzymes are extremely sensitive to the pH, and generally their activities lessen if the medium varies from a certain pH value defined as the *optimum pH.*

A graph illustrates how enzymic reaction rates depend on pH; bell-shaped curves are obtained such as that in Fig. 2.13. Such curves result from different pH effects which can alter (1) the steric structure of the protein molecule following the variation of dissociations of protic groups ($-NH_3^+$, $-COOH$, *etc.*); (2) the affinity of the substrate (and of its cofactors) for the enzyme; (3) the reactivity of the substrate bound to the enzyme.

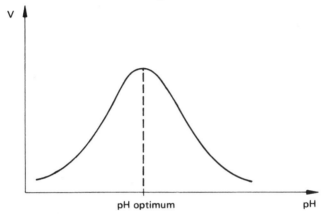

Fig. 2.13 Example of the pH dependence of an enzymic reaction rate.

Certain compounds can bind themselves to the enzyme without undergoing any chemical modification. Their presence on the protein can modify the catalytic activity of the enzyme in a positive or negative way. Those compounds which, after binding to an enzyme, increase the rate of a corresponding enzyme reaction are called *activators,* whilst the ones which act in an opposite way are called *inhibitors.* Some small molecules, and more often some inorganic ions (Zn^{2+}, Co^{2+}, Mg^{2+}) belong to the first class. The activation mechanisms, although not very well characterised, can be divided into two general types:
1. Interaction of the activator with the free enzyme, which is stabilised in the conformation most favourable for maximum catalytic activity.
2. Interaction of the activator with the substrate and formation of a complex which is more easily accepted in the active site of the enzyme. This last mechanism is typical of enzymic reactions which involve di- and triphosphate nucleosides (ADP, ATP, etc.); the activator in such cases is the Mg^{2+} ion.

Inhibition, in this textual discussion is limited to the qualitative differences between the various types of inhibitors. Inhibition is, by definition, reversible. The enzyme activity returns once the inhibitors are removed from the medium, by dialysis, electrophoresis, etc. *Inactivators* are those substances which *irreversibly* deactivate the enzyme.

1. *Competitive inhibitors.* Such molecules are generally similar to the natural substrate so that the latter is replaced by it in the active site. The inactive complex, inhibitor-enzyme (EI) is formed instead of the complex, ES.

2. *Partially competitive inhibitors.* These can combine with the enzyme, but the resulting EI complex can combine further with the natural substrate and give EIS; the same ternary complex (EIS) can be determined if I is combined with ES. The inhibition is due to the lower affinity of EI towards S and not to a slower collapsing rate (towards product, P) of the EIS complex compared to ES.

3. *Non-competitive inhibitors.* These bind to sites different from that of the substrate. They do not modify the affinity of the enzyme towards the substrate, and so S combines with EI as well as with E (and I combines with E as well as with ES); nevertheless the ternary complex EIS does not produce the product P.

4. *Partially non-competitive inhibitors.* They differ from the previous ones, since the EIS complex can, in fact, give the product, but at a slower rate than ES.

5. *Uncompetitive inhibitors.* They can combine with ES only, giving an inactive ternary complex.

Lineweaver-Burk graphs for these types of inhibition are given in Fig. 2.14.

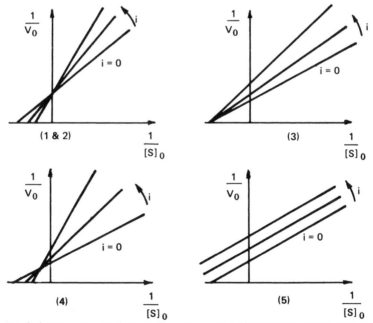

Fig. 214 Lineweaver–Burk curves for various types of inhibition, varying inhibitor (i) concentration. (1) and (2), competitive and partially competitive inhibition; (3) non-competitive inhibitor; (4) partially non-competitive inhibitor; (5) uncompetitive inhibitor.

2.7 CONTROL OF METABOLIC PROCESSES

At the beginning of this chapter we compared the living cell to a complex laboratory, and the enzymes to very clever, highly specialised 'bench chemists'. According to this picture, we can compare the metabolic pathways giving the various natural products to actual production lines along which the various 'chemists' (the enzymes) are placed.

In order for the various syntheses to proceed in a controlled and neat manner, according to the needs of the cell, it is necessary to constantly control the whole organisation by appropriate mechanisms which are capable of modulating the various enzymic activities. For instance, in a process such as $A \rightarrow B \rightarrow C \rightarrow D \rightarrow E$, if the concentration of E suddenly drops off for some reason, a quick recovery is necessary, which means a faster transformation of A into E.

Such modulation of metabolic processes can take place in two ways:

1. By increasing or decreasing the concentration of certain enzymes, corresponding to certain key passages in metabolic sequences.

2. By increasing or decreasing the catalytic activities of some of the more fundamental enzymes. The enzymes kept under control are called *regulator enzymes,* and the metabolites which start the modulation mechanisms are called *regulator metabolites.*

The first mechanism quoted above is explained by Monod and Jacob as follows: A special molecular structure, the 'operator' controls the action of the various chromosome 'structural' genes, which regulate the synthesis of the enzymes implied in a certain metabolic route, (for instance, the path from the precursor A to the compound D (Fig. 2.15)). The operator, strictly bound to its genes, is associated both with a special protein molecule, called a 'repressor', (R) — whose synthesis is directed by a particular gene (gene R), located in a position quite distant from the chromosome — and with the complex between R and a regulator metabolite. This metabolite can be one of the intermediates between A and D, or D itself. The interaction of the operator with this complex (RD) is different from the one with R alone, and might block the synthesis of the regulator enzymes (1, 2, etc.), promoted by the corresponding genes, and so interrupt the $A \rightarrow D$ chain.

The above case is a metabolic regulation by *repression.* The opposite phenomenon, or *induction,* involves an increase in the synthesis of the regulator enzymes when a metabolite concentration increases (for instance A). This derives from a different behaviour of the operator, whose action is blocked by R, and unblocked by RA.

The second mechanism for controlling enzymatic processes takes place through *direct* interactions of regulator metabolites on the regulator enzymes. These interactions can be of the type already described (activation or inhibition) or of a different kind, called *allosteric.*

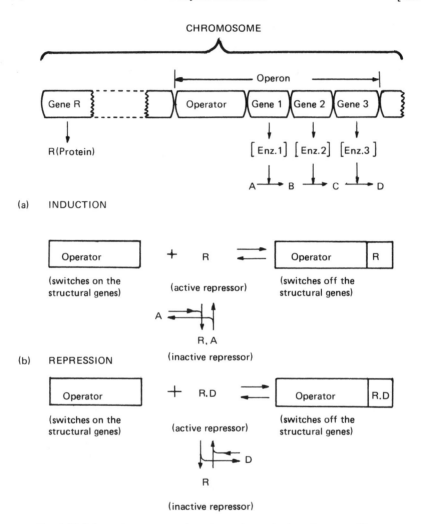

Fig. 2.15 Schematic representation of genetic mechanisms for controlling metabolism by induction and repression methods.

The interactions between *allosteric effectors* and *allosteric proteins* have been and still are under active investigation. The commonly accepted interpretive model, briefly described here, is a simplification of real phenomena, but fits the experimental data (which cannot be interpreted with classical Michaelis kinetics) quite well. In order to explain the action of allosteric proteins it has to be assumed that in their complex structure certain enzymes form an oligomeric structure in which there are, besides the active site(s) for the substrate, some regulator sites (allosteric sites), which can accept the allosteric effectors. In binding to the enzyme, a substrate or allosteric effector exerts a conformational change on the enzyme which normally (in the unbound state) can equilibrate between various conformationsl forms, only one of them having affinity for the substrate. The interaction of an (oligomeric) enzyme system with an effector molecule (different from the substrate) will stabilise one form, with the consequent formation, or disappearance of a large number of active sites. This effect is called *heterotropic.** The inhibition exerted by isoleucine on the enzyme catalysing the conversion of threonine into α-ketobutyric acid is an example of allosteric interaction. With such regulating mechanisms living organisms can control linked biosynthetic processes according to their metabolic requirements.

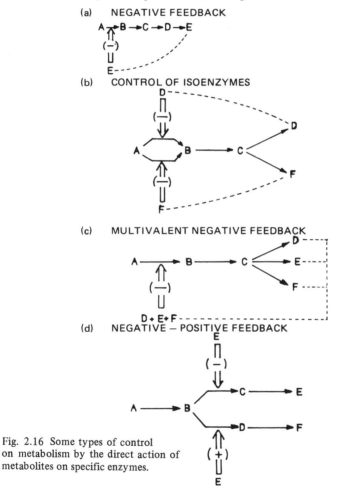

Fig. 2.16 Some types of control on metabolism by the direct action of metabolites on specific enzymes.

*There is also a homotropic effect when a molecule which is bound to an enzyme system influences the binding of other molecules identical to itself. Such molecules can be the enzyme substrate itself. The 'cooperation' of molecules undergoing enzymic transformations is very common in Nature.

The final product of a linear chain can, for instance, inhibit the first enzyme in a reaction sequence (*negative feedback* or *retroinhibition*) (Fig. 2.16(a)). In an analogous manner an enzyme situated along the path towards a certain metabolite can be activated by a degradation product of such a compound (*positive feedback* or *retroactivation*). Other situations include an intermediate metabolite preceding such a regulator enzyme which can activate either it (activation by a precursor) or an intermediate of a parallel pathway (parallel activation). If a biosynthetic route has side-branches, it is necessary to regulate both the main path and the various branches. This can be achieved in four ways:

1. Utilising, for the first stage of the main path, a series of isoenzymes, each of which is specifically inhibited by one of the final products from the various branches (Fig. 2. 16(b));
2. Only one enzyme is used to catalyse the first reaction and which is only inhibited by the simultaneous presence of all the final products (multivalent negative feedback) (Fig. 2.16(c)).
3. Inhibition of the biosynthesis of the enzyme by a coordinated action of all the final products (multivalent repression).
4. The final product of one of the branches can inhibit the first enzyme of such a branch and activate the first enzyme of another branch of the chain (negative-positive feedback) (Fig. 2.16(d)); the synthesis of a specific enzyme for a branch can be induced by the final product of another branch (induction) or by the products of many branches (multivalent induction).

2.8 CLASSIFICATION OF ENZYMES

The necessity for a rational classification and nomenclature for enzymes become overpowering by 1960, when it was realised that, compared with the 80 known enzymes of 1930, there were then over 1300. This number has since rapidly increased.

In 1961 the International Union of Biochemistry published a collection of rules for the systematic nomenclature of enzymes. These rules, now universally accepted, can be summarised as follows.

All enzymes are divided into six large classes: (1) *Oxidoreductases,* which catalyse oxidation–reduction reactions, and so transfer electrons from a donor (which is oxidised) to an acceptor (which is reduced); (2) *transferases,* which catalyse the transfer of atomic groups from a donor to an acceptor; (3) *hydrolases* which catalyse the hydrolysis of C-O and C-N bonds; (4) *lyases,* which promote either the removal (non-hydrolytic) of a group of atoms from the substrate, to subsequently form a double bond, or the addition of a group of atoms to a double bond; (5) *isomerases,* which transform the substrate into an isomer; (6) *ligases,* also known as *synthetases,* which catalyse the union of two molecules by breaking a pyrophosphate bond in ATP or a similar triphosphate.

Every enzyme is given four numbers, arranged in a well defined sequence and separated by dots. For instance phenylalanine–ammonia lyase is distinguished as E.C.4.3.1.5.

E. C. means 'enzyme classification'; 4 indicates the main class to which the enzyme belongs (which is, according to the preceding classification, the lyases); the second number indicates the subclass; the third one the sub-subclass, and the fourth one the enzyme series number within such a sub-subclass. Classes, sub-classes, and sub-subclasses are listed in Appendix 5. The subclass is assigned according to the following criteria: for oxidation-reduction reactions the nature of the oxidised group has been considered (alcohol, aldehyde, etc.); the trans-ferred group has been taken into account for transferases; the type of hydrolysed bond for hydrolases; for the lyases it is considered to be the type of bond broken when the 'leaving group' separates from the substrate, (for instance, C-N for phenylalanine-ammonia lyase); the type of isomerisation is accounted for in isomerases, and the type of bond formed for ligases.

As far as the sub-subclasses are concerned, the elements considered have been: the electron acceptor (NAD^+, $NADP^+$, O_2 etc.) for oxidoreductases; a further specification of the transferred group, or the acceptor, for transferases; a further specification of the hydrolysed bond for hydrolases; the nature of the leaving group for lyases (ammonia, for phenylalanine–ammonia lyase): a specifi-cation of the nature of the isomerised compound in isomerases; the nature of the synthesised substance for ligases.

This rational enzyme nomenclature complements the older, trivial but traditional one. Thus the major enzymes which have been known for a long time are still called by their old names, generally with their code number also quoted (for instance the enzymes of Chapter 3).

2.9 ENZYME REACTION MECHANISMS

In most reactions which take place *in vivo,* at least one of the reagents (beside the enzyme) is an organic compound.

From a mechanistic point of view therefore, these reactions follow the principles and rules of organic chemistry. In Appendix 1 the general mechanisms of organic reactions are discussed and, at this point only those reactions (oxida-tions, alkylations) which most frequently occur in metabolic processes are detailed.

2.9.1 Oxidations

Four fundamental types of biological oxidations are known (Fig. 2.17). The enzymes which catalyse them are called, substrate-acceptor oxidoreductases (see Appendix 5). They are commonly referred to by their type of oxidative behaviour, such as dehydrogenases, oxidases, mono-oxygenases, dioxygenases, etc. *Dehydrogenases* formally extract two hydrogen atoms from the substrate, but do not utilise molecular oxygen as a hydrogen acceptor, instead donating them to other organic molecules. The hydrogen acceptor can also be an electron acceptor. The *oxidases* release the substrate electrons to molecular oxygen, or, such as in *peroxidases*, to hydrogen peroxide, after dehydrogenating the substrate. The *monooxygenases* donate to the substrate just one atom from an oxygen molecule, the other atom being reduced to water by an appropriate hydrogen donor (AH_2). With the *dioxygenases* both atoms of the O_2 molecule are transferred to the substrate. *Dehydrogenases* can be divided into two categories, according to the coenzyme utilised: the latter can either be a pyridine based nucleotide (NAD or NADP) or a flavin (FMN or FAD) one (Fig. 2.7). In the first case the coenzyme acts as both a hydrogen acceptor and carrier ($K_{diss.}$ of the coenzyme-apoenzyme complex is $> 10^{-7}$ M); in the second case the flavin nucleotide represents a typical prosthetic group (the flavoprotein K_{diss} is $< 10^{-8}$ M) which remains bound to the apoenzyme and transfers the hydrogen to other acceptor molecules, for instance intermediates of the respiratory chain, some disulphides (oxidised glutathione), pyridine nucleotides, etc.

    (a) Dehydrogenases:

$$SH_2 + A \rightleftharpoons S + AH_2$$

    (b) Oxidases:

        (a) $SH_2 + \frac{1}{2}O_2 \longrightarrow S + H_2O$

        (b) $2SH_2 + H_2O_2 \longrightarrow 2SH^\bullet + 2H_2O$

    (c) Monooxygenases:

$$S + O_2 + AH_2 \longrightarrow SO + A + H_2O$$

    (d) Dioxygenases:

$$S + O_2 \longrightarrow SO_2$$

Fig. 2.17. Types of biological oxidation. S = substrate; A (or A^+) = hydrogen acceptor AH_2 (or $AH + H^+$) = hydrogen donor.

    The two types of dehydrogenases differ in their coenzymes, in the mechanism, and to a certain extent in the type of functions they attack. The NAD(P)-dependent enzymes usually oxidise hydroxyl groups to carbonyl groups, and amino-groups to imino-groups. Flavoproteins oxidise simple bonds $R_2CH-CHR_2$ to double bonds $R_2C=CR_2$, Fe^{2+} to Fe^{3+}, and various other substrates.

Recently, Hamilton proposed a rationalisation of dehydrogenation processes promoted by dehydrogenases and by oxidases. It is based on the fact that the following six mechanisms are possible *a priori*:

(1) Transfer of both hydrogens as atoms ($H^{\cdot} + H^{\cdot}$): mechanism AA;

(2) Transfer of one hydrogen as an atom, of the other one as hydride ion, and recovery of one electron ($H^{\cdot} + H^{-} - e^{-}$): mechanism AHy.

(3) Transfer of one hydrogen as an atom, of another as a proton, and of one electron ($H^{\cdot} + H^{+} + e^{-}$): mechanism AP.

(4) Transfer of both hydrogens as hydride ions and recovery of two electrons ($H^{-} + H^{-} - 2e^{-}$): mechanism HyHy.

(5) Transfer of one hydrogen as a proton, and of another as a hydride ($H^{+} + H^{-}$): mechanism PHy.

(6) Transfer of both hydrogens as protons and of two electrons ($H^{+} + H^{+} + 2e^{-}$): mechanism PP.

The most common mechanisms for enzymic oxidations are PP and PHy. The reason for these preferences is explained by the ease of proton transfer in aqueous media and at room temperature, relative to the transfer of either hydride ions or hydrogen atoms. The acid-base catalysis observed in most enzymic processes is also attributed to this factor. Major questions arise on considering the sequences by which the protons and electrons are exchanged between the substrate, the aqueous medium, and the coenzyme. Hamilton has hypothesised the following sequences in the more important dehydrogenases and oxidases. The oxidation of an alcohol group to a carbonyl group by NAD(P) — dependent dehydrogenases can be defined as a PHyC mechanism. The first hydrogen, the hydroxyl one, is lost as proton, whilst the second one, bound to the carbon, is presumably lost as a hydride ion; an intermediate is then formed, resulting from the conjugation (symbol C) between the pyridine nucleus of the coenzyme and the substrate. The specific mechanism of Fig. 2.18(b) accounts for the high stereospecificity of dehydrogenases (1) towards the stereoheterotopic (see Appendix 2) hydrogen in the 4-position of the the pyridine coenzyme; (2) towards the configuration of the substrate alcohol-bearing carbon (if chiral) and (3) towards the two stereoheterotopic hydrogens α to the hydroxyl group of primary alcohols. The specificity towards the substrate selected is not extremely pronounced (for instance towards the substituents of group C_2, see Fig. 2.18). The various specificities can be explained by considering the geometry of the intermediates formed at the active site of the enzyme (the intermediate is kept in the appropriate conformation for a hydride ion transfer by the enzyme itself).

Some dehydrogenases specifically exchange the hydrogen H_A of the dihydropyridine ring with the substrate, as in Fig. 2.18. (A-stereospecificity: for instance horse liver alcohol dehydrogenases (LADH) and the yeast dehydrogenase (YADH)); some other enzymes exchange the H_B hydrogen (B-stereospecificity: for instance the yeast D-glucose-6-phosphate dehydrogenase (YG6P DH)).

(a)

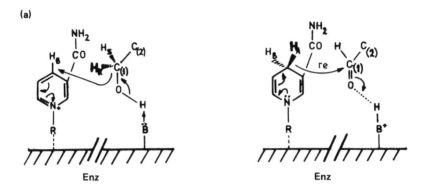

(b)   (PPC mechanism)

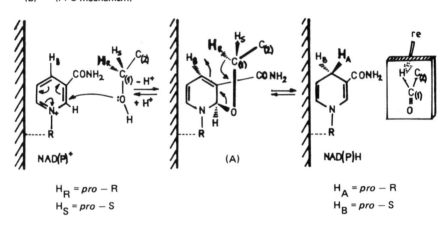

$H_R = pro - R$

$H_S = pro - S$

$H_A = pro - R$

$H_B = pro - S$

(c)   LADH, YADH

Fig. 2.18 Possible mechanisms of action for NAD(P)–dependent dehydrogenations

The stereospecificity towards the geminal hydrogens of the primary alcoholic group, or of the two faces of the corresponding aldehyde, is illustrated in Fig. 2.18 and exemplified in Fig. 2.18(c) for LADH and YADH. Such stereo-specificity is often used for determining the absolute configuration of primary alcohols which are stereospecifically labelled with ^2H or ^3H in the enantiotopic hydrogen α to the hydroxyl group, or for synthesising such alcohols. The flavine coenzymes utilise oxidation mechanisms of either the PPC type (Fig. 2.19(a)), or more often the PPR type. By considering the actual sequence in the loss of protons and electrons from the substrate, as indicated in Fig. 2.19(b), and the subsequent formation of intermediate radical species e.g. (7), Hamilton suggested the whole process should be indicated by the symbol PPR where R stresses the partial radical character of the mechanism.

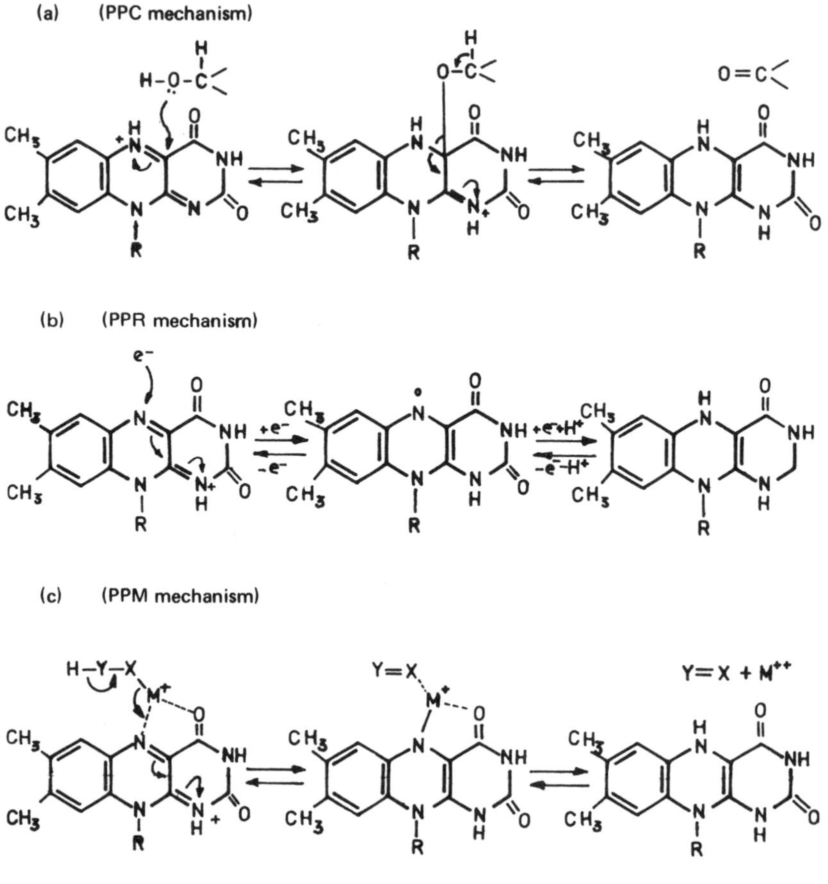

(a)   (PPC mechanism)

(b)   (PPR mechanism)

(c)   (PPM mechanism)

X = O, NH, S
Y = C, etc
M = metal cation.

Fig. 2.19 Mechanisms of action of flavoenzymes, according to Hamilton (with FAD or FMN as the prosthetic group).

Amongst the most important oxidases are ascorbic oxidase, catalysing the oxidation of ascorbic acid to dehydroascorbic acid: the cytochrome oxidase (reduced cytochrome c to oxidised cytochrome c); tyrosinase (o-diphenols to o-quinones); and laccase (p-diphenols to p-quinones) (Fig. 2.20). All these oxidases contain copper, and their catalytic activities depend on the presence of cupric and cuprous ions, which are appropriately complexed within the molecule and which transfer electrons from the substrate to the oxygen thus changing their oxidation state in a reversible manner.

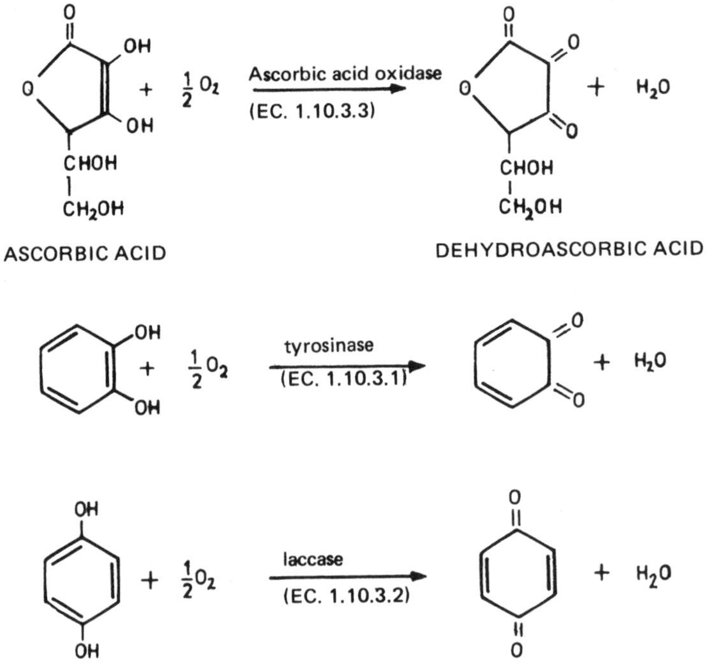

Fig. 2.20 Some oxidases

The mode of action of tyrosinase can be represented, according to Hamilton, as in Fig. 2.21, and can be assigned the symbol PPM, here M refers to the active participation of a metallic ion in the electron transfer phase. The term tyrosinase commonly refers to a large group of enzymes, often called by other names (phenol oxidase, polyphenol oxidase, catechol oxidase, phenolase, etc.); they can be distinguished from each other by their substrate specificity. Tyrosinases are notable for their double catalytic role; they behave both as oxidases, catalysing the oxidation of o-diphenols and as monooxygenases helping the hydroxylation of monophenols (such as tyrosine) to diphenols (such as dopa). It was previously thought that tyrosinases consisted of multienzyme complexes ('phenolase complexes'), but it has now been established that both catalytic activities are associated with the same enzyme species.

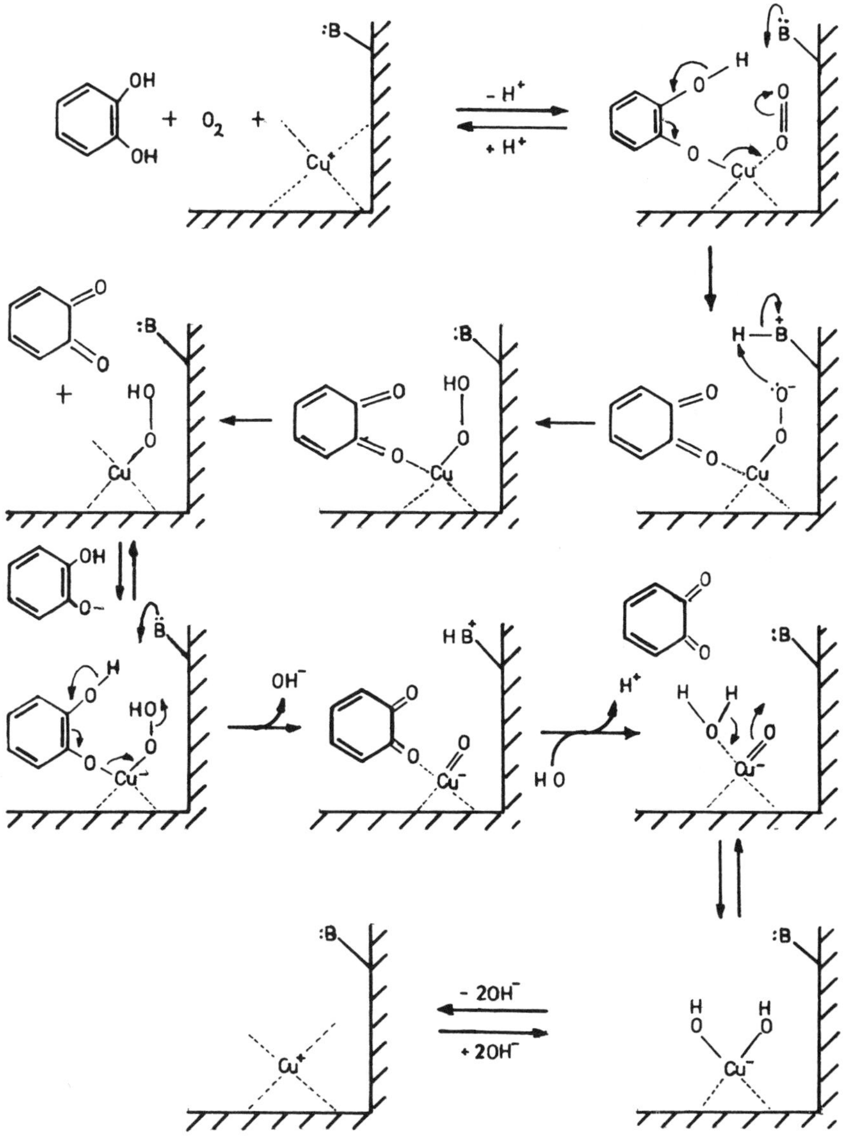

Fig. 2.21 Mechanism of action of tyrosinase according to Hamilton.

The laccases are a group of enzymes which catalyse the oxidation of a large range of substrates, including *p*-diphenols, *o*-diphenols, monophenols, p-phenylenediamines. The name laccase originates from the fact that the first extract capable of oxidising phenols by atmospheric oxygen was isolated from *Rhus vernicifera*, a Japanese tree producing lac. The best source of laccase is now obtained from the mushroom *Polyporus versicolor*. The reaction mechanism of the laccases is schematised in Fig. 2.22. They differ from the action of tyrosinases in that free radicals are formed. Unmistakeable evidence for the *in vitro* formation of intermediate phenoxy radicals have been obtained: such intermediates can react with each other, instead of undergoing further oxidation, or with other molecules, without subsequent mediation of the enzyme and the result is the formation of a large number of compounds. (Instead of reaction (b) in Fig. 2.22 a disproportionation between two phenoxy radicals can take place, leading to the formation of a quinone and a hydroquinone molecule). *In vivo* a single recombination product between phenoxy radicals is formed (see the oxidative coupling of phenols, p. 98). Many secondary metabolites originate from phenol oxidation catalysed by laccases or related phenol oxidases.

Peroxidases utilise hydrogen peroxide as the oxidising agent. The enzymes produced by plants have iron protoporphyrin IX units as prosthetic groups; the animal peroxidases are similar to those in plants, but very little is known about them. One peroxidase (*Streptococcus faecalis* peroxidase) has a flavin based prosthetic group instead of an iron porphyrin one. The best known peroxidase is horse radish peroxidase (ex. *Nasturtium armoracia*) and its probable mechanisms of action are summarised in Fig. 2.22. It is almost certain that peroxidases form a complex with hydrogen peroxide, which is subsequently split into two hydroxyl radicals. The extraction of two substrate hydrogen atoms by the hydroxyl radicals may follow. As the two hydrogen atoms are extracted successively a radical intermediate forms, which can then decay, more or less spontaneously, in many different ways, as in the case of the laccases. Peroxidases can also induce the oxidative coupling of phenols in the presence of hydrogen peroxide (see lignification processes, p. 388).

Peroxidases can introduce a hydroxyl group into substrate benzene rings, that is, as aromatic hydroxylases. A possible mechanism is reaction (g) of Fig. 2.22. Aromatic rings carrying one or more halogen atoms (Cl, Br, I) are frequently found amongst natural substances, generally as metabolites from fungi (for example the tetracyclines), or lower marine animals. Halogen atoms are introduced into aromatic rings by halogeno-peroxidases, utilising halide ions and hydrogen peroxide. The processes involved are not well characterised but possible reaction sequences are schematised in reactions (h) and (i) of Fig. 2.22. Certain halogeno-peroxidases have been isolated from microorganisms and the rat thyroid gland. They contain ironprotoporphyrin IX as the prosthetic group, as in the plant peroxidases.

Until 1955 it was thought that all biological oxidations were catalysed either by dehydrogenases (after substrate activation: the theory of Wieland and of his Münich school), or by oxidases (after oxygen activation: theory of Warburg and of his Berlin school). According to these theories, water donated the new oxygen atoms when a substrate increased the number of its oxygen atoms by

Laccase:

(a) 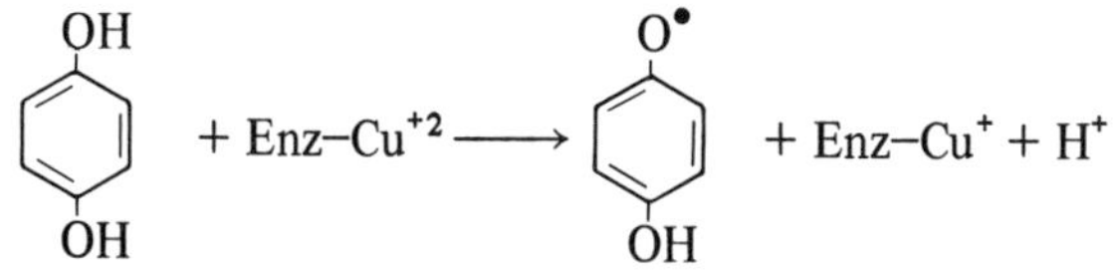 [benzene ring with OH (top) and OH (bottom)] $+ \text{Enz--Cu}^{+2} \longrightarrow$ [benzene ring with $\text{O}^{\bullet}$ (top) and OH (bottom)] $+ \text{Enz--Cu}^+ + \text{H}^+$

(b) 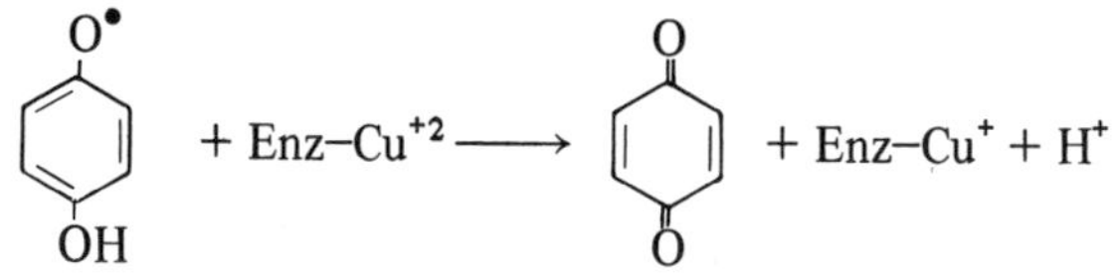 [benzene ring with $\text{O}^{\bullet}$ (top) and OH (bottom)] $+ \text{Enz--Cu}^{+2} \longrightarrow$ [benzoquinone ring with O (top) and O (bottom)] $+ \text{Enz--Cu}^+ + \text{H}^+$

(c) $2 \text{ Enz--Cu}^+ + \tfrac{1}{2}\text{O}_2 + 2\text{H}^+ \longrightarrow 2 \text{ Enz--Cu}^{+2} + \text{H}_2\text{O}$

Peroxidase:

(d) $\text{Enz} + \text{H}_2\text{O}_2 \rightleftharpoons \text{Enz} \text{ - - - } \text{H}_2\text{O}_2 \longrightarrow \text{Enz} \big\langle^{\text{OH}^{\bullet}}_{\text{OH}^{\bullet}}$

(e) $\left(\text{Enz} \big\langle^{\text{OH}^{\bullet}}_{\text{OH}^{\bullet}} + \text{SH}_2 \longrightarrow \text{Enz} \text{ - - - } \text{OH}^{\bullet} + \text{SH}^{\bullet} + \text{H}_2\text{O} \right.$

(f) $\left. \text{Enz} \text{ - - - } \text{OH}^{\bullet} + \text{SH}^{\bullet} \longrightarrow \text{Enz} + \text{S} + \text{H}_2\text{O} \right.$

(g) $\text{Enz} \big\langle^{\text{OH}^{\bullet}}_{\text{OH}^{\bullet}} + \text{SH} \longrightarrow \text{Enz} \big\langle^{\text{OH}^{\bullet}}_{\text{- - - SH - - - OH}^{\bullet}} \longrightarrow \text{Enz} + \text{S--OH} + \text{H}_2\text{O}$

Halogen Peroxidase

(h) $\text{Enz} \big\langle^{\text{OH}^{\bullet}}_{\text{OH}^{\bullet}}$ [from reaction (d)] $+ \text{X}^- \text{ (Cl}^-, \text{Br}^-, \text{I}^-) \rightleftharpoons \text{Enz} \big\langle^{\text{OH}^{\bullet}}_{\text{X}^{\bullet}} + \text{OH}^-$

(i) $\text{Enz} \big\langle^{\text{OH}^{\bullet}}_{\text{X}^{\bullet}} + \text{SH} \longrightarrow \text{Enz} \big\langle^{\text{OH}^{\bullet}}_{\text{- - - - SH - - - X}^{\bullet}} \longrightarrow \text{Enz} + \text{SX} + \text{H}_2\text{O}$

Fig. 2.22 Probable mechanisms of action of laccase and peroxidase.

oxidation, as for example in the case RCHO → RCOOH. The direct uptake of atmospheric oxygen atoms by the substrate was not considered. Such theories were shaken when various researchers used both the couples $^{18}O_2/H_2{}^{16}O$ and $^{16}O_2/H_2{}^{18}O$, and proved the existence of enzymes capable of transferring one or both oxygen atoms to the substrate. It has since been shown that such enzymes, called oxygenases (either mono- or di-) catalyse a large number of secondary metabolic reactions; this differentiates them from the dehydrogenases and oxidases, which operate mainly in primary metabolic processes.

Monooxygenases are also called 'mixed-function oxidases' or 'oxygenases'. They behave both as oxidases, to the oxygen molecule, reducing one oxygen atom to H_2O by a two-electron donor (often NADH, NADPH or ascorbic acid and eventually by a carrier chain) and as oxygenases, transferring the other oxygen atom to the substrate. In certain cases the hydrogen donor is the substrate itself ($SH_2 + O_2 → SO + H_2O$), and the enzyme is then called an 'internal mixed function oxygenase'. An example is the aminoacid oxygenase, catalysing the oxidation of aminoacids to the corresponding amides, by loss of carbon dioxide.

The general mechanism of action of a monooxygenase is illustrated in Fig. 2.23, where the microsomal P-450 cytochrome is specifically cited. This is a haemoprotein present in the endoplasmic reticulum of heptic cells, which are responsible for many hydroxylation reactions.

As far as the mechanism of oxygen activation and of introduction of the O atom into the substrate little is known. The most plausible hypothesis is the 'oxene-type mechanism'. A neutral oxygen atom can be liberated from the oxygen molecule: this atom has six valence electrons, and it is called an oxene by analogy with the well-known isoelectronic species, the carbenes and nitrenes. Reactions of atomic oxygen are known and, like those latter species, oxene has a remarkable tendency to interact with bonds or lone pairs of electrons in order to regain the more-stable eight electron configuration.

Reaction with a non-bonding lone pair on a nitrogen or sulphur atom forms the corresponding N- or S-oxides; its reaction with carbon-carbon double bonds forms epoxides. Oxene can also interact with simple σ-bond electrons, either attacking polarised bonds, such as N-H, S-H, C-H, or carbon–carbon single bonds leading to its insertion and formation of N-OH, S-OH, C-OH and C-O-C. functions. Examples of all these possible oxene reactions are found in Nature (Fig. 2.24).

A monooxygenase of the P-450 cytochrome type has been shown to participate in some of these reactions. It should be noted that biological hydroxylations of an sp^3 hybridised carbon usually occur with retention of configuration, as indicated in reaction (f) of Fig. 2.24: this is in agreement with an insertion mechanism of one oxygen atom into the carbon-hydrogen bond.

The hydroxylation of methyl groups bound to heteroatoms [reactions (h), (i) and (j) of Fig. 2.24] is the first step in the demethylation of methylamino and methoxyl groups, as well as in the formation of methylenedioxy bridges on

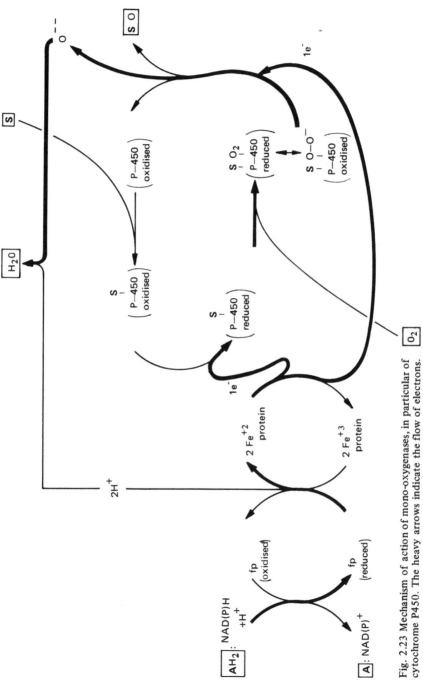

Fig. 2.23 Mechanism of action of mono-oxygenases, in particular of cytochrome P450. The heavy arrows indicate the flow of electrons.

OXENE + n – ELECTRONS (non-bonding electrons)

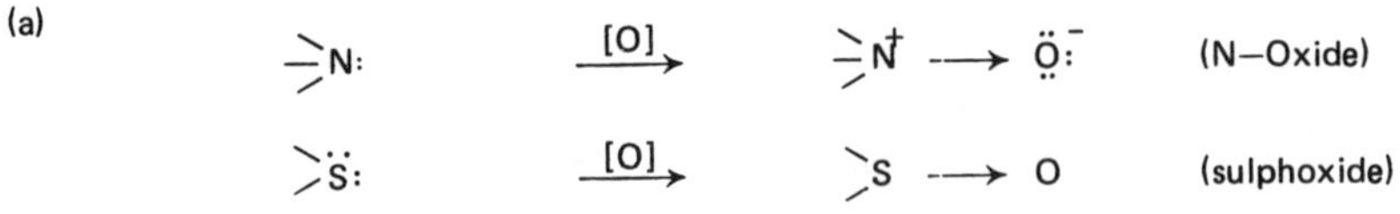

(a)

$\ce{>N:}$ $\xrightarrow{[O]}$ $\ce{>N+}$ ⟶ $\ce{\overset{..}{\underset{..}{O}}{}^{-}}$ (N–Oxide)

$\ce{>S:}$ $\xrightarrow{[O]}$ $\ce{>S}$ ⟶ O   (sulphoxide)

OXENE + π – ELECTRONS

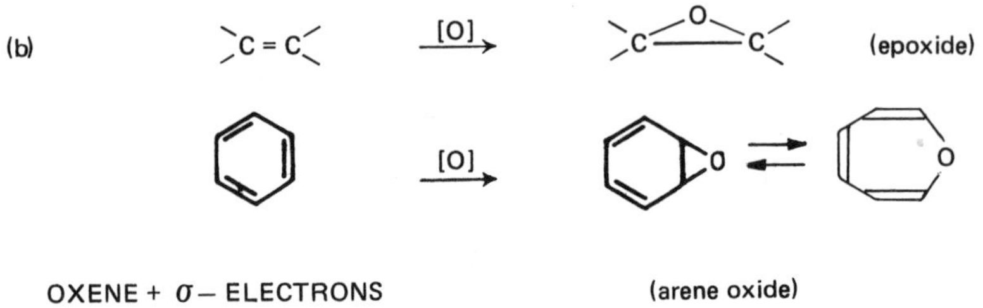

(b)

$\ce{>C=C<}$ $\xrightarrow{[O]}$ epoxide   (epoxide)

OXENE + σ – ELECTRONS   (arene oxide)

(d) $\ce{>N-H}$ $\xrightarrow{[O]}$ $\ce{>N-OH}$

(e) $\ce{-S-H}$ $\xrightarrow{[O]}$ $\ce{>S-OH}$

(f)

$$\underset{R_1}{\overset{R_3}{R_2 \cdots C - H}} \xrightarrow{[O]} \underset{R_1}{\overset{R_3}{R_2 \cdots C - OH}}$$

(g)

$$\ce{>C-\overset{O}{\overset{||}{C}}-} \xrightarrow{[O]} \ce{>C-O-\overset{O}{\overset{||}{C}}-}$$

(h) $\ce{>N-CH3}$ $\xrightarrow{[O]}$ $\ce{>N-CH2-OH}$ ⟶ $\ce{>NH}$ +   HCHO

(i) $\ce{-O-CH3}$ $\xrightarrow{[O]}$ $\ce{-O-CH2-OH}$ ⟶ $\ce{-OH}$ +   HCHO

(j)

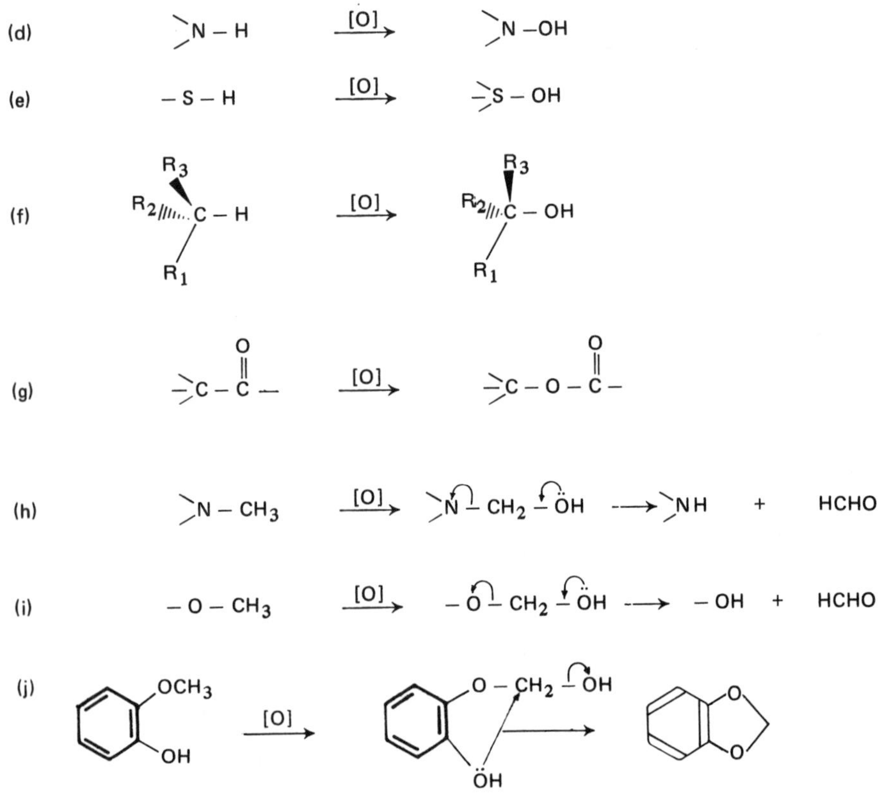

Fig. 2.24 Oxidations catalysed by mono-oxygenases.

aromatic rings. The methylenedioxy groups appear very frequently in the phenyl-propanoids (see Chapter 8) and in the isoquinoline alkaloids.

Certain oxidations of aromatic rings, with cleavage of the carbon–carbon bond adjacent to a phenolic hydroxyl group can be attributed to a mono-oxygenase action, which promotes a Baeyer-Villiger type reaction (reaction (g) of Fig. 2.24), on the carbonyl tautomeric form of the phenol.

With the exception of the particular cases mentioned above, the oxidative cleavage of an aromatic ring (and, more generally, of a carbon–carbon double bond) leading to the formation of two carbonyl groups, is usually performed by a dioxygenase. The oxygen atoms of both the resulting carbonyls come from the same molecule of oxygen. Cyclic peroxides are likely to be formed, as indicated in Fig. 2.25(a) for the general case of a carbon carbon double bond, and in Fig. 2.25(b), for the known transformation of catechol into the semi-aldehyde of *cis, cis*-α-hydroxymuconic acid semi-aldehyde (8). All dioxygenases so far isolated contain iron directly bound to the enzymic protein, or complexed by protoporphyrin IX.

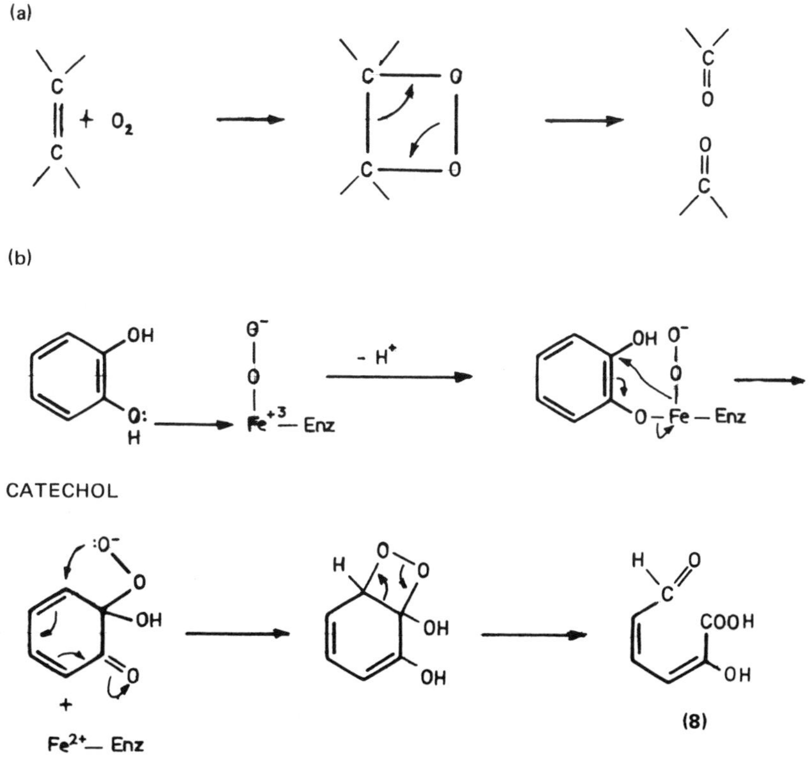

(a)

(b)

CATECHOL

Fig. 2.25 Mechanism of action of dioxygenases.

2.9.2 Arene Oxides and the N.I.H. Shift

The introduction of a hydroxyl group into an aromatic nucleus frequently occurs in nature. The first step of this process is catalysed by monooxygenase type enzymes, which allow the epoxidation of the aromatic ring with atmospheric oxygen, forming highly reactive arene oxides [Fig. 2.24, reaction (c)] such as (9)-(11) (Fig. 2.26).

The second step consists of a spontaneous rearrangement of the epoxide

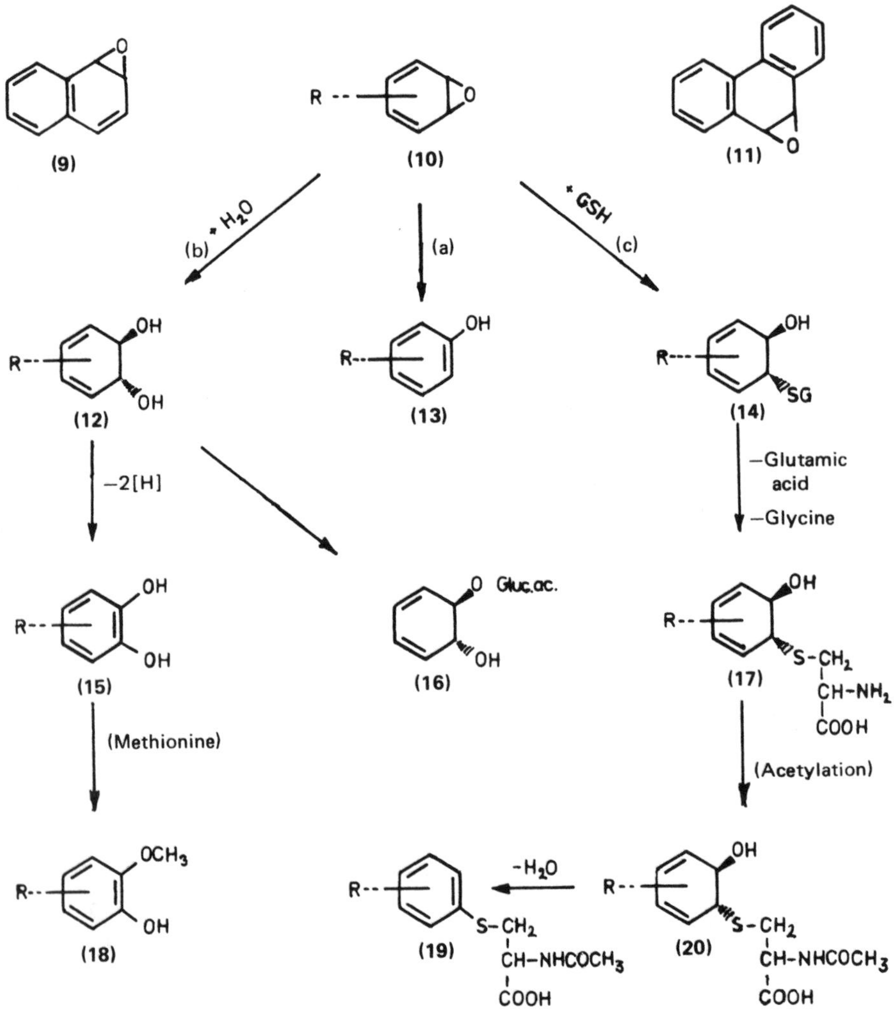

Fig. 2.26 Metabolic fates of some arene oxides.

into a phenol derivative (route (a), in Fig. 2.26). Besides being transformed into a phenol, the arene oxide can undergo an enzyme-catalysed hydrolysis in the liver of higher animals to form the corresponding *trans*-glycol (12) (route (b)); it can also be attacked by reduced glutathione (GSH), either spontaneously or enzymatically (route (c)). The glycols can be conjugated to glucuronic acid giving monoglucuronides (16), or dehydrogenated to catechols (15), which can be monomethylated to guaiacol derivatives (18). S-Glutathionyl derivatives (14) are metabolised to premercapturic acids (20), by losing the glycine and glutamic acid residues, and by N-acetylation of the resulting cysteine derivative (17). Premercapturic acids are then easily dehydrated to the mercapturic acids (19).

The mechanism of isomerisation of arene oxides to phenols has been greatly aided by a chance observation. Researchers at the National Institutes of Health (NIH) of Bethesda (Maryland), U.S.A., designed a quick, and hoped for precise, chemical test for the enzyme L-phenylalanine hydroxylase. Deficiency of the enzyme causes a disorder in the metabolism of L-phenylalanine, which cannot be transformed into L-tyrosine and hence builds up as phenylpyruvic acid. They synthesised p-[^{3}H]phenylalanine as a substrate for the enzyme but find that the resulting tyrosine showed a *retention* of up to 90% of the tritium! With p-[^{2}H] phenylalanine a 70% deuterium retention was obtained. It was found that all the isotopic hydrogen retained in the molecule was in the *meta*-position to the side chain. The label had thus undergone a 'shift' to the adjacent position during the hydroxylation process (Fig. 2.27). Since then a large number of analogous migrations (involving not only hydrogen isotopes, but other substituents such as halogens and methyl groups) have been demonstrated during enzyme-catalysed benzene-ring hydroxylations carried out both *in vivo* and *in vitro*. This phenomenon is known as the 'NIH shift', NIH being the initials of the institute in which it was discovered.

The NIH shift can be easily explained by discarding a direct mechanism (route (a) of Fig. 2.27) for converting the arene oxide into phenol and admitting the intermediate formation of a ketone (21) resulting from the arene-oxide rearrangement via 1,2-migration of the substituent X (route (b) of Fig. 2.27). If X is a deuterium or tritium atom, its high retention can be explained by a strong primary kinetic isotopic effect (see Appendix IV) at the enolisation stage ($k_H >$ k_X). The hydroxylation of *meta*-positions of monosubstituted-benzene rings are less frequent in nature, and take place with deuterium or tritium retentions quite inferior to those from *para*-hydroxylations. The mechanism is similar to the one indicated in Fig. 2.27, but there is a variation in the different epoxide-ring opening: an intermediate ketone is generated, with the carbonyl group in the meta-position to the ring substituent. The same ketone can be formed by opening the epoxide (23), an isomer of (22). The epoxides (23) and (24) can lead to *ortho*-hydroxylations, which are characterised by retentions comparable to those of the *para*-hydroxylation cases.

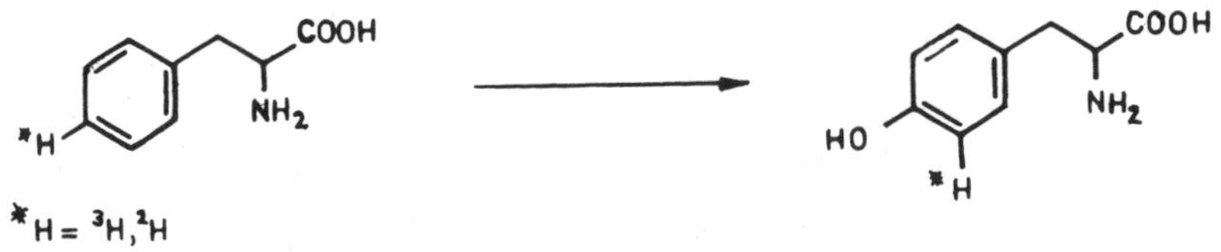

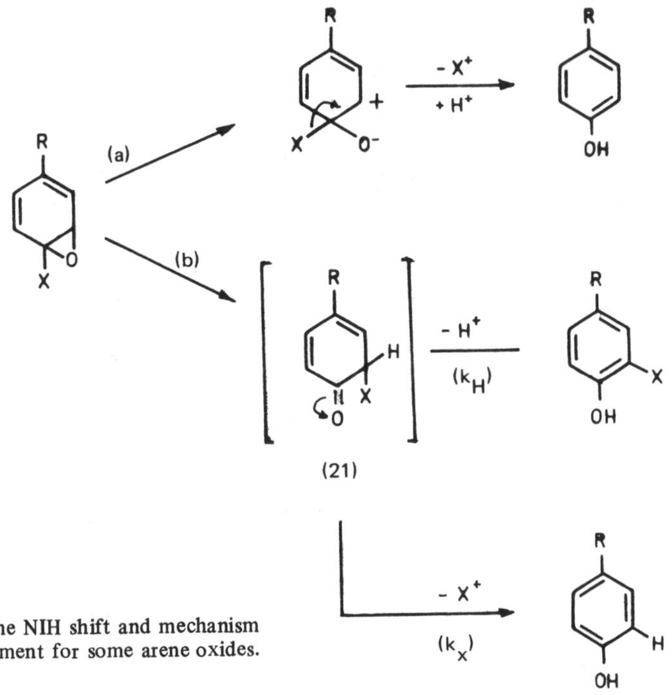

Fig. 2.27 The NIH shift and mechanism
of rearrangement for some arene oxides.

From the above mechanisms, the formation of 4-chloro-3-hydroxyphenyl-
alanine (27), tyrosine (28) and, mainly, 3-chloro-4-hydroxyphenylanine (26) is
understandable, from the treatment of 4-chlorophenylalanine (25) with phenylalanine. hydroxylase: this enzyme does not show an absolute specificity towards
the substrate, but acts on a variety of substituted phenylalanines. It has also
been observed that the introduction of a hydroxyl group in the *para-* or *ortho-*
position with respect to a preexisting hydroxyl (or amino) group, is always
followed by the total loss of the deuterium or tritium atom present in the
position undergoing hydroxylation. A possible explanation for this phenomenon
is implicit in scheme (c) of Fig. 2.28: the epoxide ring can be broken without
migration of the group X (the direct route of Fig. 2.27), the electron donating
atom Y accomplishing this process. An analogous scheme can be written when
X and Y are in relative *ortho* positions.

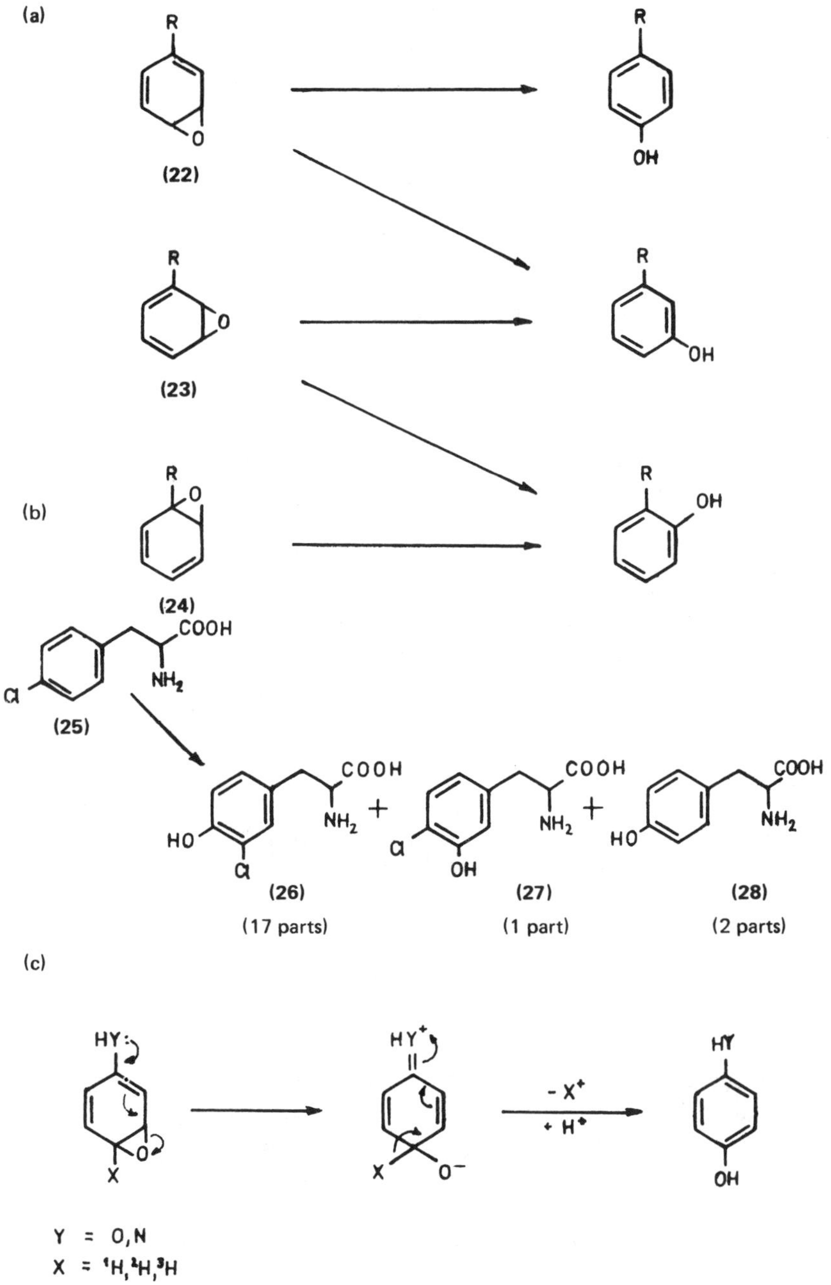

(a)

(b)

(c)

$Y = O, N$

$X = {}^1H, {}^2H, {}^3H$

Fig. 2.28 Variations of aromatic hydroxylations in relation to substituted aromatic starting materials.

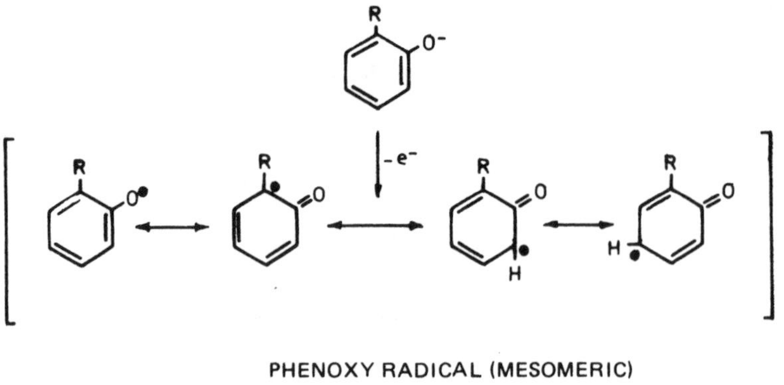

PHENOXY RADICAL (MESOMERIC)

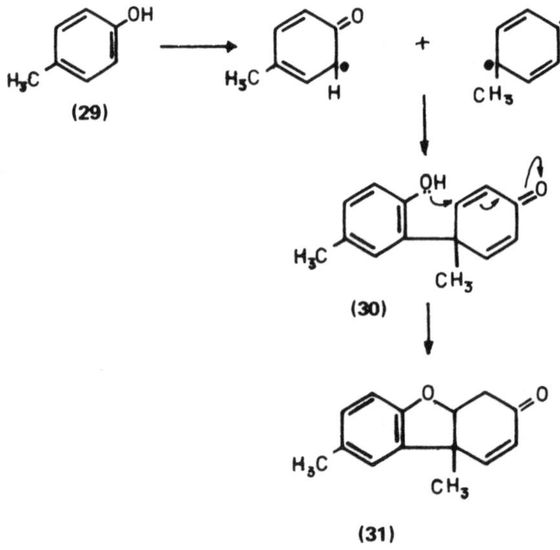

Fig. 2.29 Phenolic oxidations.

2.9.3 Oxidative Coupling of Phenols

The oxidation of a phenolate ion into a phenoxy radical (Fig. 2.29) by a one electron oxidant, for instance $K_3Fe(CN)_6$, is a well-known process in organic chemistry. Because of resonance, the phenoxy radical is quite stable in comparison to methyl or alkyl radicals. Amongst the possible reactions for these radicals (recombination, disproportionation, hydrogen atom extraction, fragmentation, etc.) they prefer to dimerise. Such dimerisations can produce a large number of dehydrodimers from phenoxy radicals, since the unpaired electron can be localised in the oxygen, or on the *ortho*- or *para*-carbon atoms. The reaction rates for the various possible combinations are very different (the peroxide formation is always very small, if not absent) and they largely depend on the nature of the substituents present in the aromatic ring. A certain regio-selectivity in the oxidative coupling of phenols can be found even without enzyme participation. When phenols with *ortho* or *para*-ring substituents are involved, certain dehydrodimers cannot readily rearomatise and they partly remain in the cyclohexadienone form. An example is in the oxidation of *p*-cresol

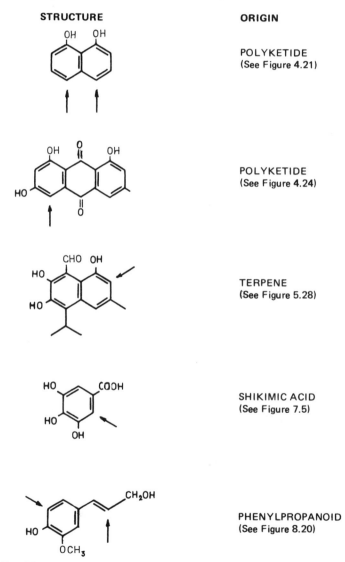

Fig. 2.30 Various types of natural products which can undergo phenolic coupling (the arrows indicate preferred sites of reaction).

(29), which gives a crystalline product, known as Pummerer's ketone, in good yield. The structure (31) of this substance and its formation through the intermediate (30) were demonstrated by Barton in 1955. Barton himself hypothesised that oxidative coupling of phenols, also known as phenol oxidation or dehydrodimerisation (DHD), takes place *in vivo* through the action of appropriate enzymes, and generates C–C and C–O bonds in a wide range of natural products: polyketides (usnic acid), terpenoids (gossipol), phenylpropanoids (for instance lignols and lignins and alkaloids.

Barton's hypothesis is nowadays universally accepted and has been confirmed in many examples using *in vivo* experiments and by the *in vitro* syntheses of natural compounds, by oxidising phenolic intermediates with radical generators (e.g. peroxidase + H_2O_2 or laccase). Noteworthy is the production, in a test tube, of lignins and lignols through dehydropolymerisation of coniferyl alcohol and analogous cinnamyl-derived alcohol mixtures (see p. 388). In Fig. 2.30 are shown some natural phenolic structures, which often undergo *in vivo* oxidative coupling: the positions which are of importance in such processes, in agreement with theoretical predictions, are indicated. One feature of natural phenol oxidations is their frequent intramolecularity. The biosynthesis of griseofulvin and of certain depside catabolites in lichens are examples. The enzymes promoting such oxidative reactions primarily account for this fact, as well as for the regiospecificity of coupling.

2.9.4 Reductions

The reductions of carbon–carbon double bonds and of carbonyls (or imines) groups to alcohols (or amines) are catalysed by dehydrogenases utilising pyridine or flavin coenzymes. These reductions are the reverse reactions of the previously-discussed reversible oxidations and they thus obey the specificity and stereospecificity requirements of the dehydrogenases. The study of such reactions poses the following questions:
(a) If an ethylenic bond is reduced through the mediation of NADH or NADPH which carbon atom binds to the coenzyme hydrogen?
(b) Which face of the double bond (*re* or *si*, see Appendix 2) does each hydrogen atom enter?
The answer to the latter question is solved by the relative direction of hydrogen addition to the double bond, i.e. *cis* or *trans*-addition.

2.9.5 Carbocations and Carbanions

Many biological reactions are commonly explained assuming the intermediacy of carbocations and carbanions. Nevertheless a doubt persists in most cases: is the carbon ion produced as a discrete molecular entity, either free or associated with an enzyme, or does it instead represent a purely formal intermediate, only useful in that it rationalises the reaction in question?

Many carbocations and carbanions are likely to be only partially formed ('incipient' ions). This is usually the case for concerted mechanisms characterised by transition states, in which marked polarisation of the heterolytic breaking (or formation) of bonds take place. The study of biogenetic mechanisms provides evidence that enzymic reactions generally involve concerted mechanisms. If they are multi-step reactions, the carbonium ions, eventually formed as intermediates, do not leave the active site of the enzyme, which thus gives them thermodynamic and configurational stability.

Potential carbocations

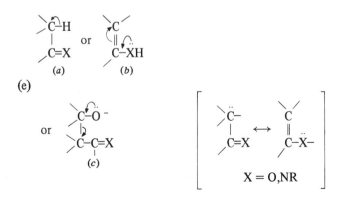

Fig. 2.31. Examples of potential carbocations and carbanions.

There is little evidence for the *in vivo* formation of free carbocations and carbanions. This is understandable on remembering that biological reactions usually take place in aqueous media at almost neutral pH and that the activation energy for carbonium ion formation is usually rather high. The thiamine pyrophosphate ion is exceptional since the thiazole ring carbanion is stabilised by the positive charge and the adjacent nitrogen and sulphur atoms. In Fig. 2.31 there are listed some of the more common functional groups constituting potential *in vivo* carbocations; in appropriate conditions they can generate incipient or free carbocations (classical and non-classical). In the same figure the more common potential carbanions are shown.

Biological reactions involving potential carbocations and carbanions generally beong to the following categories:

(a) Nucleophilic substitutions, mostly bimolecular ones (S_N2) (Fig. 2.32(a)).

(b) Addition of electrophiles to isolated double bonds, or to non-conjugated double bonds (Fig. 2.32(b)).

(c) Addition of nucleophiles to double bonds conjugated with a C=X (X=O, NR) group, as in Michael reactions (Fig. 2.32(c)).

(d) Eliminations, mostly bimolecular ones (E2) (Fig. 2.32(d)).

(e) Additions with eliminations (Fig. 2.32(e)), especially hydrolyses (Y = OH), esterifications (Y = OR, SR) and formation of amide bonds.

(f) Condensations (Fig. 2.32(f), aldol (I, Y = R), crotonyl (II, Y = R), and Claisen types (II, Y = OR, SR).

Such heterolytic reactions largely take place in living organisms in both directions, with formation and breaking of carbon–carbon, carbon–oxygen, carbon–nitrogen, and carbon–sulphur bonds. For these purposes the aforementioned, oxidative and reductive reactions, are also used (Figs. 2.17 and 2.14).

Some particular cases of the reactions (a) to (f) of Fig. 2.32 will be discussed further in the following paragraphs.

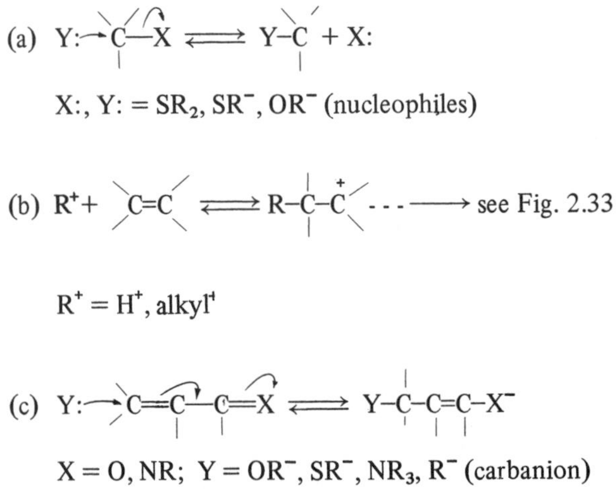

(a) $Y: \rightarrow \overset{|}{\underset{|}{C}} - X \rightleftharpoons Y - \overset{|}{\underset{|}{C}} + X:$

$X:, Y: = SR_2, SR^-, OR^-$ (nucleophiles)

(b) $R^+ + \underset{\diagdown}{\overset{\diagup}{C}} = \underset{\diagdown}{\overset{\diagup}{C}} \rightleftharpoons R - \overset{|}{\underset{|}{C}} - \overset{+}{\underset{\diagdown}{C}} \cdots \longrightarrow$ see Fig. 2.33

$R^+ = H^+$, alkyl^t

(c) $Y: \rightarrow \overset{|}{\underset{\diagup}{C}} = \overset{|}{\underset{|}{C}} - C = X \rightleftharpoons Y - \overset{|}{\underset{|}{C}} - \overset{|}{\underset{|}{C}} = C - X^-$

$X = O, NR; \; Y = OR^-, SR^-, NR_3, R^-$ (carbanion)

Fig.2.32 (contd. overleaf)

(d)

Fig. 2.32 (contd.)

$$\overset{\displaystyle X}{\underset{\displaystyle H\frown : B}{\underset{\displaystyle |}{C}}} \!\!-\!\! C \longrightarrow \; C = C \; + HB^+ + X:$$

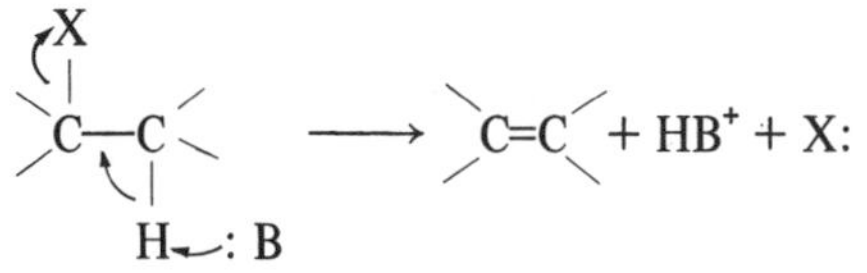

$$X: \; = SR_2, SR^-, OR^-, NR_3 \quad B: \; = base$$

(e) 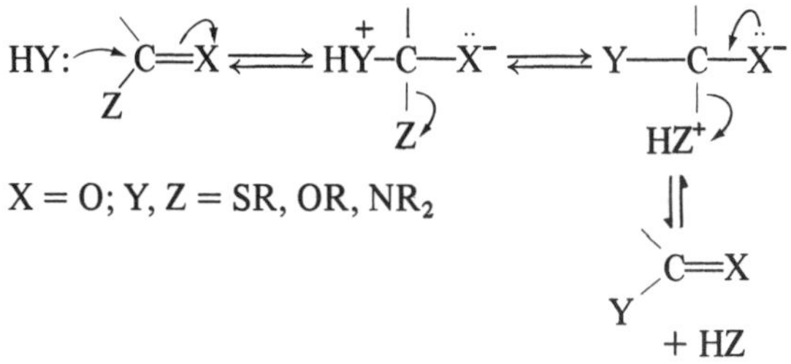 $HY: \frown C = X \rightleftharpoons \overset{+}{H\overset{|}{Y}} - \overset{|}{\underset{\underset{Z}{|}}{C}} - \ddot{X}^- \rightleftharpoons Y - \overset{|}{\underset{\underset{HZ^+}{|}}{C}} \frown \ddot{X}^-$

$$X = O; Y, Z = SR, OR, NR_2$$

$$\Updownarrow$$

$$\underset{Y}{\overset{\diagdown}{\diagup}} C = X$$

$$+ HZ$$

(f)

$$\underset{Y}{\overset{H}{\underset{|}{X = C - C:}}} \frown \overset{\diagdown}{\diagup} C = X \rightleftharpoons \underset{Y}{\overset{H}{\underset{|}{X = C - C - C - X^-}}}$$

$$+ H^+ \Big\updownarrow \; \Big| -H^+$$

$$\underset{Y}{\overset{}{\underset{|}{X = C - C = C}}} \diagup + HX^- \rightleftharpoons \underset{Y}{\overset{H}{\underset{|}{X = C - C - C - XH}}}$$

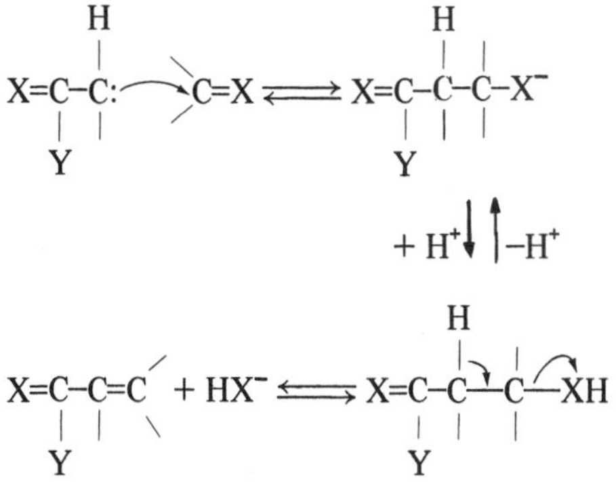

$$X = O, NR; Y = R, OR, SR$$
$$(R = H, alkyl, aryl)$$

Fig. 2.32 Types of heterolytic reactions catalysed by enzymes.

2.9.6 Fates of Carbocations

The possible fates of a carbocation, of whatever type, and arising in whatever manner are schematically summarised in Fig. 2.33. The following possibilities exist:

(1) Reaction with a nucleophilic agent, represented by an anion (OH^-, RS^-, etc.) or by a protic group (H_2O, ROH, RSH, RNH_2 etc.) or by a hydride ion donor (NADH, NADPH, etc.).

(2) Interaction with another double bond, behaving as a nucleophile. A new carbocation is then formed and may be either classical or non-classical.

(3) Elimination, as a cation, of an atom or of an atomic group adjacent to the carbocation, with simultaneous formation of a double bond. In most cases the eliminated cation is the proton. This route can produce cyclopropane rings if the eliminated group and the carbocation are in 1,3-positions.

(4) Conversion of the carbocation into another isomer through a Wagner-Meerwein type of rearrangement consisting in 1,2-shifts of hydrogen, a methyl group or alkyl radicals.

The routes (3) and (4) occur by heterolytic cleavage of bonds β- (more rarely γ-) to the carbocation.

If a cation follows either routes (1) or (3) the result is the immediate formation of a neutral product; routes (2) and (4) prolong the existence of a cationic species but finally decays according to one of the other routes. (See Appendix 3 for a discussion of non-classical carbocations.)

In terpene biogenesis (Chapter 5) there are examples of all the possible developments of classical and non-classical carbocations.

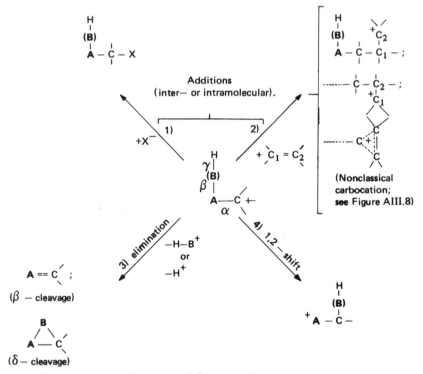

A, B = carbon atom or heteroatom; (B) can be locking

Fig. 2.33 Possible fates of carbocations.

2.9.7 Alkylations

The attack of an alkyl carbocation on to a carbon–carbon double bond or on to a hetero-atom [reactions (2) and (1) respectively of Fig. 2.33] is called an alkylation reaction.

The alkylating agents most used by living organisms are S-adenosyl-L-methionine (32) and dimethylalkylpyrophosphate (DMAPP) (34) (The dimethyl-allyl radical is often called the *prenyl* radical) (Fig. 2.34).

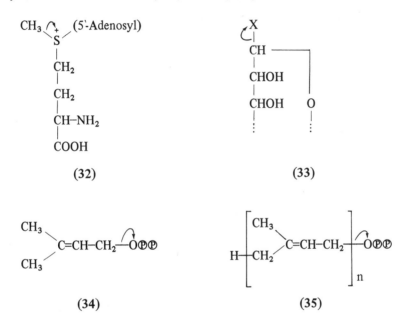

Fig. 2.34. Some natural alkylating agents.

The origin of DMAPP, its high cationic character and the principal products resulting from its action as an alkylating reagent are described in detail in Chapter 5. S-Adenosylmethionine is mostly used as a methyl donor, which is in turn taken from the C_1 unit pool. The methyl group can be transferred onto a carbon atom of a non-activated double bond, to an enolic double bond, to phenolic oxygen atoms or on to amine nitrogen atoms (for example, of alkaloids). The enzymes catalysing the various methylations (commonly called methyl transferases) are well known, and they perform a very important role in secondary metabolism. The methyl group of S-adenosylmethionine is a potential carbocation, since the positively charged sulphur atom (the sulphonium ion) strongly attracts electrons.

Glycosylation reactions can also be considered to be alkylations, in which the alkylating agent is the potential carbocation at C_1 of aldose molecules in the pyranose or furanose acetal forms (33); the most frequently used sugars are D-glucose in plants, and D-glucuronic acid in animals. Phenolic and alcoholic oxygen atoms often undergo glycosylation, and O-glycosides are formed; carbon atoms are rarely glycosylated to give C-glycosides.

2.9.8 Decarboxylations and Carboxylations

The loss or uptake of a carboxylic group by a natural molecule, by either breaking or formation of a carbon–carbon dioxide bond, and exchange of carbon dioxide with the medium, occurs in both primary and secondary metabolism. Decarboxylations can either be *non-oxidative* or *oxidative*. The former can, in most cases, be represented by the two fundamental mechanisms (36) and (37) (Fig. 2.35).

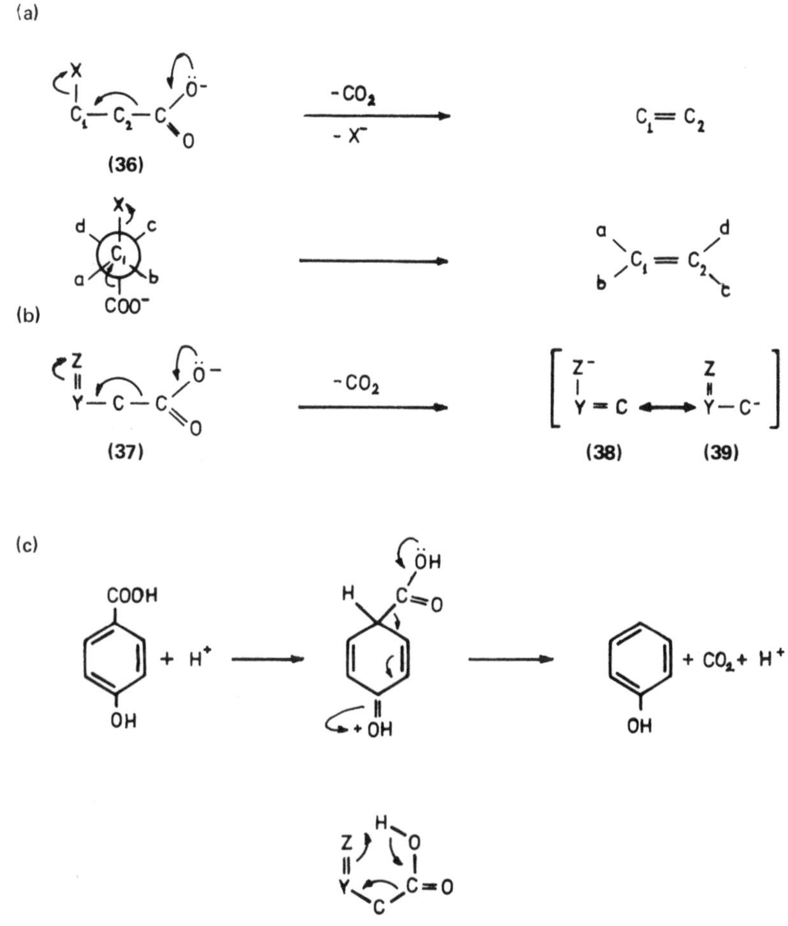

(a)

(b)

(c)

Fig. 2.35 Mechanisms for decarboxylation.

The driving force for these decarboxylations depends on the electron with-drawing strength of the group X and Z at the β-position to the carboxyl group. For (37) resonance stabilisation of the carbanion systems after CO_2 elimination favours the decarboxylation. In type (36) X is a leaving group, such as –SR, $-OH_2^+$, –OPOP, or a group belonging to the active site of an enzyme; the reaction can be considered as an elimination. In known cases such eliminations appear to be of the bimolecular type (E2). The geometry of the double bond produced is in agreement with the stereoelectronic requirements for such eliminations (see Appendix 2 and the decarboxylation of mevalonic acid, in Fig. 5.4). A vinylogous example of system (36) is given by prephenic acid (see Fig. 7.10).

The most common example of reaction (37) is given by the decarboxylation of β-ketoacids (Y = C, and Z = O), which is well known both *in vitro* and *in vivo*: for instance, acetoacetic acid to produce acetone and carbon dioxide (see the loss of the methyl groups in lanosterol at C–4, (Fig. 6.3). The first reaction product is often the enol corresponding to (38).

Vinylogues of type (37) systems are implied in the acidic decarboxylation of benzoic acids bearing electron donor groups in the *ortho-* or *para-* positions (Fig. 2.35(c)), and, possibly, when dehydroprephenic acid is converted into *p*-hydroxyphenylpyruvic acid (Fig. 7.10).

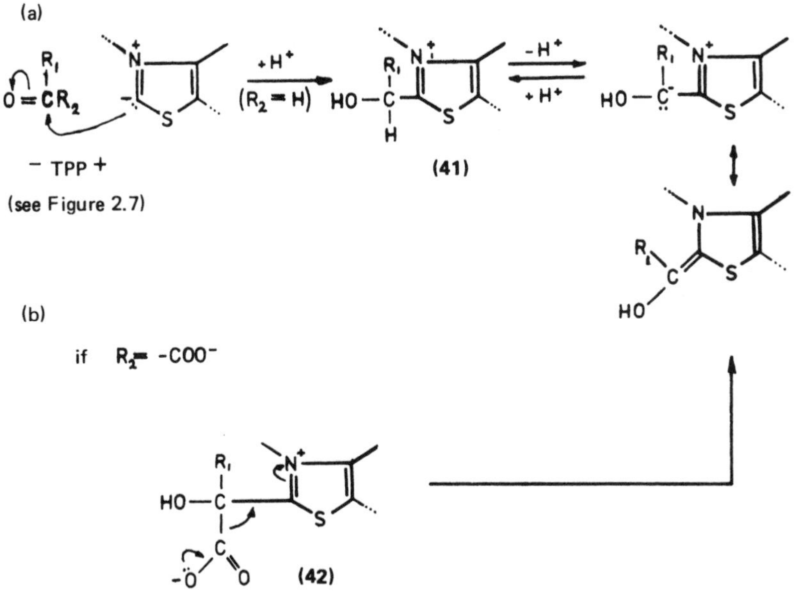

Fig. 2.36 Reactions of thiamine pyrophosphate.

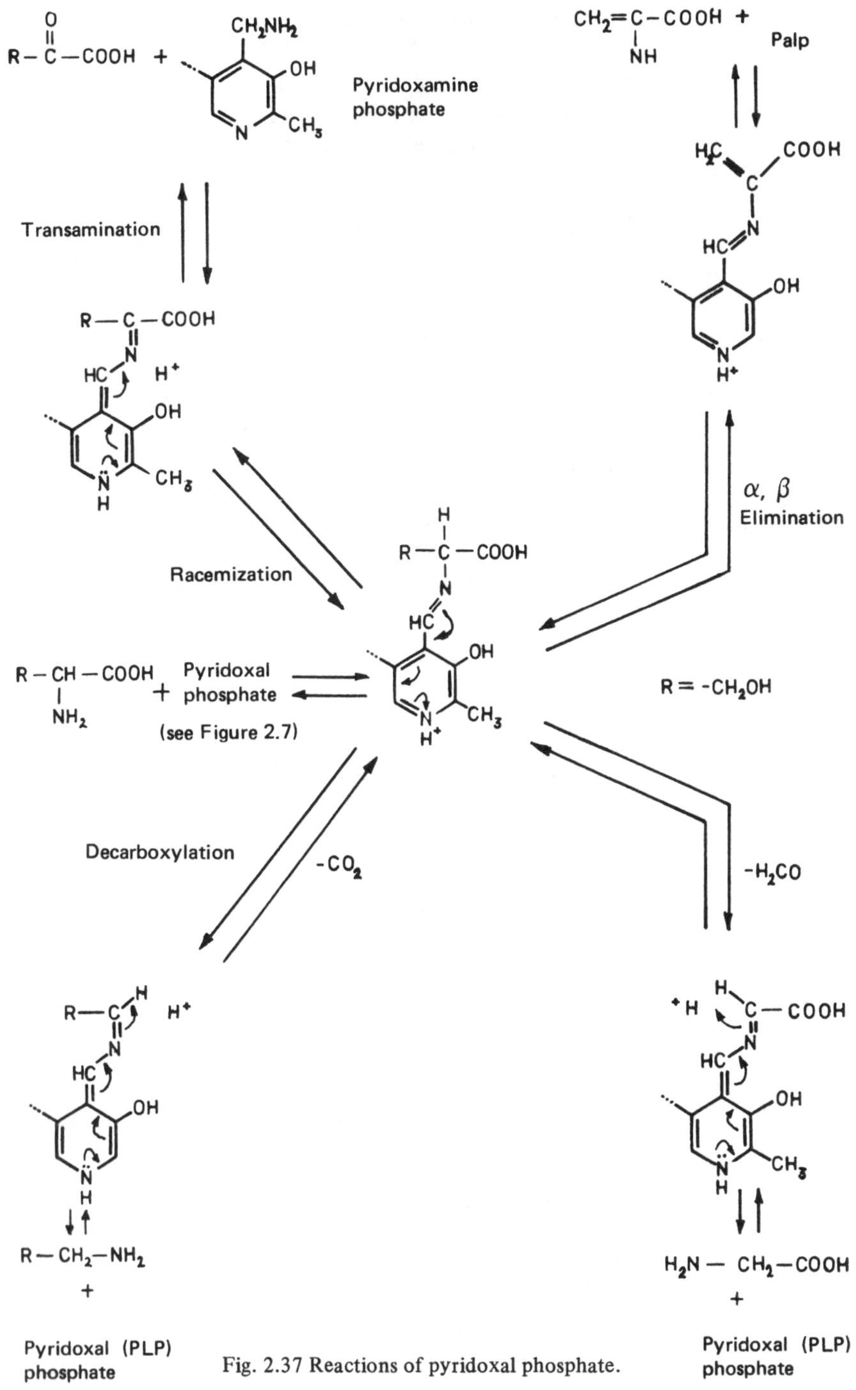

Fig. 2.37 Reactions of pyridoxal phosphate.

Although direct experimental evidence does not exist, the decarboxylations of systems of the type (37) could proceed via *in vivo* cyclic mechanisms (40), a process common for *in vitro* thermal decarboxylations of $\beta\gamma$-unsaturated acids (Y, C=C) and β-ketoacids catalysed by TPP-dependent enzymes. This process leads to the formation of the corresponding 'active' aldehydes (Fig. 2.36(b)). The thiazole nucleus of thiamine pyrophosphate (TPP) undergoes easy deprotonation, and generates an ylid, which is stable and remarkably nucleophilic with respect to carbonyl groups (Fig. 2.36(a)). Its addition product with aldehydes and ketones, for instance (41) and (42), can bear a resonance stabilised negative charge on the original carbonyl carbon, so that either proton elimination from (41) or decarboxylation from (42) are extremely easy.

The decarboxylation of α-aminoacids, catalysed by enzymes containing pyridoxal phosphate (PLP) as their prosthetic group, can be compared to system (37) (Y = N, Z = C, Fig. 2.35). In Fig. 2.37 other important reactions mediated by PLP are summarised.

Along certain biosynthetic pathways deformylation, rather than decarboxylation, is encountered. This is a relatively rare event in which a C–CHO bond is broken and formic acid is expelled.

Deformylation is similar to decarboxylation as far as mechanistic aspects and structural requirements of the substrate are concerned; the group CHOH replaces C=O in formulae (36) and (37). Two typical examples of deformylations are the methyl elimination from the steroid nucleus at C-14, when lanosterol is converted into cholesterol (see Fig. 6.30), and at C-10 in estrogen biosynthesis (Fig. 6.30).

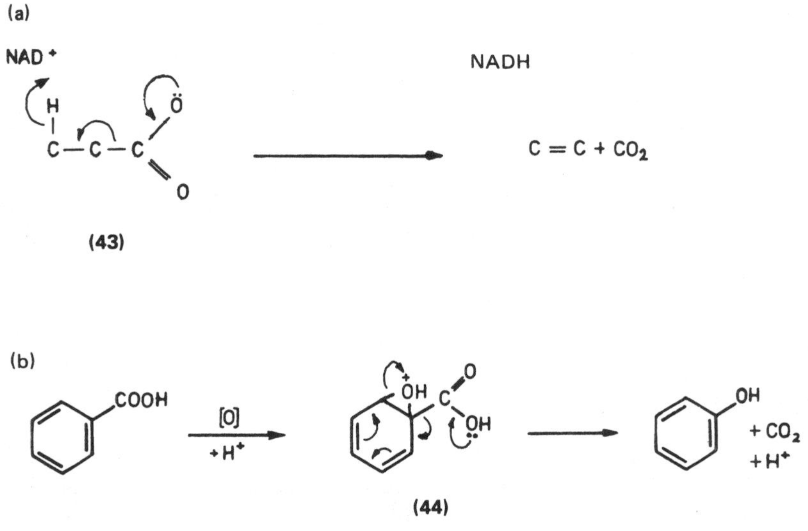

Fig. 2.38 Oxidative decarboxylation.

Oxidative decarboxylations require reduction-oxidation enzymes; their action can precede, or be simultaneous with, the decarboxylation process. A dehydrogenase can extract a hydride ion from a β-position to a carboxyl group, and thus induce a CO_2 elimination according to (43) (Fig. 2.38(a)). A formally analogous situation (the formation of an incipient carbocation β to the carboxyl group), can be generated by an oxidase, transforming an aromatic acid into an arene oxide (44); the final result is the replacement of a carboxylic group with a phenolic hydroxyl function (Fig. 2.38(b)).

Carboxylation reactions occur on carbanions derived from an active methyl or methylene group, which can bind to the positive carbon of carbon dioxide. The latter does not react in its free form, but as the carbamic ester of biotin (Fig. 2.39). The biotin coenzyme acts as a carrier of carbon dioxide. Recent studies on the carboxybiotin structure support structure (45).

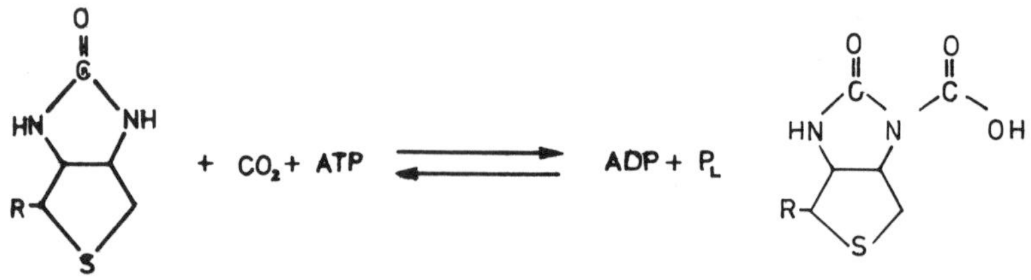

BIOTIN   (45)

(*cf* Figure 2.7)

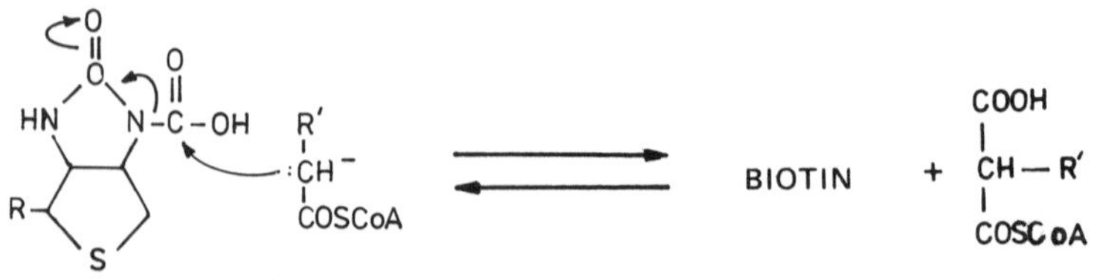

$R' = H, CH_3$

(45)

Fig. 2.39 Reactions of biotin.

2.9.9 Activation Reactions

If a product is synthesised starting from particularly stable molecules, the enzymatic catalysts can be insufficiently powerful, especially when new C–C and C–O bonds are formed simultaneously with analogous bond cleavages. In such circumstances the organisms have to initially 'activate' the reagents. Such activation processes consist in transforming the reagents into energetically less stable or more reactive derivatives. This is achieved by appropriate activating reactions and reagents (Fig. 2.40). The most common activator is adenosine triphosphate (ATP), which is produced either during respiration (p. 139) or, in green plants, during chlorophyll photosynthesis (p. 125). Adenosine triphosphate can donate either a phosphate (OP) or a pyrophosphate (OPOP) group to the hydroxyl group of a substrate. The resulting phosphoric ester shows a more marked electrophilicity on the carbon atom bearing it. The phosphoric acid group can be liberated as either the phosphate, Pi, or pyrophosphate, PPi, anion (Fig. 2.40). The remarkable leaving group character of the OP and of the OPOP residues is explained by the greater stability of the corresponding inorganic anions (see forms, Fig. 2.40). An approximate evaluation of the activation energy contained in a phosphoric ester can be given by a measurement of its hydrolysis energy (Table 2.1).

(a) Hydroxyl Group of Alcohols and Hemiacetals

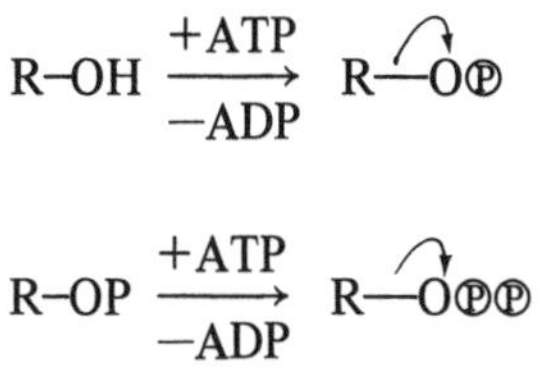

$$R{-}OH \xrightarrow[-ADP]{+ATP} R{-}O\!P$$

$$R{-}OP \xrightarrow[-ADP]{+ATP} R{-}O\!P\!P$$

(b) Carbonyl Group of Aldehydes and Ketones

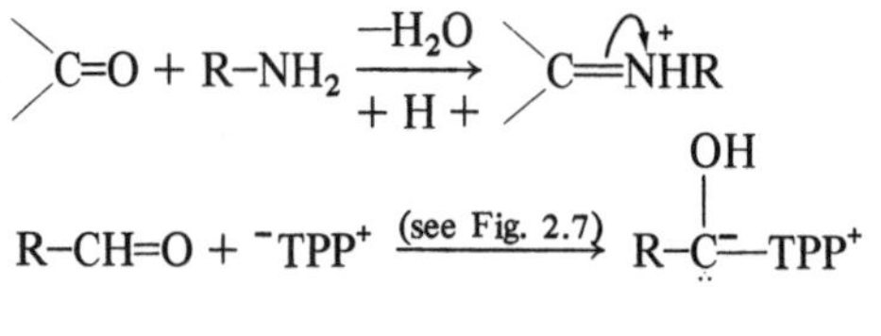

$$\begin{array}{c} \diagdown \\ \diagup \end{array}\!C{=}O + R{-}NH_2 \xrightarrow[+\,H\,+]{-H_2O} \begin{array}{c} \diagdown \\ \diagup \end{array}\!C{=}\overset{+}{N}HR$$

$$R{-}CH{=}O + {}^-TPP^+ \xrightarrow{\text{(see Fig. 2.7)}} R{-}\underset{\ddot{}}{C}\!{-}TPP^+ \quad (\text{OH above C})$$

(c) Carbonyl Group of Carboxylic Acids

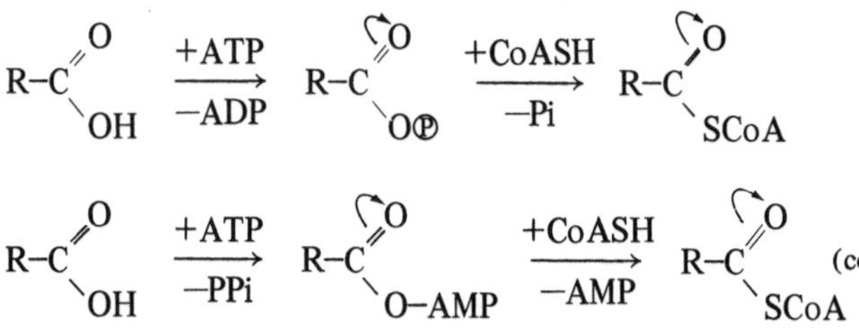

$$R{-}C\!\!\begin{array}{c}\diagup O \\ \diagdown OH\end{array} \xrightarrow[-ADP]{+ATP} R{-}C\!\!\begin{array}{c}\diagup O \\ \diagdown O\!P\end{array} \xrightarrow[-Pi]{+CoASH} R{-}C\!\!\begin{array}{c}\diagup O \\ \diagdown SCoA\end{array}$$

$$R{-}C\!\!\begin{array}{c}\diagup O \\ \diagdown OH\end{array} \xrightarrow[-PPi]{+ATP} R{-}C\!\!\begin{array}{c}\diagup O \\ \diagdown O{-}AMP\end{array} \xrightarrow[-AMP]{+CoASH} R{-}C\!\!\begin{array}{c}\diagup O \\ \diagdown SCoA\end{array}$$

Fig. 2.40
(contd. overleaf)

(d) Methyl and Methylene Groups   Fig. 2.40 (contd.)

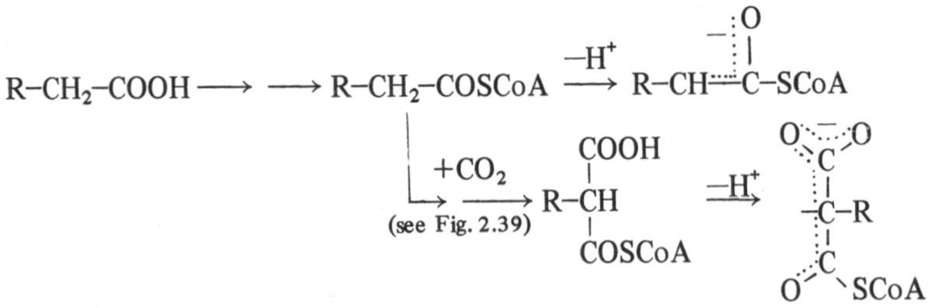

$$R-CH_2-COOH \longrightarrow \; \longrightarrow R-CH_2-COSCoA \xrightarrow{-H^+} R-CH \!=\! C-SCoA$$

with $+CO_2$ (see Fig. 2.39) giving

$$R-CH \begin{array}{c} COOH \\ | \\ | \\ COSCoA \end{array} \; \xrightarrow{-H^+} \; \begin{array}{c} O \cdots O \\ C \\ | \\ -C-R \\ | \\ C \\ O \diagup \diagdown SCoA \end{array}$$

R = H, alkyl

$$Pi = \begin{bmatrix} O \\ | \\ O-P-O \\ | \\ O \end{bmatrix}^{3-} \; ; \; PPi = \begin{bmatrix} O \quad O \\ | \quad | \\ O-P-O-P-O \\ | \quad | \\ O \quad O \end{bmatrix}^{4-} \; ; \; R-O℗ = \begin{bmatrix} O \\ | \\ R-O-P-O \\ | \\ O \end{bmatrix}^{2-}$$

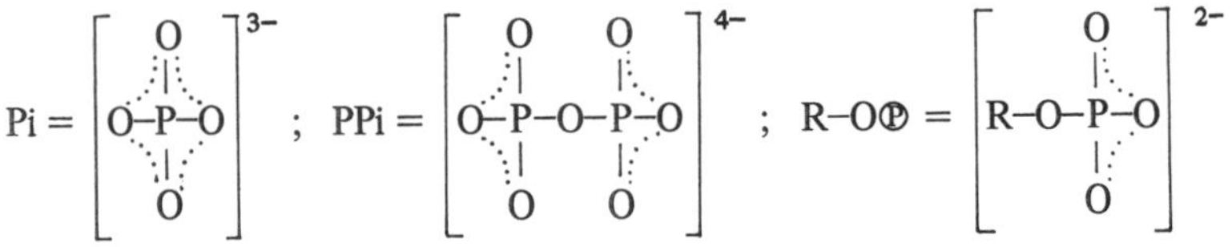

Fig. 2.40 Activation of biological molecules.

Table 2.1

Free energy of hydrolysis (kcal mol^{-1}) under standard conditions (pH7; 25°) of some phosphoric esters of biological interest.

Compound	Hydrolysis products	Energy
Phosphoenolpyruvate	Pyruvate + Pi	13.0
Acetyl phosphate	acetate + Pi	10.5
Creatinine phosphate	creatinine + Pi	9
ADP	AMP + Pi	7.6
ATP	ADP + Pi	7.4
Aldose-1-phosphate	aldose + Pi	5.0

The activation of the ketonic carbonyl group in nucleophilic reactions is often achieved *via* the formation of the corresponding imino-cation; the positive nitrogen strongly attracts electrons, inducing a marked polarisation of the π-electron bond.

Aldehydes can be activated as anions at the carbonyl carbon, which can be obtained when the carbonyl group binds with thiamine pyrophosphate.

Carboxylic acids involved in esterifications and Claisen type condensations are initially transformed into thiolesters. The esterification takes place with the –SH group of coenzyme-A (see Fig. 2.7) or with the cysteine residue of a protein. Acyl-CoA esters are formed via mixed anhydrides with phosphoric acids, or adenosine monophosphate. Thiolesters show greater reactivity at their carbonyl groups (roughly comparable to acid halides), and strongly stabilise a negative charge at the α-position, thus enhancing the acid's availability for aldol or Claisen condensations.

Sometimes the character of potential carbanions at the α-carbon of acetic or propionic acid is enhanced, by transformation of acyl- CoA esters into malonyl–CoA or methylmalonyl–CoA esters (Fig. 2.40(d)). The carboxylic group of malonic hemithiolesters is lost as CO_2, after, or during, the Claisen condensation (see the biosynthesis of fatty acids in Fig. 3.28, and of poly-ketomethylene chains in Fig. 4.7).

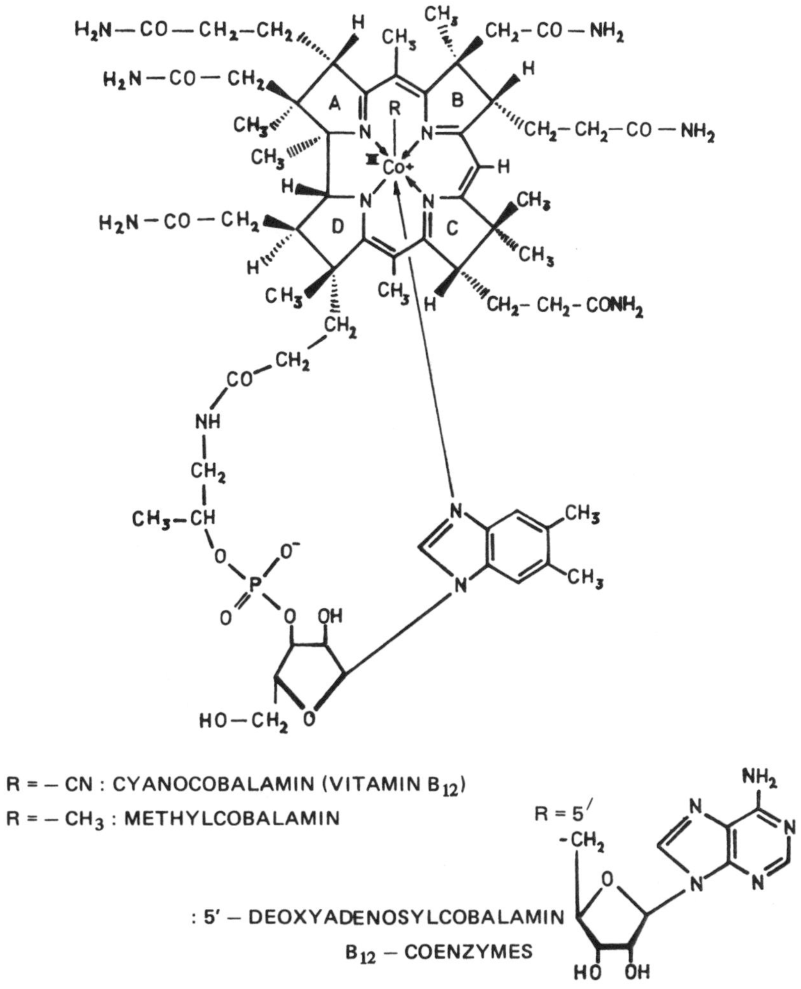

R = – CN : CYANOCOBALAMIN (VITAMIN B_{12})

R = – CH_3 : METHYLCOBALAMIN

R = 5′ –CH_2

: 5′ – DEOXYADENOSYLCOBALAMIN

B_{12} – COENZYMES

Fig. 2.41 Coenzyme B_{12}.

2.9.10 Reactions dependent upon B_{12} coenzymes

Of particular importance in biological reactions is the role played by vitamin B_{12} and its congeners (Fig. 2.41) which catalyse methylations, carboxylations and rearrangements, as illustrated in Fig. 2.42 and 2.43.

(a)   [METHIONINE SYNTHETASE]

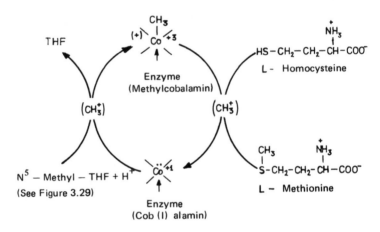

(b)   [ACETATE SYNTHETASE]

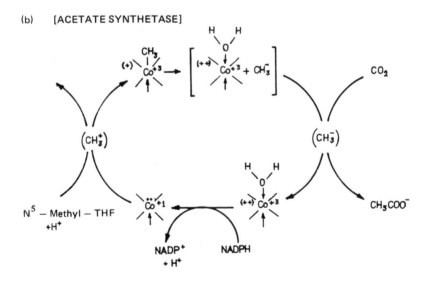

Fig. 2.42 Enzymic mechanism proposed for methylcobalamin.

(a)   HYDROGEN – TRANSFER AND C – C BOND CLEAVAGE

(a) (i) [METHYLMALONYL – CoA MUTASE]

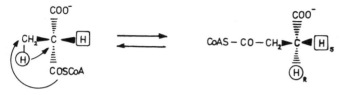

(R) – METHYLMALONYL – CoA   SUCCINYL – CoA

(ii) [METHYLASPARTATE MUTASE]

METHYLASPARTATE

GLUTAMATE

(b)   HYDROGEN TRANSFER AND C – O BOND CLEAVAGE

(b) (i) [DIOL DEHYDRASE]

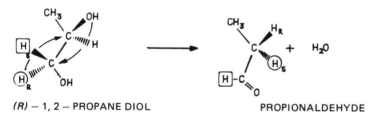

(R) – 1, 2 – PROPANE DIOL   PROPIONALDEHYDE

(ii) [GLYCEROL DEHYDRASE]

$$CH_2 - \overset{\displaystyle H}{\underset{\displaystyle OH}{C}} - \overset{\displaystyle H}{\underset{\displaystyle OH}{C}} - OH \longrightarrow CH_2 - CH_2 - CHO + H_2O$$

GLYCEROL

β– HYDROXYPROPIONALDEHYDE

Fig. 2.43 (contd. overleaf)

(c) HYDROGEN – TRANSFER AND C – N CLEAVAGE Fig. 2.43 (contd.)

(c) (i) [ETHANOLAMINE AMMONIA LYASE]

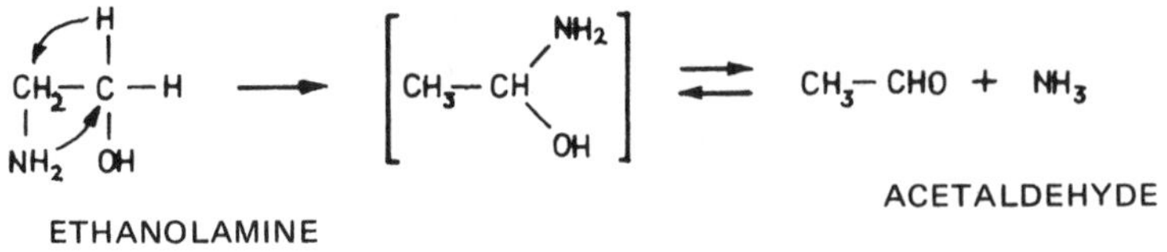
ETHANOLAMINE
ACETALDEHYDE

A RATIONALISATION OF THE GENERAL MECHANISM FOR THE ABOVE REACTIONS

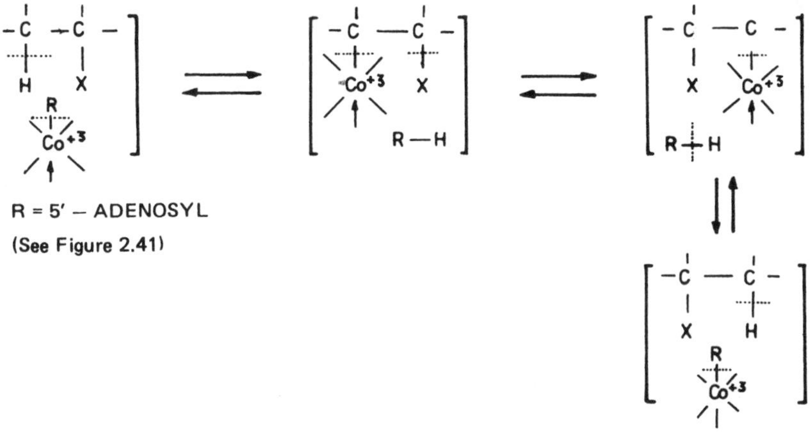
R = 5′ – ADENOSYL
(See Figure 2.41)

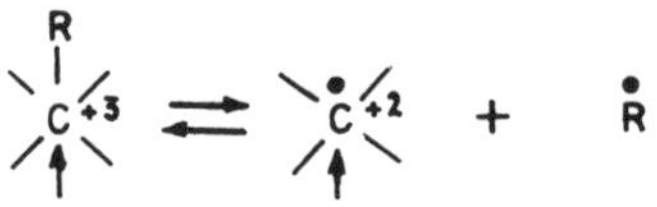

Fig. 2.43 Some processes dependent on catalysis involving the coenzyme 5′-desoxyadenosylcobalamin.

SOURCE MATERIALS AND SUGGESTED READING

(a) *Protein structure*

[1] I. Klotz, (1960), Non-covalent Bonds in Protein Structure, *Brookhaven Symp.*, **13**, 25.

[2] H. Schachman, (1963), Considerations on the Tertiary Structure of Proteins, *Cold Spring Harbor Symp. Quant. Biol.*, **28**, 409.

[3] F. Reithel, (1963), The Dissociation and Association of Protein Structure, *Adv. Protein Chem.*, **18**, 124.

[4] J. Goodwin, I. Harris and B. Hartly, (1964), *Structure and Activity of Enzymes*, Academic Press.

[5] M. F. Perutz, (1965), The Anatomy of Haemoglobin, *Chem. in Britain*, 1, 9.
[6] H. Sund and K. Weber, (1966), The Quaternary Structure of Proteins, *Angew. Chem. Int. Ed. Engl.*, 5, 231.
[7] T. L. Blundel and L. N. Johnson, (1970), Principles of Protein Structure, in *Protein Crystallography*, chap. 2, Academic Press.
[8] G. P. Hess and J. A. Rupley, (1971), Structure and Function of Proteins, *Ann. Rev. Biochem.*, 40, 1013.
[9] M. Calvin, (1974), Biopolymers: Origin, Chemistry, and Biology, *Angew. Chem. Int. Ed. Engl.*, 13, 121.
[10] H. Gutfreund, (1974), Chemistry of Macromolecules, (*International Review of Science, Biochemistry Series One*, 1), Butterworths.
[11] H. Neurath and R. L. Hill, (1976-77), *The Proteins*, 3rd ed., 1-3, Academic Press.
[12] Z. Simon, (1974), Specific Interactions – Intermolecular Forces, Steric Requirements, and Molecular Size, *Angew. Chem. Int. Ed. Engl.*, 13, 719.
[13] R. E. Dickerson and I. Geis, (1975), *The Structure and Action of Proteins*, Addison-Wesley.
[14] E. Frieden, (1975), Non-covalent Interactions, *J. Chem. Educ.*, 52, 754.
[15] J. Wilkinson, (1975), *Isoenzymes*, Chapman and Hall.

(b) *Mechanisms, stereochemistry, kinetics, and regulation of enzyme reactions*
[16] S. P. Colowick and N. O. Kaplan (Eds.), (1955-1980), *Methods in Enzymology*, Academic Press, 1-69, (A research and reference book).
[17] J. Monod, J. Changeux and F. Jacob, (1963), Allosteric Proteins and Cellular Control System, *J. Mol. Biol.*, 6, 306.
[18] H. Gutfreund, (1965), *An Introduction to the Study of Enzymes*, Blackwell.
[19] G. N. Cohen, (1965), Regulation of Enzyme Activity in Microorganisms, *Ann. Rev. Microbiol.*, 19, 106.
[20] J. Monod, J. W. Wyman and J. P. Changeux, (1965), On the Nature of Allosteric Transition, *J. Mol. Biol.*, 12, 88.
[21] A. I. Scott, (1965), Oxidative Coupling of Phenolic Compounds, *Quart. Rev.*, 19, 1.
[22] C. Walter, (1965), *Steady-state Application in Enzyme Kinetics*, Ronald.
[23] U. Henning, (1966), Multienzyme Complexes, *Angew. Chem. Int. Ed. Engl.*, 5, 785.
[24] W. I. Taylor and A. R. Battersby (Eds.), (1967), *Oxidative Coupling of Phenols*, M. Dekker.
[25] S. G. Waley, (1967), Mechanism of Enzyme Action, *Quart. Rev.* 21, 379.
[26] S. A. Bernhard, (1968), *The Structure and Function of Enzymes*, Benjamin.
[27] G. N. Cohen, (1968), *The Regulation of Cell Metabolism*, Holt, Rinehart and Winston.
[28] M. Dixon and E. Webb, (1968), *Enzymes*, Longmans Green.
[29] J. Kirschbaum, (1968), Biological Oxidations and Energy Conservation,

J. Chem. Educ., **45**, 28.

[30] T. P. Singer, (1968), *Biological Oxidations*, Wiley.

[31] B. L. Vallee and R. J. P. Williams, (1968), Enzyme Action – Views Derived from Metalloenzymes Study, *Chem. in Britain*, **4**, 397.

[32] B. E. C. Banks, (1969), Thermodynamics and Biology, *Chem. in Britain*, **5**, 514, (see also (1970), *Chem. in Britain*, **6**, 468, 472, 477, 539, 541).

[33] J. W. Cornforth, (1969), Exploration of Enzyme Mechanisms by Asymmetric Labelling, *Quart. Rev.*, **23**, 125.

[34] S. Doonan, (1969), Biological Formation and Reactions of the COOH and COOR Groups in *The Chemistry of Carboxylic Acids and Esters*, (S. Patai, Ed.), Wiley.

[35] O. Hayaishi, (1969), Enzymic Hydroxylation, *Ann. Rev. Biochem.*, **38**, 21.

[36] W. P. Jenks, (1969), *Catalysis in Chemistry and Enzymology*, McGraw-Hill.

[37] E. Lederer, (1969), Some Problems Concerning Biological C-Alkylation Reactions and Phytosterol Biosynthesis, *Quart. Rev.*, **23**, 453.

[38] A. Williams, (1969), Mechanism of Action and Specificity of Proteolitic Enzymes, *Quart. Rev.*, **23**, 1.

[39] A. Williams, (1969), *Introduction to the Chemistry of Enzyme Action*, McGraw-Hill.

[40] P. Boyer (Ed.), (1970-1976), *The Enzymes*, 3rd ed., Academic Press, 1-13, (a research and reference book).

[41] P. H. Christen, (1970), Chemical Approach to Intermediate Enzyme Catalysis, *Experentia*, **26**, 337.

[42] I. A. Rose, (1970), Enzymology of Proton Abstraction in ref. 40, 2.

[43] M. L. Bender, (1971), *Mechanism of Homogeneous Catalysis from Protons to Proteins*, Wiley.

[44] J. M. Calvo and G. R. Fink, (1971), Regulation of Biosynthetic Pathways in Bacteria and Fungi, *Ann. Rev. Biochem.*, **40**, 943.

[45] H. Eisenberg, (1971), Glutamate Dehydrogenase, Anatomy of a Regulatory Enzyme, *Accounts Chem. Res.*, **4**, 379.

[46] S. K. Erickson, (1971), Biological Formation and Reactions of the Hydroxyl Group in *The Chemistry of the Hydroxyl Group* (S. Patai, Ed.), Part 2, Wiley.

[47] C. J. Gray, (1971), *Enzyme-catalysed Reactions*, Van Nostrand-Reinhold.

[48] G. A. Hamilton, (1971), The Proton in Biological Redox Reactions in ref. 49, 1.

[49] E. T. Kaiser and F. J. Kezdy (Eds.), (1971-1976), *Progress in Bioorganic Chemistry*, **1-4**, Wiley.

[50] J. F. Riordan, (1971), Chemical Approaches to the Study of Enzymes, *Accounts Chem. Res.*, **4**, 353.

[51] G. S. Boyd and R. M. S. Smellie (Eds.), (1972), *Biological Hydroxylation Mechanisms*, Academic Press.

[52] G. W. R. Canham and A. B. P. Lever, (1972), Bioinorganic Chemistry,

J. Chem. Educ., **49**, 656.

[53] J. W. Daly, D. M. Jerina and B. Witkop, (1972), Arene Oxides and the NIH Shift: The Metabolism Toxicity and Carcinogenicity of Aromatic Compounds, *Experentia,* **28**, 1129.

[54] G. S. Fonken and R. A. Johnson, (1972), *Chemical Oxidations with Microorganisms,* M. Dekker.

[55] I. Fridovich, (1972), Superoxide Radical and Superoxide Dismutase, *Accounts Chem. Res.,* **5**, 321.

[56] E. T. Kaiser and B. Lu Kaiser, (1972), Carboxypeptidase A: Mechanistic Analysis, *Accounts Chem. Res.,* **5**, 219.

[57] R. Wolfenden, (1972), Analog Approaches to the Structure of the Transition in Enzyme Reactions, *Accounts Chem. Res.,* **5**, 10.

[58] P. D. Boyer (Ed.), (1973), *The Enzymes: Student Edition,* **1-2**, Academic Press.

[59] J. W. Cornforth, (1973), The Logic of Working with Enzymes, *Chem. Soc. Rev.,* **2**, 1 (1973).

[60] K. J. Laidler and P. S. Bunting, (1973), *The Chemical Kinetics of Enzyme Action,* Oxford University Press.

[61] R. Singleton, (1973), Bioorganic Chemistry of Phosphorus, *J. Chem. Ed.,* **50**, 538.

[62] J. S. Thayer, (1973), Biological Methylation, *J. Chem. Ed.,* **50**, 390.

[63] C. H. Wynn, (1973), *The Structure and Function of Enzymes,* E. Arnold.

[64] E. Zeffren and P. L. Hall, (1973), *The Study of Enzyme Mechanisms,* Wiley.

[65] A. Ault, (1974), An Introduction to Enzyme Kinetics, *J. Chem. Educ.,* **51**, 381.

[66] A. R. Battersby and J. Staunton, (1974), Stereospecificity of Some Enzymic Reactions, *Tetrahedron,* 30, 1707.

[67] J. W. Cornforth, (1974), Enzymes and Stereochemistry, *Tetrahedron,* **30**, 1515.

[68] A. Evans, (1974), *Glossary of Molecular Biology,* Butterworths.

[69] A. L. Fluharty, (1974), Biochemistry of the Thiol Group in *The Chemistry of the Thiol Group,* (S. Patai, Ed.), Part 2, Wiley.

[70] J. S. Fruton, (1974), The Active Site of Pepsin, *Accounts Chem. Res.,* 7, 241.

[71] G. G. Hammes, (1974), Elementary Steps in Enzyme Catalysis and Regulatio in *Pure and Applied Chemistry* – *The Official Journal of IUPAC,* **40**, Butterworths.

[72] O. Hayaishi, (1974), *Molecular Mechanisms of Oxygen Activation,* Academic Press.

[73] D. M. Jerina and J. V. Daly, (1974), Arene Oxides: A New Aspect of Drug Metabolism, *Science,* **185**, 573.

[74] I. M. Klotz, (1974), Protein Interactions with Small Molecules, *Accounts*

Chem. Res., 7, 162.

[75] J. N. Lowe and L. L. Ingraham, (1974), *An Introduction to Biochemical Reactions Mechanisms*, Prentice-Hall.

[76] L. J. Reed, (1974), Multienzyme Complexes, *Accounts Chem. Res.*, 7, 40.

[77] C. J. Suckling and K. E. Suckling, (1974), Enzymes in Organic Synthesis, *Chem. Soc. Rev.*, 3, 387.

[78] G. L. Cantoni, (1975), Biological Methylation: Selected Aspects, *Ann. Rev. Biochem.*, 44, 435.

[79] W. W. Cleland, (1975), What Limits the Rate of an Enzyme-catalyzed Reaction?, *Accounts Chem. Res.*, 8, 145.

[80] I. C. Gunsalus, T. C. Pederson and S. G. Sligar, (1975), Oxygenase-Catalysed

[81] K. R. Hanson and I. A. Rose, (1975), Interpretations of Enzyme Reaction Stereospecificity, *Accounts Chem. Res.*, 8, 1.

[82] W. P. Jenks, (1975), Binding Energy, Specificity, and Enzymic Catalysis: The Circe Effect, *Advances in Enzymology*, 43, 219.

[83] D. S. Sigman and G. Mooser, (1975), Chemical Studies of Enzyme Active Sites, *Ann. Rev. Biochem.*, 44, 889.

[84] J. Tze-Fei Wong, (1975), *Kinetics of Enzyme Mechanisms*, Academic Press.

[85] S. Blackburn, (1976), Enzyme Structure and Function, (*Enzymology Series*, 3), M. Dekker.

[86] J. Bland, (1976), Biochemical Effects of Excited State Molecular Oxygen, *J. Chem. Educ.*, 53, 274.

[87] D. M. Blow, (1976), Structure and Mechanism of Chymotrypsin, *Accounts Chem. Res.*, 9, 145.

[88] T. C. Bruice and P. Y. Bruice, (1976), Solution Chemistry of Arene Oxides, *Accounts Chem. Res.*, 9, 378.

[89] P. Cohen, (1976), *Control of Enzyme Activity*, Outline Studies in Biological Science, Chapman and Hall.

[90] J. W. Cornforth, (1976), Asymmetry and Enzyme Action, *J. Mol. Catal.*, 1, 145; *Science*, 193, 121.

[91] R. M. Denton and C. I. Pogson, (1976), *Metabolic Regulation*, Chapman and Hall.

[92] K. T. Douglas and A. Williams, (1976), Chemical and Biochemical Transfer of Acyl Groups, *J. Chem. Educ.*, 53, 544.

[93] W. Ferdinand, (1976), *The Enzyme Molecule*, Wiley.

[94] J. P. Guthrie, Enzyme Models and Related Topics in ref. 98, 2.

[95] G. A. Hamilton, Metal-Ion Dependent Redox Enzymes, in ref. 98, 2.

[96] H. A. O. Hill, (1976), Metals, Models, Mechanisms, Microbes and Medicine, *Chem. in Britain*, 12, 119.

[97] J. B. Jones and J. F. Beck, Asymmetric Synthesis and Resolution Using Enzymes, in ref. 98, 1.

[98] J. B. Jones, C. J. Sih and D. Perlman (Eds.), (1976), Application of Biochemical Systems in Organic Chemistry, (*Techniques of Chemistry*, A.

Weissberger Ed., X, erd ed.), 2 vols., Wiley-Interscience.

[99] T. Keleti, J. Ovadi, J. Batke, (1976), Catalysts and Enzymes (The Thermodynamic and Kinetic Basis of Enzyme Regulation), *J. Mol. Catal.,* 1, 173.

[100] K. Kieslich, (1976), Microbial Transformations of Non-steroid Cyclic Compounds, G. Thieme Verlag, Stuttgart.

[101] I. A. Rose and K. R. Hanson, Interpretations of Enzyme Stereochemistry in ref. 98, 2.

[102] C. F. Walter, (1976), *Enzyme Reactions and Enzyme Systems* (Enzymology Series, 4), M. Dekker, Basel.

[103] W. Albery and J. R. Knowles, (1971), Efficiency and Evolution of Enzyme Catalysis, *Angew. Chem. Int. Ed. Engl.,* 16, 285.

[104] P. C. Engel, (1977), *Enzyme Kinetics,* Chapman and Hall.

[105] J. R. Knowles and W. Albery, (1977), Perfection in Enzyme Catalysis: The Energetics of Triosephosphate Isomerase, *Accounts Chem. Res.,* 10, 105.

[106] T. Matsuura, (1977), Bio-mimetic Oxygenation, *Tetrahedron,* 33, 2869.

[107] M. I. Page, (1977), Entropy, Binding Energy and Enzymic Catalysis, *Angew. Chem. Int. Ed. Engl.,* 16, 449.

[108] W. A. Pryor (Ed.), (1977), *Free Radical in Biology,* Academic Press, 1-3.

[109] D. V. Roberts, (1977), *Enzyme Kinetics,* Cambridge University Press.

[110] K. G. Scrimgeour, (1977), *Chemistry and Control of Enzyme Reactions,* Academic Press.

[111] E. E. Van Tamelen (Ed.), (1977-78), *Bioorganic Chemistry,* 4 vols., Academic Press, 1-4.

[112] M. Akhtar and C. Jones, (1978), Some Biological Transformations Involving Unsaturated Linkages. The Importance of Charge Separation and Charge Neutralization in Enzyme Catalysis, *Tetrahedron,* 34, 813.

[113] R. Breslow, (1978), Studies on Enzyme Models and the Enzyme Carboxypeptidase in *Further Perspective in Organic Chemistry,* (Ciba Foundation Symposium 53), Elsevier.

[114] B. Feringa and H. Wynberg, (1978), Biomimetic Asymmetric Oxidative Coupling, *Bioorg. Chem.,* 7, 397.

[115] R. D. Gandour and R. L. Schowen, (1978), *Transition States of Biochemical Processes,* Plenum Publishing.

[116] C. Walsh, (1978), Chemical Approaches to the Study of Enzyme Catalysing Redox Transformations, *Ann. Rev. Biochem.,* 47, 881.

[117] M. J. Wimmer and I. A. Rose, (1978), Mechanisms of Enzyme-Catalyzed Group Transfer, *Ann. Rev. Biochem.,* 47, 1031.

[118] P. L. Luisi, (1979), Why Are Enzymes Macromolecules?, *Naturwissen schaften,* 66, 498.

[119] A. M. Mayer and E. Harel, (1979), Polyphenol Oxidases in Plants, *Phytochemistry,* 18, 193.

[120] K. H. Overton, (1979), Concerning Stereochemical Choice in Enzymic

Reactions, *Chem. Soc. Rev.*, 8, 447.

[121] J. Retey, (1979), Are the Steric Courses of Enzymatic Reactions Informative About Their Mechanisms in *Recent Advances in Phytochemistry*, ref. 57 of Chap. 1, Vol. 13.

(c) *Chemistry and biochemistry of coenzymes*

[122] E. N. Kosower, (1962), *Molecular Biochemistry*, McGraw-Hill.

[123] T. W. Goodwin, (1963), *The Biosynthesis of Vitamins and Related Compounds*, Academic Press.

[124] E. E. Snell, P. M. Fasella, A. Braunstein, and A. Rossi-Fanelli, (1963), *Chemical and Biological Aspects of Pyridoxal Catalysis*, Pergamon Press.

[125] D. W. Hutchinson, (1964), *Nucleotides and Coenzymes*, Chapman and Hall.

[126] T. C. Bruice and S. Benkovic, (1966), *Bioorganic Mechanisms*, 2 vols., Benjamin.

[127] P. Strittmatter, (1966), Dehydrogenases and Flavoproteins, *Ann. Rev. Biochem.*, 35, 125.

[128] H. C. S. Wood, (1966), The Birth of a Vitamin, *Chem. in Britain*, 2, 536.

[129] S. Chaikyn, (1967), Nicotinamide – coenzymes, *Ann. Rev. Biochem.*, 36,

[130] P. Fasella, (1967), Pyridoxal-phosphate, *Ann. Rev. Biochem.*, 36, 185.

[131] F. Lynen, (1967), The Role of Biotin-dependent Carboxylation in Biosynthetic Reactions, *Biochem. J.*, 102, 381.

[132] G. R. Penzer and G. K. Radda, (1967), The Chemistry and Biological Function of Isoalloxazines (Flavines), *Quart. Rev.*, 21, 43.

[133] D. Wellner, (1967), Flavoproteins, *Ann. Rev. Biochem.*, 36, 669.

[134] R. L. Blakley, (1969), The Biochemistry of Folic Acid and Related Pteridines in *Frontiers in Biology*, (A. Neuberger and F. L. Tatum, Eds.), 13, North-Holland Publishing Company.

[135] O. L. Krampitz, (1969), Catalytic Function of Thiamin Diphosphate, *Ann. Rev. Biochem.*, 38, 213.

[136] H. A. O. Hill, A. Roder and R. J. P. Williams, (1970), The Chemical Nature and Reactivity of Cytochrome P-450 in *Structure and Bonding*, Springer Verlag, 8.

[137] O. Isler, (1970), Developments in the Field of Vitamins, *Experientia*, 26, 225.

[138] H. Kamin (Ed.), (1971), *Flavins and Flavoproteins*, Butterworths.

[139] R. H. Wasserman and R. A. Corradino, (1971), Metabolic Role of Vitamins A and D, *Ann. Rev. Biochem.*, 40, 501.

[140] H. A. Barker, (1972), Corrinoid-dependent Enzymic Reactions, *Ann. Rev. Biochem.*, 41, 55.

[141] P. P. Nair and H. J. Kayden (Eds.), (1972), Vitamin E and Its Role in Cellular Metabolism, *Ann. N. Y. Acad. Sci.*, 203.

[142] R. H. Wasserman and A. N. Taylor, (1972), Metabolic Roles of Fat-soluble

Vitamins D, E, and K, *Ann. Rev. Biochem.*, **41**, 179.

[143] J. M. Wood and D. G. Brown, (1972), The Chemistry of Vitamin B_{12}- Enzymes in *Structure and Bonding,* Springer Verlag, **11**.

[144] S. J. Benkovic and P. W. Bullard, (1973), On the Mechanisms of Action of Folic Acid Cofactors in ref. 49, **2**.

[145] G. N. Schrauzer, (1974), 'Mechanisms of Corrin Dependent Enzymatic Reactions, in *Progress in the Chemistry of Organic Natural Products,* ref. 1 of Chap. 1, **31**.

[146] B. M. Babior, (1975), Mechanism of Cobalamin-dependent Rearrangements, *Accounts Chem. Res.,* **8**, 376.

[147] K. Yagi, (1975), *Reactivity of Flavins,* University of Tokyo Press.

[148] R. H. Abeles and D. Dophin, (1976), The Vitamin B_{12} Coenzyme, *Accounts Chem. Res.,* **9**, 114.

[149] C. T. Bruice, (1976), Models and Flavin Catalysis in ref. 49, **4**.

[150] I. M. Arias and W. B. Jacoby (Eds.), (1976), *Glutathione: Metabolism and Function,* Raven Press.

[151] P. Hemmerich, (1976), The Present Status of Flavin and Flavocoenzyme Chemistry in *Progress in the Chemistry of Organic Natural Products,* ref. 1 of Chap. 1, **33**.

[152] G. N. Schrauzer, (1976), New Developments in the Field of Vitamin B_{12}: Reactions of the Cobalt Atom in Corrins and in Vitamin B_{12} Model Compounds, *Angew. Chem. Int. Ed. Engl.,* **15**, 417.

[153] H. C. Wood and R. E. Barden, (1977), Biotin Enzymes, *Ann. Rev. Biochem.,* **46**, 385.

[154] G. I. Dimitrienko, V. Snieckus and T. Viswanatha, (1977), On the Mechanism of Oxygen Activation by Tetrahydropterin and Dihydroflavin — Dependent Monooxygenases, *Bioorg. Chem.,* **6**, 421.

[155] M. Hughes and R. H. Prince, (1977), Some Aspects of Zinc Ion Involvement in Alcohol Dehydrogenase Catalysis, *Bioorganic Chem.,* **6**, 137.

[156] A. Marquet, (1977), New Aspects of the Chemistry of Biotin and of Some Analogs, *Pure and Applied Chemistry,* **49**, 183.

[157] C. M. Visser and R. M. Kellog, (1977), Mimesis of the Biotin Mediated Carboxyl Transfer Reactions, *Bioorganic Chem.,* **6**, 79.

[158] S. J. Benkovic, (1978), On the Mechanism of Folate Cofactors, *Accounts Chem. Res.,* **11**, 314.

[159] E. J. M. Helmreich and H. W. Klein, (1980), The Role of Pyridoxal Phosphate in the Catalysis of Glycogen, *Angew. Chem. Int. Ed. Engl.,* **19**, 441.

[160] L. J. Machlin, (1980), *Vitamin E. A Comprehensive Treatise,* Marcel Dekker.

[161] C. M. Visser, (1980), Role of Ascorbate in Biological Hydroxylations: Origin of Life Considerations and Nature of the Oxenoid Species in Oxygenase Reactions, *Bioorg. Chem.,* **9**, 261.

[162] C. Walsh, (1980), Flavin Coenzymes: At the Crossroads of Biological Redox Chemistry, *Accounts Chem. Res.*, **13**, 148.

Chapter 3

Primary and Intermediate Metabolism

3.1 CHLOROPHYLL PHOTOSYNTHESIS

Green plants can convert light energy, from the sun, into chemical energy. This phenomenon is called photosynthesis: it takes place in cytoplasmic organelles rich in chlorophyll (the chloroplasts, see p. 24) and results in the conversion of carbon dioxide and water into glucose:

$$6\ CO_2(g) + 6\ H_2O(l) \xrightarrow{\ h\nu(\sim 700\ nm)\ } C_6H_{12}O_6\ (solution) + 6\ O_2(g) \qquad (1)$$

For the synthesis of one mole of glucose,

$$\Delta G^{\circ\prime} = 686\ kcal\quad [\text{i.e. for (1)}]$$

Where $\Delta G^{\circ\prime}$ is the free energy difference under standard conditions of 25°C and pH 7.0.

Six moles of ATP are also produced from six moles of ADP (reaction 2): in this manner a part of the photochemical energy is retained in a chemical form and is eventually used by the organism to fuel a large number of reactions:

$$6\ ADP + 6\ Pi \longrightarrow 6\ ATP + 6\ H_2O \qquad (2)$$

$$\Delta G^{\circ\prime} = 42\ kcal.\quad [\text{i.e. for (2)}]$$

The total amount of useful energy accumulating is up to 728 kcal per mole of glucose formed. This energy is gained after absorption of 48 moles of light quanta (einsteins) with a wavelength around 700 nm. Since one einstein has an amount of energy, dependent on the wavelength (λ), that is:

$$E(kcal) = \frac{h.c.N}{\lambda} = \frac{2.86 \times 10^4}{\lambda}$$

(where Planck's constant $h = 1{,}583 \times 10^{-37}$ kcal.s; the speed of light, c, is 3×10^{17} nm.s^{-1}; N, the Avogadro number, 6.023×10^{23} and λ the wavelength of light in nm), the total amount of light energy absorbed is,

$$E = \frac{48 \times 2.86 \times 10^4}{700} = 1961 \text{ kcal.}$$

The efficiency of photosynthesis, converting this light energy into chemical energy, according to equations (1) and (2) is thus 37%.

The major chemical paths by which glucose is synthesised have been unravelled by Calvin's school. The 'Calvin-Bassham cycle' gives a satisfactory explanation for most experimental results. Although some *in vivo* observations show that the Calvin-Bassham route (Fig. 3.3) is not uniquely utilised by photosynthetic organisms for capturing light energy, it is certainly the most important anabolic process for green plants. The appearance of starch granula in the stroma of exposed chloroplasts is the visible proof of glucose synthesis. Hydrolysis of the starch granules gives glucose which can then migrate into other parts of the plant.

Using $^{18}O_2$, it was shown that all the liberated oxygen derives from water, whilst the glucose oxygen derives from CO_2. The reduction of one CO_2 molecule, to one sixth of a hexose molecule, requires two molecules of water*, that is:

$$CO_2 + 2H_2{}^{18}O \longrightarrow [CH_2O] + {}^{18}O_2\uparrow + H_2O \tag{3}$$

A very short time (in the order of one thousandth of a second!) elapses between the photon absorption and the subsequent liberation of oxygen. The subsequent reduction of carbon dioxide, takes place with a slight delay, as is shown by $^{14}CO_2$ experiments. This step is a relatively slow one and it does not require simultaneous illumination. If a suspension of chloroplasts is irradiated in the presence of $NADP^+$, ADP and Pi, but in the absence of CO_2, NADPH and ATP are formed and oxygen is evolved. These facts suggest that chlorophyll-catalysed photosynthesis is split into two successive phases: a 'light' phase, in which the light quanta are captured and their energy utilised for separating water into oxygen and hydrogen (the latter species being bound to a carrier, such as NADP) and a 'dark' phase, in which carbon dioxide is reduced by utilising the hydrogen produced in the 'light' phase.

These two steps will now be considered in more detail.

---

*In sulphur bacteria (Thiorhodaceae) the reducing agent is not water but H_2S. The organisms thus liberate sulphur. The reduction of CO_2 occurs only under anaerobic conditions:

$$CO_2 + 2H_2S \longrightarrow [CH_2O] + 2S + H_2O$$

3.1.1 Absorption of Light Quanta. The 'Light' Phase.

The 'light' processes takes place in the quantasomes (see p. 25). *Eight* light quanta, absorbed by the photosynthetic pigments, cause the photolysis of water according to the following reactions:

H_2O photolysis and NADP reduction:

$$2NADP^+ + 2H_2O \longrightarrow 2\ NADPH + O_2 + 2H^+ \qquad (4)$$

Photosynthetic phosphorylation:

$$4\ ADP + 4\ Pi \longrightarrow 4\ ATP \qquad (5)$$

Two NADPH and three ATP molecules are stoichiometrically necessary for reducing one CO_2 molecule to $\frac{1}{6}$th of a glucose molecule in the subsequent 'dark' phase (Fig. 3.3). If this, as well as equations (4) and (5) are taken into account, one net ATP molecule is produced per carbon dioxide unit reduced; therefore six ATP molecules are formed for each glucose unit formed (equations (1) and (2)). A further amount of ATP is very likely to be formed through reaction (5) when disconnected from (4) (see following discussion).

Besides $NADP^+$ other substances may act as hydrogen acceptors. If ferricyanide is added to the isolated and irradiated chloroplasts, oxygen development is increased, whilst the added oxidising reagent is reduced; this is the well-known Hill reaction.

The process by which the light energy is captured and transformed, as chemical energy, into NADPH and ATP has not been completely unravelled. The currently accepted scheme is illustrated in Fig. 3.1.

This scheme agrees with studies on the photoexcitation of chlorophyll and with the ultrastructure of the quantasomes and their chemical composition. In this hypothesis chlorophyll *a* forms two types of 'active' associations (system I and system II) with protein and lipid molecules. In each quantasome only one molecule of chlorophyll *a* is enveloped in a *system I* association.

System I is excited by wavelengths between 680 and 700 nm. (It is also called the P_{700} system. In photosynthetic bacteria, containing bacteriochlorophyll, the pigment corresponding to P_{700} is called P_{870}).

The system II pigment can be excited either directly, by wavelengths near to 670 nm., or by energy transfer from other pigments present in the quantasome (chlorophyll *b*, carotenes, xanthophylls, biliproteins, etc.), with absorption maxima between 400 and 600 nm. The spectrum of photosynthetic action thus coincides, within good approximation, to the total absorption spectrum of the pigments (Fig. 3.2). For chlorophyll photosynthesis in higher plants light of wavelength > 700 nm. is ineffective; the useful light is at about 690 nm and below.

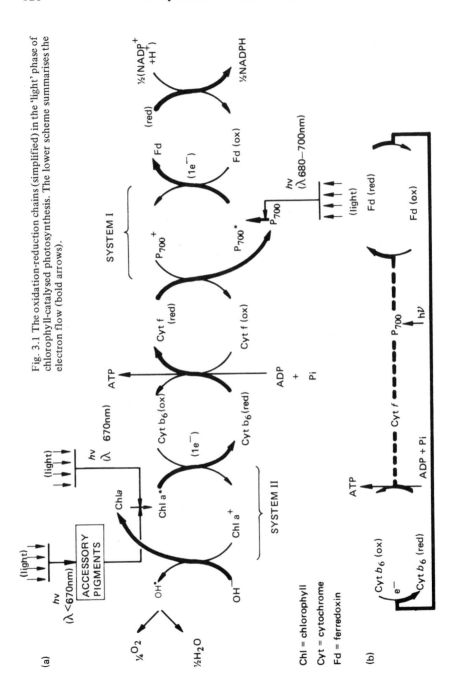

Fig. 3.1 The oxidation-reduction chains (simplified) in the 'light' phase of chlorophyll-catalysed photosynthesis. The lower scheme summarises the electron flow (bold arrows).

Chl = chlorophyll
Cyt = cytochrome
Fd = ferredoxin

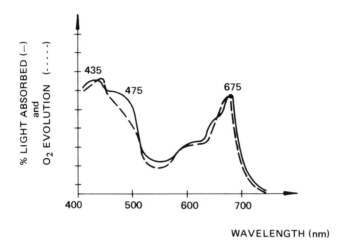

Fig. 3.2 Visible absorption in *Ulva taeniata*: - - - - - absorption causing
                                                        photosynthetic action
                                                ———— absorption of pigments

System II, when excited by a photon ($\lambda \leqslant 670$ nm) transfers one electron to a cytochrome (cytochrome b_6) through certain carriers (plastoquinones, not included in Fig. 3.1), which then transfers it to cytochrome f. The latter oxidation-reduction reaction is coupled to the phosphorylation of ADP to ATP. The electron flow continues from cytochrome f to system I [probably through a copper containing protein, the plastocyanin again not shown in the carrier chain of Fig. 3.1(a)]. The P_{700}^+ ion resulting from this process acts as an electron acceptor, in as much as it indirectly accepts the negative charge transferred to ferredoxin after interacting with another photon, of different wavelength from the former one (λ 680–700 nm). The radical nature of P_{700}^+ has been confirmed by electron spin resonance spectroscopy. The electron is finally transferred from ferredoxin to $NADP^+$, through a flavoprotein.

System II chlorophyll, when deprived of one electron, interacts with an OH^- ion to restore its neutral form (its fundamental state) liberating an $OH^{\cdot}$ radical, which disproportionates into water and oxygen gas. The outcome of the oxidation-reduction chains shown in Fig. 3.1 is that two quanta of light determine the flow of one electron, the production of a quarter of one oxygen molecule, of one ATP molecule and of half a NADPH molecule. [Eight quanta are therefore necessary for the stoichiometry of reactions (4) and (5).] It is thought that ferredoxin can also reduce cytochrome b_6 (Fig. 3.1(b)). A certain amount of phosphorylation should therefore take place independently from the photolysis of water, without developing oxygen, and with the participation of P_{700} only. In this case one photon would so produce one ATP molecule.

3.1.2. 'Dark' Reactions. The Calvin-Bassham Cycle

The ATP and NADPH molecules produced in quantasomes during the light phase migrate into the stroma, and are used there as cofactors by the enzymes catalysing the Calvin-Bassham cycle, which is the transformation of carbon dioxide into carbohydrates. The cycle, shown in Fig. 3.3, determines the fixation of three carbon dioxide molecules and the production of one D-glyceraldehyde-3-phosphate molecule, i.e. half of a glucose molecule. The balance of the inputs and outputs is in agreement with the consumption of two NADPH and three ATP moles per mole of CO_2.

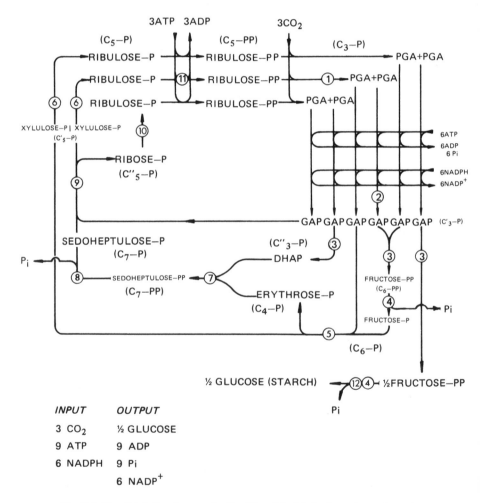

Fig. 3.3 The Calvin-Bassham cycle. See Figs. 3.3–3.9 for formulae and reaction mechanisms.

The separate reactions of the Calvin cycle are listed in Figure 3.4–3.8, with the compound formulae represented according to the Fischer convention (see Appendix 2)*. From these reactions it is apparent why, when the cycle is blocked after one round in the presence of ^{14}C-labelled carbon dioxide (for example, after irradiating chloroplasts for about 5 seconds) the radioactivity is concentrated in certain positions of the various intermediates and in glucose (Fig. 3.10).

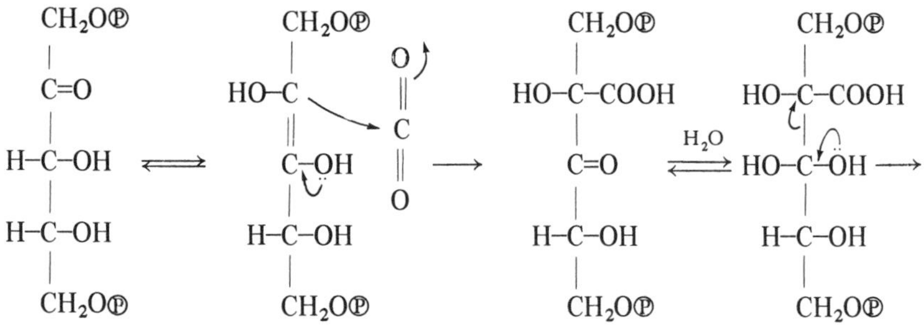

Ribulose-PP
(Ribulose-1,5-Diphosphate)

(1) [Carboxydismutase] : $(C_5\text{-}PP) + C_1 = 2(C_3\text{-}P)$

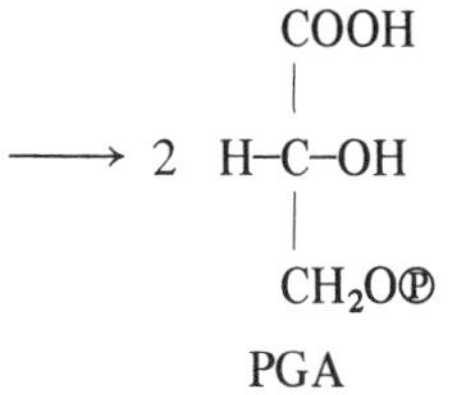

PGA
(3-Phospho-D-Glyceric Acid)

(2.1) [Phosphoglycerate Kinase]: $C_3 - P + P = C_3 - PP$

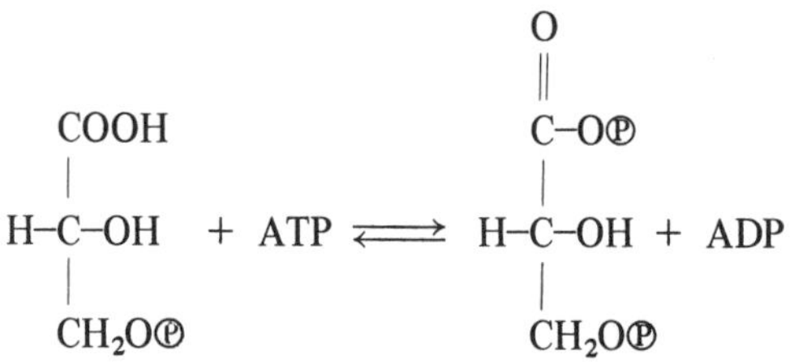

PGA   D-Glycerate-1,3-Diphosphate   Fig. 3.4 (contd. overleaf)

*The names of the enzymes mentioned in these Figures are the traditional, common names.

Fig. 3.4 (contd.)

(2.2) [D-Glyceraldehyde Phosphate Dehydrogenase]: $C_3\text{-PP} = C_3'\text{-P} + P$

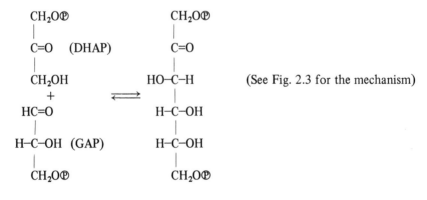

$$
\begin{array}{c}
O \\
\parallel \\
C\text{-}O\textcircled{P} \\
| \\
H\text{-}C\text{-}OH \\
| \\
CH_2O\textcircled{P}
\end{array}
+ \text{NADPH} + H^+ \rightleftharpoons
\begin{array}{c}
CHO \\
| \\
H\text{-}C\text{-}OH \\
| \\
CH_2O\textcircled{P}
\end{array}
+ \text{NADP}^+ + \text{Pi}
$$

GAP
(D-Glyceraldehyde-3-Phosphate)

Fig. 3.4. Mechanism of steps (1) and (2) of the Calvin-Bassham cycle (*cf.* Fig. 3.3).
The traditonal names of the enzymes involved are indicated in the parentheses.

(3.1) [Triosephosphate Isomerase]: $(C_3'\text{-P}) = (C_3''\text{-P})$

$$
\begin{array}{c}
CHO \\
| \\
H\text{-}C\text{-}OH \\
| \\
CH_2O\textcircled{P}
\end{array}
\rightleftharpoons
\begin{array}{c}
CH_2OH \\
| \\
C\text{=}O \\
| \\
CH_2O\textcircled{P}
\end{array}
$$

GAP   DHAP
(Dihydroxyacetone Phosphate)

(3.2) [Aldolase]: $(C_3''\text{-P}) + (C_3'\text{-P}) = (C_6\text{-PP})$

$$
\begin{array}{c}
CH_2O\textcircled{P} \\
| \\
C\text{=}O \quad (\text{DHAP}) \\
| \\
CH_2OH \\
+ \\
HC\text{=}O \\
| \\
H\text{-}C\text{-}OH \quad (\text{GAP}) \\
| \\
CH_2O\textcircled{P}
\end{array}
\rightleftharpoons
\begin{array}{c}
CH_2O\textcircled{P} \\
| \\
C\text{=}O \\
| \\
HO\text{-}C\text{-}H \\
| \\
H\text{-}C\text{-}OH \\
| \\
H\text{-}C\text{-}OH \\
| \\
CH_2O\textcircled{P}
\end{array}
$$

(See Fig. 2.3 for the mechanism)

Fructose-PP
(D-Fructose-1,6-Diphosphate)   Fig. 3.5 (contd. overleaf)

(4) [Phosphatase] : $(C_6\text{-}PP) = (C_6\text{-}P) + P$ Fig. 3.5 (contd.)

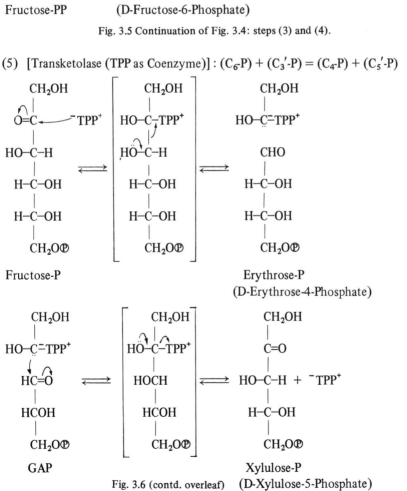

Fig. 3.5 Continuation of Fig. 3.4: steps (3) and (4).

(5) [Transketolase (TPP as Coenzyme)] : $(C_6\text{-}P) + (C_3'\text{-}P) = (C_4\text{-}P) + (C_5'\text{-}P)$

Fig. 3.6 (contd.)

(6) [Phosphoketopentose Epimerase] : $(C_5'\text{-P}) = (C_5\text{-P})$

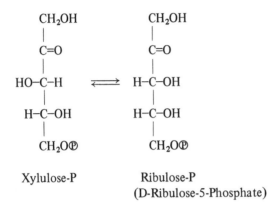

Xylulose-P   Ribulose-P
                     (D-Ribulose-5-Phosphate)

Fig. 3.6 Continuation of Fig. 3.4: steps (5) and (6).

(7) [Aldolase] : $(C_3''\text{-P}) + (C_4\text{-P}) = (C_7\text{PP})$

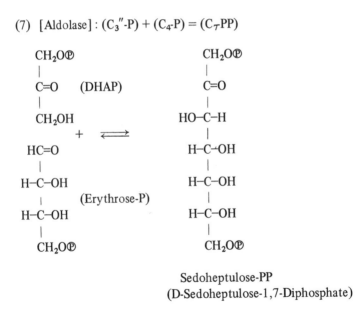

Sedoheptulose-PP
(D-Sedoheptulose-1,7-Diphosphate)

Fig. 3.7 (contd. overleaf)

Fig. 3.7 (contd.)

(8) [Phosphatase] : $(C_7PP) = (C_7P) + P$

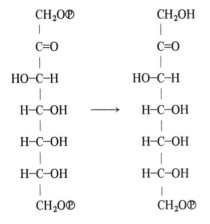

Sedoheptulose-PP   Sedoheptulose-P
                          (D-Sedoheptulose-1-Phosphate)

Fig. 3.7 Continuation of Fig. 3.4: steps (7) and (8)

(9) [Transketolase (TPP as coenzyme)] : $(C_7P) + (C_3''\text{-}P) = (C_5''\text{-}P) + (C_5'\text{-}P)$

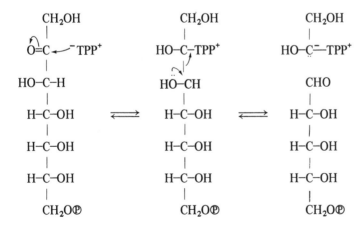

Sedoheptulose-P   (D-Ribose-5-Phosphate)

Formation of Xylulose-P occurs as described in reaction (5).

Fig. 3.8 (contd. overleaf)

Fig. 3.8 (contd.)

(10) [Phosphopentose Isomerase] : $(C_5'\text{-}P) = (C_5\text{-}P)$

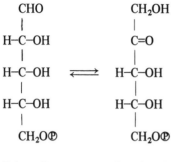

Ribose-P   Ribulose-P

(11) [Phosphopentose Kinase] $(C_5\text{-}P) + P = (C_5\text{-}PP)$

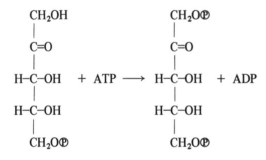

Fig. 3.8 Continuation of Fig. 3.4: steps (9) (10) and (11)

Step 1 (Fig. 3.4) is the carboxylation reaction; step 2 is an activation reaction, followed by a hydrogenation prompted by a NADP dependent-dehydrogenase. The mechanisms of the aldol reactions (step 3.2), aldose-ketose isomerisations (steps 3.1, 10 and 12a), and of the epimerisation of the alcoholic carbon adjacent to a carbonyl group (step 6) imply forming an enol bound to the enzyme, as indicated in Fig. 2.3. The transfer of a glycolaldehyde unit (steps 5 and 9) is an example of activating the aldehyde group through thiamine pyrophosphate (see p. 65).

The fructose-6-phosphate arising from operation of the Calvin cycle isomerises to glucose-6-phosphate (step 12a), which is the starting intermediate for all the metabolic uses of glucose (Fig. 3.9).

(12)

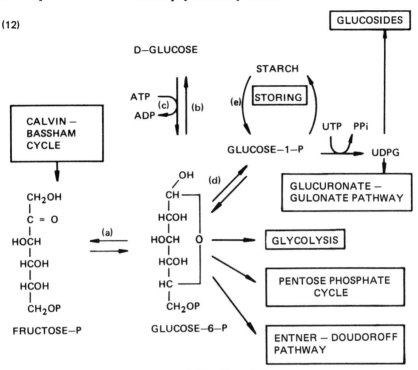

(a) : [PHOSPHOGLUCOISOMERASE] (See Figure 2.3)
(b) : [GLUCOSE−6−PHOSPHATASE]
(c) : [GLUCOKINASE]
(d) : [PHOSPHOGLUCOMUTASE]
(e) : [α−1,4−GLUCAN PHOSPHORYLASE]

Fig. 3.9 Synthesis and destinations of D-glucose-6-phosphate and D-glucose-1-phosphate.

Position of the Carbon atom chain	C-3 PGA	C-6 Fructose, Glucose	C-7 Sedoheptulose	C-5 Ribulose
1	*(82)	(3)	(2)	(11)
2	(6)	(3)	(2)	(10)
3	(6)	*(42)	*(28)	*(69)
4		*(43)	*(24)	(5)
5		(3)	*(27)	(3)
6		(3)	(2)	
7			(2)	

Fig. 3.10 Distribution of radioactivity (*) between intermediates of one Calvin-Bassham cycle after introduction of $^{14}CO_2$. The number in parenthesis represent the values, in percentages, in the various molecular positions after a total experimental term of 5 seconds.

As already mentioned, the Calvin-Bassham cycle is not the only way for fixing light energy and CO_2. Other known, or hypothesised, routes are summarised in Fig. 3.11. The glycolic acid route is probably responsible for the considerable formation of this labelled acid in cells having photosynthetic activity under a labelled carbon dioxide atmosphere. This phenomenon cannot be simply interpreted as oxidation of some active glycolic aldehyde, formed in transketolase reactions such a step (9) of Fig. 3.8. Furthermore, when there are low carbon dioxide concentrations in the air, an asymmetric distribution of the radiocarbon

(a) Glycolate Pathway

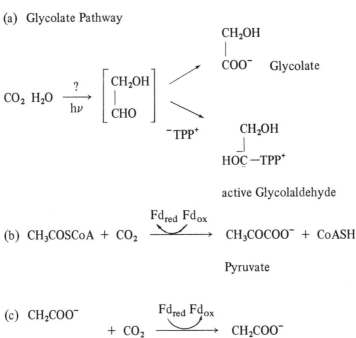

active Glycolaldehyde

(b) $CH_3COSCoA + CO_2 \xrightarrow{Fd_{red} \; Fd_{ox}} CH_3COCOO^- + CoASH$

Pyruvate

(c) $\begin{array}{l} CH_2COO^- \\ \\ CH_2COSCoA \end{array} + CO_2 \xrightarrow{Fd_{red} \; Fd_{ox}} \begin{array}{l} CH_2COO^- \\ | \\ CH_2COCOO^- \end{array} + CoASH$

α-Ketoglutarate

(d) $N_2 \xrightarrow{Fd_{red} \; Fd_{ox}} 2NH_3$

Fd_{red} = reduced Ferredoxin
Fd_{ox} = oxidised Ferredoxin

Fig. 3.11 Some possible means for the photofixation of CO_2 and N_2

in glucose is observed but not in fructose-1,6-diphosphate; C-4 becomes much more radioactive than C-3 (the Gibbs effect) and C-1 and C-2 are more labelled than C-5 and C-6 (Fig. 3.10). The supplementary formation of active glycol-aldehyde, coming from a 'pool' directly maintained by CO_2, according to reaction (a) of Fig. 3.11 could explain the Gibbs effect. Thus, active glycol-aldehyde reacts with GAP (reaction 5.2) giving rise to xylulose-5-phosphate labelled at positions 1 and 2; this pentose is then converted into glucose (via fructose-6- phosphate) through reactions 5, 10, 9 and 15, the last one being part of the pentose phosphate cycle (see Figs. 3.13 and 3.14).

In some experiments in which certain photosynthetic systems are fed with $^{14}CO_2$, alanine and glutamic acid appear to be the initial labelled products. The fixture of CO_2 has been hypothesised to take place either by reaction with acetyl-CoA, yielding pyruvic acid and subsequently alanine (Fig. 3.11 (b)), or by reaction with succinyl-CoA, forming α-ketoglutaric acid (Fig. 3.11 (c)), and finally glutamic acid by transamination. The reducing agent is probably reduced ferredoxin, (Fig. 3.1), which also participates in the fixation of atmospheric nitrogen (Fig. 3.11, (d)), and in the reduction of nitrates to nitrites and ammonia.

3.2 GLYCOLYSIS

Glucose-6-phosphate, derived from the phosphorylation of glucose or from the phosphorylative hydrolysis of starch (or of glycogen), undergoes several trans-formations in the cytosol (Fig. 3.9). Its degradation to pyruvic acid (glycolysis) is the most important of them. This metabolic path is also called the Embden-Meyerhof route, after the names of the two biochemists who made major contributions to its discovery. It is a fully anaerobic process, as indicated in Fig. 3.12. 3-Phosphoglyceric acid (PGA) is formed by a reverse sequence of the steps which lead from PGA to glucose in the chloroplast stroma (which are steps 2, 3, 4, and 12 of Fig. 3.3), thus returning chemical energy from glucose to the ATP and NADH coenzymes (whilst NADP was implied in photosynthesis).

Three moles of ATP and two of NADH are produced for every mole of glucose-6-phosphate transformed into pyruvic acid. If the glucose-6-phosphate is derived from starch (or from glycogen) the cell still gains three moles of ATP; only two moles of ATP can be gained if glucose-6-phosphate arises from the phosphorylation of glucose (see Fig. 3.9). Pyruvic acid and NADH are then consumed anaerobically (fermentation, Section 3.6) or aerobically (respiration, see Krebs cycle, Section 3.7). Three carbon-atom compounds are also important synthetic intermediates: dihydroxyacetone phosphate, after reduction by NADH, gives L-α-glycerolphosphate, utilised for the formation of fats (glycerides); the aminoacid serine is formed from D-3-phosphoglyceric acid (PGA); cysteine and glycine (the latter one with simultaneous transfer of material to the C_1 unit pool, see p. 163) are formed from serine; phosphoenolpyruvic acid (PEP) is used for synthesising aromatic aminoacids (phenylalanine, tyrosine, and tryptophan;

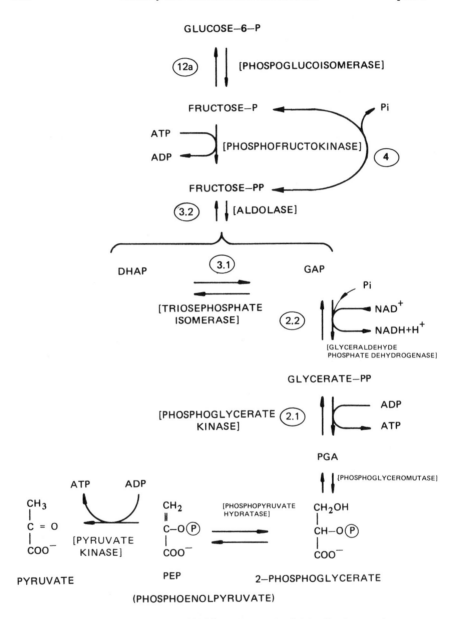

*NAD is used here in place of NADP involved in the Calvin—Bassham cycle.

Fig. 3.12 The Embden-Meyerhof route to glucose. The numbers encircled refer to the steps depicted in Figs. 3.4–3.9.

Chapter 7) and thus the phenylpropanoids and various alkaloids; pyruvic acid is the precursor of the aliphatic amino acids alanine, valine, and leucine. It should be noted, however, that whilst the steps from PEP to glucose-6-phosphate are all reversible, or easily made to proceed in both directions (for instance, fructose-1-phosphate to and from fructose-1,6-diphosphate), the transformation of PEP into pyrivic acid is practically irreversible. This means that the synthesis of carbohydrates from other metabolites (fats or aminoacids) is possible only if such substances are transformed into PEP or into the intermediates nearer the top of the glycolytic process ('anaplerosis', see Section 3.8).

3.3 THE PENTOSE PHOSPHATE CYCLE

In the cell (either animal or vegetable) cytosol, glucose-6-phosphate can also be degraded by the pentose phosphate (or phosphogluconic acid) cycle (Fig. 3.13). Glucose-6-phosphate is initially oxidised to 6-phosphoglucono-δ-lactone by the enzyme, glucose-6-phosphate dehydrogenase, which utilises $NADP^+$ as the hydrogen acceptor (step 13 in Fig. 3.13 and 3.14). 6-Phosphogluconolactone is hydrolysed (enzyme-accelerated), producing 6-phosphogluconic acid, which decarboxylates oxidatively to ribulose-5-phosphate, liberating CO_2, and producing a new NADPH molecule (step 14, Fig. 3.14). Three molecules of ribulose-5-phosphate are then transformed into two molecules of glucose-6-phosphate, and into one of glyceraldehyde-3-phosphate (GAP), through a path tracing back many of the steps of the Calvin cycle from fructose-6-phosphate, to ribulose-5-phosphate (steps 5, 6, 9, and 10 of Fig. 3.3). The novelty is the transaldolation reaction between sedoheptulose-7-phosphate and glyceraldehyde-3-phosphate, forming fructose-6-phosphate, and erythrose-4-phosphate (reaction 15 of Fig. 3.14).

The final balance of the complete cycle indicates that the cell recovers the chemical energy from one glucose molecule (completely transformed into CO_2) in the form of 12 NADPH molecules, a reducing agent utilised in numerous syntheses, in particular for the formation of fatty acids. Another advantage offered by the pentosephosphate pathway is the transformation of glucose-6-phosphate into ribose-5-phosphate, proceeding either by the practically, irreversibleoxidative phase of the cycle, or the non-oxidative reversible one. Ribose-5-phosphate is an essential compound for the production of nucleosides (NAD, ATP, etc.), nucleic acids, and the pyrrole ring of tryptophan (see page 000). Another intermediate of remarkable synthetic value is erythrose-4-phosphate, used by plants for producing the benzene nucleus of phenylalanine, tyrosine, and tryptophan (see Chapter 7). Lastly the pentose phosphate cycle offers an alternative route for glycolysis for producing glyceraldehyde-3-phosphate, and thus phosphoenolpyruvic and pyruvic acids.

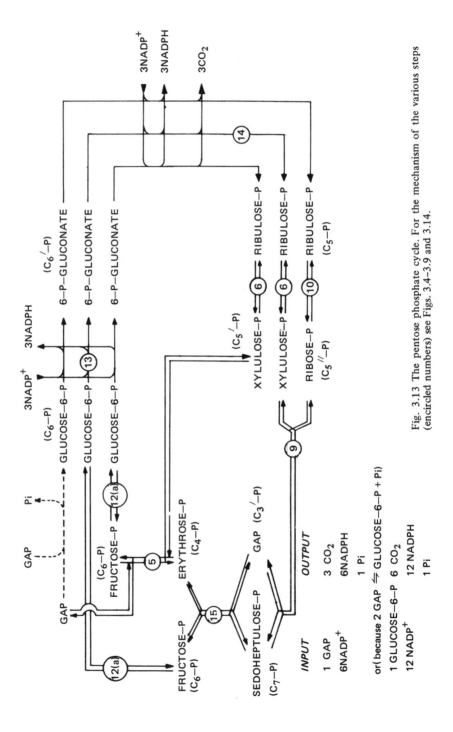

Fig. 3.13 The pentose phosphate cycle. For the mechanism of the various steps (encircled numbers) see Figs. 3.4–3.9 and 3.14.

(13) [Glucose-6-Phosphate Dehydrogenase (a); Gluconolactonase (b)]; C_6-P $= C_6'$-P

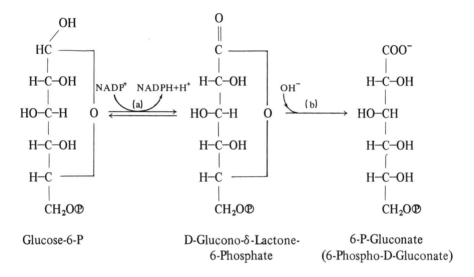

Glucose-6-P   D-Glucono-δ-Lactone-   6-P-Gluconate
                            6-Phosphate   (6-Phospho-D-Gluconate)

(14) [Phosphogluconate Dehydrogenase]: C_6'-P $= C_5$-P $+ C_1$

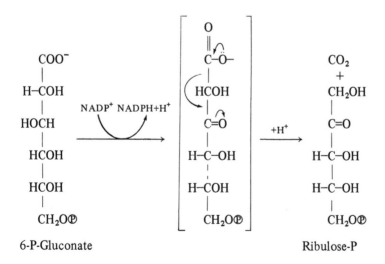

6-P-Gluconate   Ribulose-P

Fig. 3.14 (contd. overleaf)

(15) [Transaldolase]: $(C_7P) + (C_3'\text{-}P) = (C_4P) + C_6\text{-}P$ Fig. 3.14 (contd.)

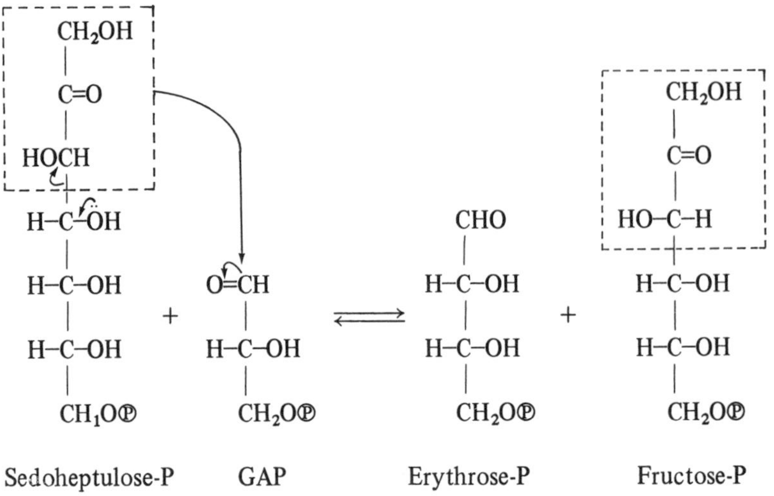

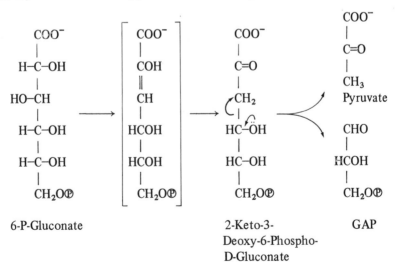

Sedoheptulose-P	GAP	Erythrose-P	Fructose-P

Fig. 3.14 Further reactions involved in the pentose phosphate cycle (*cf* Fig. 3.13).

3.4 THE ENTNER-DOUDOROFF PATHWAY

A third degradation process of glucose-6-phosphate is represented by the Entner-Doudoroff route. This path is undoubtedly a minor one with respect to the previous routes and it has only been found with certainty in one genus of bacteria, the *Pseudomonas* type. It is outlined in Fig. 3.15.

6-P-Gluconate		2-Keto-3-Deoxy-6-Phospho-D-Gluconate	GAP

Fig. 3.15 The Entner-Dondoroff route (in *Pseudomonas* spp.)

The first compound formed is 6-phosphogluconic acid, produced in step 13.1(a) of Fig. 3.14. By a retroaldol reaction it produces pyruvic acid, the methyl group of which derives from C-3 (and C-6) of glucose. In the pyruvic acid produced by glycolysis the methyl group arises from C-1 and C-6 of glucose.

3.5 D-GLUCURONIC AND L-GULONIC ACIDS, GLUCOSIDES, GLUCURON-IDES AND ASCORBIC ACID

α-D-Glucose-1-phosphate, derived from the isomerisation of glucose-6-phosphate, reacts with uridine triphosphate (UTP), giving uridine diphosphate glucose (UDPG) which is the active form of glucose for the synthesis of glucosides.

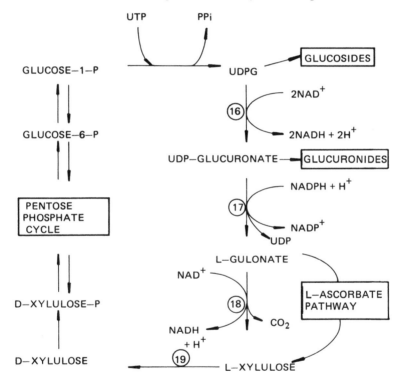

Fig. 3.16 The routes to D-glucuronic and L-gulonic acids. For the mechanisms of the various steps (encircled numbers) see Figs. 3.17 and 3.18.

All the natural glycosides derive from similar intermediates differing only in the sugar group. Some of these intermediates, for example the uridine di-phosphate-α-D-glucose (Fig. 3.16), can also undergo oxidation of the primary alcohol group at C-6 to the corresponding carboxylic acid. This oxidation involves the consumption of two NAD^+ moles, and gives uridine diphosphate-glucuronic acid (UDP-glucuronic acid). In mammals UDP-glucuronic acid is utilised for numerous detoxification processes, as it donates D-glucuronic acid to alcohols and phenols, when appropriate glucuronyl-transferases are present: such conjugation products (glucuronides), in which a hemiacetal bond is formed between the glucuronic acid at position 1 and the phenolic or alcoholic groups, are easily excreted in the urine.

(16) [UDP-D-Glucose Dehydrogenase]

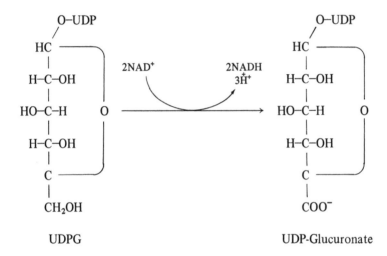

(17) [UDP-D-Glucuronyl Transferase]; [Glucuronate Reductase]

Fig. 3.17 Reactions of steps (16) and (17) involved in the route to D-glucuronic and L-gulonic acids (cf. Fig. 3.16)

In mammals and plants, some of the UDP-glucuronic acid is also transformed into L-gulonic acid (Figs. 3.16 and 3.17). L-Gulonic acid decarboxylates to give L-xylulose, which subsequently isomerises to D-xylose (Figs. 3.16 and 3.18). At this point there is an overlap into the non-oxidative stage of the pentose phosphate cycle. The main difference between the D-glucuronic route and that of the pentose phosphate way lies in the choice of the carbon atom eliminated as CO_2: the primary carbon, C-6, is lost in the glucuronic acid route whilst C-1 is eliminated in the pentose phosphate pathway. As a consequence, D-pentoses formed by the pentose phosphate route (*via* oxidation-decarboxylation) have, as C-1, the corresponding C-2 of glucose or fructose, whilst the ones synthesised from the D-glucuronic and L-gulonic acid have the same C-1 as in the original hexoses.

(18)

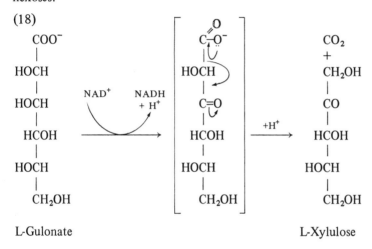

L-Gulonate   L-Xylulose

(19)

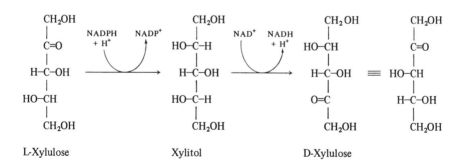

L-Xylulose   Xylitol   D-Xylulose

Fig. 3.18 Reactions of steps (18) and (19) involved in the route to
D-glucuronic and L-gulonic acids (*cf.* Fig. 3.16).

The importance of the D-glucuronic and L-gulonic acid pathway also arises from its intimate connection with the biosynthesis and catabolism of ascorbic acid (Fig. 3.19). In man, primates, and guinea-pigs this biosynthesis is forbidden by a genetic, metabolic block between L-gulonolactone and 2-keto-L-gulono-lactone, so that for such species ascorbic acid is an essential factor in diet, i.e. the vitamin C.

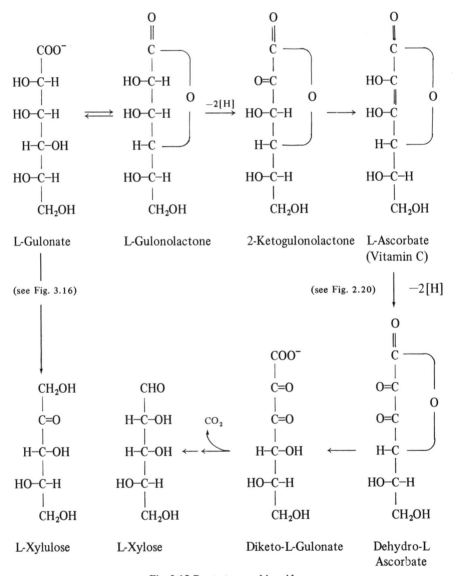

Fig. 3.19 Route to ascorbic acid.

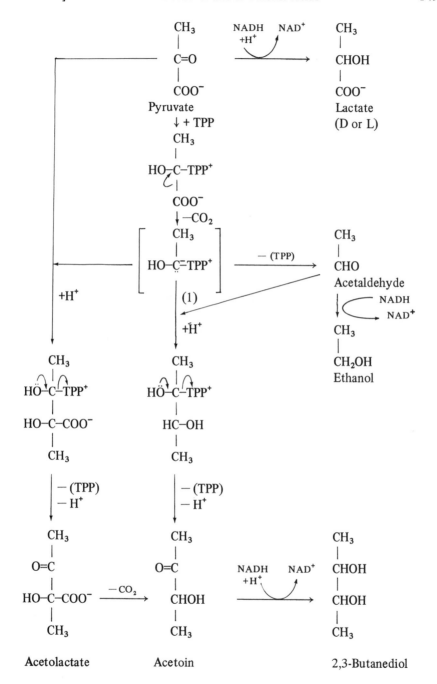

Fig. 3.20 The fermentation route to pyruvic acid.

3.6 FERMENTATION

Pyruvic acid arising from the glycolytic route undergoes a very different fate according to whether the cell is kept in anaerobic or aerobic conditions. In the first case we speak of *fermentation,* in the second one, of *respiration.*

In fermentation, the principal metabolic routes of pyruvic acid are its reduction to lactic acid, the formation of ethanol, and the formation of acetoin (Fig. 3.20).

The reduction to lactic acid (through a NAD-dependent dehydrogenase action) is a reversible process, taking place in vertebrate muscles, and in many microorganisms (*Lactobacilli, Streptococci, Clostridia, etc.*). Whilst lactic acid of animal origin is invariably in the L-(+) form, that of bacterial origin can have both configurations.

Ethanol is the classical product of yeasts (*Saccharomyces cerevisiae*) from fermentation and brewing and is produced by the reduction of acetaldehyde, which in turn originates from the decarboxylation of pyruvic acid. The reduction is catalysed by the familiar enzyme yeast alcohol-dehydrogenase (YADH), the stereospecificity of which has been discussed on p. 83. The decarboxylation of pyruvic acid takes place on its adduct with thiamine pyrophosphate (TPP), giving an active form of acetaldehyde (the activated form is analogous to that of glycolic aldehyde, formed in transketolisation reactions, see Fig. 3.8). The acetaldehyde-TPP adduct can also react, as a carbanion (Fig. 3.20), with another acetaldehyde molecule (acting as an electrophilic species), giving an acetoin molecule. Such acyloin condensations are frequent in bacteria and plants. In certain microorganisms, such as *Aerobacter aerogenes* and *Proteus vulgaris,* the activated acetaldehyde species attacks the carbonyl group of pyruvic acid, forming acetolactic acid, which then decarboxylates to acetoin, by the action of a specific enzyme. When an appropriate dehydrogenase is present, the acetoin can be reduced to 2,3-butanediol.

In many living organisms under anaerobic conditions, the acetaldehyde-TPP adduct ((1), Fig. 3.21(a)) is oxidised by an electron acceptor (A), through appropriate carriers (ferredoxin?), and forms of acetyl-TPP ((2), Fig. 3.21(a)) which, by reaction with nucleophiles (XH) can generate active forms of acetic acid, for instance acetyl-CoA ($CH_3COSCoA$) and acetylphosphate. The whole process, known as the Clastic reaction, varies from organism to organism as does the nature of the final electron acceptor (when A is $2H^+$, molecular hydrogen is liberated and, when A is CO_2, formic acid is formed) and the nature of the acetyl acceptor (CoASH, etc.).

3.7 THE TRICARBOXYLIC ACID CYCLE (TCA CYCLE)

In aerobic organisms (microorganisms, animals, and plants) the adduct between pyruvic acid and TPP is metabolised by enzymic complexes, called 'pyruvate dehydrogenase complexes' (PDC) placed inside the mitochondria of animal and

(a)

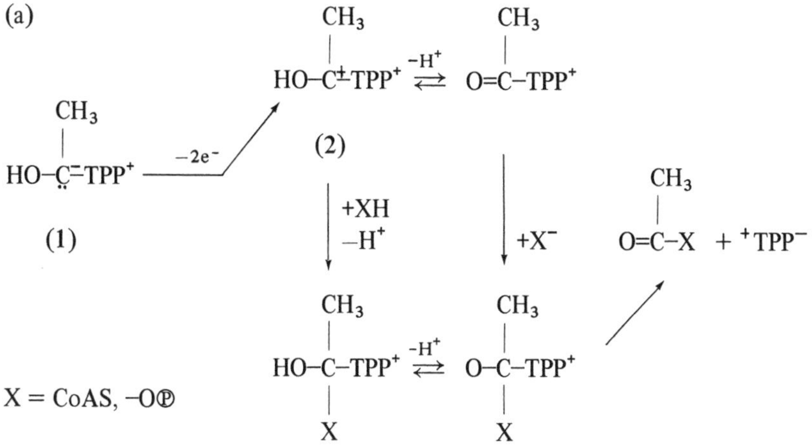

$$CH_3 \qquad CH_3$$
$$HO-C^{\pm}TPP^+ \xrightleftharpoons[\ ]{-H^+} O=C-TPP^+$$

$$CH_3$$
$$HO-\underset{\cdot\cdot}{C}=TPP^+ \xrightarrow{-2e^-} \qquad (2)$$

(1)

$$\downarrow \begin{array}{c} +XH \\ -H^+ \end{array} \qquad \qquad \downarrow +X^- \qquad O=C-X \ + \ ^+TPP^-$$

$$CH_3$$

$$X = CoAS, -O\text{℗}$$

$$CH_3 \qquad CH_3$$
$$HO-C-TPP^+ \xrightleftharpoons[\ ]{-H^+} O-C-TPP^+$$
$$\quad | \qquad\qquad\quad |$$
$$\quad X \qquad\qquad\quad X$$

(b)

$$CH_3 \qquad\qquad CH_3 \qquad\qquad (CH_2)_4.CO-Enz$$
$$CO \ + \ ^-TPP^+ \xrightarrow[+H^+]{-CO_2} HO-\underset{\cdot\cdot}{C}=TPP^+ \qquad \begin{array}{c} \diagup\diagdown \\ S{-}S \end{array} \text{(Enzyme bound Lipoic Acid)}$$
$$COO^- \qquad\qquad\qquad\qquad (1) \qquad\qquad\qquad\qquad +H^+$$

Pyruvate

$$CH_3 \qquad\qquad\qquad\qquad CH_3$$
$$H\ddot{O}-C-TPP^+ \xrightarrow[-H^+]{-TPP} O=C \qquad\qquad \xrightarrow{CoASH}$$
$$\quad | \qquad\qquad\qquad\qquad\quad |$$
$$\quad S{-}(CH_2)_5.CO-Enz \qquad\qquad S{-}(CH_2)_4.CO-Enz$$
$$\quad HS \qquad\qquad\qquad\qquad\qquad HS$$

$$\longrightarrow CH_3COSCoA \ + \qquad (CH_2)_4.CO-Enz \qquad\qquad FAD \qquad\quad NADH + H^+$$
$$\quad (3) \qquad HS \quad SH \qquad\qquad\qquad \text{(Flavoprotein)}$$
$$\qquad\qquad\qquad\qquad (CH_2)_4.CO-Enz \qquad\quad FADH_2 \qquad\quad NAD^+$$
$$\qquad\qquad\qquad S{-}S$$

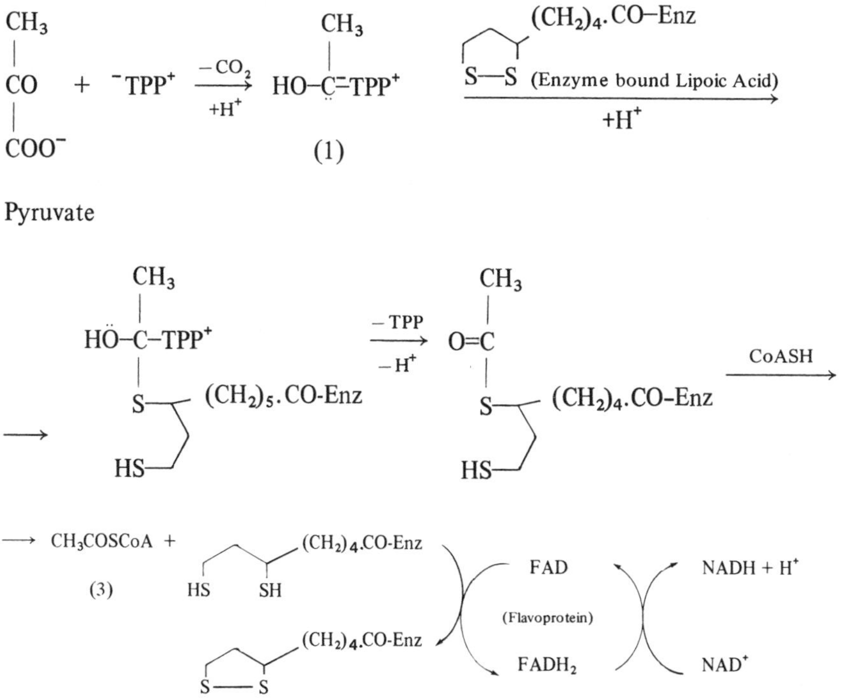

Fig. 3.21 (a) The Clastic reaction; (b) mitochondrial oxidative decarboxylation route to pyruvic acid.

plant cells, or associated with the plasmatic membranes of bacteria. Acetyl CoA
((3), Fig. 3.21(b)) is obtained through a reaction sequence which shows remarkable similarities to the decarboxylation step of fermentation processes (Fig. 3.20)
and the oxidative step of Clastic reactions. Lipoic acid, through the FAD of
flavoproteins, carries electrons to NAD^+, which is in turn re-oxidised in the
respiratory chain.

The acetyl CoA derived from pyruvic acid (and thus from carbohydrates)
merges in the mitochondria with that arising from other sources, such as the
fatty acids and amino-acids (proteins), forming a unique mitochondrial pool.
This pool mainly feeds the respiration process, through which the cell burns
acetic acid to CO_2 with molecular oxygen, and utilises the combustion energy as
chemically energetic ATP molecules.

The first part of this respiratory process is called the tricarboxylic acid cycle,
otherwise known as the citric acid, or the Krebs' cycle. It consists of a sequence
of reactions in which the final product loop into the original step. Through this
closed chain of reactions the acetyl unit is gradually oxidised to carbon dioxide.
The cycle is outlined in Fig. 3.22. The reduced forms of the hydrogen acceptor
coenzymes are oxidised through the respiratory chain to which ADP phosphorylation is coupled. 3 Moles of ATP are obtained from each mole of NADH and
2 moles of ATP from one mole of $FADH_2$ (Appendix 1, Fig. A1.2). Thus the
complete combustion of 1 mole of acetyl CoA supplies the organism with 12
ATP molecules; 11 are obtained through the respiratory chain and one directly
by reaction 24. The overall recovery of energy $(\Delta G^{\circ'})$ is of 15 moles of ATP
with respect to one mole of pyuric acid and of 38 moles of ATP with respect to
one mole of glucose burnt to CO_2 *via* glycolysis and the TCA cycle.

The reactions of the tricarboxylic acid cycle require some comments from a
mechanistic point of view. The condensation between oxaloacetic acid and the
acetyl CoA, forming citric acid, is a stereospecific reaction, since the methylene
carboxylic group derived from acetyl CoA (marked in the Fig. 3.22 as ●)
exclusively assume the *pro-S* position with respect to the central carbon of citric
acid. (A C_{aabc} prochiral carbon, see Appendix II). This configurational result
requires a *si* attack of acetyl CoA to the carbonyl group of oxaloacetic acid
(reaction 20, Fig. 3.23).

When citric acid is isomerised into isocitric acid, the enzyme chooses between
the two enantiotopic methylenecarboxylic chains: the *pro-R* group is the one
involved in the transposition of hydrogen and hydroxyl groups (reaction 21,
Fig. 3.22). As a consequence of the subsequent reactions, the two CO_2 molecules
formed after a whole cycle, derive from the oxaloacetic acid. This sequence is in
perfect agreement with the labelling observed in α-ketoglutaric acid: for instance,
the specific formation of [5-^{13}C] α-ketoglutaric acid by using [1-^{13}C] acetic
acid (Fig. 3.22). Ogston, in 1948, was the first scientist, to give an interpretation
of such selective labelling which is now universally accepted. He state that 'an
asymmetric enzyme, attacking a symmetric compound, is able to distinguish

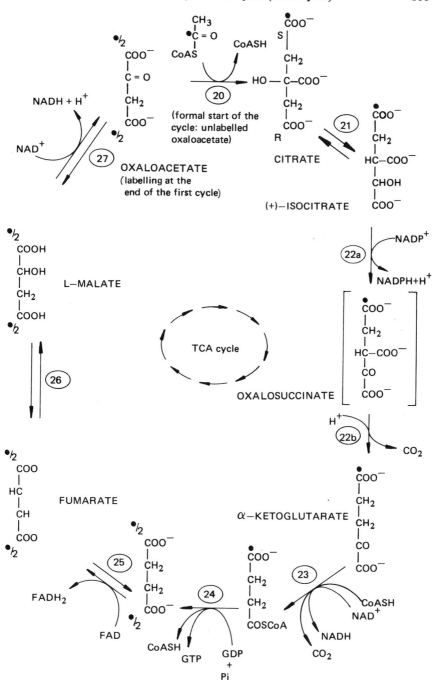

Fig. 3.22 The tricarboxylic acid (TCA) cycle. Details of the steps (encircled numbers) are given in Figs. 3.23 and 3.24.

between identical groups'. He also gave a theoretical explanation for such enzymatic stereospecificity known as the 'Ogston effect', which is treated more fully in Appendix II.

The enzyme transforming citric acid into isocitric acid (aconitase) also shows other interesting peculiarities: it replaces the central carbon hydroxyl of citric acid with the same *pro-R* hydrogen abstracted from the *pro-R* methylene, and replaces the hydroxyl group in the hydrogen's place, without inverting the configurations of the carbon atoms involved in these substitutions. Furthermore, it also catalyses the formation of some free *cis*-aconitic acid, which can be subsequently transformed into isocitric acid but, in this case, the *pro-R* hydrogen of the *pro-R* arm of citric acid is exchanged with those of the medium. Mechanism 21 of Fig. 3.23 interprets such facts, as evidenced by stereospecifically-labelled deuteriated or tritiated substrates. Aconitase, at least formally, is a bifunctional enzyme, since the proton extracted from the substrate in the first step can be transferred to another position of an intermediate according to an E2 and a retro-E2 process respectively. In this isomerisation reaction *cis*-aconitic acid appears to be remarkably mobile within the active site, and to have a relatively low affinity towards the enzyme.

Reaction 22(a), Fig. 3.24, is a simple dehydrogenation, in which hydrogen from position 2 of isocitric acid is given to NAD^+. (This hydrogen corresponds to the ex *pro-S* hydrogen from the *pro-R* arm of citric acid.)

Oxalosuccinic acid has got three hydrogens: the methylene pair is derived from acetyl CoA, and the third one corresponds, in part, to the ex *pro-R* of the *pro-R* chain of citric acid and, in part, comes from the medium. The 'detection' of the latter hydrogen arises since isocitric acid, as has been mentioned above, can form either by direct intramolecular transposition of citric acid, or from *cis*-aconitic acid. The decarboxylation of oxalosuccinic acid (step 22(b), Fig. 3.24) implicates acquisition of a new hydrogen atom from the medium, and placed at the C–3 *pro-S* position of α-ketoglutaric acid. The conversion (step 23, Fig. 3.22) of α-ketoglutaric acid into succinyl CoA has no stereochemical implications and has an identical mechanism to the pyruvic acid – acetyl CoA conversion (see Fig. 3.21(b)). The loss of CoA with GTP production by reactions opposite to those involved in carboxylic group activation, produces a perfectly symmetrical molecule – the free succinic acid – in which the two methylene carboxylic portions are equivalent and are thus indistinguishable even by chiral reagents (see Appendix II). The stereospecific dehydrogenation of succinic acid to fumaric acid is schematised in reaction 25, Fig. 3.24. An elimination of a pair of antiperiplanar hydrogens in the most stable conformation of succinic acid takes place. The two hydrogens of fumaric acid are chemically and stereochemically indistinguishable.

The addition of water to the double bond of fumaric acid, reaction 16, takes place with proton attack to the *re,re* face, and a hydroxyl attack to the *si,si* face: L-malic acid is thus formed, which is then oxidised to oxaloacetic acid by a NAD-dependent dehydrogenase. Thereafter the cycle is repeated.

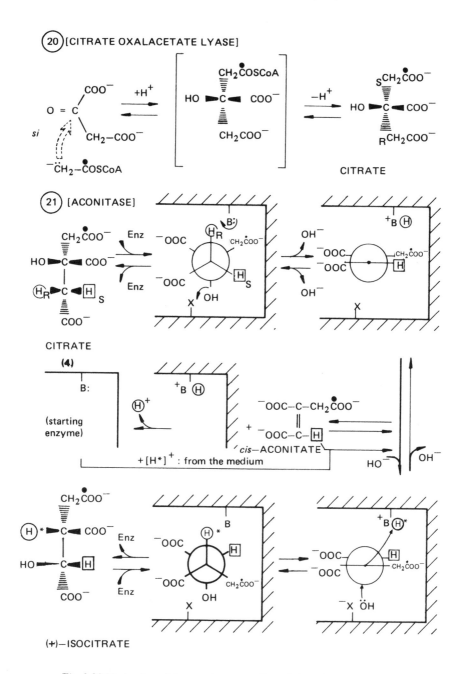

Fig. 3.23 Mechanism of the reaction steps 20 and 21 of the TCA cycle (*cf* Fig. 3.22).

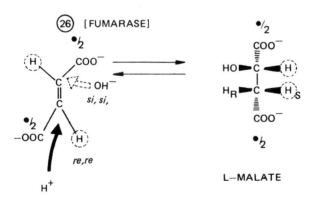

FUMARATE   Fig. 3.24 Mechanism of the reaction steps 22, 25 and 26 of the TCA cycle (*cf* Fig. 3.22).

The importance of the TCA cycle is not only due to its combustion action, in which acetic acid is burnt and ATP formed *via* the respiratory chain but also to the utilisation of some of the intermediates in other processes, such as:

(i) Transport of acetyl CoA from the inside to the outside part of mitochondria. Since the mitochondrial membrane (see p. 19) is rather impermeable to acetyl CoA this should result in the existence of two distinct C_2 pools. Actually only one pool is present in the cell, because acetic acid is easily exchanged between the cytosol and the mitochondria in at least three ways (Fig. 3.25 (a)-(c)). One involves citrate ion which can freely pass through the mitochondrial membrane; the two others utilise acetate ion and acetyl-carnitine as diffusible forms.

(a) [Citrate Lyase]

Citrate + ATP + CoASH $\rightleftharpoons$ Acetyl-SCoA + Oxaloacetate + ADP + Pi

(b) [Acetate Thiokinase]

Acetate + ATP–CoASH $\rightleftharpoons$ Acetyl-SCoA + AMP + PPi

(c) [Acetyl coenzyme A − Carnitine Acetyl Transferase]

$$CH_3CO\text{-}SCoA + (CH_3)_3\overset{+}{N}\text{-}CH_2\text{-}CH\text{-}CH_2\text{-}COO^- \rightleftharpoons$$

        |
        OH
    Carnitine

$$(CH_3)_3\overset{+}{N}\text{-}CH_2\text{-}CH\text{-}CH_2\text{-}COO^-$$

        |
        $OCOCH_3$ + HSCoA

    Acetyl Carnitine

(d) [Phosphoenolpyruvate Carboxykinase]

    COO⁻
    |
    CH₂   CH₂
    | + GTP ⇌   ‖   + GDP + CO₂
    CO   C–O℗
    | |
    COO⁻   COO⁻

Oxaloacetate   PEP

Fig. 3.25 Other uses of the TCA cycle: (a)–(c) production of acetyl-CoA; (d) retroglycolysis.

The first path also allows the exit of oxaloacetic acid from the mitochondria.
(ii) The synthesis of sugars. Oxaloacetic acid, also produced from malic acid diffusing out of mitochondria, can give rise to phosphenol-pyruvic acid in the ioloplasm (reaction d Fig. 3.25). The latter reactions catalysed by a phospho-enolpyruvate-carboxykinase utilising ITP, or GTP as coenzymes. Retro-glycosis can proceed from PEP (Fig. 3.12). Since oxaloacetic acid can also be considered a glucose catabolite, via pyruvic acid (see Fig. 3.26(b)), the irreversibility of the reaction PEP to pyruvic acid may be overridden through process (c), Fig. 3.25, which thus ensures that glycolysis can be a reversible process.
(iii) The synthesis of various metabolites. α-Ketoglutaric and oxaloacetic acids can, for instance, be used for producing glutamic and aspartic acids respectively, whilst succinyl CoA is used for the synthesis of tetrapyrrole pigments (porphyrins, chlorophylls, etc.)

3.8 ANAPLEROTIC PROCESSES

According to the tricarboxylic acid cycle (Fig. 3.22) it appears that every dicarboxylic acid is produced and successively used up, without accumulation. In fact, many of these acids are utilised by the cell for producing various primary and secondary metabolites. The organism must synthesise such dicarboxylic acids through independent ways, called anaplerotic processes, in order to keep the pool of such acids at its optimum level. In this way C_1 (such as CO_2) and C_2 (such as CH_3COOH or $CHOCOOH$) units are fed into the tricarboxylic acid (TCA) pool, through reactions schematised in Fig. 3.26.

The method for recovering carbon dioxide exists in nearly all cells; it produces oxaloacetic acid, and is connected to the retroglycolysis step in forming PEP, according to reaction (c) Fig. 3.25. The enzyme catalysing the carboxylation of pyruvic acid requires biotin as a CO_2 carrier (p. 65).

If [^{14}C] carbon dioxide is used, oxaloacetic acid is initially labelled at the C-4 carboxyl group and, subsequently, uniformly at both carboxyl groups, since interconversions of oxaloacetic acid to malic acid to fumaric acid (the 'dicarboxylic acid shuttle') is rapid.

Consequently PEP resulting from reaction (c) Fig. 3.25, is to some extent labelled in its carboxyl group. For the same reason pyruvic acid specifically labelled at C-2 or C-3 produces PEP which is randomly labelled at these positions.

The second anaplerotic pathway, or glyoxylate cycle (the Krebs-Kornberg cycle) only operates in plants and microorganisms, and involves the synthesis of two L-malic acid molecules (which become oxaloacetic molecules) starting from *one* molecule of oxaloacetic acid, and *two* molecules of acetyl CoA. Overall, there is a net synthesis of TCA intermediates using C_2 units. The cycle offers to plants the possibility of utilising glyoxylic acid (derived from the photosynthetic oxidation of glycolic acid) as a source of various metabolites and energy.

(a)   (a) RECOVERY of CO_2 : [ATP–DEPENDENT PYRUVATE CARBOXYLASE]

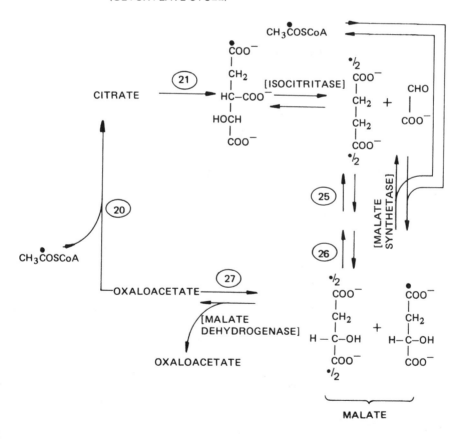

(b)   CONVERSION of C_2–UNITS INTO C_4–UNITS.
       (GLYOXYLATE CYCLE)

Fig. 3.26 Anaplerotic processes.

3.9 METABOLISM OF FATTY ACIDS

The synthesis of fatty acids takes place in three different sites of the cell, and is mediated by three enzymatic systems.

(i) The mitochondrial system, catalysing the lengthening of the aliphatic chain of a fatty acid, according to the scheme in Fig. 3.27. Acetic acid units (as acetyl CoA) are successively added, followed each time by reduction of the resulting β-carbonyl group to a saturated methylene group.

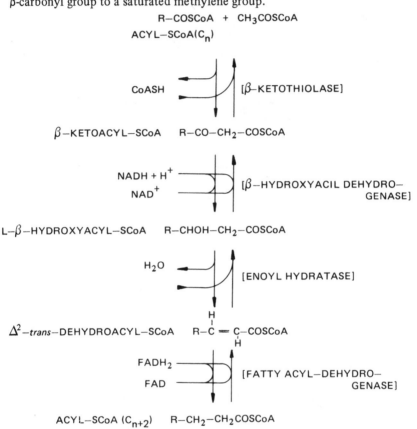

Fig. 3.27 Synthesis and mitochondrial degradation (β-oxidation) of fatty acids.

This biosynthetic route is like the reverse of the most important degradative pathway of fatty acids, β-oxidation, of importance in the production of acetyl CoA (n/2 moles) starting from saturated fatty acids with even-numbered carbon chains $[CH_3-(CH_2)_{n-2}-COOH]$ in mitochondria. A difference is in the step, saturated $\rightleftharpoons$ unsaturated acyl CoA, catalysed by a NADP-dependent enoyl reduction in the synthetic route and by a FAD-dependent dehydrogenase in the degradative one.

(ii) The cytoplasmic system converts acetyl CoA into long chain fatty acids; it requires ATP, CO_2, Mn^{++} and NADPH. Part of the acetyl CoA is transformed into malonyl CoA by an acetyl carboxylase, which utilises biotin as CO_2 carrier. Malonyl CoA can also be formed by activation of malonic acid produced by a different route.

The condensation of acetyl CoA with malonyl CoA and the successive lengthening of the fatty acid chain takes place according to the reactions shown in Fig. 3.28(a).

It is of note how the carrier protein (acyl carrier protein, ACP) and the enzymes catalysing each step form a multienzymatic complex (fatty acid synthetase). The substrates never detach from the complex during the whole series of transformations which determine the lengthening of the aliphatic chain by two carbon atoms. The progressive lengthening of the fatty acid only stops when the aliphatic chain has reached the critical value of about 16-20 carbon atoms; the reasons for limit are still unknown. A schematic representation of fatty acid synthetase is shown in Fig. 3.28(b). According to Lynen's hypothesis the enzymatic complex should consist of 6 macromolecules, set around a central protein (ACP), from which the 4'-phosphopantetheine chain should protrude; such a chain should terminate with a thiol (–SH) group, acting as a 'hook' for the carboxylic group of the lengthening fatty acid. This chain is long enough to allow the substrate to reach into the active sites of the surrounding enzymes. Every complete rotation of such an ACP chain should correspond to a two-carbon lengthening of the fatty acid. Other acyl CoA's (isobutyryl CoA, isovaleryl CoA, etc), besides acetyl CoA can initiate the building up of a fatty acid.

(a)

I CH_3CO–S–CoA + Protein–SH $\longrightarrow$ CH_3-CO-S–Protein + CoASH

$$(5)$$

II CH_2–CO-S-CoA + ACP-SH $\longrightarrow$ CH_2CO-S-ACP + CoASH + CO_2
    | |
   COOH   COOH   (6)

III (5) + (6) $\longrightarrow$ CH_3-CO-CH_2-CO-S-ACP + Protein-SH

$$(7)$$

IV (7) + NADPH + H^+ $\longrightarrow$ CH_3-CH-CH_2-CO-S-ACP + $NADP^+$
                                            |
                                           OH   (8)

V (8) $\longrightarrow$ CH_3-CH=CH–CO-S-ACP + H_2O

$$(9)$$ Fig. 3.28 (contd. overleaf)

VI (9) + FMNH$_2$ → CH$_3$-CH$_2$-CH$_2$-CO-S-ACP + FMN   Fig. 3.28 (contd.)

(10)

VII (10) + CoASH → CH$_3$-CH$_2$-CH$_2$-CO-S-CoA + ACP-SH

The series of reactions then
continues, as from I, with
chain elongation

ACP = Acyl Carrier Protein

(b)

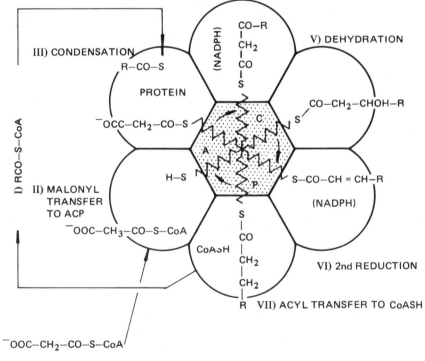

Fig. 3.28 Biosynthesis of fatty acids: (a) reaction sequence; (b) model of fatty-acid synthetase.

(iii) The third site of fatty acid synthesis, the microsomal system, can lengthen the chain of both saturated and unsaturated fatty acids, utilising an acyl CoA, malonyl CoA and NADPH, according to mechanisms very similar to the ones implicated in the cytoplasmatic synthesis, except that the acyl carrier protein is absent.

The mitochondrial and microsomal pathways lengthen aliphatic (saturated or unsaturated) chains of no less than ten initial carbon atoms whilst the cytoplasmic system synthesises fatty acids from scratch.

3.10 ONE-CARBON METABOLISM

During the biosynthesis of some compounds the gain or loss of a one-carbon unit often occurs.

Such single carbons, generically called the C_1 unit, can be incorporated either directly from, or to, the medium, or by the mediation of a donor or acceptor; they can also exist in different oxidation states relative to formaldehyde (O), e.g. methanol (−2), formic acid (+2) and carbon dioxide (+4). Methylations, hydroxymethylations, formylations, and carboxylations are possible.

The C_1 unit pool may be viewed as comprised of one-carbon atom molecules (CO_2, HCOOH *etc.*) and certain substances capable of transferring such units to various substrates. The transfer substances bind such units from the medium, or from primary metabolites, and transfer them to other metabolites, especially secondary metabolites, sometimes with a change of their oxidation number. We have already mentioned biotin, as a CO_2 carrier, and S-adenosylmethionine, as a methylating agent in Chapter 2.

Here we mention the fundamental role played by tetrahydrofolic acid (THF, or FH_4) in the metabolism of the C_1 unit. THF can bind a carbon atom in such oxidation states as $[^+CH_3]$, $[^+CH_2OH]$, $[^+CHO]$ (Fig. 3.29). Serine is an important C_1 donor to THF, and itself transforms into the simpler amino acid glycine. The N^5, N^{10}-methylene-THF can be reduced to N^5-methyl-THF; the latter can transfer its methyl group to homocysteine, to produce methionine via a methylcobalamin-catalysed reaction (see p. 113). Methionine is thus converted into S-adenosyl-methionine, after further reaction with ATP. Formic acid bound to the glutamic acid or glycine nitrogen, or as a formimino group, can be transformed into the methionine methyl group, via N^{10}-formyl-THF, N^5-formyl-THF, and N^5-formimino-THF: all such species may be converted into N^5, N^{10}-methenyl-THF. N-formimino − as well as N-formylglutamic acids arise from the degradation of histidine in mammals, whilst N-formylglycine is the metabolic end-product of purine catabolism in microorganisms. Formic acid can derive from the decarboxylation of oxalic acid arising *via* glyoxylic acid oxidation. Through this process the C_1 unit pool is connected to the C_2 acetate unit (see Fig. 3.26); in photosynthetic organisms it is also connected to the pool of glycolic acid, which directly derives from CO_2 and is in a reduction − oxidation equilibrium with glyoxylic acid.

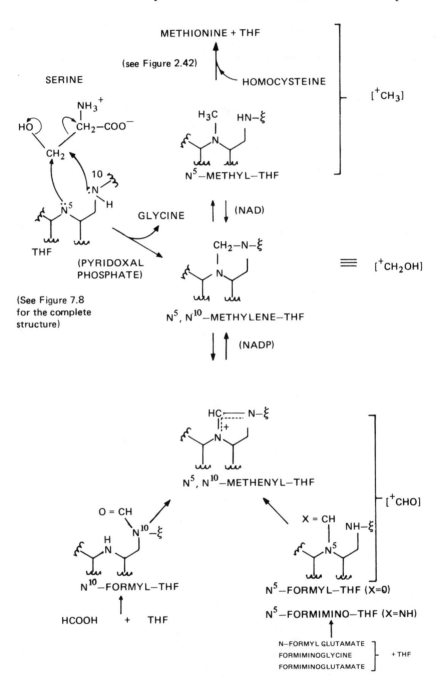

Fig. 3.29 Interconversions of some C_1-THF adducts.

3.11 THE C$_3$ UNIT

The C$_3$ (propionic) unit sometimes substitutes into the C$_2$ acetate system.
This is especially common amongst the secondary metabolite product of fungi, for example in the biosynthesis of macrolide antibiotics.

The active form of the C$_3$ unit is methyl malonyl CoA. This species can be formed either by carboxylation of propionyl CoA or via a chemically unusual cobalamin-catalysed rearrangement of succinyl CoA. Another route to methylmalonyl CoA is by the degradation of valine. Propionyl CoA is also the product of metabolism of odd-numbered fatty acids.

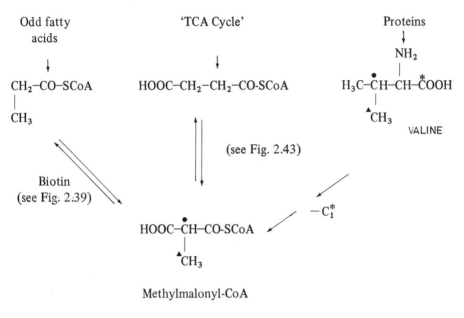

Fig. 3.30 Origin of C$_3$ units.

SOURCE MATERIALS AND SUGGESTED READING

[1] J. A. Bassham and M. Calvin, (1957), *The Path of Carbon in Photosynthesis*, Prentice-Hall.

[2] M. Calvin and J. A. Bassham, (1962), *The Photosynthesis of Carbon Compounds*, Benjamin.

[3] M. D. Kamen, (1963), *Primary Processes in Photosynthesis*, Academic Press.

[4] T. P. Hilditch and P. N. Williams, (1964), *The Chemical Constitution of Natural Fats*, 4th ed., Chapman and Hall.

[5] R. K. Clayton, (1965), *Molecular Physics in Photosynthesis,* Blaisdell.

[6] B. L. Horecker, (1965), Pathways of Carbohydrate Metabolism and Their Physiological Significance, *J. Chem. Educ.,* **42**, 244.

[7] D. I. Arnon, (1966), The Photosynthetic Energy Conversion Process in Isolated Chloroplasts, *Experentia,* **22**, 273.

[8] J. A. Bassham, (1966), Photosynthesis in *Survey of Progress in Chemistry* (A. F. Scott, Ed.), 3, Academic Press.

[9] T. W. Goodwin (Ed.), *The Biochemistry of Chloroplasts,* Academic Press, (1966), 1; (1967), 2.

[10] H. R. Mahler and E. H. Cordes, (1966), *Biological Chemistry,* Harper and Row.

[11] J. B. Mudd, (1967), Fat Metabolism in Plants, *Ann. Rev. Plant. Physiol.,* **18**, 229.

[12] F. D. Gunstone, (1967), *An Introduction to the Chemistry and Biochemistry of Fatty Acids and Their Glycerides,* 2nd ed., Chapman and Hall.

[13] M. E. Pullman and G. Schatz, (1967), Mitochondrial Oxidations and Energy Coupling, *Ann. Rev. Biochem.,* **36**, 539.

[14] K. P. Strickland, The Biogenesis of the Lipids in *Biogenesis of Natural Compounds,* ref. 20 of Chap. 1.

[15] J. H. Wang, (1967), The Molecular Mechanism of Oxidative Phosphorylation, *Proc, Nail. Acad. Sci. U. S.,* **58**, 37.

[16] A. J. James, (1968), Biosynthesis of Unsaturated Acids by Plants, *Chem. in Britain,* **4**, 484.

[17] K. Bloch, (1969), Enzymatic Synthesis of Monounsaturated Fatty Acids, *Accounts Chem. Res.,* **2**, 193.

[18] H. A. Lardy and S. M. Ferguson, (1969), Oxidative Phosphorylation in Mitochondria, *Ann. Rev. Biochem.,* **39**, 991.

[19] J. M. Lowenstein, (1969), *Citric Acid Cycle – Control and Compartimentation,* Marcel Dekker.

[20] E. Rabinowitch and Govindjee, (1969), *Photosynthesis,* Wiley.

[21] P. K. Stumpf, (1969), Metabolism of Fatty Acids, *Ann. Rev. Biochem.,* **38**, 691.

[22] R. W. McGilvery, (1970), *Biochemistry: A Functional Approach,* W. B. Saunders.

[23] W. J. Lennarz, (1970), Lipid Metabolism, *Ann. Rev. Biochem.,* **39**, 359.

[24] N. I. Bishop, (1971), Photosynthesis: the Electron Transport System of Green Plants, *Ann. Rev. Biochem.,* **40**, 197.

[25] G. Cohen, (1971), *Le métabolisme cellulaire et sa regulation,* Hermann.

[26] R. M. Delvin and A. V. Barker, (1971), *Photosynthesis,* Van Nostrand Reinhold Co.

[27] C. Hitchcock and B. W. Nichols, (1971), *Plant Lipid Biochemistry,* Academic Press.

[28] D. O. Hall, (1972), *Photosynthesis,* E. Arnold.

[29] W. J. Lennarz, (1972), Studies in the Biosynthesis and Functions of Lipids in Bacterial Membranes, *Accounts Chem. Res.,* 5, 361.

[30] J. H. Wang and S. I. Tu, (1972), Primary Energy Conversion Reactions in Photosynthesis in *Recent Advances in Phytochemistry,* ref. 57 of Chap. 1, 5.

[31] D. F. Wilson, P. L. Dutton, M. Erecinska, G. Lidsay, and N. Sato, (1972), Mitochondrial Electron Transport and Energy Conservation, *Accounts Chem. Res.,* 5, 234.

[32] W. L. Butler, (1973), Primary Photochemistry of Photosystem II of Photosynthesis, *Accounts Chem. Res.,* 6, 177.

[33] J. B. Pridham (Ed.), (1974), *Plant Carbohydrate Biochemistry,* Phytochemical Society Symposia Series No. 10, Academic Press.

[34] E. Quagliarello, F. Palmieri and T. P. Singer (Eds.), *Horizons in Biochemistry and Biophysics,* Addison Wesley, (1974), 1, (1976), 2.

[35] J. T. Warden and J. R. Bolton, (1974), Light-Induced Paramagnetism in Photosynthetic System, *Accounts Chem. Rev.,* 7, 189.

[36] L. P. Whittingham, (1974), *The Mechanism of Photosynthesis,* E. Arnold.

[37] A. J. Bearden and R. Malkin, (1975), Primary Photochemical Reactions in Chloroplast Photosynthesis, *Quarterly Rev. Biophys.,* 7, 2.

[38] F. Hucho, (1975), The Pyruvate Dehydrogenase Multienzyme Complex, *Angew. Chem. Int. Ed. Engl.,* 14, 591.

[39] P. Karlson, (1975), *Introduction to Modern Biochemistry,* 4th ed., Academic Press.

[40] A. Lehninger, (1975), *Biochemistry,* 2nd ed., Worth.

[41] L. Stryer, (1975), *Biochemistry,* W. H. Freeman and Co.

[42] N. C. Van Hummel, (1975), Chemistry and Biochemistry of Plant Galactolipids in *Progress in the Chemistry of Organic Natural Products,* ref. 1 of Chap. 1, 32.

[43] IZelitch, (1975), Pathways of Carbon Fixation in Green Plants, *Ann. Rev. Biochem.,* 44, 123.

[44] R. C. Bohinski, (1976), *Modern Concepts in Biochemistry,* 2nd ed., Allyn and Bacon.

[45] E. E. Conn and P. K. Stumpf, (1976), *Outlines of Biochemistry,* 4th ed., John Wiley.

[46] F. Gabrielli, (1976), Glucogenesis: A Teaching Pathway, *J. Chem. Educ.,* 53, 86.

[47] C. W. Jones, (1976), *Biological Energy Conservation,* Chapman and Hall.

[48] W. H. Kunau, (1976), Chemistry and Biochemistry of Unsaturated Fatty Acids, *Angew. Chem. Int. Ed. Engl.,* 15, 61.

[49] J. D. Watson, (1976), *Molecular Biology of the Gene,* 3rd ed., Addison Wesley.

[50] R. P. F. Gregory, (1977), *Biochemistry of Photosynthesis,* 2nd ed., Wiley.

[51] D. O. Hall, (1977), *Photosynthesis,* 2nd ed., E. Arnold.

[52] D. E. Metzler, (1977), *Biochemistry: The Chemical Reactions of Living*

Cells, Academic Press.

[53] R. E. Blankenship and W. W. Parson, (1978), The Photochemical Electron Transfer Reactions of Photosynthetic Bacteria and Plants, *Ann. Rev. Biochem.,* 47, 635.

[54] M. Calvin, (1978), Simulating Photosynthetic Quantum Conversion, *Accounts Chem. Res.,* 11, 369.

[55] R. A. Heller and D. R. Person, (1978), The Other Photosynthesis, *J. Chem. Educ.,* 55, 233.

[56] J. Staunton, (1978), *Primary Metabolism,* Oxford University Press.

[57] J. Barber, (1979), Primary Processes of Photosynthesis: Structural and Functional Aspects, *Photochem. Photobiol.,* 29, 203.

Chapter 4

The Polyketides

4.1 INTRODUCTION

The C$_2$ unit of acetic acid is one of the most common 'building bricks' used by living organisms for generating complex molecules. Besides being used in fatty acid synthesis (see p. 160), acetic acid (in its activated forms of acetyl CoA and malonyl CoA) is the most important carbon atom source for two other large classes of natural substances, the *polyketides* and the *terpenes*.

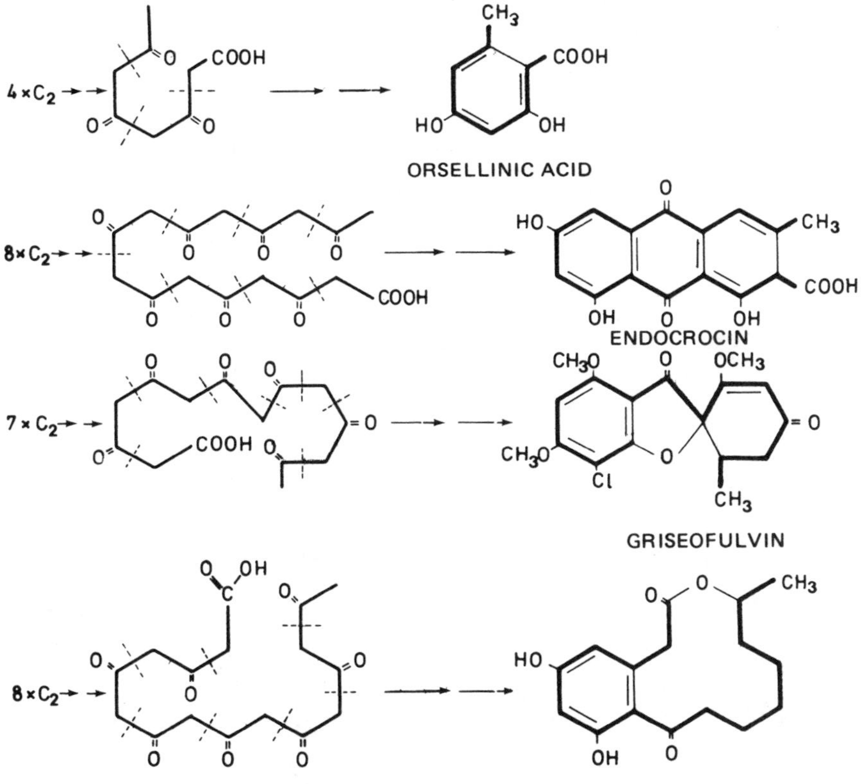

Fig. 4.1 The origin of some polyketides.

ORSELLINIC ACID

ENDOCROCIN

GRISEOFULVIN

CURVULARIN

The biosynthesis of polyketides is of minor significance in plants,* but is particularly well developed in bacteria, fungi, and lichens. This route is largely responsible for the great variety of secondary metabolites produced by such organisms. Examples of polyketides include orsellinic acid (a common metabolite from fungi and lichens), endocrocin (an anthraquinone pigment produced by the lichen *Centralia endocrocea* and the fungus *Claviceps purpurea*), griseofulvin (an important antifungal antibiotic generated by *Penicillium griseofulvum*), and curvularin (a macrolide, synthesised by some *Curvularia* species) (Fig. 4.1).

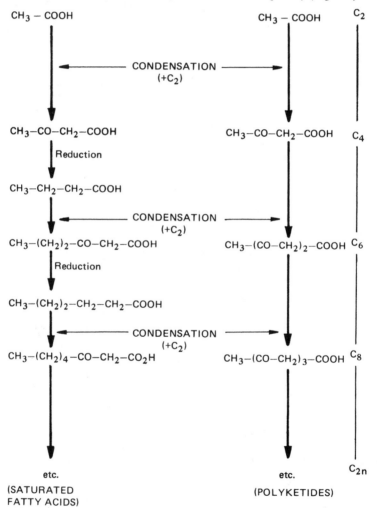

Fig. 4.2 Assembly of acetate units during the biosynthesis of fatty acids and polyketides.

*The flavanoids which are ubiquitous plant products, are only partially polyketides: they result from a mixed shikimic-polyacetic acid biosynthesis and are discussed in Chapter 8.

These compounds are called 'polyketides' and arise from polyketomethylenic chains $[-(CH_2-CO)_m-]$, formally ketene polymers. There is a close parallel between the biosynthesis of fatty acids and polyketides, since, in both cases, the formation of linear chains proceeds by the addition of C_2 units (Fig. 4.2). Nevertheless, whilst in fatty acid biosynthesis every C_2 unit is added to the growing chain only after reduction of the previous carbonyl unit to a methylene group, the growth of a polyketide chain does not usually require such prior reduction. Instead poly-β-ketoacids are formed: such compounds are very reactive since they contain both active methylene groups (potential nucleophiles) and carbonyl groups (potential electrophiles).

Intramolecular reactions are common and crotonic and Claisen type condensations are particularly favoured: in the simplest cases a polyphenolic aromatic nucleus is formed (Fig. 4.3). The benzene nucleus can be formed by two different paths: type (a) condensations (crotonic) produces resorcinol derivatives [2, 4-dihydroxy-6-alkylbenzoic acids (1)], also called *orsellinic acids* the parent orsellinic acid being (1, R = CH_3); type (b) condensation (Claisen), leads to acylphloroglucinols, such as acetylphloroglucinol (2, R = CH_3). Reactions of type (a) can also involve more central portions of the polyketomethylenic chain: this often happens, and polycyclic compounds, such as endocrocin (Fig. 4.1), are formed. Single condensations in chains having the potential for polycyclic ring formation are rare, although exceptions, such as curvularin (Fig. 4.1) are known.

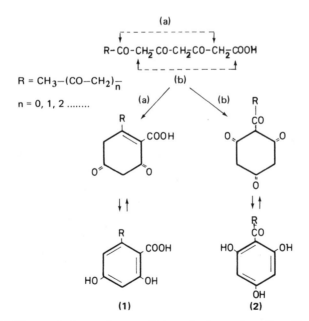

Fig. 4.3 The two fundamental types of intramolecular condensation of β-polyketo-acids.

Polyketomethylenic chains have a marked tendency for intramolecular cyclisation: oxygen bridges between carboxylic and hydroxyl groups (lactonisation), or between two hydroxyls are depicted in Fig. 4.4. The lactonisation, reaction (c), is sometimes the only cyclisation process, predominating over reactions of type (a) and (b). This happens in oligoketides such as triacetic acid lactone (3, R = CH$_3$) and tetracetic acid lactone (3, R = CH$_3$COCH$_2$-) produced by *Penicillium stipitatum*. Simple etherifications, reaction (d), are not favoured, but are generally associated with other cyclisations: the γ-pyrone ring (4) tends to be condensed to at least one benzene ring, such as in chromones (6). The isocoumarin nucleus (5) is formed when the lactonisations reaction (c) is combined with reactions of type (a). Other empirical rules appear to regulate the cyclisations of polyketomethylenic chains *in vivo*, and will be summarised later on (see p. 186). In all the structures determinable from reactions (a) to (d) the phenolic hydroxyls (or the carbonyls) end up in alternate positions. This gives the typical oxygenation pattern of polyketide compounds and allows a general distinction between aromatic rings derived from acetate and from other sources, such as that from the shikimic acid path. (See Chapter 7).

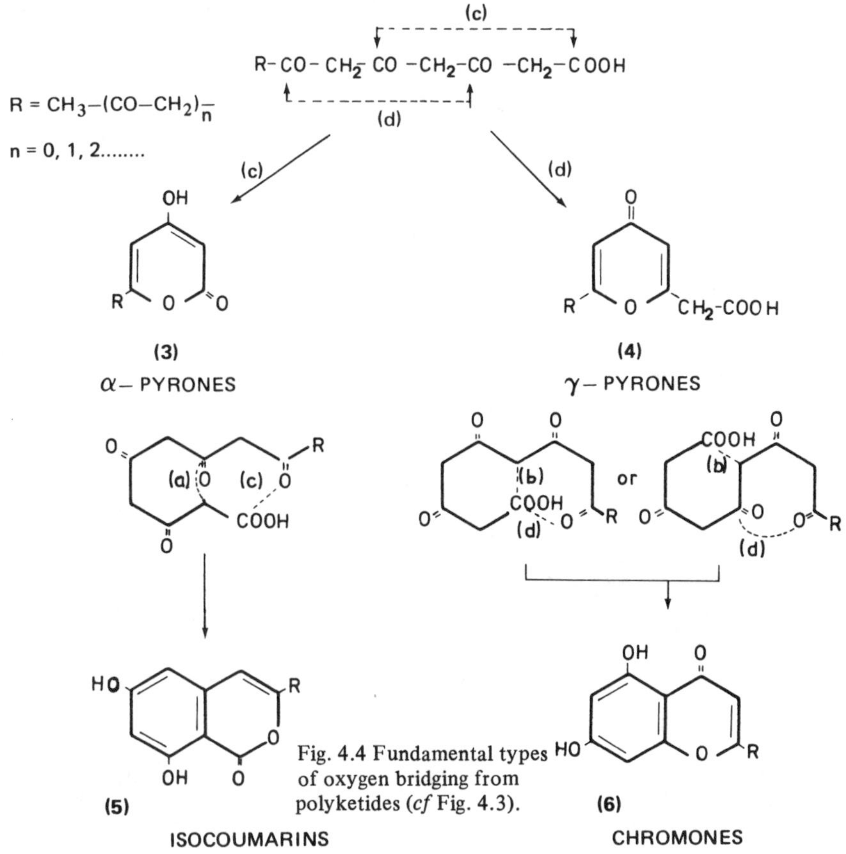

Fig. 4.4 Fundamental types of oxygen bridging from polyketides (*cf* Fig. 4.3).

The hydroxylation pattern, however, cannot always be used for biogenetic diagnoses, since the typical alternate substitution pattern can be radically altered by oxidative and reductive processes, which may take place after, or even before, the cyclisation process. Despite this qualification there are a large number of structural indications and experimental results entirely consistent with the hypothesis that such compounds are really formed from polyketide intermediates. Collie (1893) first postulated the polyacetic origin of many natural products. He studied the chemistry of polyacetyl (ketomethylenic) compounds and noticed that products identical with, or similar to, others already known in nature, could be obtained from these ketomethylenic substances using reactions catalysed by base. Orsellinic acid was obtained in this manner from dehydroacetic acid (8), which was itself obtained by the pyrolysis of acetoacetic ester (7). Similarly, dihydroxynaphthalene (10) was produced from 1,3-diacetylacetone (9) (Fig. 4.5).

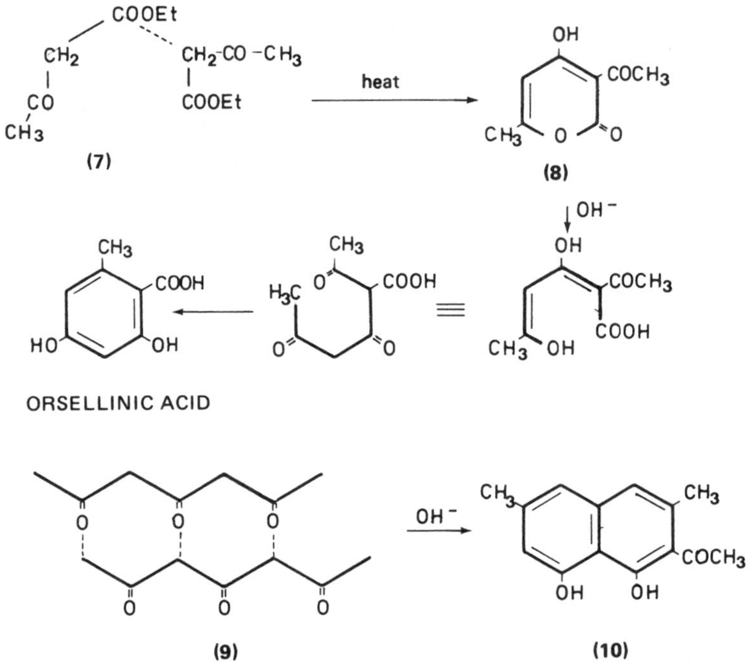

Fig. 4.5 *In vitro* reactions considered by Collie as possible biological models.

Collie's idea was not followed up until 1953, when Birch rediscovered it and proposed it as a general biogenetic hypothesis. Initial circumstantial evidence in favour of this theory was its success in accounting for the structure of a large number of natural substances. Such examples, which are always important for the support of a biogenetic hypothesis, were soon confirmed by experimental proof, collected in abundance by both Birch's group and by others. These experiments utilised the incorporation of labelled precursors, mainly specifically labelled [^{14}C]- or [^{18}O]acetic acid, into products from microorganisms. The reactions used for determining the distribution of the labelled atoms in the griseofulvin obtained from [1-^{14}C] acetic acid are given as an example in Fig. 4.6(a). The orsellinic acid biosynthesised by *Chaetomium cochlioides*, in which

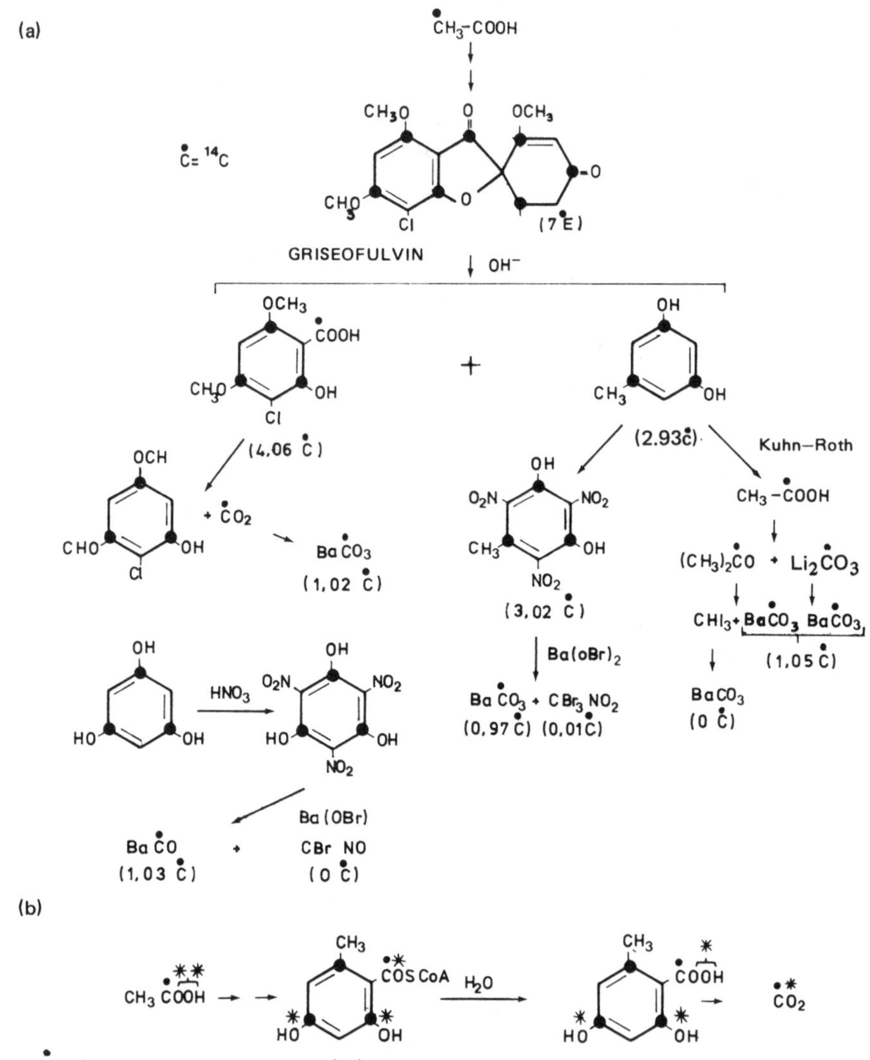

(a)

(b)

$\overset{\bullet}{C} = {}^{14}C$

$\overset{\bullet}{O} = {}^{18}O$

Fig. 4.6 Distribution of the labels in polyketides produced from [1-^{14}C, ^{18}O] acetic acid.

doubly labelled [^{14}C] and [^{18}O] acetic acid (on the carboxylic group) was used (Fig. 4.5(b)) was analysed in a similar way. The agreement between the labelling patterns obtained by chemical degradation, and those predicted by the Birch hypothesis (schematised in Fig. 4.1) are perfect. In particular, the isotope abundance ratio of ^{14}C/^{18}O in the carboxyl group of the orsellinic acid is equal to $\frac{1}{2}$ of the ratio in the original acetic acid carboxyl (containing two oxygen atoms) and to $\frac{1}{2}$ of that in the phenolic (C–OH) groups in the aromatic ring. This result is in agreement with the formation and the hydrolysis of an orsellinic acid ester, which could well be orsellinyl–CoA (11) (Fig. 4.6).

(Note that the hydrolysis of RCO*SCoA produces R–CO*₂H i.e. dilution of the labelled oxygen by a half).

Birch's biogenetic acetate rule states that the synthesis of polyketides usually takes place in four successive steps:

(i) The addition of the C₂ units (chain assembly).
(ii) Oxidations, reductions and alkylations of the polyketide chain.
(iii) Stabilisation of such a chain, by intramolecular cyclisations.
(iv) Secondary modifications of the functions or of the mono- or polycyclic skeleton resulting from step (iii).

Steps (ii) and (iv) do not always occur; step (ii) can mix with step (i). A subsequent addition (step (i)), after a step (iii) or (iv), or else a step (iii) taking place through an intermolecular cyclisation (for instance between two polyketomethylenic chains, as considered in Collie's earlier hypothesis, see Fig. 4.5) can be considered as very unusual, if not impossible. For certain biosyntheses both 'one chain' and 'two chain' pathways can be formulated, but, whenever such cases have been experimentally tested, according to the procedures and criteria described herein, the single chain process has almost always been found. This is true, for example for 6-methylsalicylic acid and for the anthraquinones (p. 195). It is therefore generally accepted that the biosynthesis of polyketides *normally* involves only one polyketide chain. Citromycetin (38) is one of the very few exceptions to this generalisation. It should be noted that this compound possesses a very unusual branched chain (see Fig. 4.25).

4.2 ADDITION OF C₂ UNITS

The addition of C₂ units in polyketide chains probably takes place through repeated Claisen-type condensations, very similar to the ones implicated in the cytoplasmic synthesis of the fatty acids (see p. 162). An acyl group bound to a thiol group of a multienzyme complex reacts with a malonyl unit, bound to the same complex by a second sulphur-bridge, situated in an appropriate position. The whole process can be represented as in Fig. 4.7 (steps (a)–(i)).

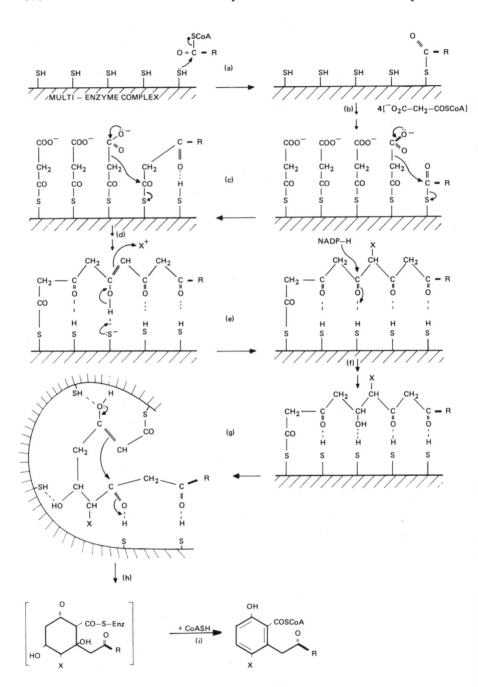

Fig. 4.7 Model for the enzyme system for the biosynthesis of polyketides.

The polyacyl synthetase is hypothesised as a modified fatty acid synthetase (Fig. 3.28): numerous active sites for the Claisen type condensation step should be present, as well as rarer sites, catalysing other reactions, such as alkylations (X^+, as an alkylating agent), reductions (with NADH and NADPH), and oxidations (steps (e)–(f) of Fig. 4.7). This representation is supported by the following facts:

(i) The active form of the first C_2 unit is acetyl CoA, whilst malonyl CoA, formed by carboxylation of acetyl CoA, is used for subsequent C_2 units. This utilisation of two different units by the enzyme accounts for the different distribution of radioactivity in the polyketide molecules obtained after incorporating either labelled acetic acid or malonic acid. For instance, the localisation of radioactivity in 6-methylsalicylic (12) acid isolated from *Penicillium urticae,* gave the results shown in Fig. 4.8: in one case diethyl [2-¹⁴C] malonate

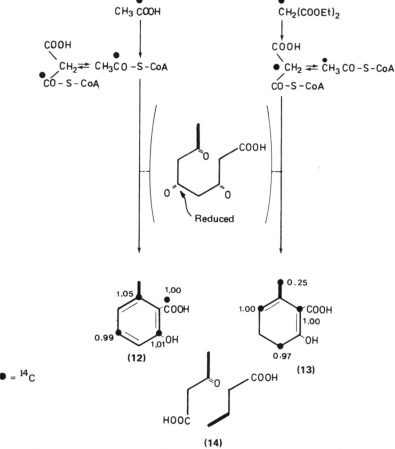

Fig. 4.8 Incorporation of (1-¹⁴C) acetic acid and (2-¹⁴C) melonic acid into 6-methylsalicylic acid.

was fed and, in another experiment, [1-^{14}C] acetic acid was used. The positions corresponding to the first C_2 unit (derived from acetyl CoA) and the ones corresponding to the successive C_2 units (derived from malonyl CoA) are labelled to different extents.

This difference may be accounted for since the rate of utilization of the two labelled products competes with the rate at which they are diluted into the common C_2 unit pool. This behaviour of the microorganism is general and illustrates how 6-methylsalicylic acid is formed by a one-chain process. A two-chain synthesis would produce the same specific activity for both the methyl-carbon atom and for the para-ring carbon to the methyl group (14), Fig. 4.8 which is inconsistent with the experimental results given by (13).

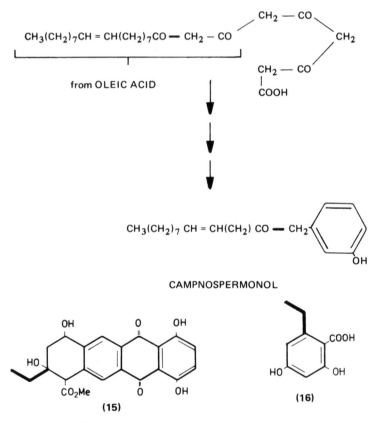

Fig. 4.9 Some polyketides generated by conjugation with various starters (the conjugating bond is indicated).

(ii) Just as for the biosynthesis of fatty acids via malonyl CoA, the starting acyl group can sometimes be different from the acetyl group. This frequently takes place in plants, in which cinnamic acids (see the biosynthesis of flavanoids, p. 400), benzoic acids (p. 395) and fatty acids can act as 'starters'. Oleic acid is for instance utilised for the synthesis of campnospermonol (Fig. 4.9), which is a constituent of Tigaso oil (genus, *Campnosperma*; family, Anacardiacae). In certain fungi C_3 units, such as monomalonamide (see tetracycline, p. 207) and propionic acid act as initiators. The latter unit is incorporated into rutilantinone (or ε-pyrromycinone) (15), produced by a *Streptomyces* sp. and in homo-orsellinic acid (16) produced, together with orsellinic acid, by *Penicillium boarnense*, cultured in the presence of propionate.

(iii) The polyketomethylene chains must somehow be protected during their growth, since they are chemically very unstable and spontaneous, intra-molecular reactions can occur *in vitro* in the absence of protection. The multi-enzyme complex which forms such chains probably fulfils the protection role itself; hydrogen-bonding or metal-chelation may be involved. When the polyketide chain reaches a suitable length, a well-defined folding of the enzyme-polyketo-acid complex takes place, followed by stabilisation reactions (condensations, etc.) and final release of the product polyketide from its matrix (steps g-i of Fig. 4.7).

(iv) Structural modifications to the polyketide chain can sometimes be recognised in the final product, such as alkylations, reductions, etc. These modifications usually take place before final stabilisation as indicated by the results from certain incorporation experiments, some of which are mentioned below.

4.3 MODIFICATIONS OF POLYKETIDE CHAINS

Some examples of the attack of reducing agents (e.g. NADH or NADPH), oxidants (e.g. activated oxygen), or alkylating agents (e.g. S-adenosylmethionine, dimethyl-allylpyrophosphate) on polyketide chains are illustrated in Fig. 4.10.
Such reactions may take place when the polyketide product has either formed or is in the process of formation. This duality of biosynthetic pathways makes the sequential description of the earlier steps difficult. In most cases the exact point where such modifications takes place can not be determined (see Fig. 4.11).

Labelling experiments *can* be used, however, to determine when some of these transformations take place. Thus, the incorporation of appropriate pre-cursors into the final product can be determined. Such precursors should have the basic skeleton, but not the final structure, of the molecule under study. Clavatol (20), for instance, produced by *Aspergillus fumigatus*, incorporates [methyl-14C] methionine into the two methyls bound to the benzene ring, but does not incorporate either [1-14C] 2,4-dihydroxyacetophenone (21) or [1-14C]-2,4-dihydroxy-3-methylacetophenone (22) (Fig. 4.12). This result means that the two methyls are introduced into the polyketomethylene chain and not into the aromatic ring. Both (21) and (22) are incorporated by *A. fumigatus* but they are transformed into metabolites other than clavatol. This result serves to confirm the previous conclusions, as it cannot be argued that the two compounds are hindered from passing through the plasma membrane and are therefore not utilised.

From numerous experiments of this type, the general conclusions summarised in Table 4.1 can be made.

Table 4.1

Frequency of reactions introducing structural variations into polyketides.

Reactions	On product polyketide	Before formation of polyketide product
(1) Reduction	Common (F)[a]	Rare
(2) Oxidation	rare	common (F, P)[a]
(3) C-methylation	very frequent (F)	infrequent
(4) O-methylation	–	very frequent (F, P)
(5) C-prenylation	rather frequent (F, P)	common (F, P)
(6) O-prenylation	–	common (P)
(7) C-glycosylation	–	rare (P)
(8) O-glycosylation	–	common (P)

[a]F = fungi; P = higher plants.

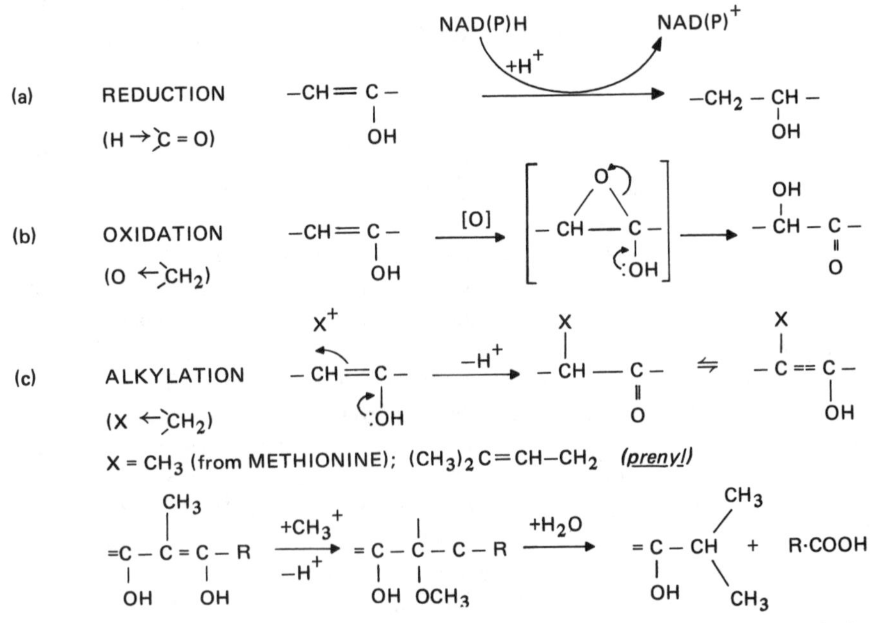

Fig. 4.10 Principal modifying reactions on polyketomethylene chains.

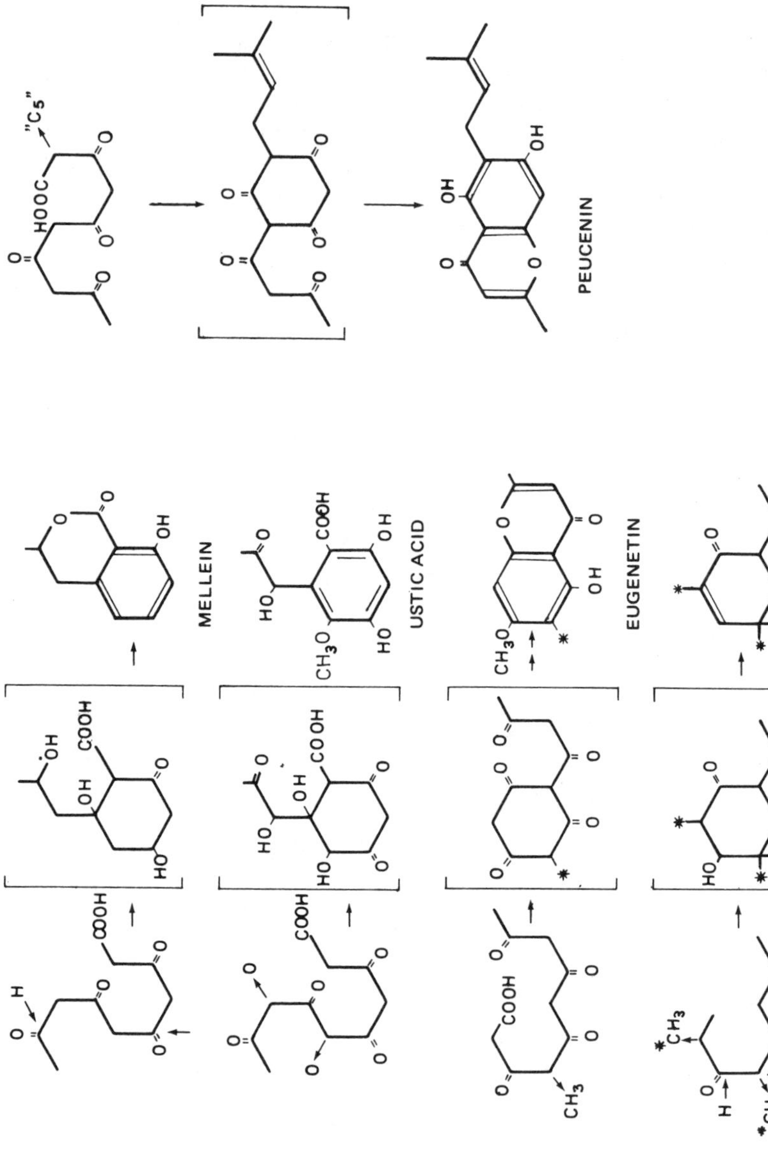

Fig. 4.11 Some compounds derived from modified polyketomethylene chains (* = methyl from methionine).

If the methylation reaction occurs twice on the same carbon atom (Fig. 4.10(c), a system of the type (17) is obtained and the conjugation of the chain's enolic groups becomes interrupted, making the hydrolytic fission into fragments (18) and (19) easier. Many examples of such methylations, for example leading to the formation of isobutyric groups, are known in nature, with both the geminal methyl groups arising from methionine. The reduction of a ketone group to a secondary alcohol, followed by its loss during aromatisation seems to occur without any preference along a polyketide chain. As a general rule it has been found that the oxygen atom placed β to the terminal, polyketoacid carboxyl group is *always* present in the final product when the latter results from a condensation involving the methylene group α to the acid function (Fig. 4.12). This is probably due to the fact that the carboxyl group itself, either bonded or free, does not activate the adjacent methylene group sufficiently for the necessary crotonic type condensation reaction.

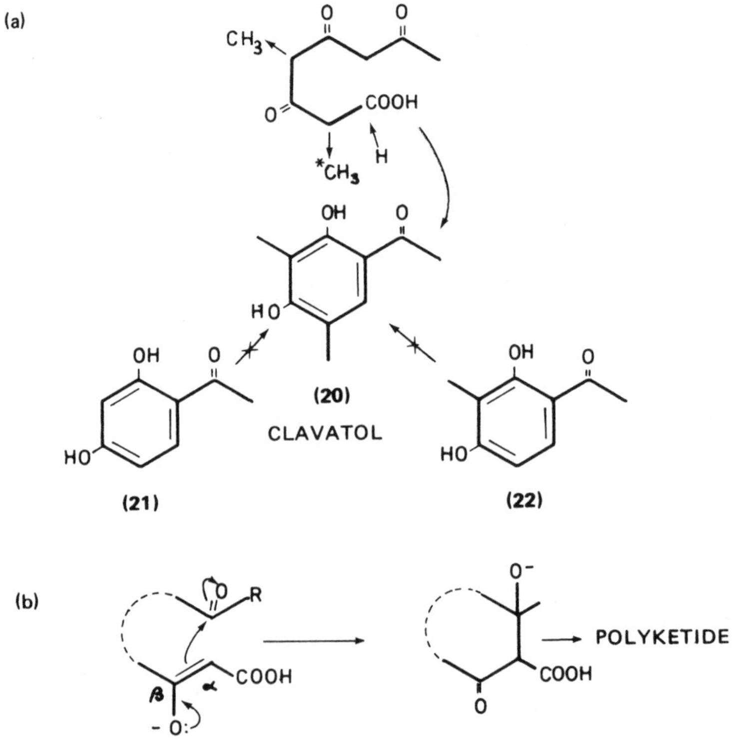

Fig. 4.12 (a) Biosynthesis of clavatol; (b) importance of the β-carbonyl group (enol) in condensations α to the carboxyl function.

4.4 REACTIONS OF THE POLYKETOMETHYLENE CHAIN

The polyketomethylene chains, temporarily bound to the enzyme, can react via processes depicted in Figs. 4.3 and 4.4. Many resulting structures are thus possible, these depending on the length and functionality of the starting chain. The cyclisation reactions outlined can also take place and combine with each other in different sequences. It is found, however, that a particular polyketide precursor does *not* seem to produce a host of the expected products, some specificity of reaction being observed. Such specificity depends on the enzymic matrix, which tends to keep the polyketide chain in a precise conformation. The fact that a large number of matrixes, related to each other, do not exist in any one organism can be explained by natural evolution and selection! The most common polketide skeletons are shown in Fig. 4.13. The corresponding polyketide (polyacetate) chain is schematically indicated beside the natural compound. The heavier segments in the formulae correspond to the acetyl CoA starter unit, whilst the dotted lines correspond to crotonic or Claisen-type condensations.

Tetra-, penta-, hepta-, and octaketides are very numerous; tri-, hexa-, nona-, and decaketides are less frequently encountered.

TETRAKETIDES :

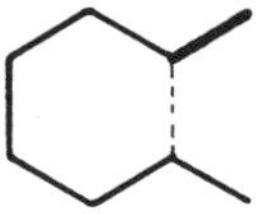

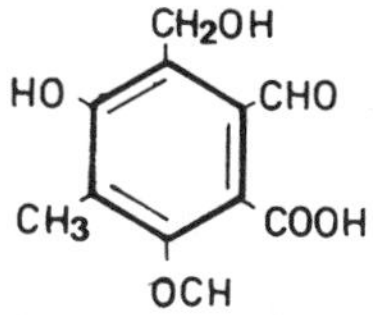

CYCLOPOLIC ACID
(P. cyclopium)

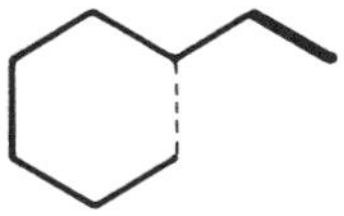

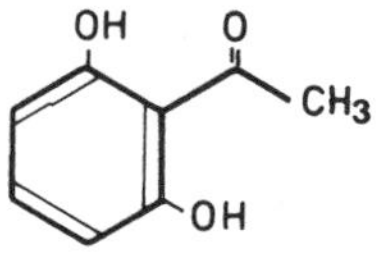

2,6 – DIHYDROXYACETOPHENONE
(Daldinia concentrica)

PENTAKETIDES :

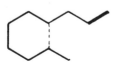

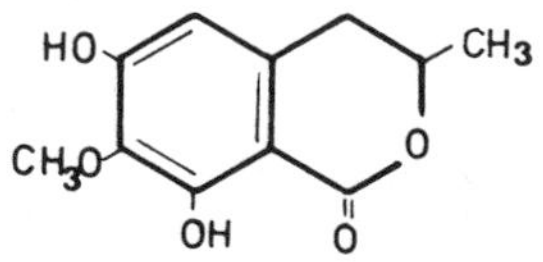

RETICULOL
(Streptomyces rubrireticulae).

Fig. 4.13 (contd. overleaf)

Fig. 4.13 (contd.)

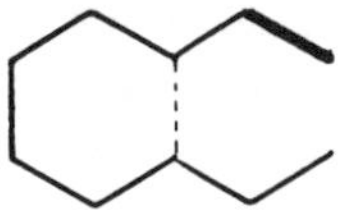

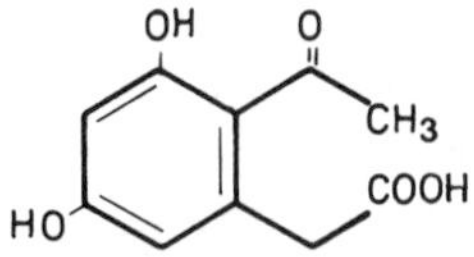

CURVULINIC ACID
(Curvularia siddiqui)

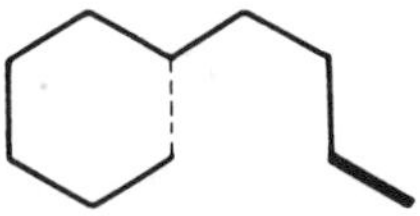

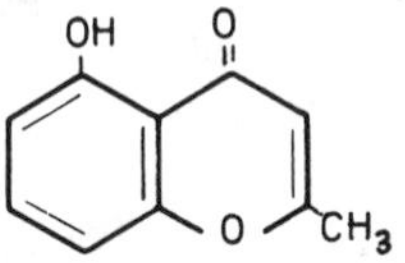

5–HYDROXY–2–METHYLCHROMONE
(Daldimo coucentrica)

HEXAKETIDES :

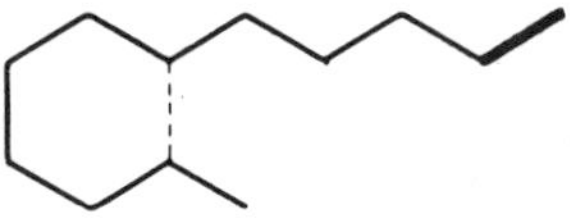

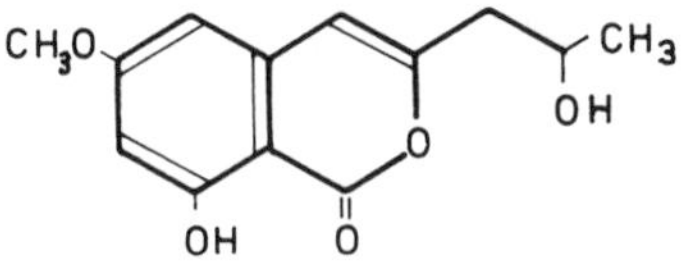

DIAPORTHIN
(Endothia parasitica)

HEPTAKETIDES :

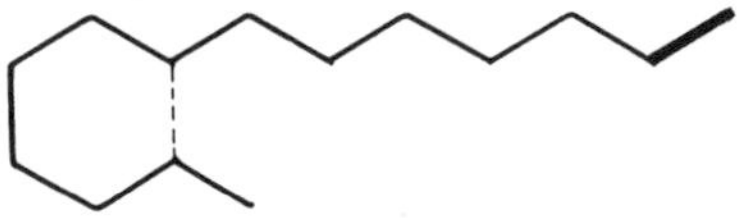

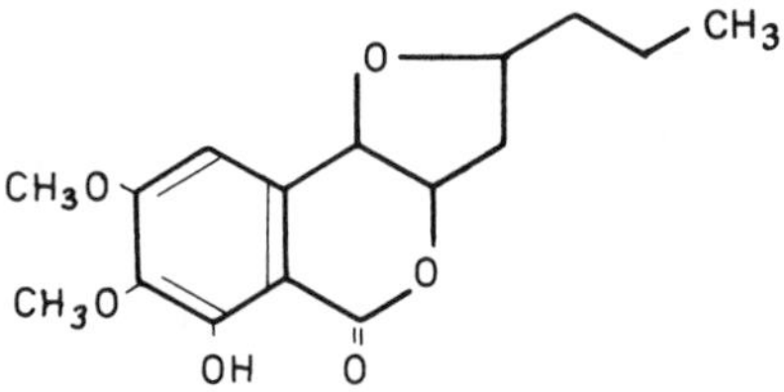

MONOCERIN
(Helminthosporium monoceras)

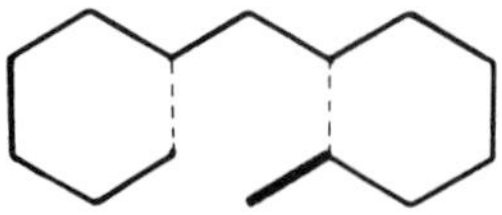

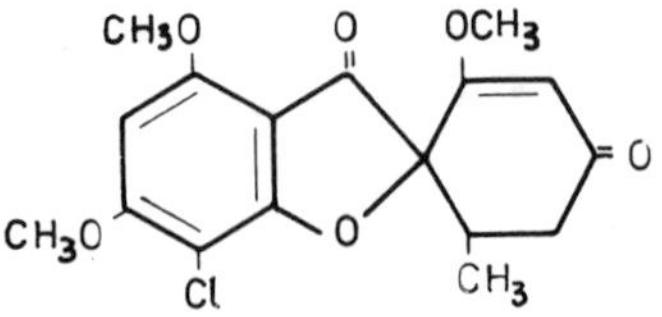

GRISEOFULVIN
(See Figure 4.1)

Fig. 4.13 (contd. overleaf)

Fig. 4.13 (contd.)

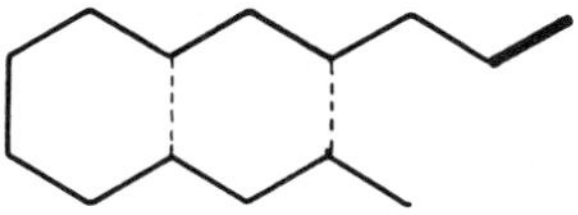

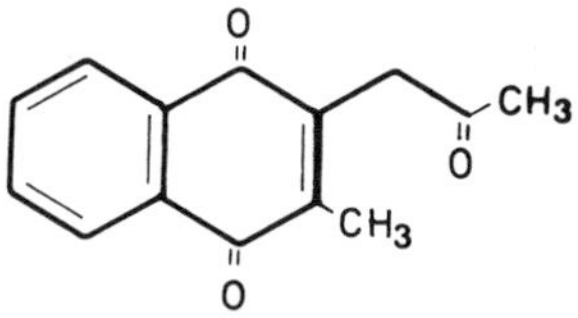

JAVANICIN
(Fusarium javanicum)

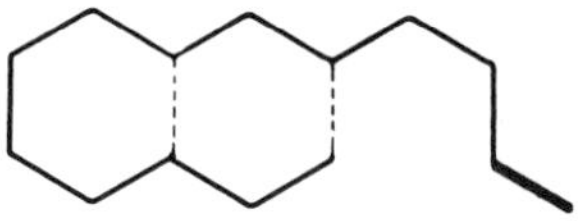

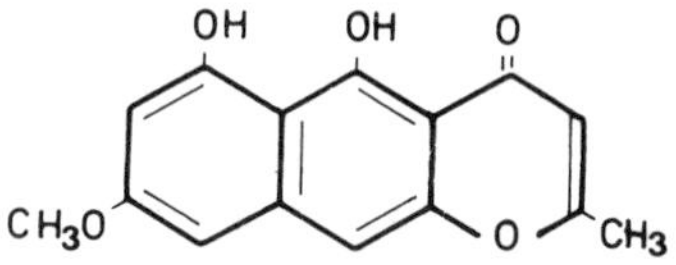

RUBROFUSARIN
(Fusarium culmorum)

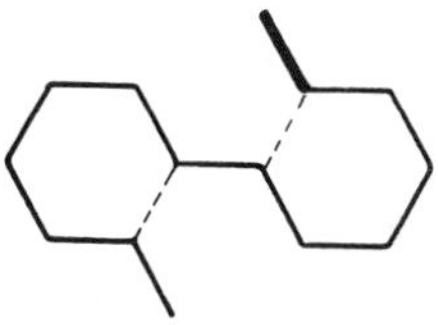

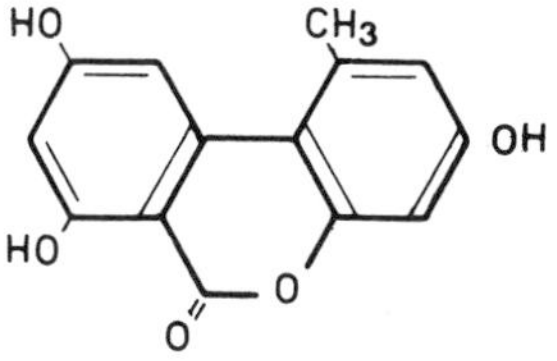

ALTERNARIOL
(Alternaria tenuis)

OCTAKETIDES :

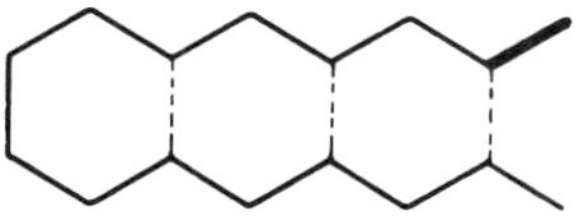

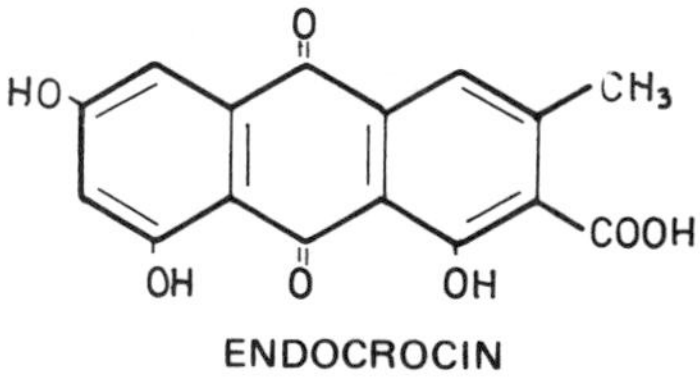

ENDOCROCIN
(See Figure 4.1)

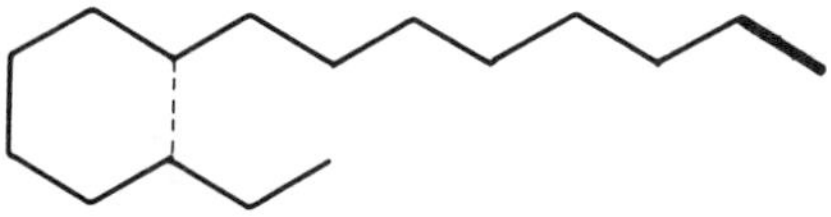

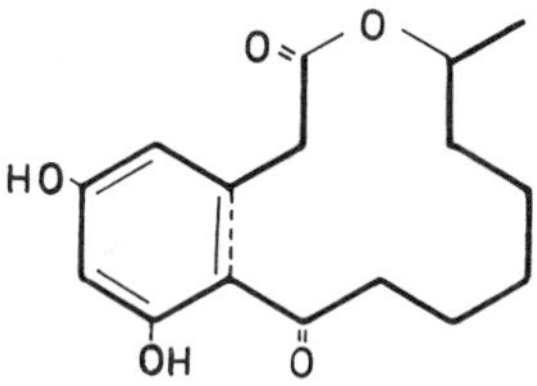

CURVULARIN
(See Figure 4.1)

Fig. 4.13 (contd. overleaf)

Fig. 4.13 (contd.)

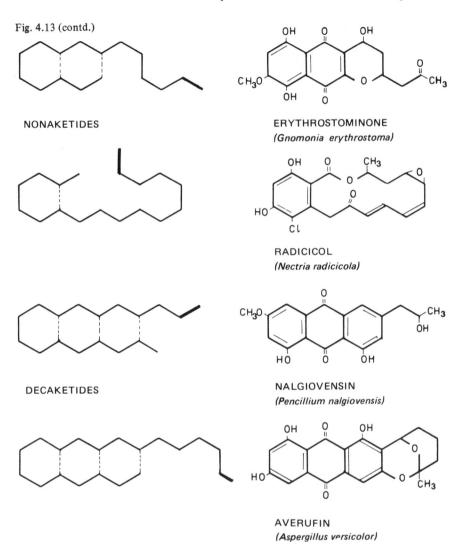

NONAKETIDES

ERYTHROSTOMINONE
(Gnomonia erythrostoma)

RADICICOL
(Nectria radicicola)

DECAKETIDES

NALGIOVENSIN
(Pencillium nalgiovensis)

AVERUFIN
(Aspergillus versicolor)

Fig. 4.13 Some of the more common polyketide skeletons.

4.5 SECONDARY MODIFICATIONS OF POLYKETIDES

After the initial cyclisations, polyketides can undergo many modifications, via various reactions, which are common to those of most secondary metabolites of whatever biogenetic origin. Most secondary transformations are oxidative, thus helping the general catabolic phase of the biosynthetic products. One of the most frequent processes is the conversion of an aromatic ring methyl group into a carboxylic acid, through successive oxidation reactions (Fig. 4.14); the carboxyl can then be lost and replaced with a hydrogen atom (non-oxidative decarboxylation) or with a hydroxyl group (oxidative decarboxylation). The first, irreversible, oxidation step takes place through a mixed function oxygenase whilst the interconversions primary alcohol, to aldehyde, to carboxylic acid are catalysed by NAD(P)-dependent dehydrogenases or by flavoproteins.

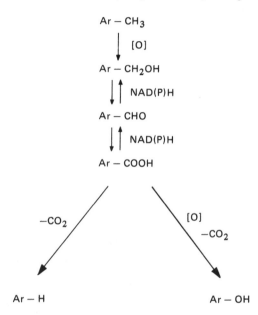

Fig. 4.14 Catabolism of aromatic methyl groups.

Mono- and dicarboxylic aromatic acids can be formed through reactions as exemplified in Fig. 4.15 by routes to gentisic acid. Two distinct routes to this acid are known; one involves derivation from glucose via the shikimic acid pathway and the other via acetic acid. The latter route has been confirmed by incorporation experiments in *Penicillium urticae*. Two metabolic ways were proposed: the first one starts from 4-hydroxy-2-methylbenzoic acid (23) and the second one from 6-methylsalicylic acid. Figure 4.15 also illustrates some secondary transformation frequently observed in fungi from the polyketide products orsellinic and 6-methylsalicylic acids.

The biosynthesis of griseofulvin in fungi (Fig. 4.16) is characterised by two rather common processes: the introduction of one halogen atom into an aromatic ring (see p. 88, as well as the biosynthesis of tetracyclines, p. 207), and the oxidative coupling of phenolic rings (see p. 98). Such coupling is intramolecular for griseofulvin, but it is more commonly intermolecular, such as the formation of usnic acid in lichens (Fig. 4.17).

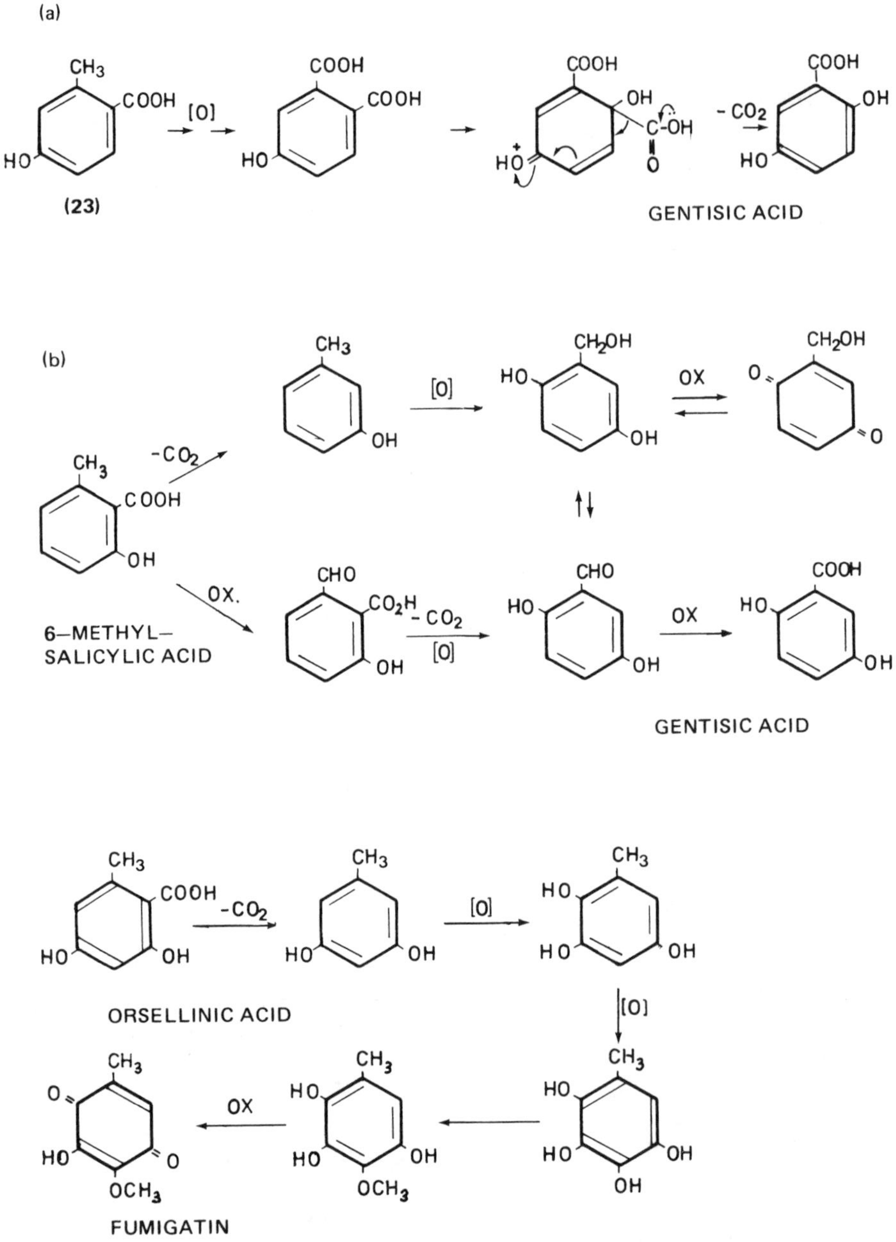

(a)

(23)   GENTISIC ACID

(b)

6—METHYL—
SALICYLIC ACID

GENTISIC ACID

ORSELLINIC ACID

FUMIGATIN

Fig. 4.15 Biosynthesis of gentisic acid and some secondary transformations of 6-methylsalicylic and orsellinic acids.

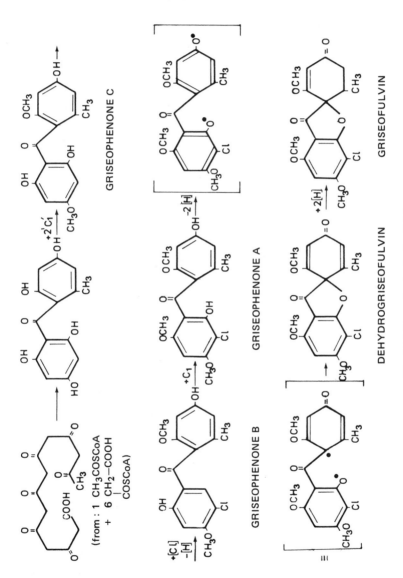

Fig. 4.16 Biosynthesis of griseofulvin.

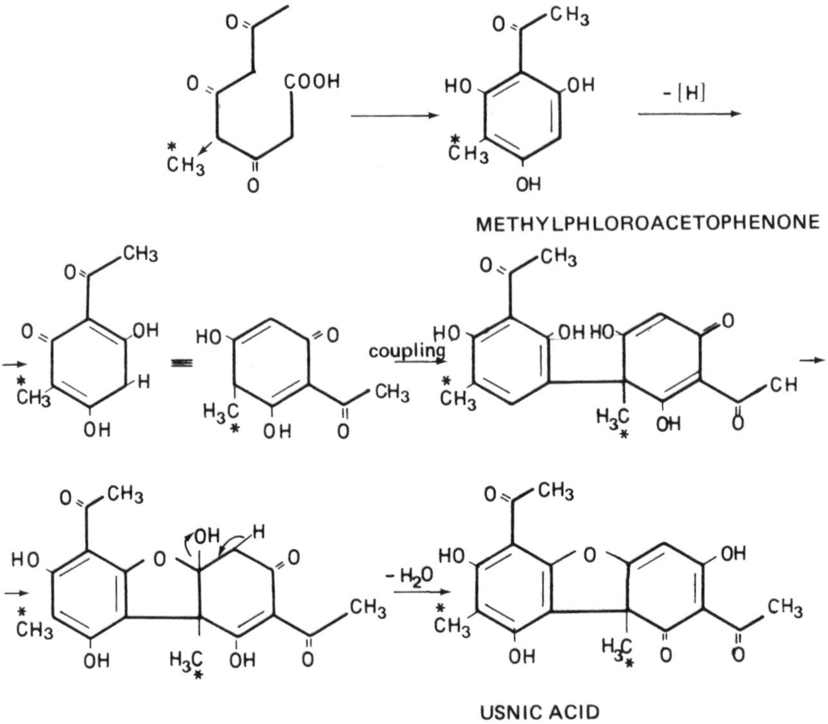
COOH
- [H]
METHYLPHLOROACETOPHENONE
coupling
- H_2O
USNIC ACID

Fig. 4.17 Biosynthesis of usnic acid via oxidative coupling.

(a)

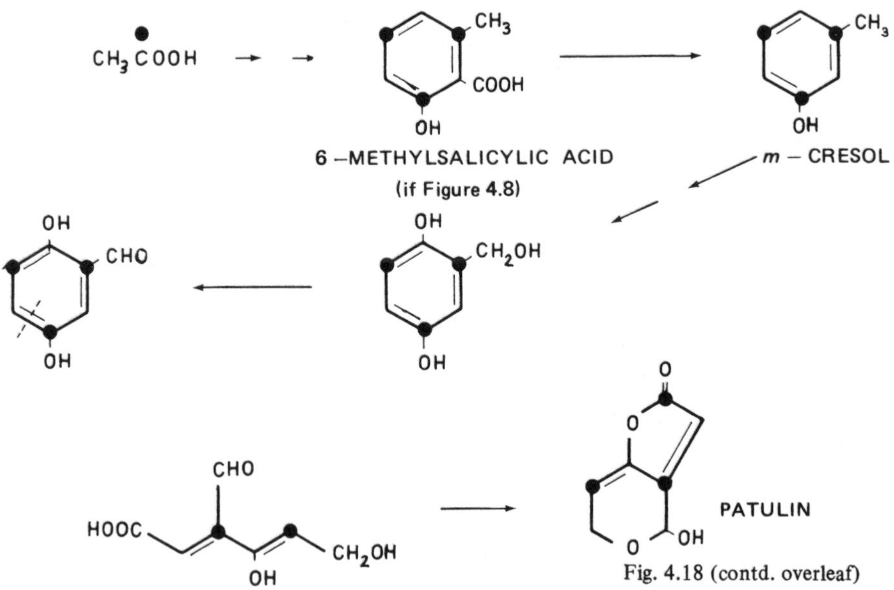
CH3COOH
6-METHYLSALICYLIC ACID
(if Figure 4.8)
m - CRESOL
CHO
CH2OH
HOOC
PATULIN

Fig. 4.18 (contd. overleaf)

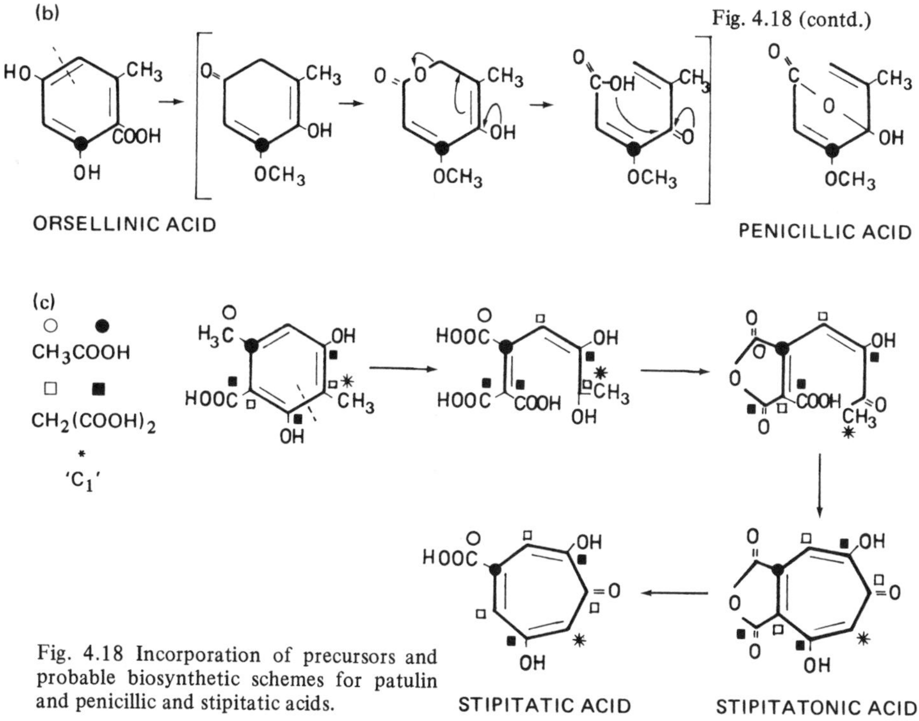

(b)   Fig. 4.18 (contd.)

ORSELLINIC ACID   PENICILLIC ACID

(c)

○ ●
CH₃COOH

□ ■
CH₂(COOH)₂

*
'C₁'

Fig. 4.18 Incorporation of precursors and
probable biosynthetic schemes for patulin
and penicillic and stipitatic acids.   STIPITATIC ACID   STIPITATONIC ACID

The action of an oxygenase on the aromatic polyketide rings can result in ring
cleavage: the formation of patulin in *Penicillium patulum,* of penicillic acid in
P. puberulum and of stipitatic acid in *P. stipitatum* (Fig. 4.18) are examples.
The tropolone ring characterises the structure of stipitatic acid and of its
derivatives found in plant metabolites, however then tropolone compounds arise
from completely different metabolic routes.

4.6 QUINONES

1,4-Quinones structures often appear in nature as the final product of oxidation
of mono- and polycyclic-aromatic nuclei. The carbon atoms of such rings can
derive from acetic acid, (as in the polyketide quinones described in this chapter)
from mevalonic acid (terpene quinones; Chapter 5), or from glucose, through the
shikimate and aromatic amino-acid pathway (Chapter 8). The facile interconversion of quinones (Q) with hydroquinones (H_2Q) according to equation (1), is
often of advantage to plants in helping to carry electrons to and from certain
substrates:

$$H_2Q \rightleftharpoons Q + 2e^- + 2H^+ \qquad (1)$$

One classical example of such a process is the ubiquinone (coenzyme Q) involved
in the respiratory chain.

A chemical feature of quinones is their tendency to add nucleophiles. Quinones, formed in large amounts by soil microorganisms and/or by the autoxidation of pyrogallol derivatives readily add phenols, amines, amino-acids, etc. By further oxidative processes some brown, high molecular weight compounds (humic acids) are obtained. Such substances form the major part of organic material present in 'humus'-rich soils.

4.6.1 Benzoquinones

Besides fumigatin and hydroxymethyl-p-benzoquinone (24) (Fig. 4.15), many other p-benzoquinones have been isolated from various fungal species. The biosynthesis of shanorellin (25, Fig. 4.19), a yellow pigment from *Shanorella spirotricha*, an Ascomyte, has recently been studied and the incorporation data are in agreement with the pathway summarised.

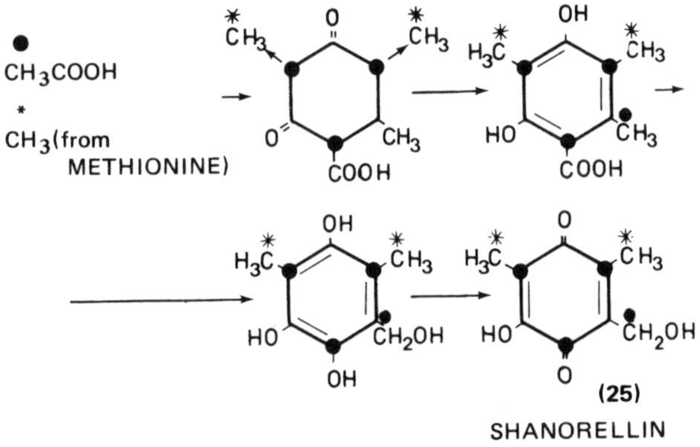

Fig. 4.19 Biosynthesis of shanorellin.

Certain quinomethide structures such as citrinin (26) and fuscin (27) are closely related to p-benzoquinones. Citrinin is a very common mould metabolite and shows a remarkable antibiotic activity *in vitro*; it has also been isolated from a higher plant, *Crotolaria uripata*. Fuscin (from *Oidiodendron fuscum*) possesses a C_5 unit of mevalonic origin. The probable biosynthesis of both polyketides are summarised in Fig. 4.20.

p-Benzoquinone and some of its simpler derivatives have been identified in numerous arthropods, milliapodes and insects in which they probably act as defence substances. The biosynthesis of these quinones is still unknown. The quinone nuclei of volucrisporin (p. 366), the ubiquinones (see p. 302) and of the plastiquinones (p. 303) are certainly not of the polyketide type.

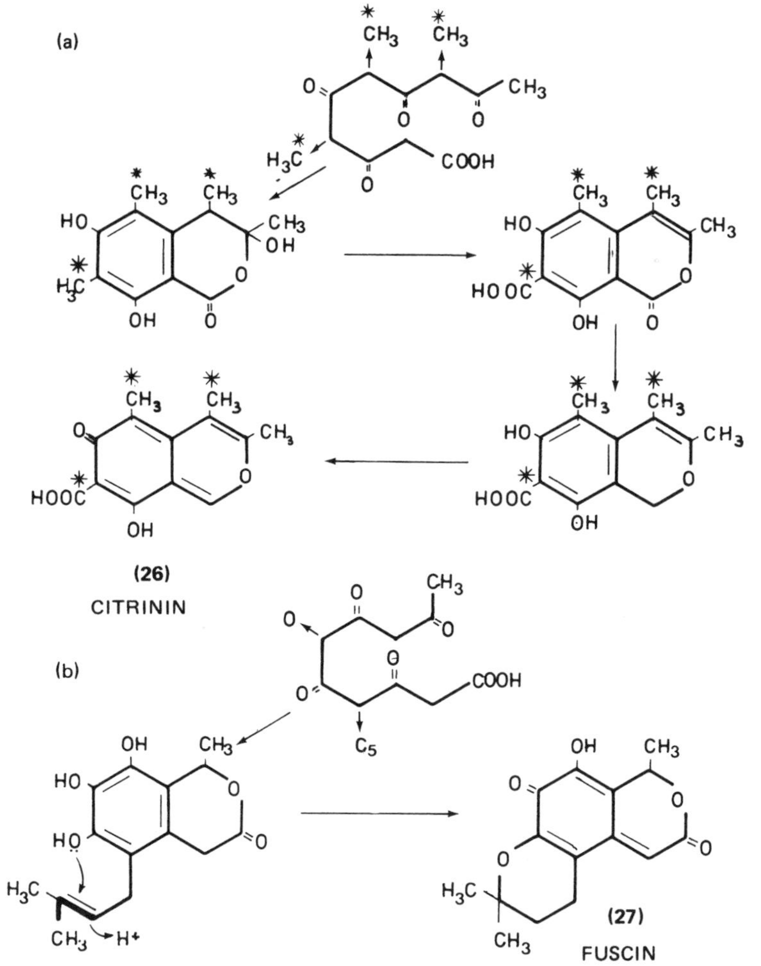

Fig. 4.20 Biosynthesis of some quinomethides.

4.6.2 Naphthoquinones

The polyketide route to the naphthalene nucleus and consequently of napththoquinones, is well established in microorganisms but it appears to be quite rare in higher plants, where other biosynthetic pathways are preferred. In Fig. 4.21 are illustrated some typical naphthalene and naphthoquinone compounds produced by fungi. Metabolites (29) (binaphthyl) and (30) (4,9-dihydroxyperilene-3,10-quinone), from *Daldinia concentrica* presumably arise via 1,8-dihydroxynaphthalene (28) by oxidative processes.

There are many uncertainties about the type of folding of the parent penta-ketide chain for the naphthalene system. Potential cyclisation schemes for such systems are illustrated (Fig. 4.21).

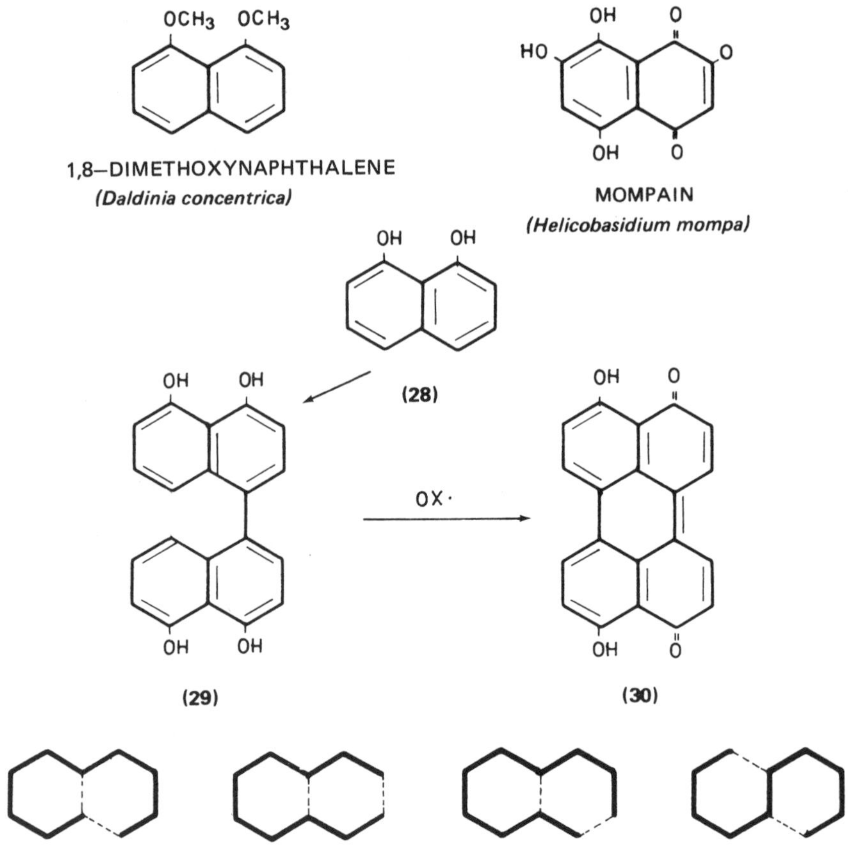

Fig. 4.21 Some naphthalene and naphthoquinone derivatives of polyketide origin which are produced by fungi.

A further biosynthetic possibility is the hypothesis that the ten carbon atoms of the naphthalene nucleus derive from a hexaketide, from which two carbon atoms are lost after cyclisation. Degradative processes of this kind are common in the biosynthesis of anthraquinones. It has recently been demonstrated that the two naphthoquinones, plumbagin (31) and 7-methyljuglone (32) (Fig. 4.22), synthesised by plants of the *Drosera* and *Plumbago* genera, arise from a hexaketide rather than the shikimic acid route as occurs for juglone and menadione.

In fungi, naphthaquinones from heptaketides, such as mavanicin, and actaketides, such as erythrostominone (Fig. 4.13), are known. Certain naphthoquinone pigments, the echinochromes and spinochromes (Fig. 4.23) found in the sexual organs and spines of sea urchins for instance *Paracentrotus lividus,* also appear to derive from acetic acid.

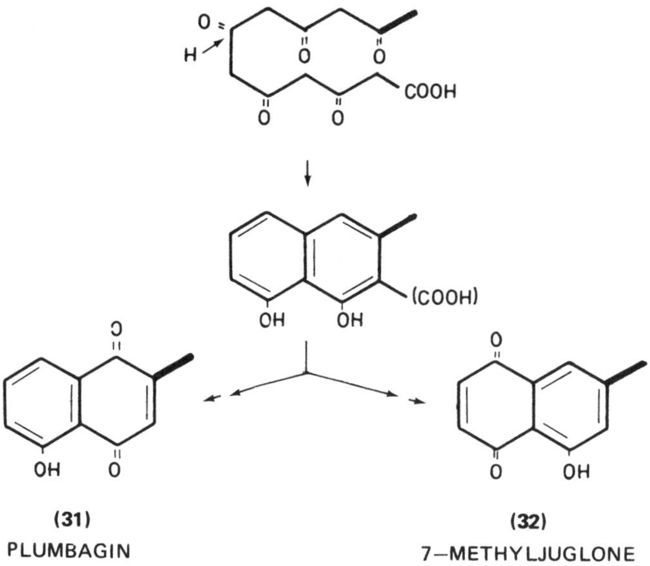

Fig. 4.22 Biosynthesis of some polyketide-derived naphthoquinone produced by plants.

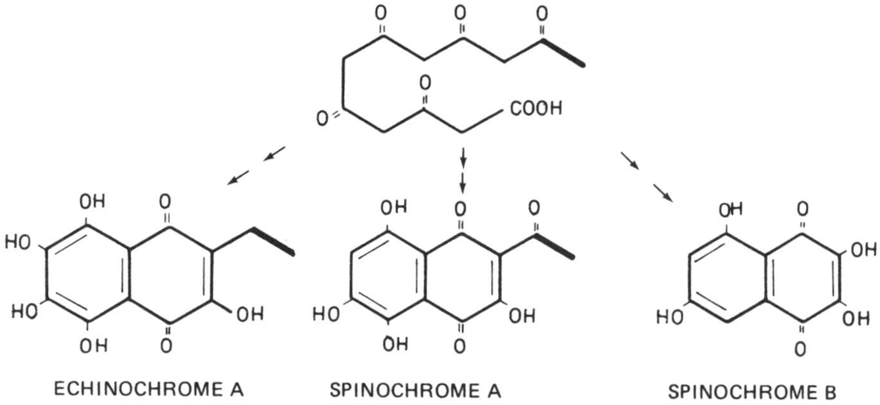

Fig. 4.23 Probable biosynthesis of some pigments derived from marine organisms.

4.6.3 Anthraquinones and Anthrones

Anthracenes, mostly at the quinone oxidation level are commonly found in microorganisms, plants, and lower animals. Such compounds are mainly of polyketide origin (other sources also exist, see p. 361) and are derived from the cyclisation of an octaketide chain in all the cases so far examined, (e.g. endocrocin, Fig. 4.1). The tricyclic skeleton can lose the 3-carboxylic group, produc-

ing anthracene derivatives with 15 carbon atoms (Fig. 4.24). Such derivatives
are common and are found in a wide range of fungi, often with anthrones,
such as (33) and (34), the anthraquinone dimers, such as (36). The former are
biosynthetic intermediates leading to the anthraquinone structure, and the latter
result from oxidative couplings. A particularly rich source of anthraquinones and
their dimers is *Penicillium islandicum.*

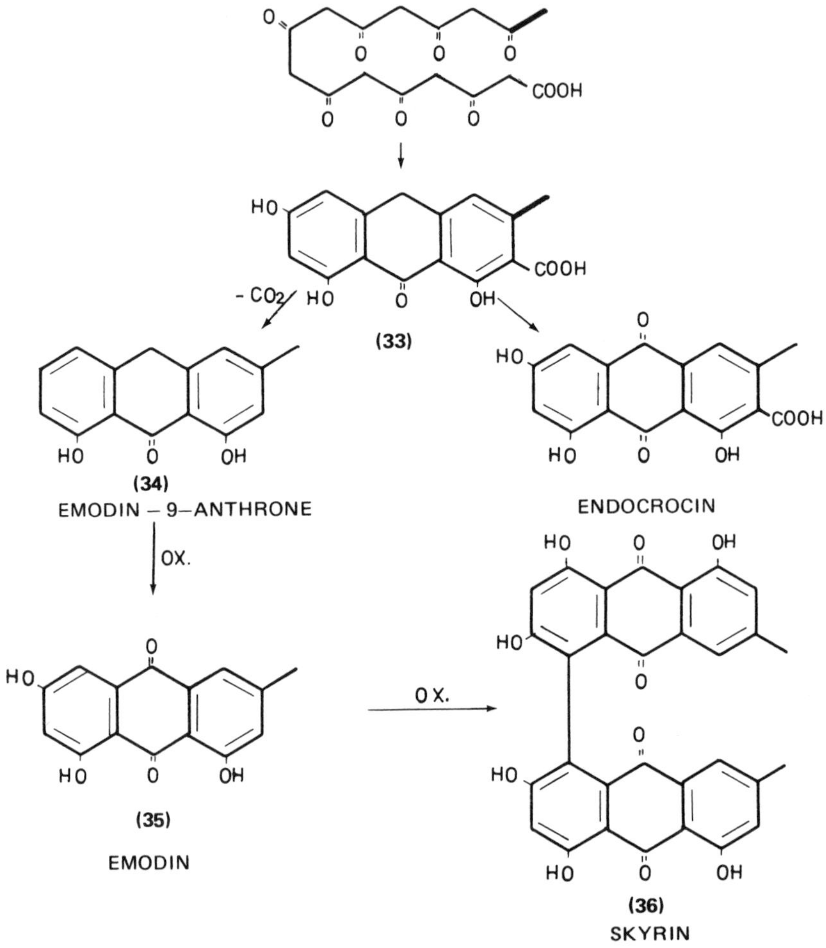

Fig. 4.24 Biosynthesis of polyketide-derived anthrones and anthraquinones.

Nalgiovensin and averufin are illustrated in Fig. 4.13 as examples of nona-
and decaketides. Other interesting anthraquinones can be found amongst the
anthracyclines, which are a group of antibiotic glycosides produced by various
Streptomyces species: rutilantinone (15) (Fig. 4.9) is a typical example.

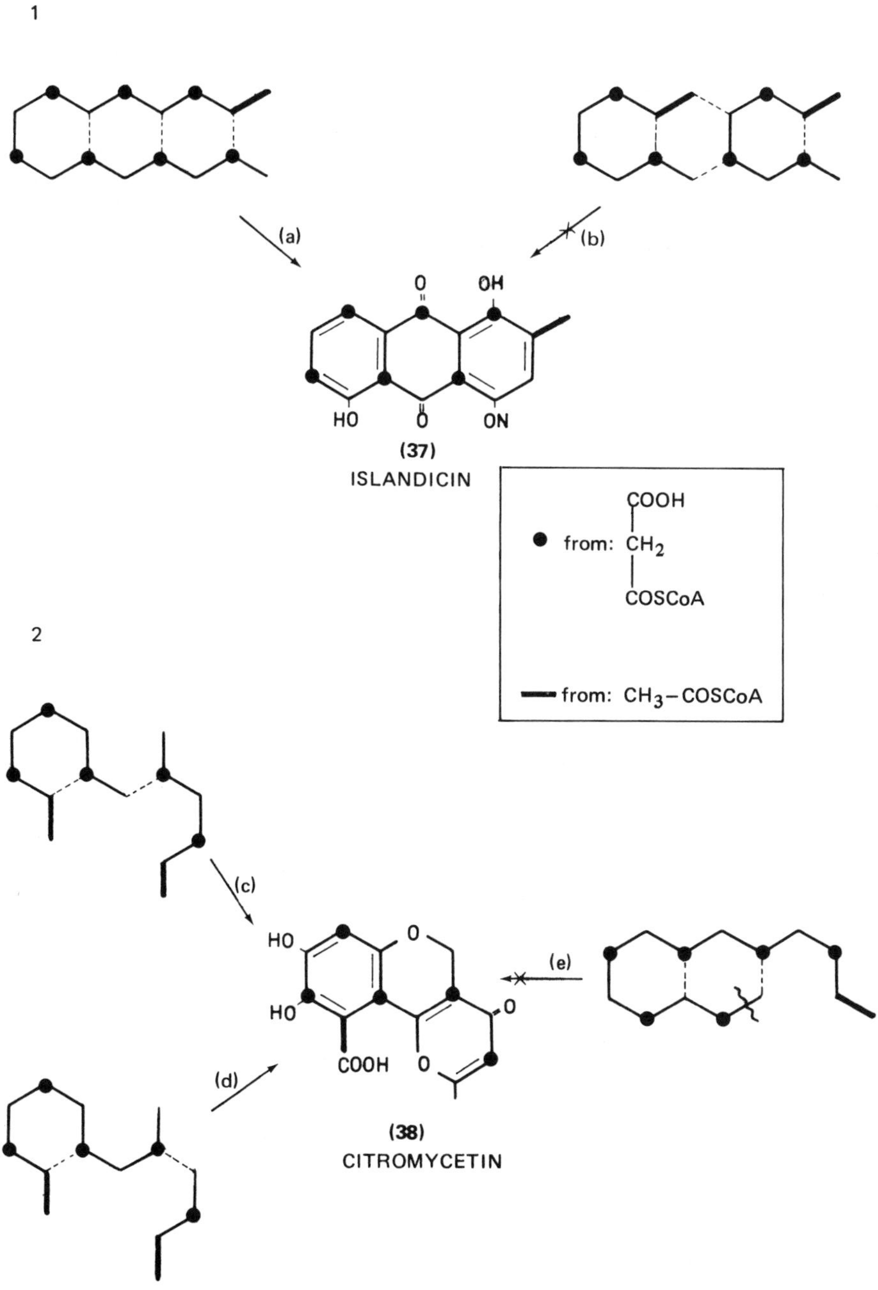

1

(a)

(b)

O   OH

HO   O   ON

(37)
ISLANDICIN

| COOH
| ● from: CH₂
| COSCoA
|
| ━━ from: CH₃−COSCoA

2

(c)

HO

HO

COOH   O

(e)

(d)

(38)
CITROMYCETIN

Fig. 4.25 Possible schemes for the biosynthesis of islandicin acid and citromycetin.

Emodin (**35**) (Fig. 4.24) is present in many imperfect fungi and higher plants (such as *Rhamnus frangula*), in which it is present as a glycoside. Incorporation experiments have shown that the same polyketide route to emodin exists in both systems. Fungal anthraquinones can probably arise from either a single, or more than one, polyketide chain. The situation has been resolved for the quinone islandicin (**37**) (Fig. 4.25) following the same method and reasoning adopted for 6-methylsalicylic acid (Fig. 4.8). The radioactive islandicin, obtained after incorporation of diethyl [2-^{14}C] malonate, demonstrates how a specific starter unit (acetyl CoA) is utilised by the microorganism for building the anthraquinone molecule. This supports the one chain route (path (a), Fig. 4.25). An entirely different example, although it refers to a non-anthraquinone molecule, is encountered for citromycetin (**38**). By use of diethyl [2-^{14}C] malonate in incorporation experiments the single chain route (path (e)) has been discarded leaving either the paths (c) or (d) as biosynthetic routes (Fig. 4.25).

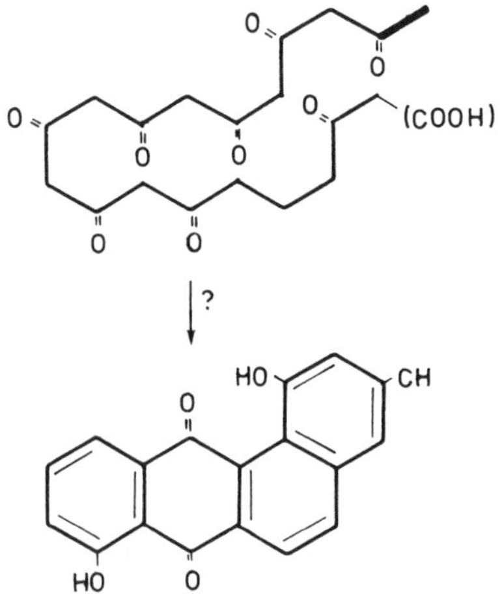

(**39**)

TETRANGOMYCIN

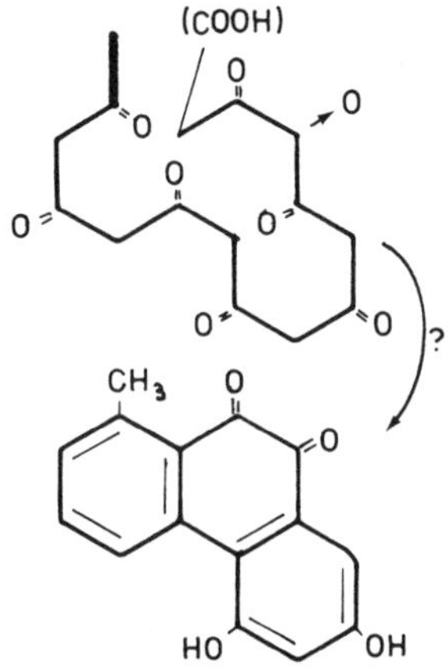

(**40**)
DENTICULATOL

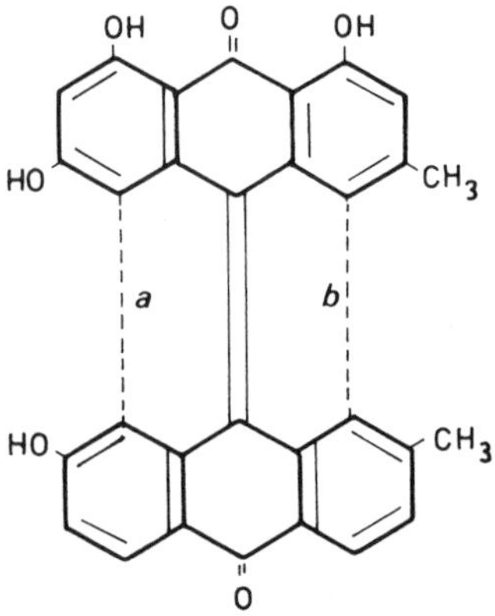

41 : no bonds at **a** and **b**
        (PROTOHYPERICIN I)
42 : no bond at **a**; bond at **b**
        (PROTOHYPERICIN II)
43 : bonds at **a** and **b**
        (HYPERICIN)

Fig. 4.26 Polycyclic quinones.

4.6.4 Other Quinones

Many other polycyclic natural quinones are known besides the naphtho- and anthraquinones. They constitute the major group of fungal and bacterial pigments, whilst being relatively rare in plants. Tetrangomycin (39) (see Fig. 4.26) produced by *Streptomyces rimosus*, is probably a decaketide; denticulatol (40) is the only phenanthranequinone found in nature; the protohypericins (41), (42) and hypericin (43), from *Hypericum perforatum*, arise from oxidative coupling processes, involving the 4-, 5-, and 10-positions of anthracene precursors (Figs. 4.24 and 4.27). Oxidative dimerisations of naphthalenes form perylene-quinones, such as (30) (Fig. 4.21).

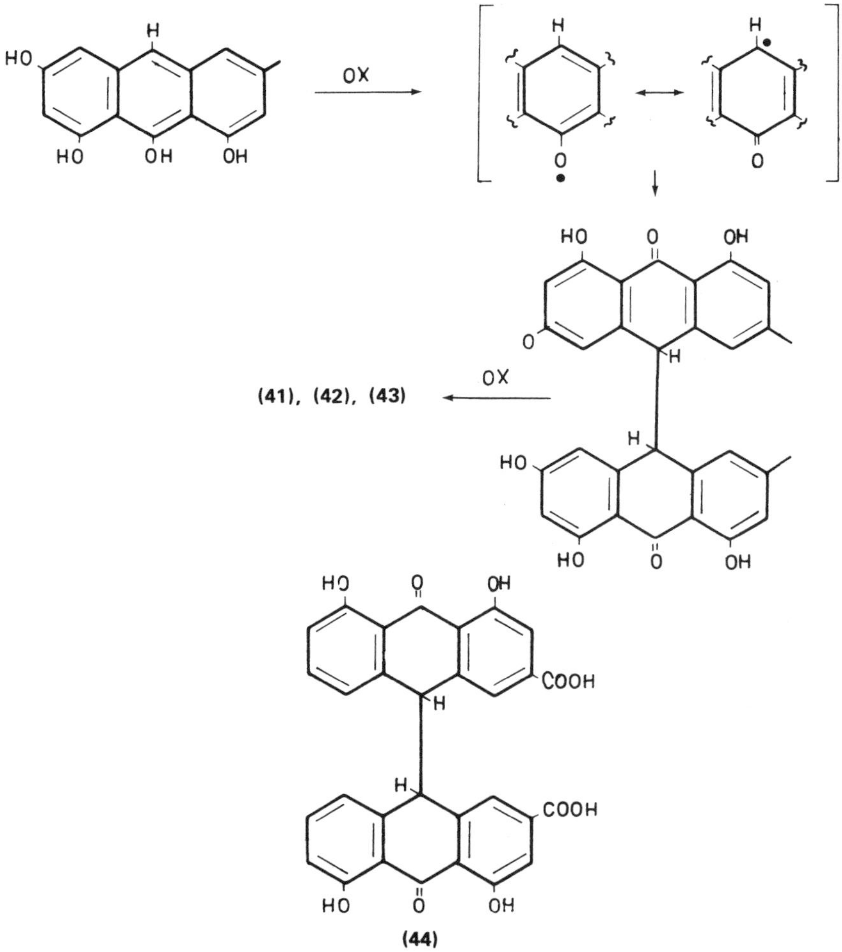

(41), (42), (43)

(44)
SENNIDINES A and B
(stereoisomers)
Fig. 4.27 Oxidative dimerisation of 9-anthrones.

4.7 BENZOPHENONES AND XANTHONES

Three routes have been established for the biosynthesis of the benzophenones; two of which involve polyketides and operate mainly in microorganisms and lichens. The first route is exemplified by the formation of various griseophenones (C_{14}) (Fig. 4.16), whilst the alternative route for sulochrin (**46**) formation was discovered in *Aspergillus terreus* (Fig. 4.28). The 9–9a bond of an emodin type anthraquinone (Fig. 4.24) is cleft oxidatively. This latter route was demonstrated via a [^{14}C]-labelled experiment involving the incorporation of questin (**45**) into sulocrin.

The third route to benzophenone systems, such as protocotoin (**48**), mainly takes place in higher plants; it involves the cyclisation (acylphloroglucinol-type) of a polyketomethylene chain, initiated by benzoic acid, and built up through three C_2 units (malonyl CoA) (Fig. 4.28(b)). This mixed biosynthesis can usually be recognised by the number of skeleton carbon atoms (C_{13} as against C_{14} or C_{15} of the former two ways) and by the different hydroxylation pattern of the two benzene rings: the triacetic ring (A) possesses oxygen atoms in alternate positions (2, 4, 6) whilst the ring of shikimate origin (B) has them in adjacent positions (3', 4' and 5'). (A fourth route to benzophenones and xanthones has recently been proposed which is of wholly shikimate origin and operates in plants via the catabolism of 4-aryl-coumarins).

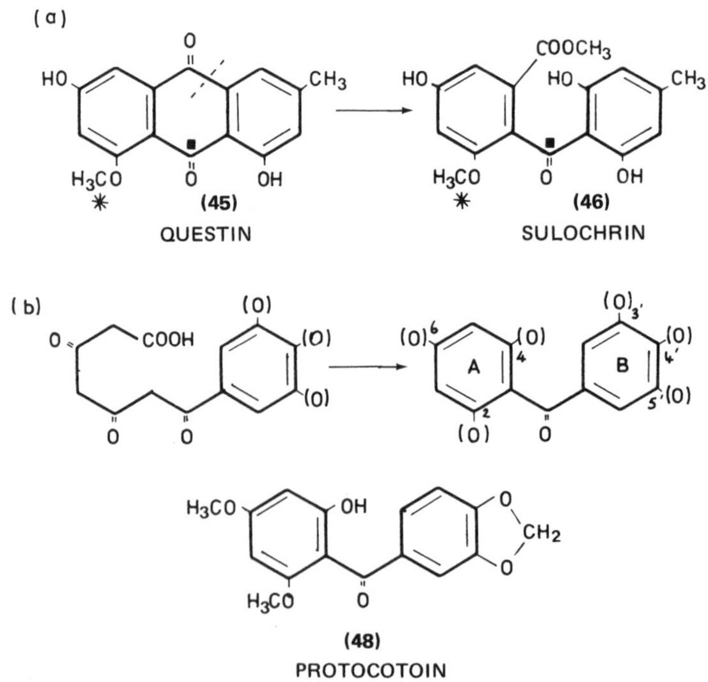

Fig. 4.28 Biosynthetic paths to some naturally-occurring benzophenones.

About a hundred different xanthones have so far been isolated and they raise interesting biogenetic questions, especially about the formation of the ether bridge. Xanthones can, *a priori*, derive from the corresponding benzophenones, either through an intramolecular phenol oxidation (via (a)) or through elimination of water from two phenolic hydroxyls in the 2- and 2'-positions (via (b)) (Fig. 4.29). Path (a) requires the presence of a hydroxyl group *meta* to the carbonyl function of the benzophenone and a second, *ortho*-hydroxyl group on the other ring, whilst route (b) requires two *ortho*-hydroxyls on both rings.

Path (a) can explain the formation of the C_{13} skeleton of xanthones of plant origin, such as jacarubin and mangiferin. Since they derive from a mixed biosynthesis the *meta*-hydroxyl in ring B can either be already present in the starting benzoic acid, or be introduced on hydroxylating the benzophenone precursor already becoming a *para*-hydroxyl group. Route (b) appears to be more likely for xanthones of the lichexanthone type (51), produced by microorganisms: the two hydroxyl groups in the *ortho*-positions of rings A and B probably arise as a consequence of the polyketide origins of the molecule.

(a)

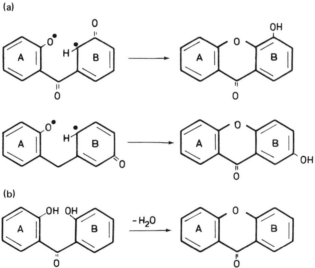

(b)

Fig. 4.29 Possible paths for the formation of some xanthones.

Secoanthraquinones (52) and (54) can be considered as intermediates of either actual xanthones, such as pinselin (53), isolated from certain microorganisms (e.g. *Penicillium amarum*) and from higher plants (e.g. *Cassia occidentalis*), or of the ergochromes from the pigments of *Claviceps purpurea*. Pinseline probably derives from a non-oxidative cyclisation of type (b) (Fig. 4.29). The ergochromes, on the other hand, probably form by a nucleophilic addition of the A ring hydroxyl to an arene oxide of B ring, followed by an intermolecular (2, 2' or 4, 4') phenol oxidation, and finally by successive reduction-oxidation steps in the ring carrying the methyl group (Fig. 4.30).

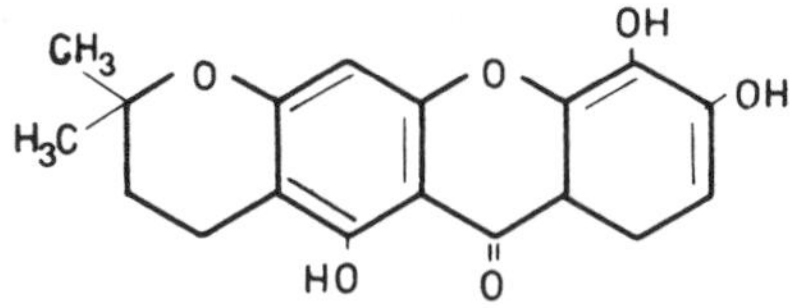

(49)
JACAREUBIN

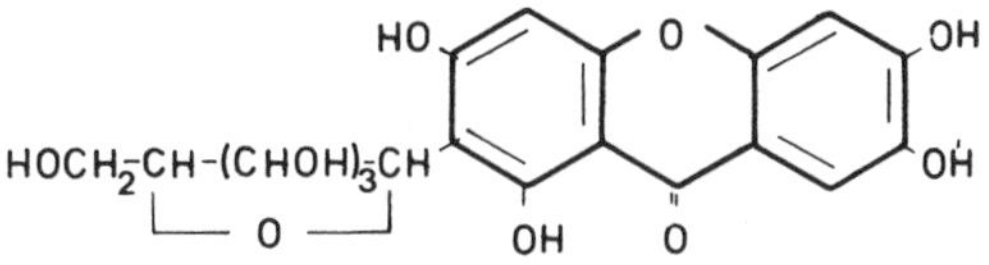

(50)
MANGIFERIN

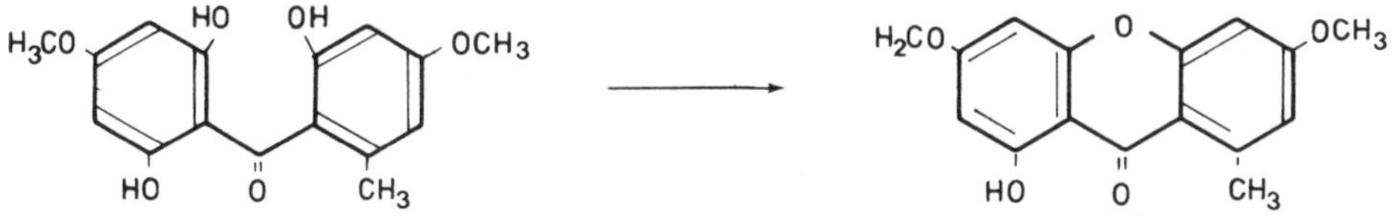

(51)
LICHEXANTHONE

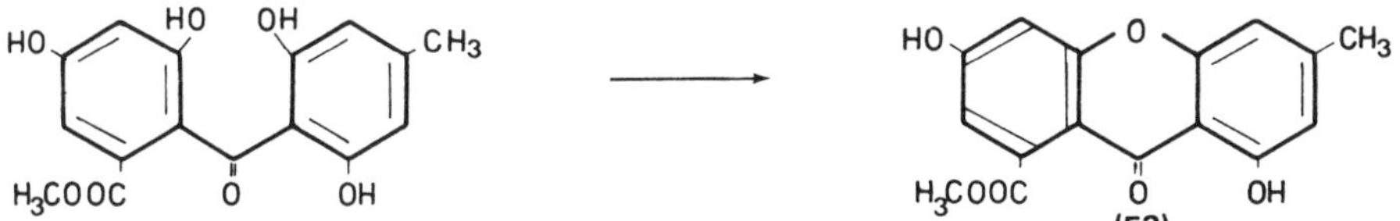

(52)

(53)
PINSELIN

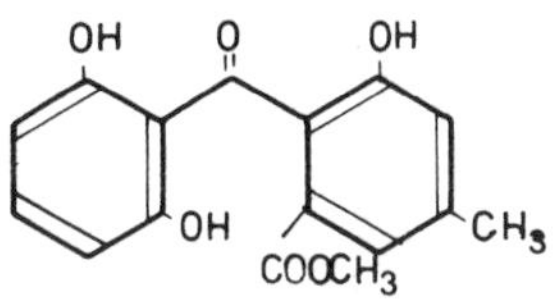

(54)

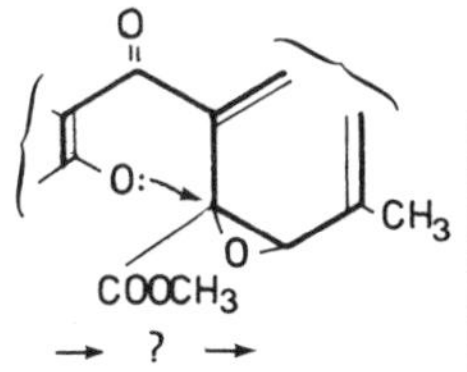

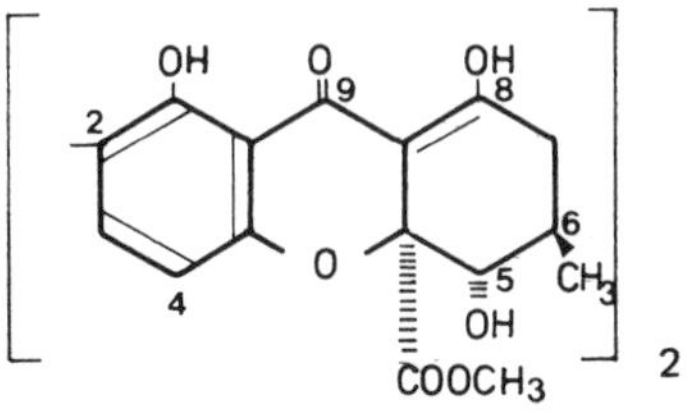

ERGOCHROME AA(4,4')

Fig. 4.30 Various types of xanthones and ergochromes.

4.8 DEPSIDES AND DEPSIDONES

Depsides result from the union of two or more molecules of di- or trihydroxy-benzoic acids and possess ester bonds between the carboxylic group of one unit and the phenolic group of another acid molecule. Depsides are typical lichen metabolites (Fig. 4.31). The di- and trigallic acids, present in tannin (p. 420), should also be considered as depsides.

The polyketide nature of the lichen depsides is evident from their structure, since the monomeric acid unit is represented by orsellinic acid (e.g. **55**), its simple derivatives, or one of its homologues, e.g. (**56**). The polyketide origin has also been experimentally demonstrated through incorporation experiments. Such incorporation experiments are very difficult in lichens, owing to the extremely slow metabolic processes of such organisms.

DEPSIDES :

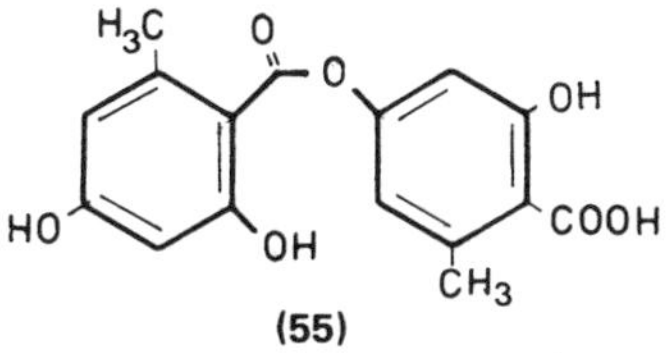

(55)
LECANORIC ACID

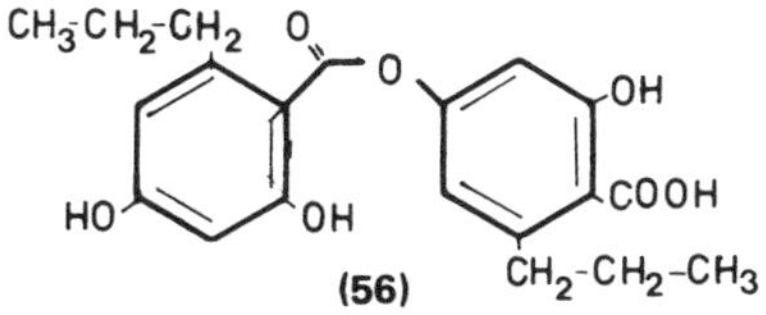

(56)
DIVARICATIC ACID

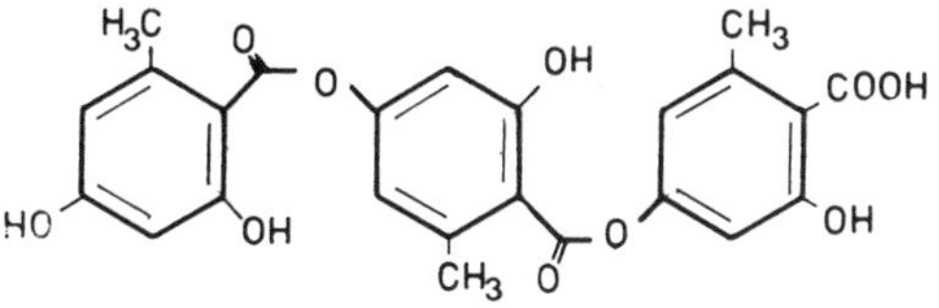

(57)
GYROPHORIC ACID

DEPSIDONES :

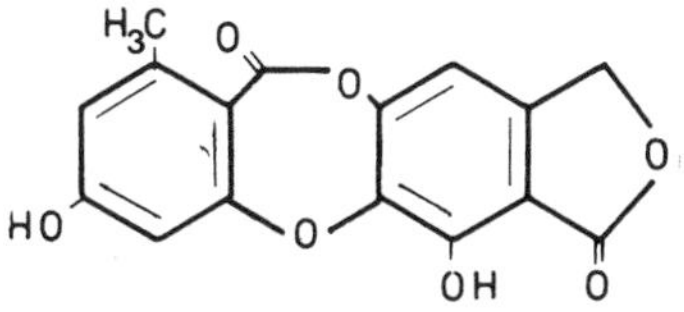

(58)
VARIOLARIC ACID

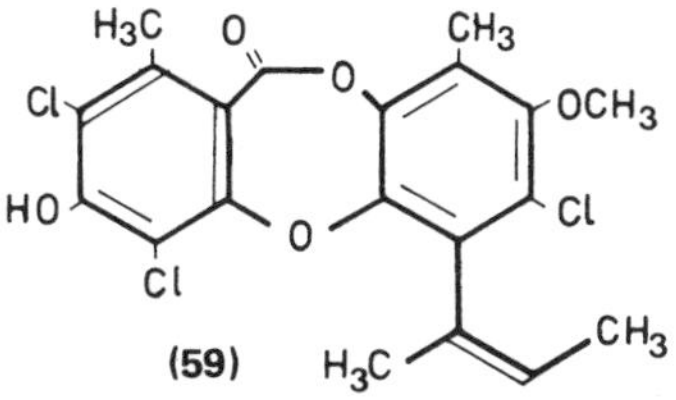

(59)
NIDULIN

Fig. 4.31 Depsides and Depsidones.

Atranorin (61) (Fig. 4.32) requires two 'extra' carbon atoms (asterisked). As predicted, the uptake of these extra carbons takes place in the polyketo-methylenic chain and not at the intermediate orsellinic or lecanoric acid stages. Labelled orsellinic acid is incorporated into lecanoric acid, but not into atranorin, whilst 2,4-dihydroxy-3,6-dimethylbenzoic acid (60) is incorporated into atranorine but not into lecanoric acid. These results as well as others suggest that the biosynthesis of depsides takes place through the following steps:

(a) Synthesis of aromatic monomeric units, via orsellinic acid-type cyclisations, starting from intact or modified polyketide chains.

(b) Condensation of the various units, through reaction of a phenolic hydroxyl with an activated carboxyl group.

Some experiments with lichens have been performed in which the fungus (mycobiont) is grown separately from its natural symbiont (the alga or phycobiont). These experiments indicate that the former synthesises the monomeric aromatic acids (phase a), whilst the latter carries out the intermolecular condensations (phase b). This distribution of tasks, although consistent with the peculiar metabolic nature of the symbionts, requires much more experimental confirmation.

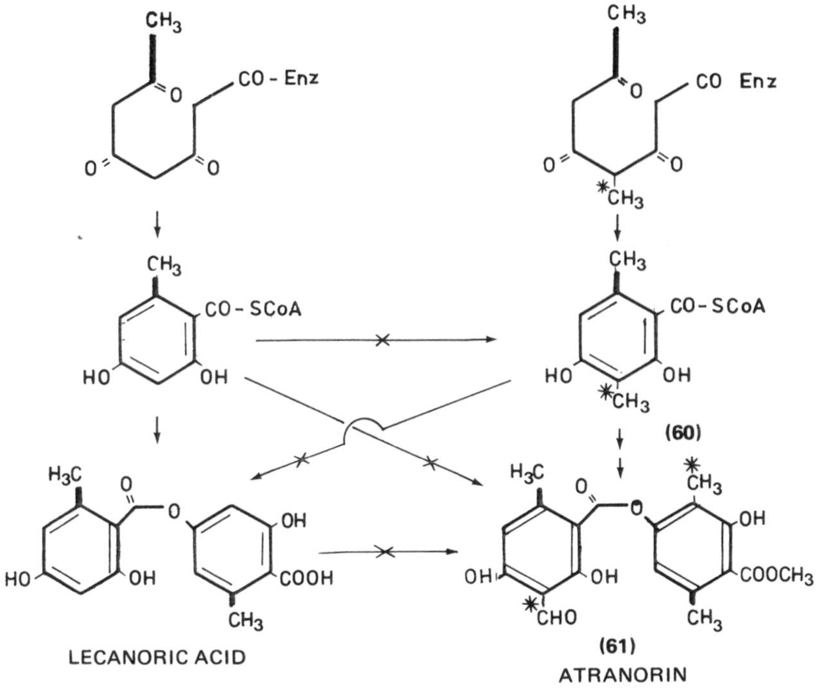

Fig. 4.32 Biosynthesis of depsides.

Depsidones are also present in lichens. These compounds have one ether bridge between two consecutive aromatic rings of a depside sequence. Structures such as those in Fig. 4.31 are obtained, and in which a dioxepin ring condensed with two benzene rings is present.

The close structural nature and the coexistence of the depsides and depsidones in the same organisms strongly support their close biogenetic relationship. The depsidones probably derive from the depsides through phenol oxidation, as indicated in Fig. 4.33. About thirty depsides and depsidones are currently known and they are all produced by lichens, except for nidulin (59) and its derivatives which are isolated from the imperfect fungus, *Aspergillus nidulans*.

4.9 AFLATOXINS

Aflatoxins are a group of fungal metabolites, with homogenous structural, biogenetic and toxicological properties. Their most evident common feature is the presence in the molecule of two tetrahydrofuran rings, condensed with each other across the 2,3 bonds (Fig. 4.34). The remainder of the molecule can be an anthraquinone such as in (62), a xanthone such as in (63), or coumarin such as in (64) and (65).

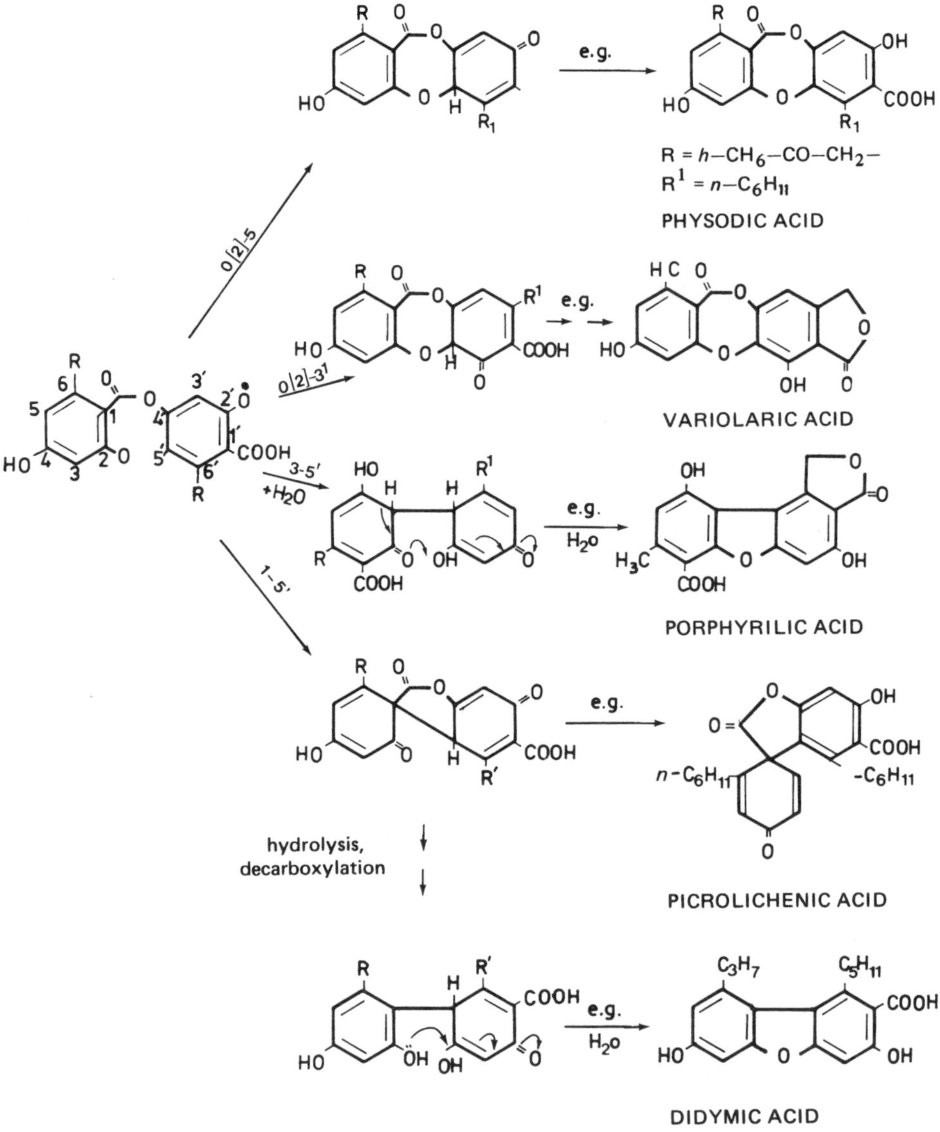

Fig. 4.33 Biosynthesis of depsidones and related metabolites by the phenolic oxidation of depsides.

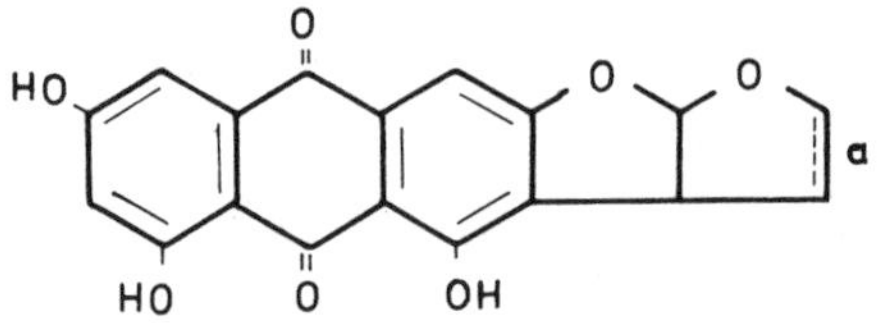

HO

O

O   O

a

HO   O   OH

(62)

a = double bond :
VERSICOLORIN A

b = single bond :
VERSICOLORIN B

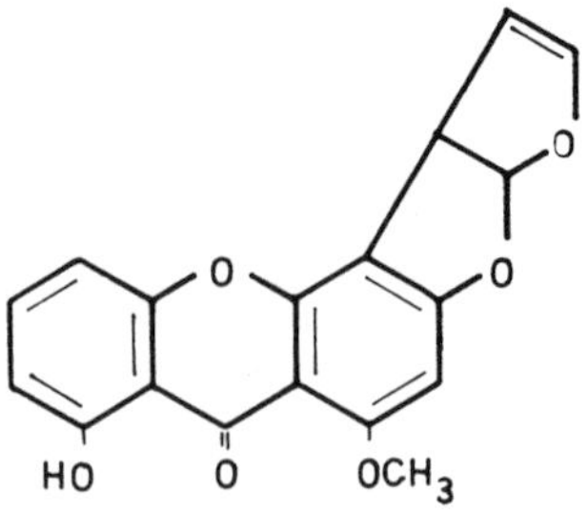

O

O

HO   O OCH_3

(63)
STERIGMATOCYSTIN

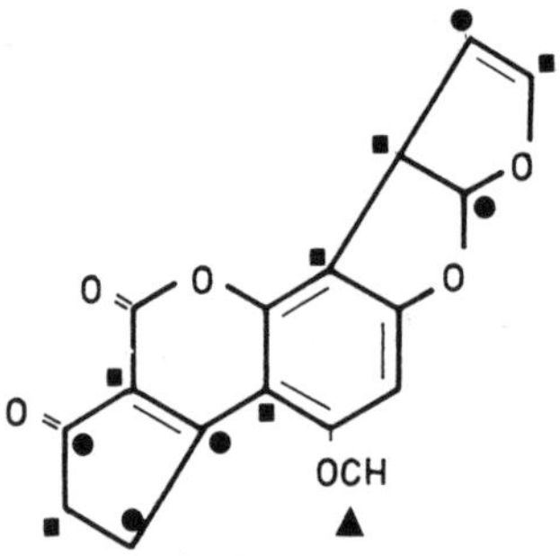

O

O   O

O

O

OCH

(64)
AFLATOXIN B_1

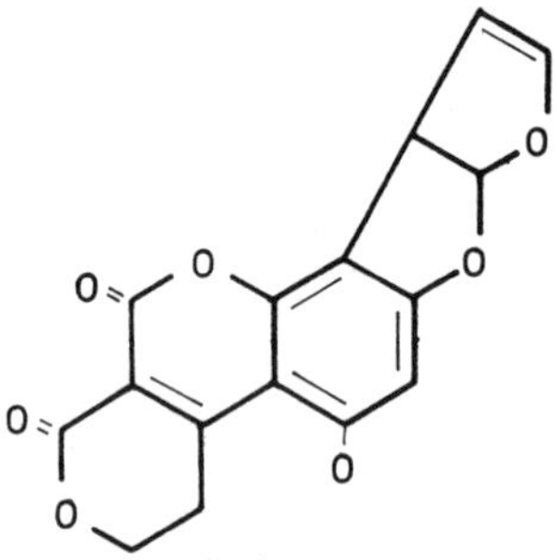

O

O

O

O

O

(65)
AFLATOXIN G_1

CH_3
|
COOH

C_1

Fig. 4.34 Aflatoxins and the origin of the carbons in aflatoxin B_1.

Aflatoxins are produced by imperfect fungi of the *Aspergillus* type (*A. versicolor* and *A. flavus*) and are all highly toxic to animals, which may ingest them on eating mouldy food. They are amongst the most carcinogenic agents known, being active in rats at a dose level of 1 μg per day.

The biosynthesis of the aflatoxins has been studied by Büchi and coworkers. They were able to establish the origin of most carbon atoms in aflatoxin B, (Fig. 4.34), by using elegant incorporation and degradation products. Fig. 4.35 illustrates the biosynthetic scheme for different aflatoxin structures.

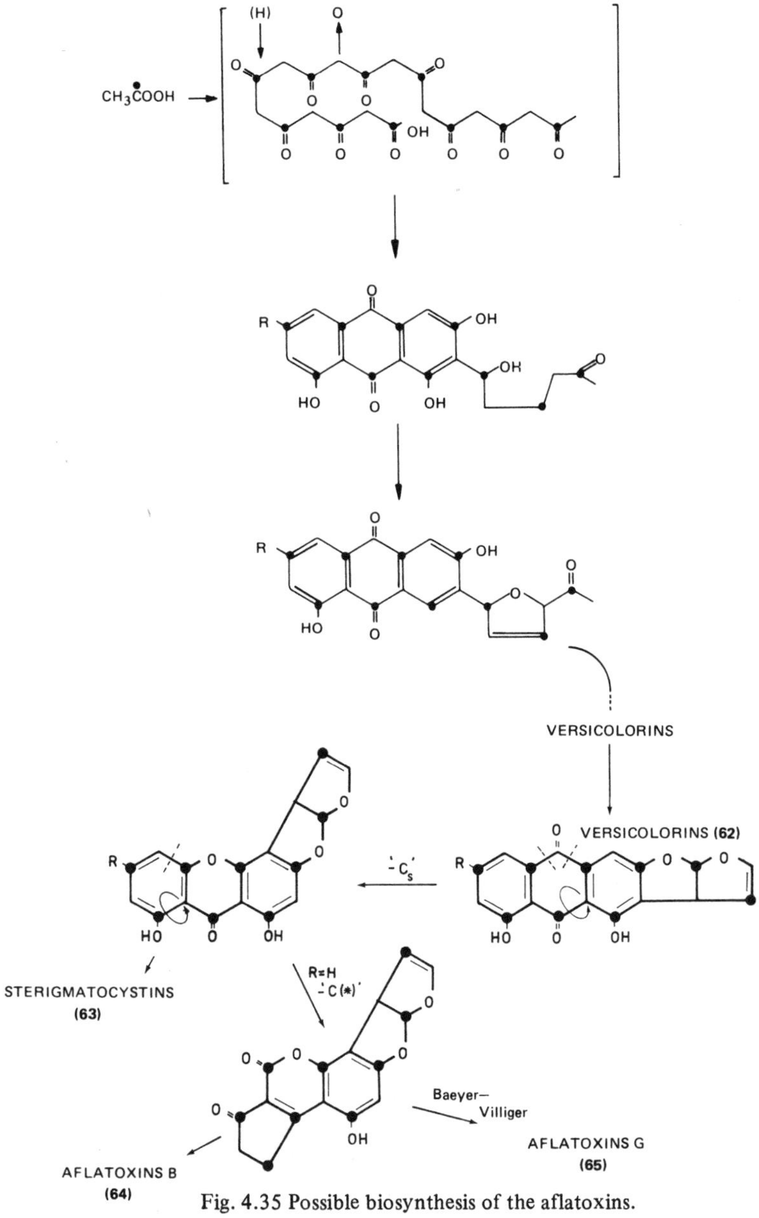

Fig. 4.35 Possible biosynthesis of the aflatoxins.

4.10 THE TETRACYCLINES

Tetracyclines are antibiotics with a wide range of antibacterial activity (bacterio-static activity); they are produced by various *Streptomyces* species, and are characterised by a partially hydrogenated C_{18}-naphthacene skeleton (Fig. 4.36).

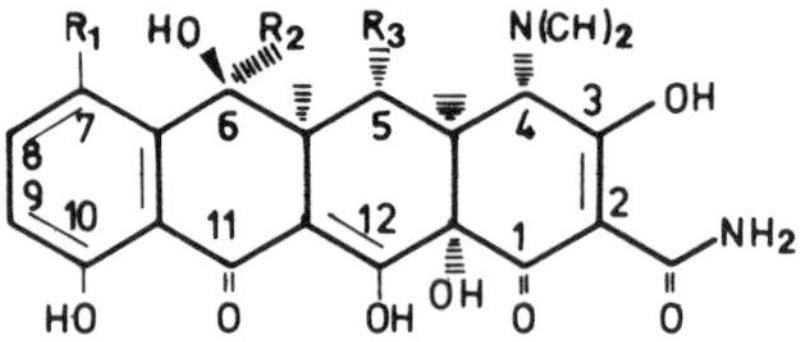

	R_1	R_2	R_3	
(66)	H	H	H	6–DEMETHYLTETRACYCLINE
(67)	H	CH_3	H	TETRACYCLINE
(68)	Cl	H	H	7–CHLORO–6–DEMETHYLTETRACYCLINE
(69)	H	CH_3	OH	5–HYDROXYTETRACYCLINE (TERRAMYCIN)
(70)	Cl	CH_3	H	7–CHLOROTETRACYCLINE (AUREOMYCIN)

Fig. 4.36 Tetracyclines.

The biosynthesis of the tetracyclines (Fig. 4.37) has been confirmed by experiments in terms of an initial cyclisation of a non-aketide chain (with or without prior modification). In this manner key intermediates are produced, such as pretetramid and 6-methylpretetramid, which contain all of the carbon atoms of the final products. The starter unit for the polyketomethylene chain of the tetracyclines is malonamoyl CoA. The tetracyclines of the 6-demethyl series, e.g. (66) and (68), are derived from pretetramide, through reactions very similar to those included in Fig. 4.37 for 6-methylpretetramid. The introduction of the chlorine atom into the aromatic ring A of intermediate (72) produces a dichotomy in the biosynthetic path leading to the 7-chloro, and 7-H (normal) series. A second dichotomy arises from the intermediate (73) leading to the 5-hydroxy and 5-H (normal) series.

4.11 MACROLIDE ANTIBIOTICS

The macrolide antibiotics, like the tetracyclines, are produced by various species of *Streptomyces*. They often possess remarkable bacteriostatic properties, so that they have been widely used as therapeutic agents. The term 'macrolide' derives from a structural feature common to all such antibiotics: a macrocyclic lactone is present in the molecule. The generally saturated and branched aliphatic chain is composed of more than ten carbon atoms as in the erythromycins (Fig. 4.38). Another common structural feature is the presence of from one to three sugar units, bound through glycosidic links to hydroxy-substituents on the lactone ring. The sugars are generally deoxyhexoses, in the pyranose form, often with methyl, dimethylamino, or methoxy substituents, e.g. (76)–(78).

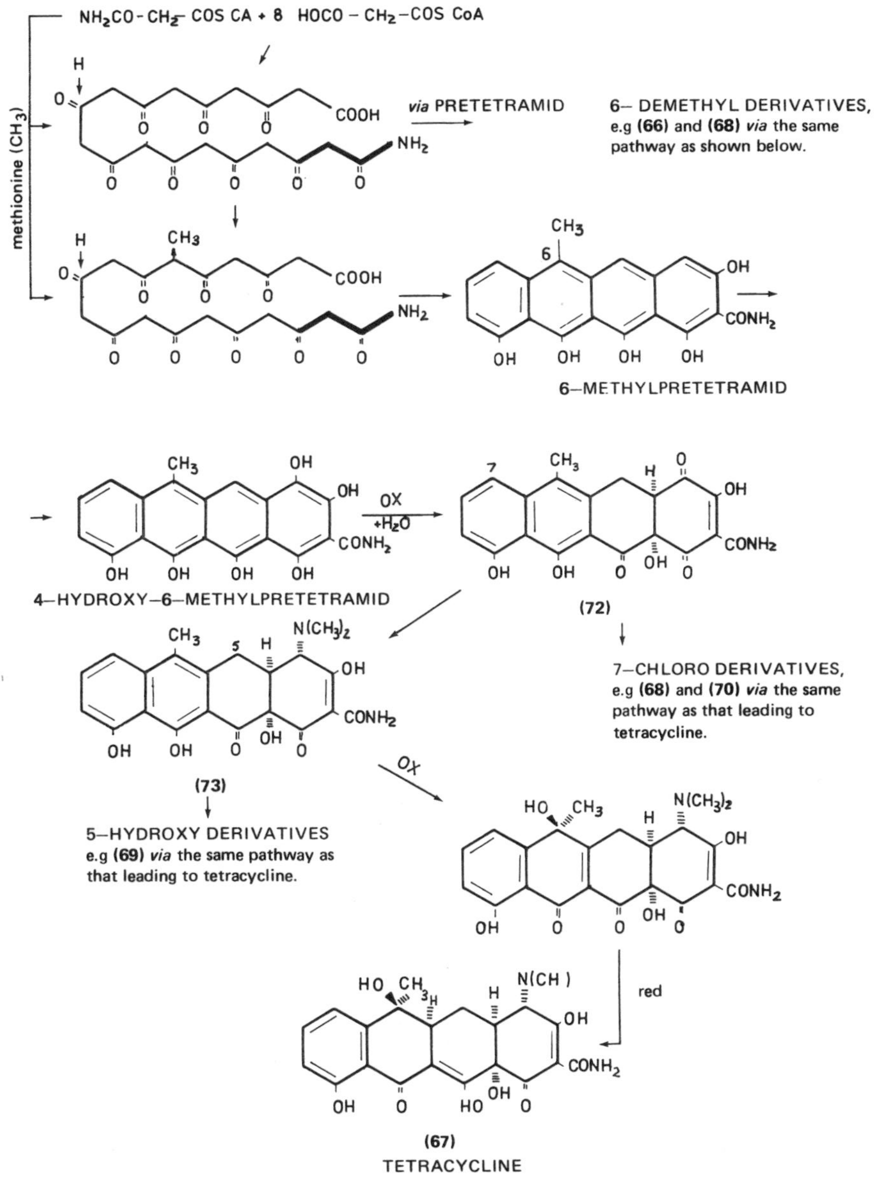

Fig. 4.37 Biosynthesis of the tetracyclines.

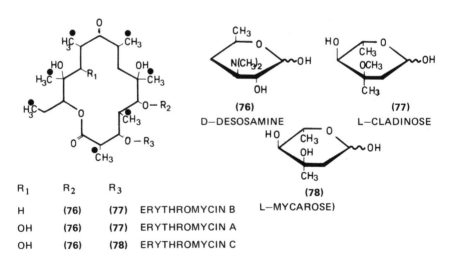

Fig. 4.38 Biosynthesis of some macrolide antibiotics.

SOURCE MATERIALS AND SUGGESTED READING
[1] J. N. Collie, (1893), The Production of Naphthalene Derivatives from Dehydroacetic Acid, *J. Chem. Soc.*, **63**, 329T.
[2] J. N. Collie, (1907), Derivatives of the Multiple Ketene Group, *J. Chem. Soc.*, **91**, 1806T.
[3] A. J. Birch and F. W. Donovan, (1953), Studies in Relation to Biosynthesis. I. Some Possible Routes to Derivatives of Orcinol and Phloroglucinol, *Austral. J. Chem.*, **6**, 360.
[4] A. J. Birch, (1957), Biosynthetic Relations of Some Natural Phenolic and Enolic Compounds in *Progress in the Chemistry of Organic Natural Products*, ref. 1 of Chap. 1, **14**.
[5] A. J. Birch, (1962), Some Pathways in Biosynthesis, *Proc. Chem. Soc.*, 3.
[6] J. F. Grove, (1964), Griseofulvin and Some Analogues in *Progress in the Chemistry of Organic Natural Products*, ref. 1 of Chap. 1, **22**.
[7] A. J. Birch, (1966), Biosynthetic Intermediates of Aromatic Compounds in *Biosynthesis of Aromatic Compounds*, (G. Billek, Ed.), Pergamon Press.
[8] E. Jones, (1969), Natural Polyacetylenes and Their Precursors, *Chem. in Britain*, **2**, 6.
[9] Lord Todd, (1966), Some New Developments in the Chemistry of Natural Colouring Matters, *Chem. in Britain*, **2**, 428.
[10] J. F. Snell (Ed.), (1966), *Biosynthesis of Antibiotics*, Academic Press.
[11] F. Bohlmann, (1967), Biogenetische Beziehungen der natürlichen Acetylenverbindugen in *Progress in the Chemistry of Organic Natural Products*, ref. 1 of Chap. 1, **25**.
[12] A. J. Birch, (1967), Biosynthesis of Polyketides and Related Compounds, *Science*, **156**, 202.
[13] P. Gottlieb and P. D. Shaw, (1967), *Antibiotics*, Springer Verlag, **II**.
[14] S. Shibata, (1967), Some Recent Studies on the Metabolites of Fungi and Lichens, *Chem. in Britain*, **3**, 110.
[15] W. B. Whalley, The Biosynthesis of Fungal Metabolites in *Biogenesis of Natural Compounds*, ref. 24 of Chap. 1.
[16] K. Mosbach, (1969), Biosynthesis of Lichen Substances, Products of a Symbiotic Association, *Angew. Chem. Int. Ed. Engl.*, **8**, 240.
[17] T. Money, (1970), Biogenetic-Type Synthesis of Phenolic Compounds, *Chem. Rev.*, **70**, 553.
[18] S. huneck, (1971), Chemie und Biosynthese der Flechtenstoffe in *Progress in the Chemistry of Organic Natural Compounds*, ref. 1 of Chap. 1, **29**.
[19] R. H. Thomson, (1971), *Naturally Occurring Quinones*, 2nd ed., Academic Press.
[20] W. B. Turner, (1971), *Fungal Metabolites*, Academic Press.
[21] M. O. Moss, Aflatoxin and related Mycotoxins in *Phytochemical Ecology*, ref. 69 of Chap. 1.

[22] K. L. Rinehart, (1972), Antibiotics with Ansa Rings, *Accounts Chem. Res.*, 5, 57.

[23] F. Bohlmann, T. Burkhardt, and C. Zdero, (1972), *Naturally Occurring Acetylenes*, Academic Press.

[24] C. W. J. Chang, (1973), Marine Natural Products, Pigments, *J. Chem. Educ.*, 50, 102.

[25] B. Frank and H. Flasch, (1973), Die Ergochrome (Physiologie, Isolierung, Struktur und Biosynthese) in *Progress in the Chemistry of Organic Natural Products*, ref. 1 of Chap. 1, 30.

[26] N. M. Packter, (1973), *Biosynthesis of Acetate-Derived Compounds*, Wiley.

[27] T. Money, Biosynthesis of Polyketides in *Biosynthesis*, ref. 30 of Chap. 1, (1973) 2; (1976), 4.

[28] R. Bentley and I. M. Campbell, (1974), Biological Relations of Quinones in *The Chemistry of Quinonoid Compounds*, Part 2, (S. Patai, Ed.), Wiley.

[29] T. M. Harris, C. M. Harris, and K. B. Hindley, (1974), Biogenetic-Type Syntheses of Polyketide Metabolites in *Progress in the Chemistry of Organic Natural Products*, ref. 1 of Chap. 1, 31.

[30] J. C. Roberts, (1974), Aflatoxins and Sterigmatocystins, in *Progress in the Chemistry of Organic Natural Compounds*, ref. 1 of Chap. 1, 31.

[31] R. Bentley, (1975), Biosynthesis of Quinones in *Biosynthesis*, ref. 30 of Chap. 1, 3.

[32] W. Dürckheimer, (1975), Tetracyclines: Chemistry, Biochemistry and Structure-Activity Relations, *Angew. Chem. Int. Ed. Engl.*, 14, 721.

[33] J. V. Rodricks (Ed.), (1976), Mycotoxins and Other Fungal Related Food Problems, *Advances in Chemistry Series, No. 149, Amer. Chem. Society.*

[34] R. H. Thomson, (1976), Quinones, Nature, Distribution and Biosynthesis and Miscellaneous Pigments in *Chemistry and Biochemistry of Plant Pigments*, ref. 39 of Chap. 1, 1.

[35] S. Masamune, G. S. Bates and J. W. Corcoran, (1977), Macrolides. Recent Progress in Chemistry and Biochemistry, *Angew. Chem. Int. Ed. Engl.*, 16, 585.

[36] T. J. Simpson, (1977), Biosynthesis of Polyketides in *Biosynthesis*, ref. 30 of Chap. 1, 5.

[37] G. E. Evans, M. J. Garden, D. A. Griffin, F. J. Leeper and J. Staunton, (1978), Biomimetic Syntheses of Phenols from Polyketides, in *Further Perspectives in Organic Chemistry*, (Ciba Foundation Symposium 53), Elsevier.

[38] J. G. Heathcote and J. R. Hibbert, (1978), *Development in Food Science-1: Aflatoxins, Chemical and Biological Aspects*, Elsevier.

[39] J. S. Glasby, (1979), *Encyclopaedia of Antibiotics*, 2nd ed., Wiley.

[40] M. U. S. Sultanbawa, (1980), Xanthonoids of Tropical Plants, *Tetrahedron*, 36, 1463.

Chapter 5

The Isoprenoids

Natural products often possess a carbon framework comprised of units of the five carbon arrangement (1). Such compounds are called terpenes. The mono-meric unit (1) is called 'isoprenic', because of its relationship to the diene, isoprene (2); it is commonly indicated by the symbol, C_5. Most terpenes possess a carbon content in multiples of this five carbon arrangement.

Before the common biosynthesis of this class of products was recognised the term terpene was introduced for those compounds containing ten carbon atoms and this base is still used for the modern classification of such natural products. This classification divides terpenes into *hemiterpenes* ($1 \times C_5$), *monoterpenes* ($C_{10} = 2 \times C_5$), *sesquiterpenes* ($C_{15} = 3 \times C_5$), *diterpenes* ($C_{20} = 4 \times C_5$), *sesterpenes* ($C_{25} = 5 \times C_5$) and *triterpenes* ($C_{30} = 6 \times C_5$).

An example of each group is shown in Fig. 5.1. The C_5 units can be easily recognised and they are generally limited by 'head to tail' bonds as well as supplementary bonds (which are indicated by dotted lines). These structural features are a consequence of the common biosynthetic origin of the terpenes. The examples illustrated in Fig. 5.1 have a regular pattern. One often finds related structures possessing irregularities and these are the result of subsequent bond rearrangements. For example, one or more head-to-tail bonds may be missing or part of the carbon skeleton may not possess isoprenic character. In some cases, because of subsequent loss of, or gain of, carbon atoms the terpenes may not contain a simple multiple of five carbon atoms. The family of regular and irregular terpenoids is very large and of great importance.

Research into the biosynthesis of the terpenoids may be divided into three parts:

(a) On the structure and origin of the parent isoprenic units.
(b) On study of the enzymic process by which the C_5 units are assembled, and the fundamental skeletons are formed.
(c) On the study of the sequences and nature of processes by which modified terpenic skeletons and the introduction of functional groups occur.

As yet such studies have not revealed a complete picture of all *in vivo* terpenic syntheses. Nevertheless, a general pattern of the key biosynthetic

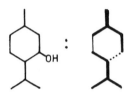

(1)   **(2)**

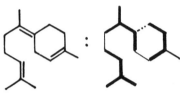

$\gamma,\gamma-$ DIMETHYLALLYL
ALCOHOL

MONOTERPENES   *SESQUITERPENES*

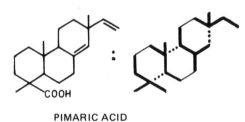

MENTHOL $\gamma-$ BISABOLENE

DITERPENES   *SESTERTERPENES*

CHO

O

HO   OPHIOBOLIN – A

PIMARIC ACID

TRITERPENES

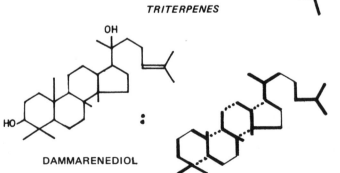

OH

HO

DAMMARENEDIOL

Fig. 5.1
Isoprene units
and terpene
structures.

process has emerged. This pattern, based on solid experimental data, has led to formulation of the biogenetic isoprene 'rule'. The isoprene rule was originally proposed by Ruzicka in 1953-1954 and has been successively improved and extended to cover the increasing number of terpenic compounds discovered in nature.

The isoprene rule was a rationalisation, through plausible chemical reactions and probable biological intermediates, of the old 'structural isoprenic rule' of Wallach (1887) and Robinson (1920), based upon the frequency and regularity of isoprenic features in the terpenes.

5.1 BIOSYNTHESIS OF THE C₅ UNIT

Earlier chemists hypothesised a direct participation of isoprene in the *in vivo* synthesis of terpenes.

This hypothesis was supported by the possible formation of 'dipentene' from two isoprene units by a simple Diels-Alder process (Fig. 5.2) and by the wide occurrence, amongst essential oils, of compounds with the dipentene structure.

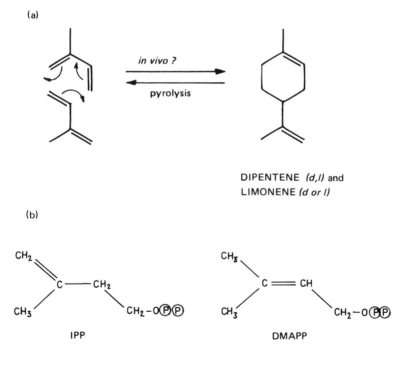

Fig. 5.2 (a) The first postulated method for uniting C₅ units; (b) the real active forms of the C₅ units.

Dipentene found this way is racemic. It is particularly abundant in turpentine oil. Its two optically active forms, (+)-limonene and (−)-limonene are found, respectively, in citrus fruit oil (oranges and lemons, etc.) and peppermint oil. Other diterpene derivatives are widely distributed in nature. An objection to this early postulate, however, was that isoprene itself did not appear to be present in nature and could only be obtained by the pyrolysis of certain monoterpenes.

Elucidation of the structure of the real, natural C_5 precursor unit, utilised by organisms for the synthesis of terpenes, was achieved by J. W. Cornforth in 1959. In his work on the biosynthesis of steroids (see Chapter 6) he characterised two active forms of isoprene, isopentenyl pyrophosphate (IPP) and dimethylallyl-pyrophosphate (DMAPP) (Fig. 5.2). Such intermediates are obligatory for the synthesis of plant terpenes. The incorporation of these intermediates into terpenes is catalysed by various enzymes. Enzymes catalysing the reactions illustrated in Fig. 5.3, 5.4 and 5.5 have been identified and isolated from a variety of vegetable sources. The intermediates, IPP and DMAPP, arise from (+)-mevalonic acid (MVA) and an enzyme complex which effects this conversion in high yields (over 30%) has been isolated from the latex of the rubber plant. Acetic acid, or

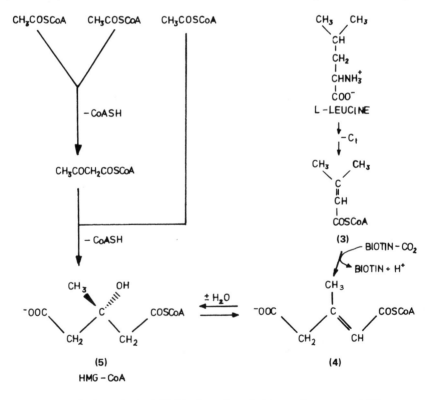

Fig. 5.3 Biosynthesis of (S)-3-hydroxy-3-methylglutaroyl-coenzyme A (5).

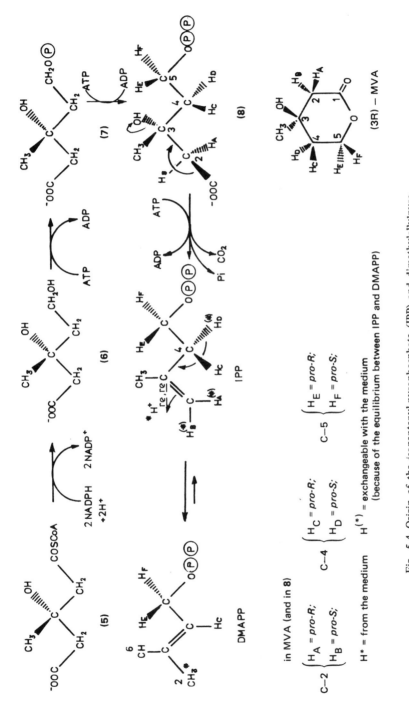

Fig. 5.4 Origin of the isopentenyl-pyrophosphate (IPP) and dimethylallylpyro-phosphate (DMAPP) units.

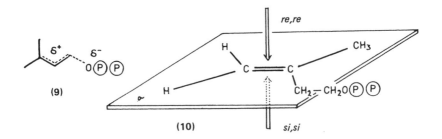

Fig. 5.5 Electronic characteristics of DMAPP (9) and steric sites of IPP (10)

its derivative acetyl CoA, is the only carbon atom source of mevalonic acid and, hence, the intermediates IPP and DMAPP. In the process leading to these intermediates (Fig. 5.3 and 5.4) a key intermediate is S-3-hydroxy-3-methyl-glutaryl CoA (5). This can be formed from either acetic acid (major route) or L-leucine. (Fig. 5.3). Thereafter a unique pathway leads to IPP and DMAPP (Fig. 5.4). The enzymes involved in the biosynthesis of the active forms of the C_5 unit are highly specific; they have been isolated from many different sources and the reactions they catalyse have been studied in detail.

Compound (8) is the biological equivalent of R-(+)-mevalonic acid lactone. MVA-lactone is a crystalline substance, chemically stable, which is easily converted into compound (8) *in vivo*, by opening of the lactone ring and esterifying its primary alcoholic group with pyrophosphoric acid. MVA-lactone is by far the most used precursor for selective incorporation experiments on biosynthesis of the terpenes and is commercially available in many different [3]H- and ^{14}C-specifically labelled forms.

Only the R form is utilised by organisms for producing terpenes, whilst the S form is metabolically inert. This is fortunate since the optical resolution of the racemate obtained by synthesis (at least eleven different synthetic routes have been described) is extremely difficult.

After formation of the 3-phosphate ester the 5-pyrophosphate of mevalonic acid (8) undergoes decarboxylation, induced by the incipient formation of a tertiary carbocation at position 3 formed by loss of this phosphate group. The mechanism schematised in Fig. 5.4 can be considered as an E2 elimination reaction with antiperiplanar leaving groups. (See Appendices 1 and 2). This process agrees with the following experimental facts:
(1) the enzyme catalysing the decarboxylation step requires ATP;
(2) the oxygen originally present at C-3 cannot be found in the aqueous medium but in phosphate ions (Pi) (experiments with ^{18}O);
(3) no hydrogen atoms from the medium are incorporated into the reaction product.

IPP can undergo an isomerisation into DMAPP. The consequence of this isomerisation is to transform a relatively unreactive substance into a reactive molecule (see formula (9)) capable of attacking nucleophilic species, such as a double bond. This reactivity is fully exploited in combining with other C_5 units and prenylation reactions.

The isomerisation of IPP into DMAPP is one of the few reversible reactions observed in terpene biosynthesis. Mechanistically the prototropic shift of the double bond in the isoprenic skeleton implies the loss of, and appearance of, the C-4 prochiral centre of IPP. As far as the two enantiotopic hydrogens at these positions are concerned the enzyme involved selectively removes H_D (4-*pro-S*). The reprotonation takes place from the *re,re*- face of IPP (Fig. 5.5). (Also see Appendix 2). These steric aspects of DMAPP formation were brilliantly confirmed by Cornforth. In the reverse isomerisation, only the *pro-E* methyl group of DMAPP is involved in the proton transfer. The hydrogen atoms originally present at position 2 of mevalonic acid, initially well distinguishable, become scrambled between positions 2 and 4 in the pyrophosphate esters, IPP and DMAPP, as the isomerisation between them occurs, i.e. between 4-*pro-S* (HD), 2-*pro-S* (HB) and 2-*pro-R* (HA). This scrambling explains why a certain amount of tritium randomisation is observed when $(2R)$-2-$^3H_1$- and $(2S)$-2-^{3}H-mevalonic acids are incorporated into carotenes in the tomato plant. In most cases, however, the DMAPP formed by IPP is quickly utilised *in vivo* and so limit such complications in interpreting the data from the incorporation of stereospecifically C(2)-tritiated mevalonic acid. Whilst isopentenylpyrophosphate and dimethyl-allyl-pyrophosphate are ubiquitous compounds in nature, the corresponding alcohols are rarely found. They can be considered as the principal members of the hemiterpenoid group. It is to be noted that certain compounds with an isoprenic skeleton of non-mevalonic acid origin, such as tiglic and angelic acid, (see p. 237) must be excluded from this very small group of terpenes.

5.2 COMBINATION OF C₅ UNITS

A DMAPP molecule can condense in a head-to-tail manner with IPP to produce geranyl pyrophosphate (Fig. 5.6). This type of reaction can be repeated, by further reaction of the product with IPP, and a series of pyrophosphate esters of aliphatic alcohols is obtained. Such systems are called prenylogues, since the dimethylallyl radical is known as the prenyl group.

DMAPP thus acts as the foundation stone upon which are added the building bricks of IPP units. Such additions are possible since the product obtained from each prior C_5-addition has the same tail structure and thus the same reactivity as DMAPP.

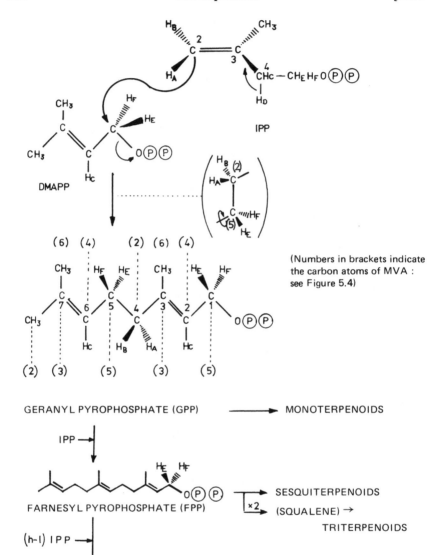

(Numbers in brackets indicate
the carbon atoms of MVA :
see Figure 5.4)

GERANYL PYROPHOSPHATE (GPP) ─────────► MONOTERPENOIDS

FARNESYL PYROPHOSPHATE (FPP) ┌─► SESQUITERPENOIDS
                              │ ×2
                              └─► (SQUALENE) →
                                       TRITERPENOIDS

                                   n=2 ─► DITERPENOIDS
                                    │ ×2
                                    └─► (PHYTOENE) →
                                           CAROTENOIDS

Fig. 5.6 Assembly of isoprenes.   n=3 ─► SESTERTERPENOIDS

n = 2 : GERANYL–GERANYL PYROPHOSPHATE (GGPP)

n = 3 : GERANYL–FARNESYL PYROPHOSPHATE (GFPP)

(a)

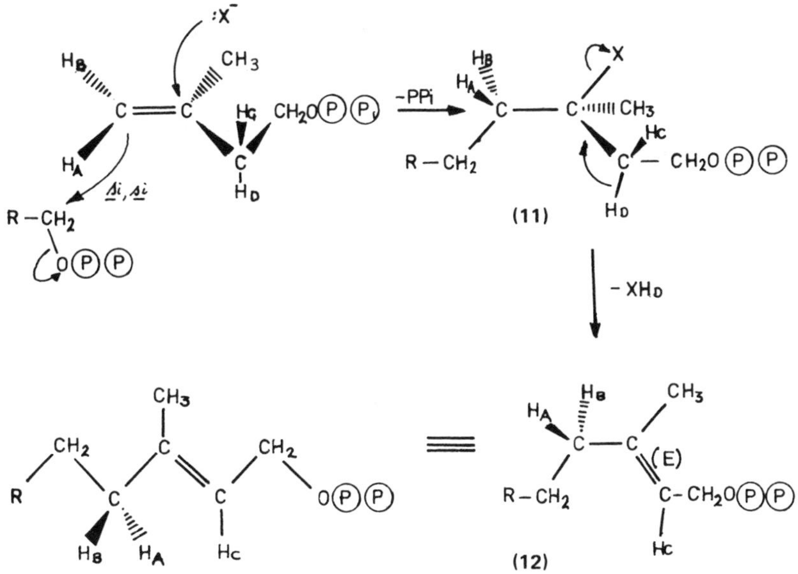

(b)

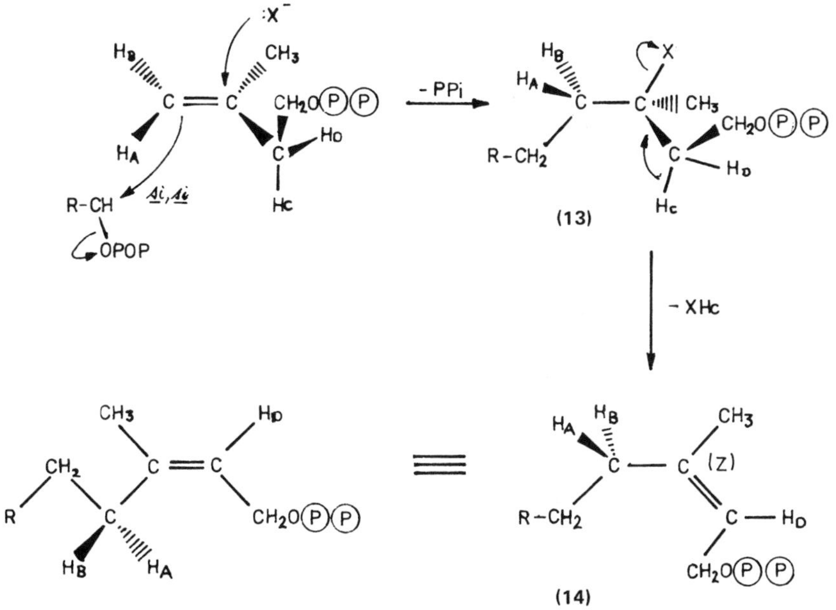

[H_C = **4—*pro-R*** of MVA; H_D = **4—*pro-S*** od MVA; x⁻ = nucleophile]

Fig. 5.7 The two mechanisms proposed by Cornforth for the coupling of isoprenyl units.

In the combination of IPP with DMAPP, which is enzymatically controlled, the C-5 carbon atom of geranyl pyrophosphate, corresponding to the allylic alcohol group of DMAPP, undergoes inversion of configuration in agreement with a formal S_N2 mechanism. The configurations of the product geranyl ester at C-4 and the 2,3 E – double bond (see Fig. 5.6) are accompanied by the retention of H_C pro-R-H from C-4 of mevalonic acid in IPP. These steric results are not consistent with a concerted Ad_E–E2 process following the initial displacement of the phosphate group. If such a concerted reaction took place, the proton leaving IPP at C-4 should be Hc, following attack of the incipient allylic carbocation on the si, si face of the IPP double bond. (Compare the analogous concerted mechanism postulated for the isomerisation of IPP to DMAPP; Fig. 5.4) Cornforth suggested the two step mechanism, represented in Fig. 5.7, to explain the observed result. The first step is an antiperiplanar addition, in which X is an unknown nucleophile, for example a possible -SH group of the enzyme, and the second one is a further antiplanar elimination, in which X acts as a leaving group.

In nature various types of prenyl transferases exist, each of them able to catalyse one or more of the reactions illustrated in Fig. 5.6. By use of such enzymes cells can discriminate between the synthesis of pyrophosphate esters with different numbers of carbon atoms. For example, one prenyl transferase, isolated from the liver, could promote the formation of either geranylpyrophosphate (GPP) or farnesylpyrophosphate (FPP) but not of higher prenylogues, probably because its active site could not accept allylic reagents larger than GPP.

The names of the free pyrophosphate esters formed from the isoprene units are geranyl pyrophosphate, farnesyl pyrophosphate, geranylgeranyl pyrophosphate (GGPP) and geranylfarnesyl pyrophosphate (GFPP), and are obligatory intermediates for synthesis of all the terpenes and carotenes. The structural variety of the mono to sesterpenes arises from elaborations (cyclisations, etc.) of these five open-chain intermediates. The higher terpenes, the triterpenes (C_{30}) and carotenes (C_{40}) derive instead from two hydrocarbons, squalene and phytoene, respectively produced from FPP and GGPP through complicated tail-to-tail dimerisation processes.

The polymerisation of isoprene chains above C_{25} (Fig. 5.6) is a rather common process in nature. It leads to the formation of the natural rubbers.

5.3 MONOTERPENES
The fundamental skeletons of monoterpenes can be grouped into four categories: (a) open chains, (b) cyclohexanes (mono- and polycyclic), (c) cyclopentanes (the iridoids and secoiridoids), (d) irregular systems.
5.3.1 Open Chain Monoterpenes
Some examples of monoterpenes are pictured in Fig. 5.8. They derive from GPP by hydrolysis, isomerisation, reduction, oxidation and dehydrations, etc. and these reactions do not change the number of carbon atoms.

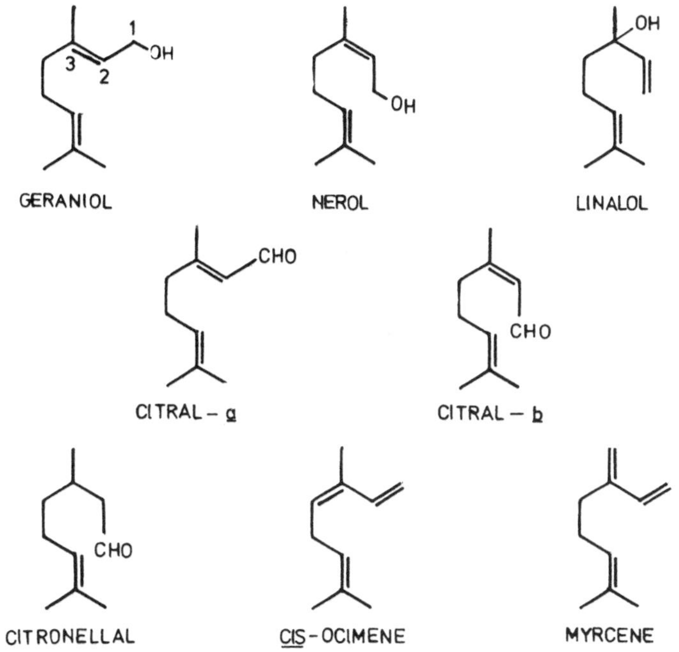

Fig. 5.8 Principal acyclic monoterpenes.

From a biogenetic point of view the *in vivo* interconversion processes of the three alcohols geraniol, nerol, linalool are particularly important. The plant uses these materials for the production of more complex cyclic monoterpenes. The allylic isomerisations of geraniol into linalool, and of linalool into nerol, probably take place via their pyrophosphate esters. They can be considered as anionotropic reactions, with the pyrophosphate anion undergoing either intramolecular or intermolecular transposition (Fig. 5.9). Because of restricted rotation about the C-2, C-3 bond in the allylic carbocations, (15) and (17), the formation of linalyl pyrophosphate (LPP) or of an intermediate (16) bound to an enzyme must be a necessary step in interconverting GPP and NPP (Fig. 5.9).

Another hypothesis for the formation of NPP is by a stereoselective coupling process involving DMAPP and IPP. If this were so, the mechanism would be similar to the one implied in the biosynthesis of India rubber (Fig. 5.7(b)), in which a double bond in the Z-configuration is formed. This latter hypothesis appeared to be in contrast with successive incorporation experiments of (3R, 4S)-[2-^{14}C, 4-^{3}H] MVA and (3R, 4R)-[2-^{14}C, 4-^{3}H] MVA in rose petals; the first precursor gave geraniol and nerol with complete loss of tritium in both alcohols, in good agreement with mechanism a, but not with mechanism b (Fig. 5.9). It should nevertheless be noticed that the direct formation of the terminal double bond having a Z-configuration is not necessarily excluded by the latter experi-

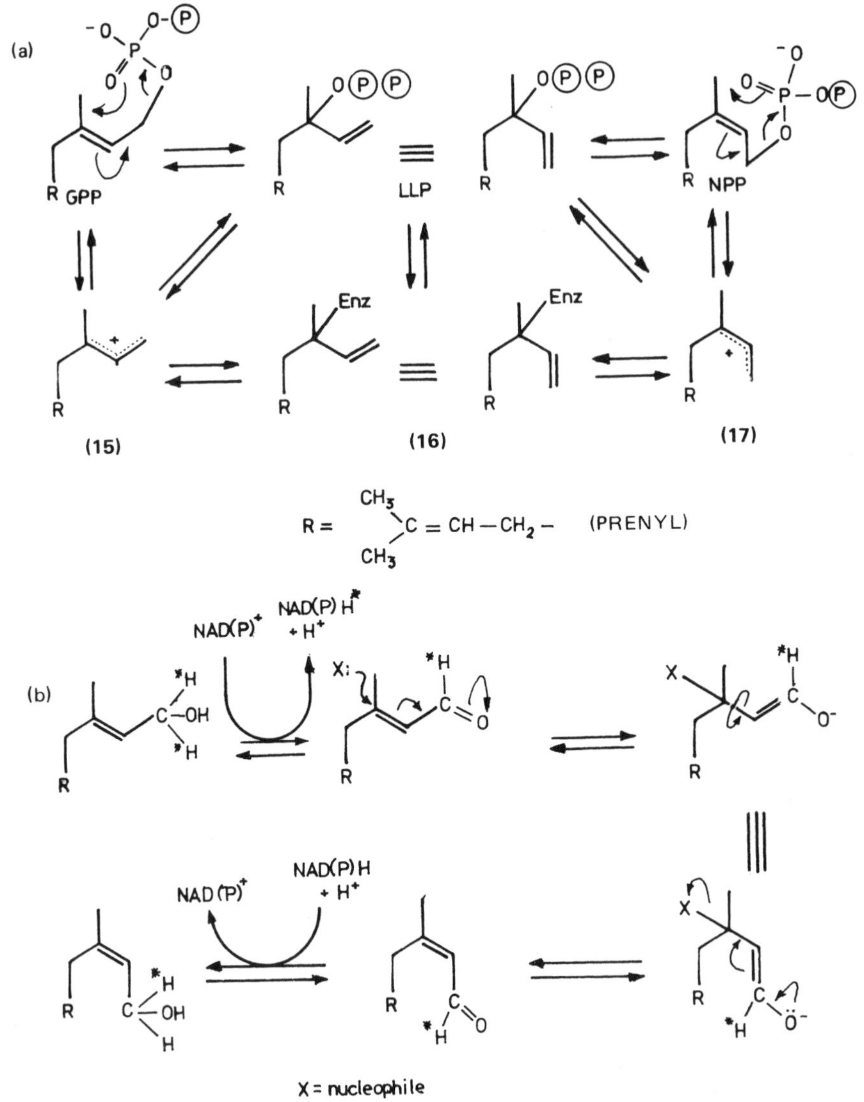

Fig. 5.9 Possible mechanisms for the interconversions of geraniol, linalool and nerol.

mental results: the retention of the 4-*pro-R* hydrogen of mevalonic acid can be explained by assuming that plants either use stereochemically different prenyl-transferases or enzymes with two mechanistically different active sites for assembling the C_5-units, in order to give isomeric C_{10} and C_{15} compounds. In recent work with a cellular preparation either of *Trichothecium roseum* or tissue from *Andrographis paniculata* the loss of one tritium atom out of the six was observed when incorporating $[2\text{-}^{14}C, 5\text{-}^{3}H_2]$ MVA into $(2Z, 6E)$ farnesol (see Fig. 5.23; (57), with H instead of OPOP), more commonly called cis-farnesol, which is an acyclic sesquiterpene alcohol, prenylogous to nerol. For the isomer $(2E, 6E)$-farnesol (58), (with H instead of OPOP), more commonly called trans-farnesol), which was isolated together with its isomer, a full retention of tritium was observed. Subsequently it was demonstrated that, on interconnecting trans-into cis- farnesol, using soluble enzymes of Andrographis paniculata, one of the two hydrogen atoms of the primary alcoholic group at C-1 is exchanged with the medium. These results add plausibility to the sequence for the observed geometrical isomerisations (Fig. 5.9(b)).

5.3.2 Cyclohexane Monoterpenes

Figure 5.10 presents a biogenetic scheme in which all the fundamental skeletons of the cyclohexyl monoterpenes are related according to mechanistically likely routes.* These biosynthetic paths diverge from a unique cation (18), originating either by cyclisation of cation (17) or directly from NPP or LPP through a concerted mechanism (S_N2 at C-1 with OPOP$^-$ as a leaving group). Apart from rare exceptions, it cannot be said whether cations such as (17) and (18) really exist in vivo, or merely represent the biogenetic equivalent of other active forms, such as the pyrophosphate esters, thiol esters or thiol ethers. Their discrete existence is not required for concerted reaction processes. The biogenetic picture of Fig. 5.10, although biosynthetically and chemically plausible, needs more experimental data to place it on a solid foundation. Alternative biosynthetic routes have been hypothesised for certain monoterpenes (Fig. 5.11). The final choice can only be made if data from previous experiments with labelled precursors are available. Unfortunately, incorporation levels of mevalonic acid and related precursors are generally low ($< 0.01\%$) for this group of monoterpenes, which often precludes the collection of such data. A peculiarity observed in the biosynthesis of the non-cyclopentane monoterpenes is the remarkable asymmetry of labelling regularly noticed in products obtained after feeding leaves and stems (but not petals!) with $[2\text{-}^{14}C]$ MVA (Fig. 5.12). The labelling is localised almost exclusively on only one carbon atom of the ring (the carbon corresponding to the C(2) of IPP) with no labelling at the two methyls of the isopropyl group, which arise from isoprenyl unit corresponding to DMAPP. Three possible explanations can be given for this asymmetry: (a) a DMAPP pool exists in plants, and it reacts with the IPP obtained from MVA before the IPP isomerises into DMAPP; (b) the DMAPP might originate from molecules other than from MVA; (c) compartmentation effects can take place, so that $[^{14}C]$ IPP, but neither $[^{14}C]$ MVA nor $[^{14}C]$ DMAPP can pass through the membrane of the cell containing the biosynthetic site for monoterpene formation.

*The classical or non-classical carbocations (see Appendix 3), pictured in Fig. 5.10 and in all the following biosynthetic schemes of terpenoids, should not be interpreted as discrete chemical species, but rather as formal intermediates, useful for rationalising the evolution of a certain skeleton along more or less branched pathways.

Cyclohexane and acyclic monoterpenes are especially present in the oleipher glands of leaves, and can also be found in the low boiling fractions of essential oils.

MONOTERPENES (SKELETA)

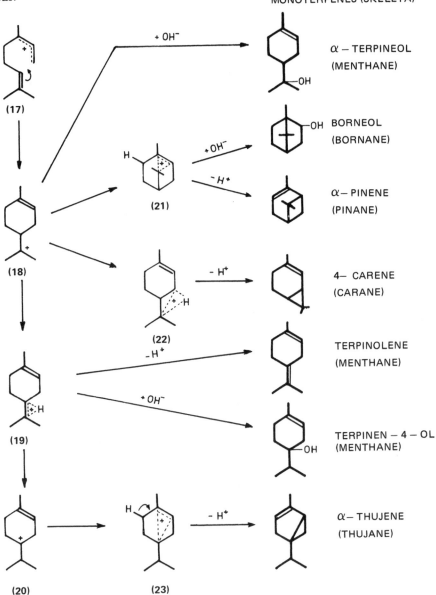

Fig. 5.10 Biogenesis of the cyclohexyl monoterpenes.

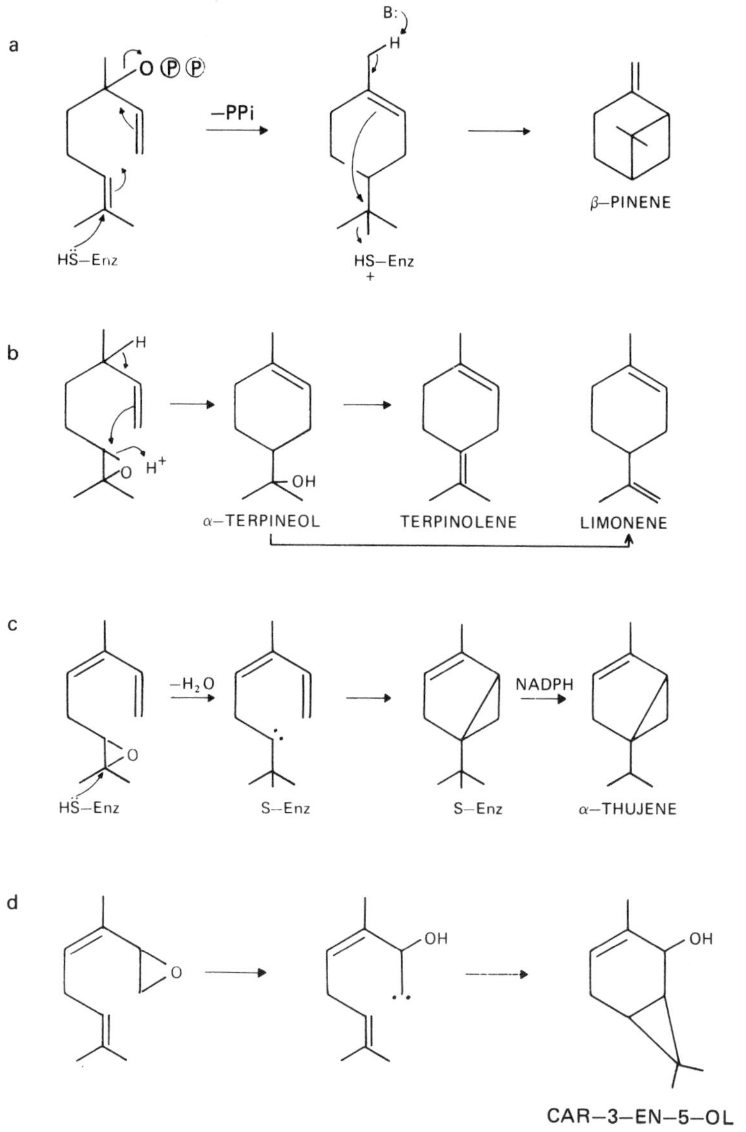

Fig. 5.11 Possible biosynthetic mechanisms for the function of some monoterpenes.

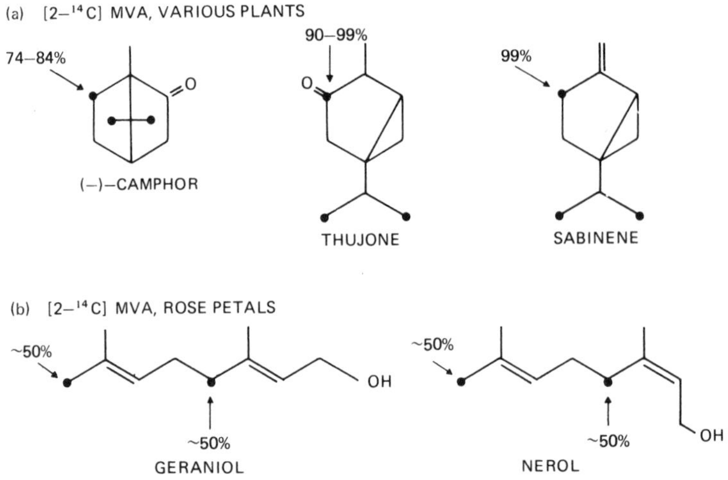

(a) [2–¹⁴C] MVA, VARIOUS PLANTS

(–)-CAMPHOR

THUJONE   SABINENE

(b) [2–¹⁴C] MVA, ROSE PETALS

GERANIOL   NEROL

Fig. 5.12 Distribution of radioactivity in monoterpene molecules after incorporation of [2-¹⁴] mevalonic acid. The symbol • indicates the position of the label.

5.3.3 Iridoids and Seco-Iridoids

The subgroup of the iridoids comprises of several hundreds of monoterpenes, all with a cyclopentane skeleton of the type (24). In most cases a hemiacetal bridge links positions 1 and 9, forming an α-hydroxytetrahydro-, or dihydro-, pyrane ring, condensed to the cyclopentane ring, whilst the hydroxyl group at position 1 is conjugated to a glucose molecule, e.g. (25).

Thus the iridoids are almost invariably found in nature as 1-D-glucosides, soluble in water. The cyclohexane monoterpenes, being insoluble in water, are accumulated in special oleipher glands of leaves, whilst iridoids are largely diffused in plant tissues (leaves, seeds, bark and roots etc.), especially in the dicotyledons. This difference in distribution is accompanied by much higher incorporation yields of MVA into the iridoids compared to that into the cyclo-hexyl monterpenes.

The name 'iridoid' originates from the fact that the first members (iridomyrmecin and anisomorphal, Fig. 5.13) were first isolated from the secretions of ants of the genes *Iridomyrmex*.

The biosynthesis of the iridoids has attracted the attention of many research teams for two main reasons: first, because good results can be obtained on incorporating labelled precursors; second, because of the close biogenetic relation-ship between the iridoids and many indole and isoquinoline alkaloids, in which 10 (or 9) carbon atoms of the iridoid skeleton are bound.

An important place amongst the iridoids is held by loganin, which is the commonest precursor of the so-called secoiridoids, for instance of sweroside (Fig, 5.13), and of the terpene portion of the above-mentioned alkaloids.

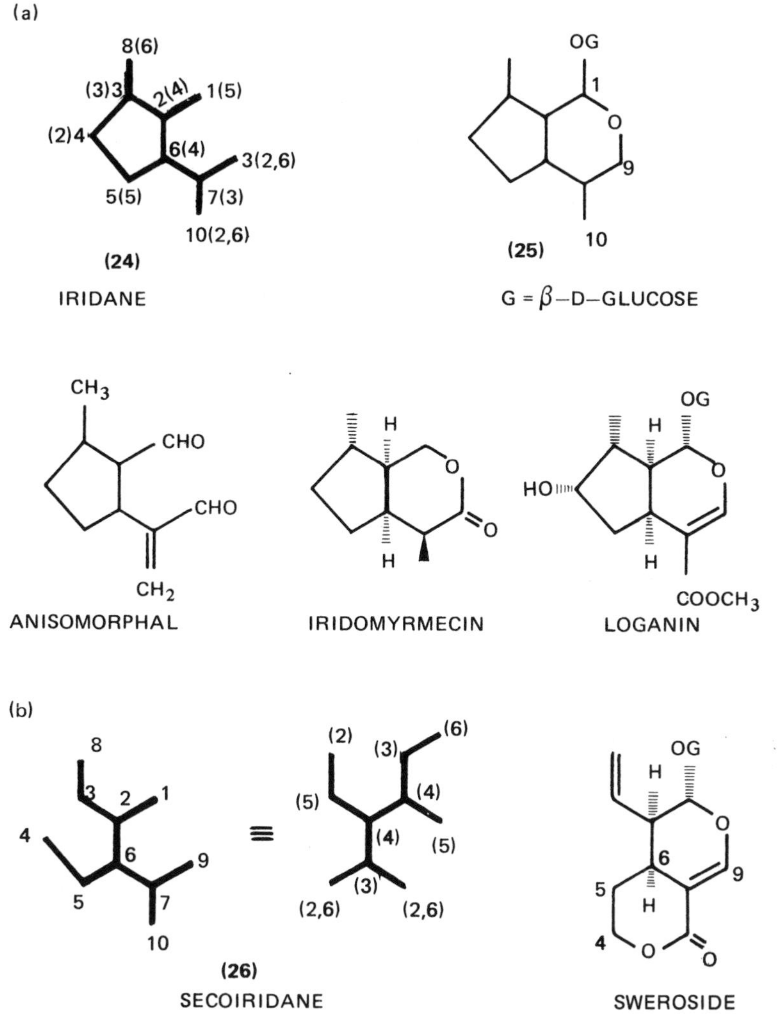

(a)

(24)

IRIDANE

(25)

G = β—D—GLUCOSE

CH₃

ANISOMORPHAL   IRIDOMYRMECIN   LOGANIN

(b)

(26)

SECOIRIDANE   SWEROSIDE

Fig. 5.13 (a) Iridoids and (b) secoiridoids. The numbers in parentheses indicate the carbon atom numbering in the starting molecules of MVA.

Biosynthetic studies carried out up to the present indicate that the most probable sequence from nerol to loganin is as shown in Fig. 5.14.

The identification in loganin of the hydrogen atoms originally carried by geraniol and MVA has been performed using (2R)-, (2S)-, (4R)-, (4S)- and (5R)-[³H]MVA as labelled precursors. Nerol has been hypothesised as an intermediate, since 10-hydroxynerol (27) is a more efficient precursor than 10-hydroxygeraniol.

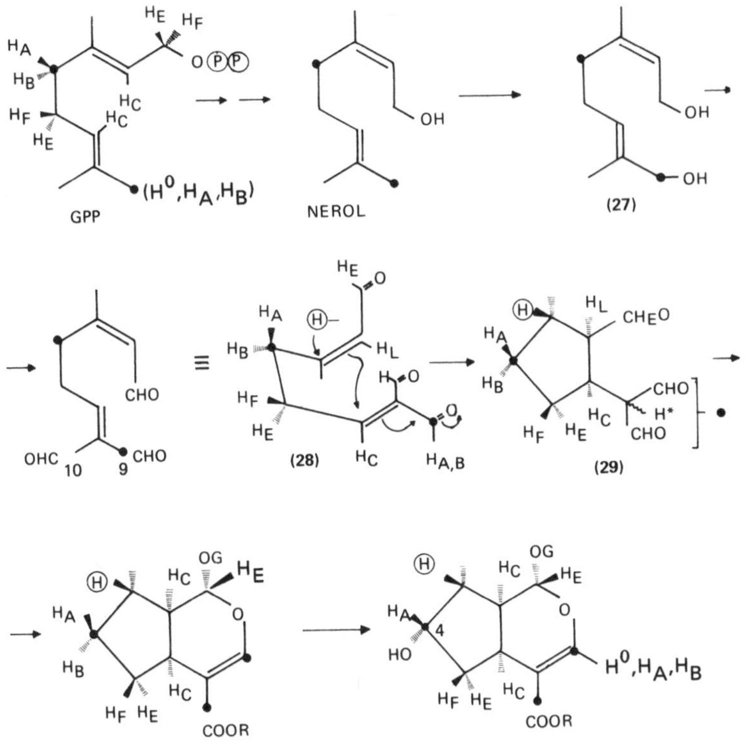

(30, R = H)
(31, R = CH_3 : from METHIONINE)

(32, R = H) LOGANIC ACID
(33, R = CH_3) LOGANIN

(•) = C–2 of MVA
H_A, H_B, H_C, H_E, H_F : see Fig 5.4 and 5.6
H* : from the medium
Ⓗ : from NAD(P)H?

Fig. 5.14 Biosynthesis of loganin.

D. Arigoni postulated the formation of the trialdehyde (28) to explain, amongst other facts, the randomisation of the label between positions 9 and 10 of loganin, obtained from *Vinca rosea* after feeding experiments with [2-^{14}C] mevalonic acid. A label localised on carbon 9 has been found in some iridoids, or in compounds strictly correlated to them, as in verbenaline or skytanthine (Fig. 5.15).

Figure 5.15 illustrates the distribution of labelling in verbenaline and skytanthine resulting from incorporation experiments with [2-^{14}C] mevalonic acid in plants of different age. Two points arise from such data. The first is that, in mature plants, the five-membered ring is probably formed from an acyclic monoterpene, which has different functions at the two terminal carbons, e.g. C-9 as an aldehyde and C-10 as an acid in (28). An enzyme might also be able to distinguish between two aldehyde functions in the non-equivalent positions at 9 and 10 (see Appendix 2). The second point is that, in the iridoids which have incorporated [2-^{14}C] mevalonic acid, there is no asymmetric labelling observed between positions of the molecule deriving from IPP and DMAPP, whilst such an asymmetry is generally observed for the cyclohexane-type monoterpenes.

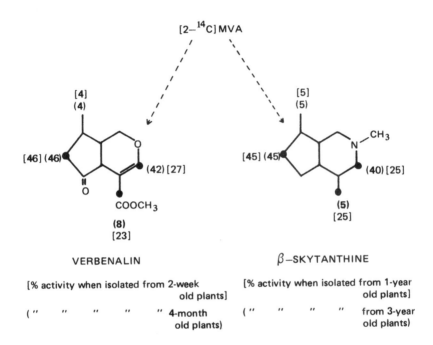

Fig. 5.15 Distribution of radiolabels in iridoids, highlighting the degree of variation observed after feeding with [2-^{14}C] mevalonic acid. The symbol ● indicates the position derived from carbon atom C-2 of MVA.

In loganic acid (32), isolated after feeding *Swertia caroliniensis* with [2-^{14}C, 2-^{3}H$_2$] MVA, a slightly smaller ^{3}H/^{14}C ratio was found at C-4 than one should expect according to the scheme of Fig. 5.14. This deviation, of the same order of magnitude as that found in the biosynthesis of steroids, can be attributed to use of DMAPP by the prenyl transferases, which competes kinetically with the dynamic equilibration between DMAPP and IPP catalysed by IPP isomerase (see p. 220).

The skeleton of the secoiridoids (26) is formally derived from the iridoids (24) by breaking the C-3–C-4 bond. The exact mechanism by which loganin is converted into secologanin (35) is not known; Fig. 5.16 illustrates one possibility.

LOGANIN (34)

(35)

SECOLOGANIN GENTIOPICROSIDE

X = O(P)(P)?

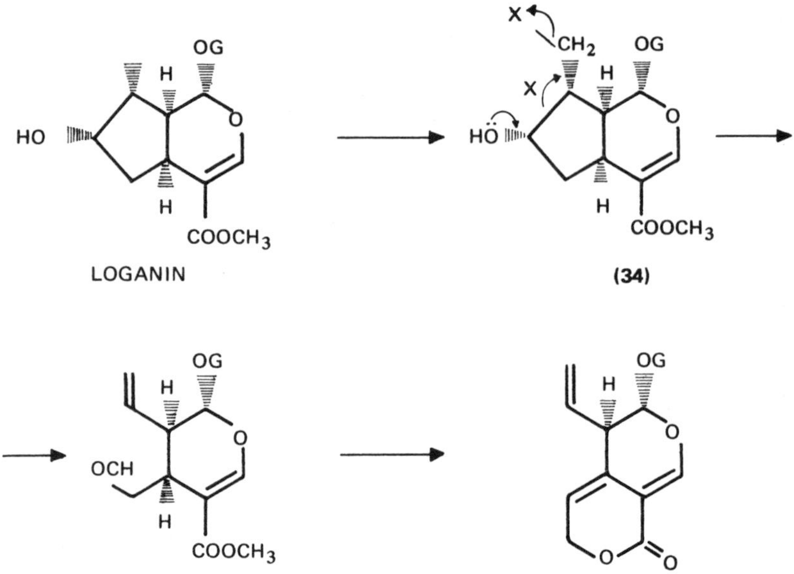

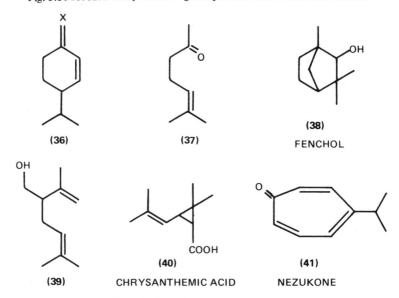

Fig. 5.16 Probable biosynthesis of gentiopicroside and related secoiridoids.

(36)

(37)

(38)
FENCHOL

(39)
LAVANDULOL

(40)
CHRYSANTHEMIC ACID

(41)
NEZUKONE

Fig. 5.17 Some irregular monoterpenes.

5.3.4 Irregular, Apparent and False Monoterpenes

Examples of irregular monoterpenes are illustrated in Fig. 5.17. Those with a skeleton of less than ten carbon atoms, such as cryptone (**36**, X = 0) or 2-methyhept-2-en-6-one (**37**), are likely to arise from regular terpene precursors through degradative processes. Cryptone almost certainly derives from the *in vivo* oxidation of β-phellandrene (**36**, X = CH₂), and ketone (**37**), which is a defence substance from certain insects, very likely derives from the oxidative cleavage of the C-2–C-3 bond of an acyclic monoterpene.

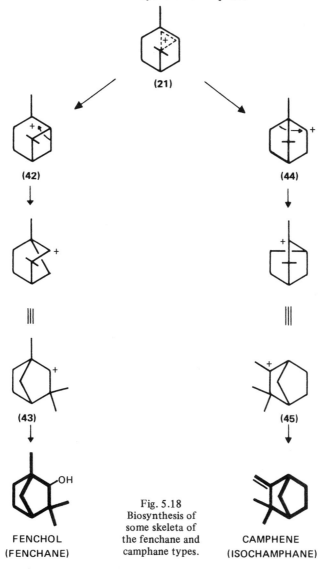

Fig. 5.18
Biosynthesis of
some skeleta of
the fenchane and
camphane types.

It is more difficult to find a biogenetic explanation for the structures of compounds **(38)-(41)**; they contain ten carbon atoms but they do not follow the
isoprene rule. Some probable hypotheses about their origins are schematised, in
Fig. 5.18 for the isocamphane and fenchane structures and, in Fig. 5.19, for
artemisia ketone, lavandulol, santolina triene, and chrysanthemic acid. The
hypotheses summarised in Fig. 5.19, involving the formation of cyclopropane
intermediates of type **(47)**, are supported by some incorporation data of [2-^{14}C]
MVA into artemisia ketone (in *Artemisia annua*) and into chrysanthemic acid (in

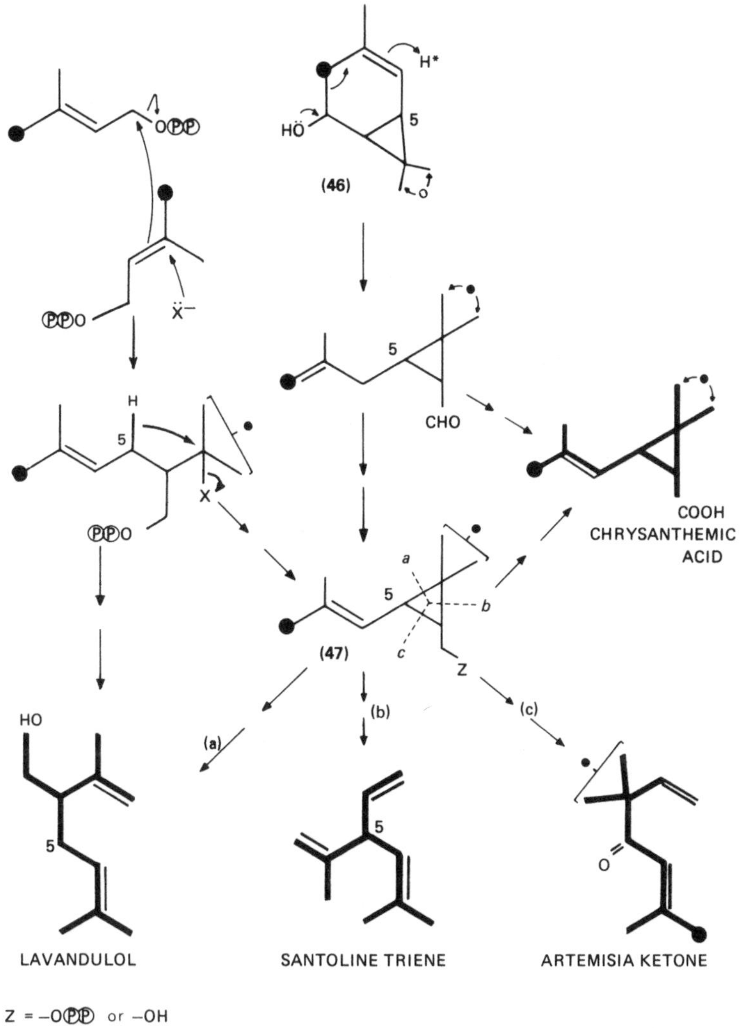

Fig. 5.19 Possible biosynthetic routes to the irregular monoterpenes. The symbol
• indicates the position of the label derived from [2-^{14}C] MVA.

Chrysanthemum cinerariaefolium). Such intermediates could arise by breaking the C-2-C-3 bond of a carene type skeleton, for instance (46), or by the attack of an isopentenylidene carbene onto the double bond of a DMAPP molecule, by a mechanism analogous to the one operating in presqualene formation (see Fig. 5.55).

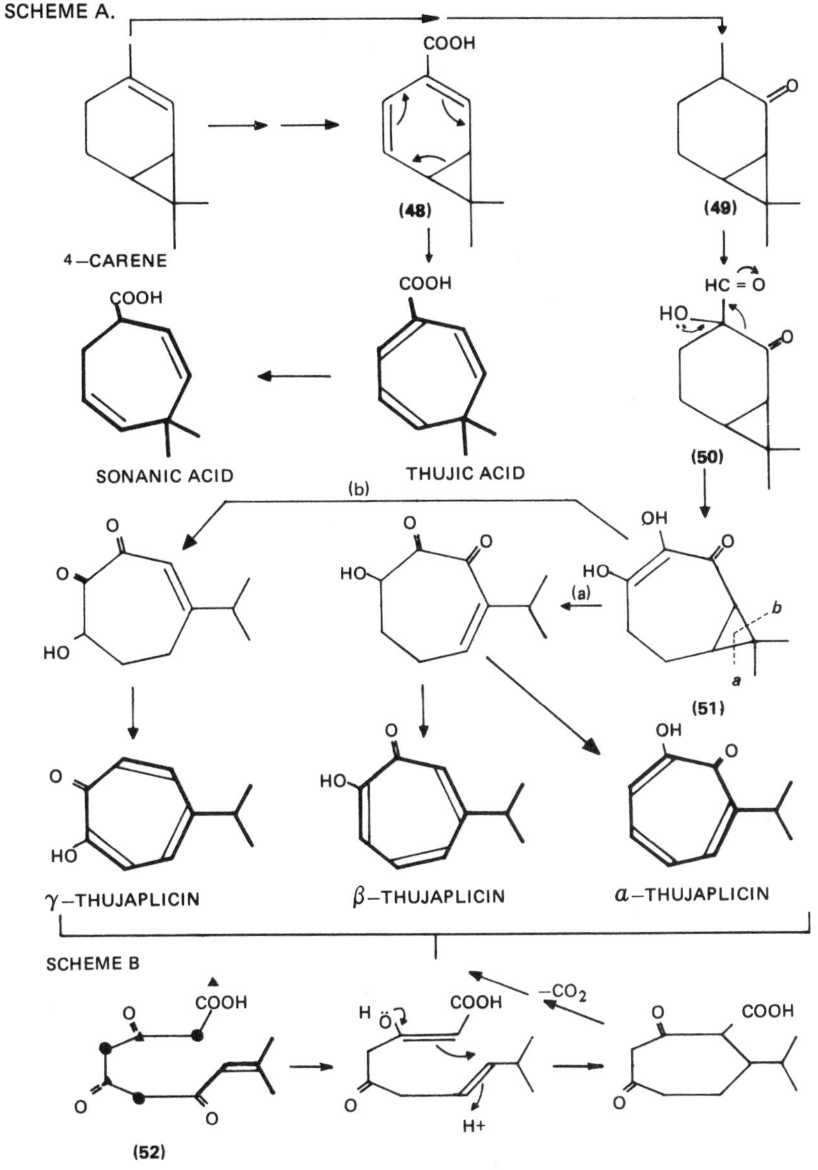

SCHEME A.

Fig. 5.20 Possible routes for the formation of cycloheptyl monoterpenes.

The compounds shown in Fig. 5.20 (isolated from the wood of numerous Gymnospermae) are commonly considered monoterpenoids, although no experimental evidence for their derivation from MVA exists. The presence of an isopropyl group or of two geminal methyl groups in such molecules cannot be considered sufficient evidence of a terpenic precursor. Other cycloheptane, or tropolone compounds, which are produced by many fungi (see p. 170) are known to have a polyketide origin. The terpene-derived tropones, tropolones and cycloheptane-carboxylic acids are often derived from 4-carene (scheme A, Fig. 5.20). Another biogenetic path can also be envisaged (scheme B), by growth of a polyketo-methylene chain (52) through C_2-units (malonyl CoA) using $\beta\beta$-dimethylacryloyl-5 CoA as a 'starter'.

The formation of monoterpenes with a 1,1,2,3-tetramethylcyclohexane skeleton (53–54) via a cyclisation of GPP of the type (55) is very unlikely (Fig. 5.21). The frequency of such cyclisations is related to the length of the linear precursor; it is infrequent with farnesol, more prevalent with geranylgeraniol, and very common in carotenoids, where a single cyclohexane ring is formed for each end of the chain. For this reason safranal (53) and its analogues are more likely to be degradation products of carotene molecules (see p. 287) and hence are only apparent monoterpenes. (Monoterpenes, by definition, derive from the union of two isoprene units).

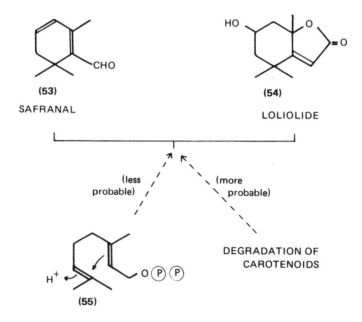

Fig. 5.21 Apparent monoterpenes (1,1,2,3-tetramethylcyclohexanes).

The 'false' monoterpene-nature of some compounds can be identified by incorporation experiments. The senecic acids, e.g. (56), have thus been shown to be 'false' monoterpenes. In nature they are combined through their carboxylic groups to the nitrogen nuclei of the pyrrolizidine alkaloids. These acids, as well as certain false hemiterpenes, such as tiglic and angelic acids, are in reality the metabolic products of certain aliphatic amino-acids (valine, leucine, and isoleucine) (Fig. 5.22).

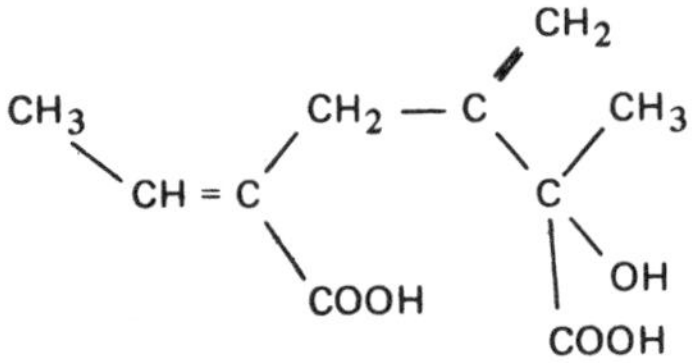

(56)

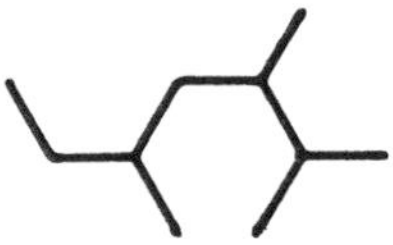

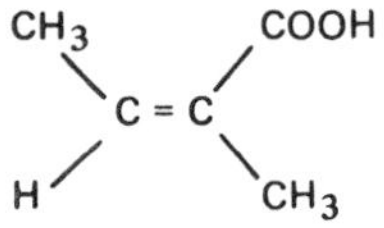

ANGELIC ACID

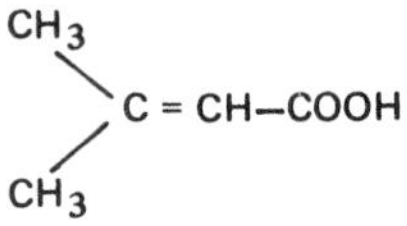

SENECIOIC ACID

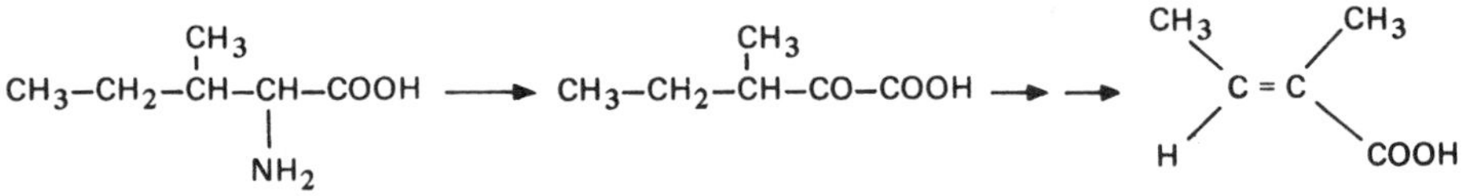

ISOLEUCINE

TIGLIC ACID

Fig. 5.22 False mono- and hemiterpenes.

5.4 SESQUITERPENES

The sesquiterpene family of compounds provide a seemingly inexhaustible supply of biogenetic problems and variations. A rich range of sophisticated reaction processes are utilised by nature, including the use of non-classical cations, molecular rearrangements, hydride ion or methyl group shifts and anti-Markownikoff additions! About 1,000 sesquiterpenes are known and these possess over 100 different carbon skeletons. They arise mainly from plants, although the lower animals (such as the coelenterates, molluscs and arthropods) and certain fungi, also contain members of this group of terpenes. The progress in the experimental knowledge of their biogenesis has hardly kept pace with the rate of their discovery. One reason for the rather slow elucidation of biogenetic patterns in this family of compounds is the difficulty encountered in incorporating labelled precursors when working with plants, for similar reasons to those explained for the monoterpenoids.

The carbon atom skeleta of almost all the known sesquiterpenoids can be derived from *trans*-farnesylpyrophosphate (58) and the *cis*-isomer (57), through appropriate cyclisations and rearrangements. *Cis*-FPP should derive from *trans*-FPP via (reversible) mechanisms very similar to the ones connecting GPP to NPP (Fig. 5.9). The schemes for sesquiterpene biogenesis, summarised in Figs. 5.23–5.37 has many similarities to those for the cyclohexane monoterpenes and comments made earlier about the interpretation of such schemes are also pertinent here.

The departure of the pyrophosphate anion from (57) and (58) leads to the six possible monocyclic cations (62)–(67) through the nonclassical cations (53) and (61) (Fig. 5.23). The stabilisation of such classical carbocations, either through loss of a proton from the adjacent carbon atoms, or attack of a hydroxide ion, produces compounds referred to as primary skeletal sesquiterpenes. Compounds of this group, as well as those modified by further functionalisation of the skeleton but not in the sequence of the carbon atoms or the configuration of the endocyclic bonds, form from the cations (62) (bisabolene group, e.g. γ-bisabolene); (66) (germacrane group, e.g. germacrone); (67) (humulane group, e.g. humulene). The fact that primary derivatives of the cations (63), (64) and (65) are not known probably relates to the ease with which they form further C–C bonds rearrangements, intramolecular additions, etc.

The cations (62)–(67) and their primary skeletal compounds can give rise to the other classes of sesquiterpenes (the secondary skeletal sesquiterpenes). Herein their description will be limited to the most common of these types and those that are of special interest from the biogenetic angle.

The biogenetic scheme proposed for the formation of bergamotane and of α- and β-santalane (Fig. 5.24) is related to those proposed for the biosynthesis of the monoterpene skeletons of pinane, bornane and isocamphane (with a dichotomy between the cation (68) and rearrangement of the cation (69), see Figs. 5.10 and 5.18).

γ-Bisabolene is considered to be the precur or to a number of sesquiterpene systems (Fig. 5.25), often all present in the same plant. For instance, in the essential oil from *Chamaecyparis taiwanensis* leaves, seventeen sesquiterpene constituents have been isolated: they belong to the subgroups of cuparane, cedrane, chamigrane, thujopsane, and widdrane. This fact provides circumstantial evidence supporting a close biogenetic relationship between these systems, as indicated in Fig. 5.25.

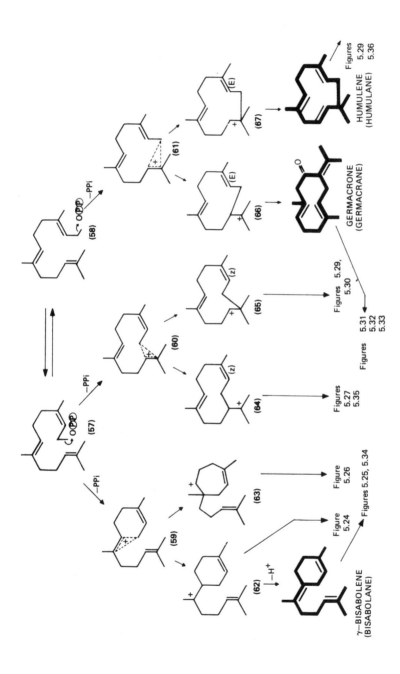

Fig. 5.23 Cyclisation of *cis*- and *trans*-farnesylpyrophosphate Sesquiterpene skeleta.

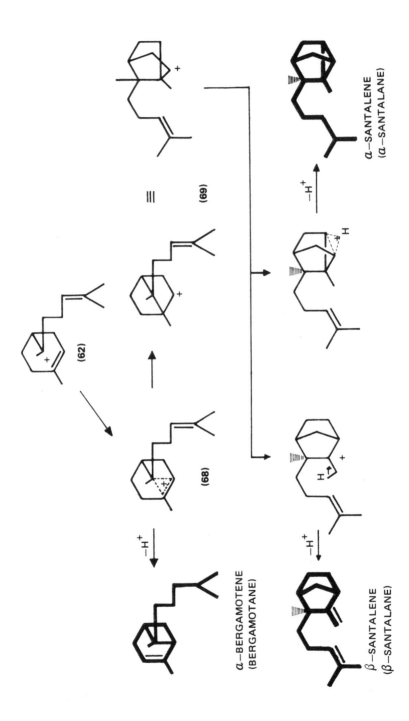

Fig. 5.24 Biosynthesis of some structures of the bergamotane, and α- and β-santalane types.

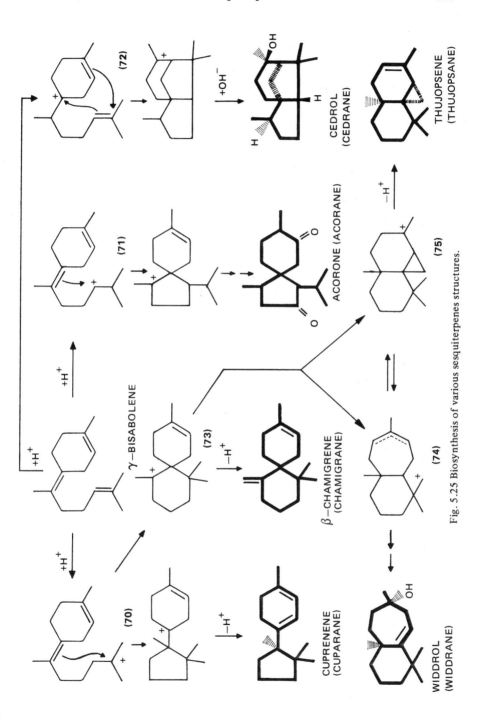

Fig. 5.25 Biosynthesis of various sesquiterpenes structures.

The skeleton of carotane (Fig. 5.26) is probably formed from cation (63), through an anti Markowninoff addition and a 1,3-hydride shift. The labelling pattern from [1-^{14}C] acetate is in agreement with this hypothesis.

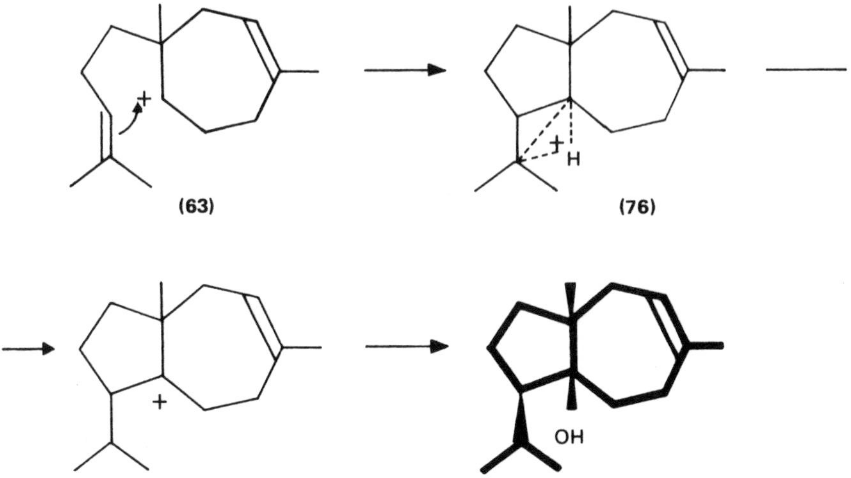

Fig. 5.26 Biosynthesis of the carotane skeleton.

The cation (64) presumably gives rise to both the sesquiterpenes of the cadinane subgroup and numerous related but rather complex structures such as picrotoxin, copacamphene, etc. (Fig. 5.27). The formation of the various structural units depends on the stereochemistry in the cyclisation process of (78), on the configuration at the carbon bearing the isopropyl group in (79), on the usual biosynthetic dichotomy in the intramolecular interaction between carbocations and double bonds in α-terpineyl systems (compare (18) in Fig. 5.10 to (79) in Fig. 5.27) and on the rearrangement of the bornyl carbocation into the isocamphane skeleton, (compare (44) in Fig. 5.18 with (80) in Fig. 5.27).

Compounds with the picrotoxane skeleton are extremely interesting for their very high toxicity. Turine and coriamyrtin (Fig. 5.28), present in *Coriaria* species of plants (Coriariaceae), are responsible for the death of large numbers of cattle in New Zealand. Their toxic effects are caused by an anomalous stimulation of the central nervous syctem, particularly of the respiratory, vasomotor and cardiac centres. Mellitoxin is a toxic constituent found in the honey produced by bees which visit the flowers of *Coriaria arborea* Lindsy. (In this particular case the hymenoptera do not digest mellitoxin directly from the plant but they often obtain it by feeding upon the sugary excretions of another insect, *Scolypopa australis,* which does feed from *Coriaria.*)

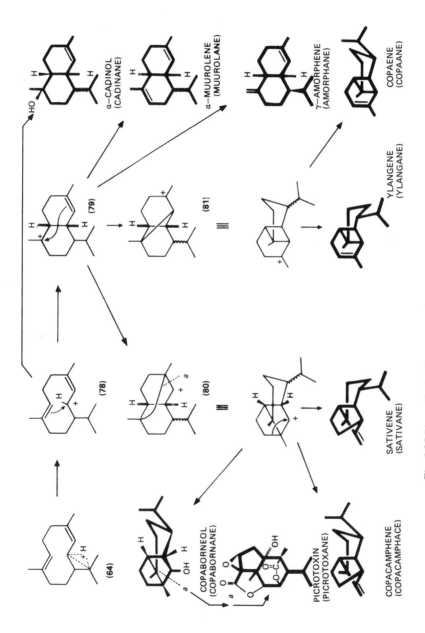

Fig. 5.27 Biosynthesis of some sesquiterpene structures.

Another very toxic sesquiterpene is gossypol, (Fig. 5.28), a yellow pigment contained in the seeds from cotton plants. It is derived from a cadinane-type intermediate, by aromatisation of the cyclohexane rings and oxidative dimerisation (compare the oxidative coupling of phenols).

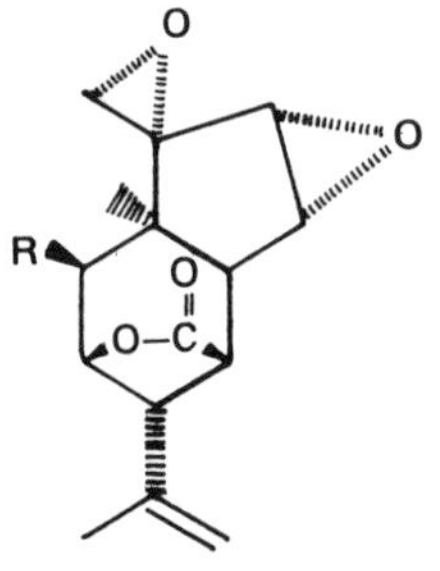

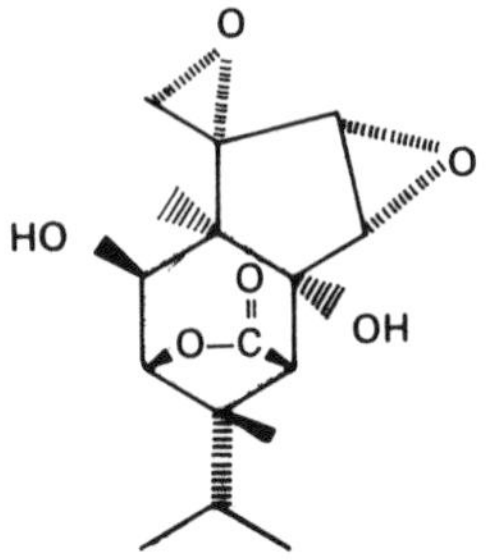

R = H, CORIAMYRTIN   MELLITOXIN
R = OH, TUTIN

GOSSYPOL

C_{15}-INTERMEDIATE $\xrightarrow{OX}$ GOSSYPOL

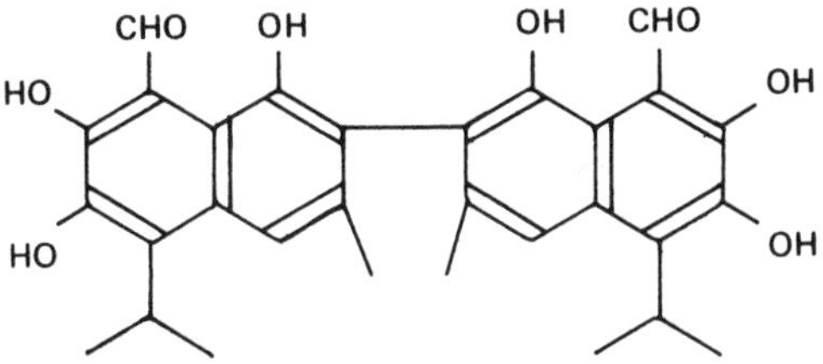

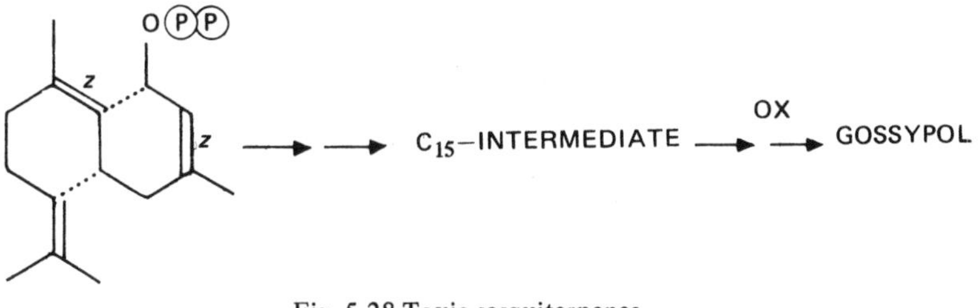

Fig. 5.28 Toxic sesquiterpenes.

2−cis−6−cis−FPP

There is a biogenetic relationship between humulene and caryophyllene (Fig. 5.29). The two hydrocarbons are often found together in essential oils and this should indicate a closer relationship than the one illustrated in Figs. 5.23 and 5.29; the same cation, (65) or (67), could produce both. Their stereochemical relationship is, however, more easily explained by assuming humulene, with all its double bonds in the E-configuration is derived from *trans*- and the caryophyllene arises from *cis*-farnesol. It has also been hypothesised that caryophyllene could form from humulene as indicated in Fig. 5.29 (via route (b)).

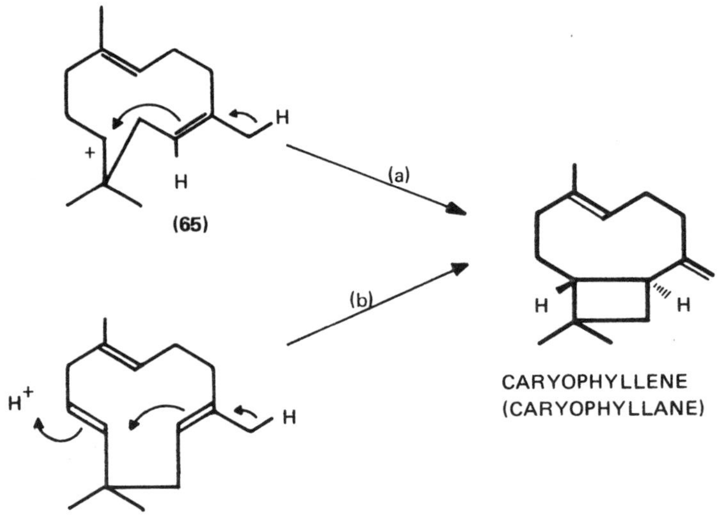

HUMULENE   Fig. 5.29 Possible routes for the biosynthesis of caryophyllene.

In Fig. 5.30 are illustrated some skeleta derived from the cation (65). The sesquiterpenoids of the germacrane subgroup are presumably formed from cation (66) (Fig. 5.23). If the double bonds of the macrocyclic ring were left out of consideration, they could also derive from the cation (64). The presence of both double bonds in the E-configuration for all of the cyclodecane sesquiterpenes indicates their probable production from *trans*-farnesylpyrophosphate (58). Germacradiene structures are not so important for their range and wide occurrence but because they play a fundamental role as intermediates for a variety of bicyclic structural types. The two double bonds are held in an appropriate position and configuration in the ten-membered ring to allow intramolecular electrophilic cyclisations, to yield bicyclic products. The final structure of the products depends on the intimate structure of and initial conformation adopted by the macrocyclic ring, taking into account the orientation of the isopropyl group and the stereoelectronic features of the cyclisation reaction (Figs. 5.31, 5.32, 5.33). Enantiomeric structures to the ones shown in the above figures are also known: their formation can be explained by assuming that the germacradiene intermediates are able to adopt enantiomeric conformations prior to cyclisation.

Figure 5.31 depicts some possible cyclisations of germacradiene intermediates in various conformations; the double bonds are always in the same positions and with the (*E*)-configuration typical of the primary cation (66). Electrophilic addition follows the Markownikoff rule for both double bonds. When on oxidative (OH$^+$) attack is required, it appears to follow the pattern set by opening the corresponding epoxide, with Markownikoff ring opening. Such 'epoxide' intermediates themselves originate by attack of an oxidase on the double bond (see also triterpene biosynthesis, p. 266).

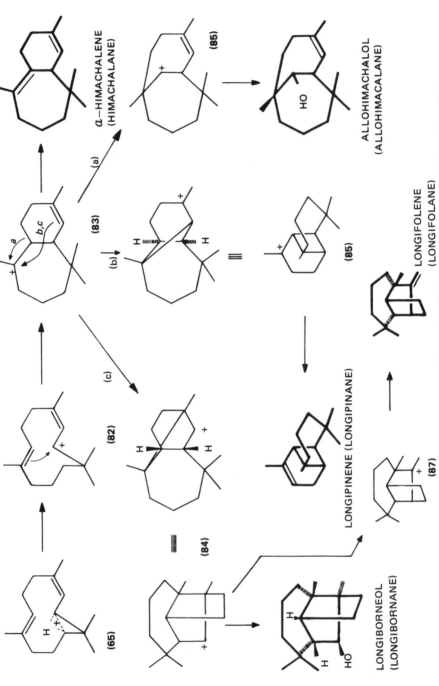

Fig. 5.30 Biosynthesis of the skeleta of the longibornane, longipinane, longifolane, himachalene and allohimachalane types.

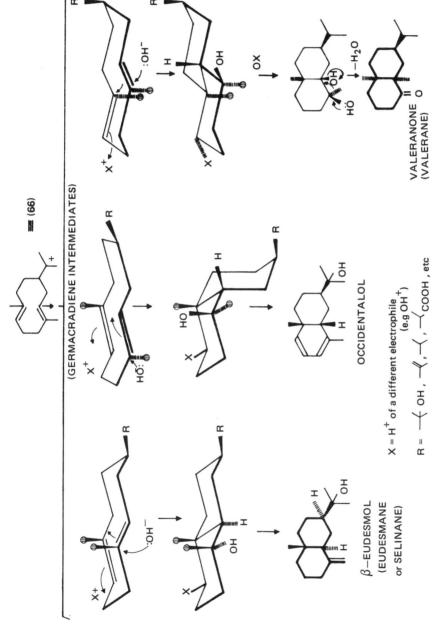

Fig. 5.31 Biosynthesis of some structures of the eudesmane and valerane types derived from germacradiene types.

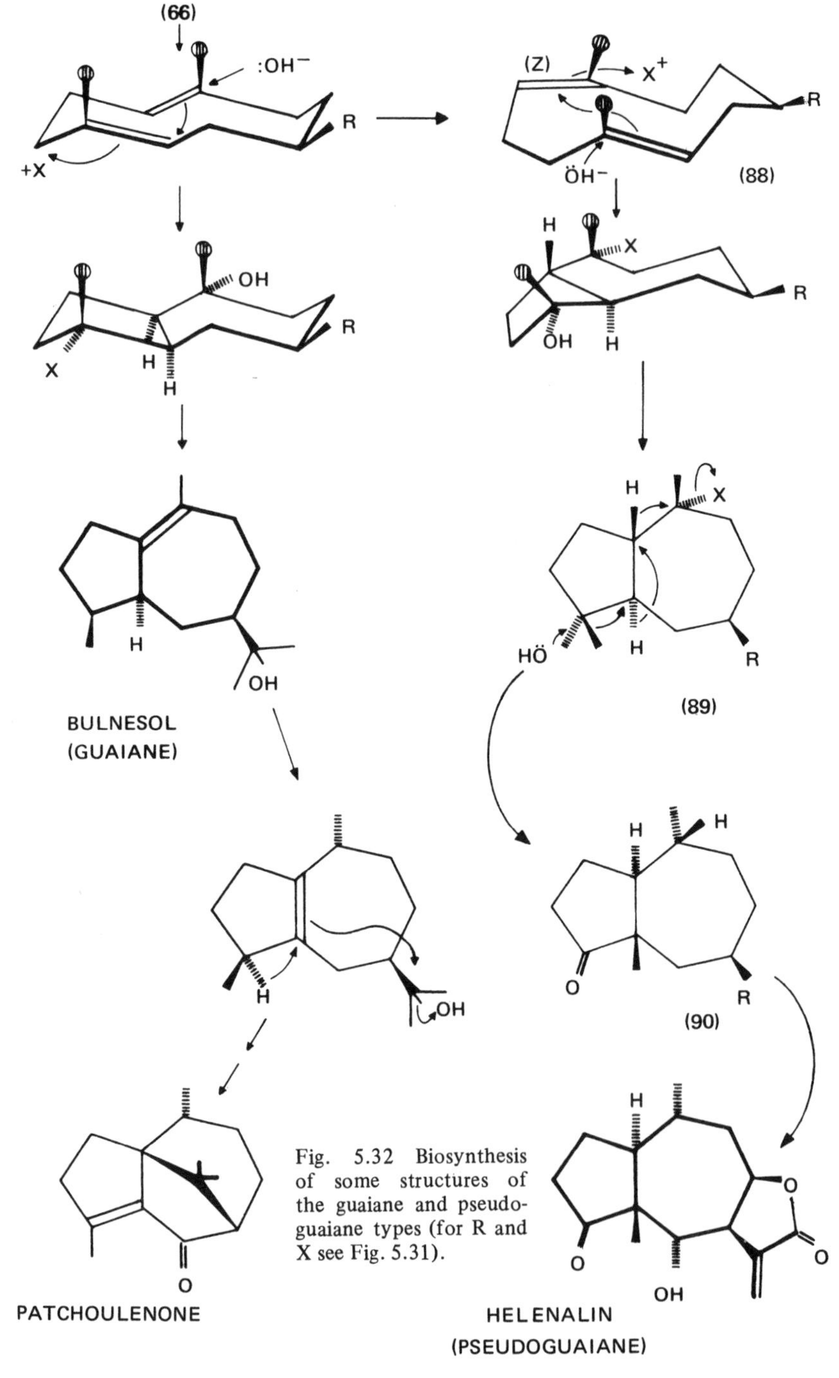
(66)
:OH⁻
R
+X
(Z)
X⁺
ÖH⁻
(88)
OH
X
H
BULNESOL
(GUAIANE)
HÖ
(89)
(90)
O
PATCHOULENONE
HELENALIN
(PSEUDOGUAIANE)

Fig. 5.32 Biosynthesis of some structures of the guaiane and pseudoguaiane types (for R and X see Fig. 5.31).

The structures of Fig. 5.32 originate by an anti-Markownikoff attack of electrophiles either on the usual germacradiene derivative, having both its double bonds in the (E)-configuration, or to an isomerisation product such as (88), having one double bond in the (Z)-configuration.

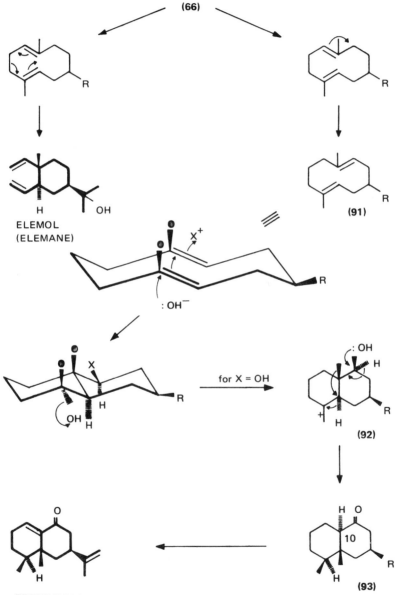

Fig. 5.33 Biosynthesis of some structures of the elemane and eremophilane types (for R and X see Fig. 5.31).

The biosynthesis of eremophilone (Fig. 5.33) requires the prior migration of one double bond of a primary germacradiene intermediate, forming the isomer (91), with both double bonds in the (E)-configuration. Subsequent cyclisation, induced by the usual electrophilic addition, then takes place (Markownikoff type on both double bonds).

5.4.1 Sesquiterpenes from Fungi
About thirty sesquiterpenes are known to be produced by fungi (Imperfect Fungi and Basidiomycetes). They mostly belong to the three subgroups of trichothecane, helminthosporane, and illudane. The biosynthesis of the first two structures, schematised in Fig. 5.34 and 5.35, agrees with the many incorporation experiments carried out using specifically labelled mevalonic acid. In vitro cultures of microorganisms generally prove more efficient than plants for incorporation experiments.

(80)
(see Figure 5.27)

HELMINTHOSPORAL
(Helmintosporium sativum)

$[2-{}^{14}C]$ MVA : ●

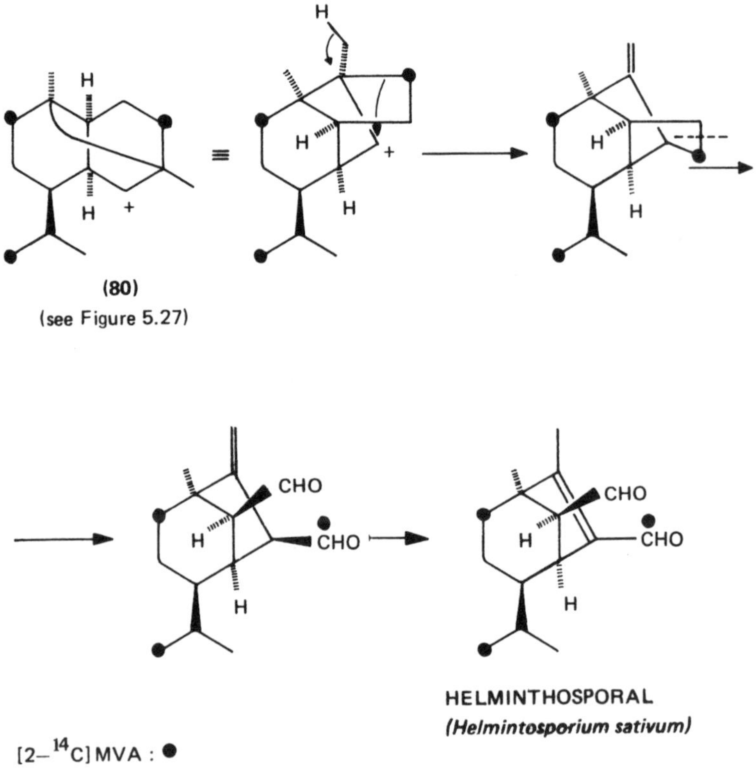

Fig. 5.35 Biosynthesis of Helminthosporal.

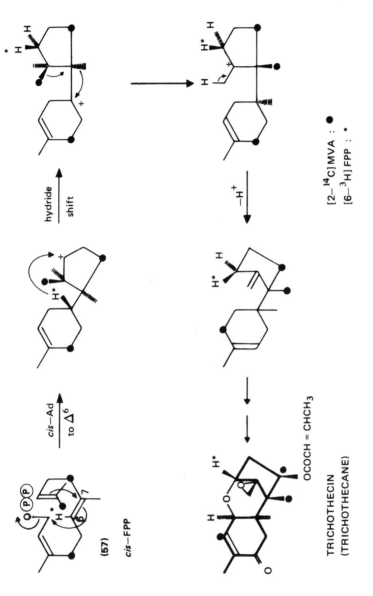

Fig. 5.34 Biosynthesis of the tricothecane skeleton.

Compounds showing potent antibacterial properties belong to the trichothecane subgroup. Illudin M and marasmic acid (illudin group) are antibacterial, whilst formannosinosin is a phytotoxin. All of these three sesquiterpenes are derived from a common protoilludane skeleton (94), as shown in Fig. 5.36.

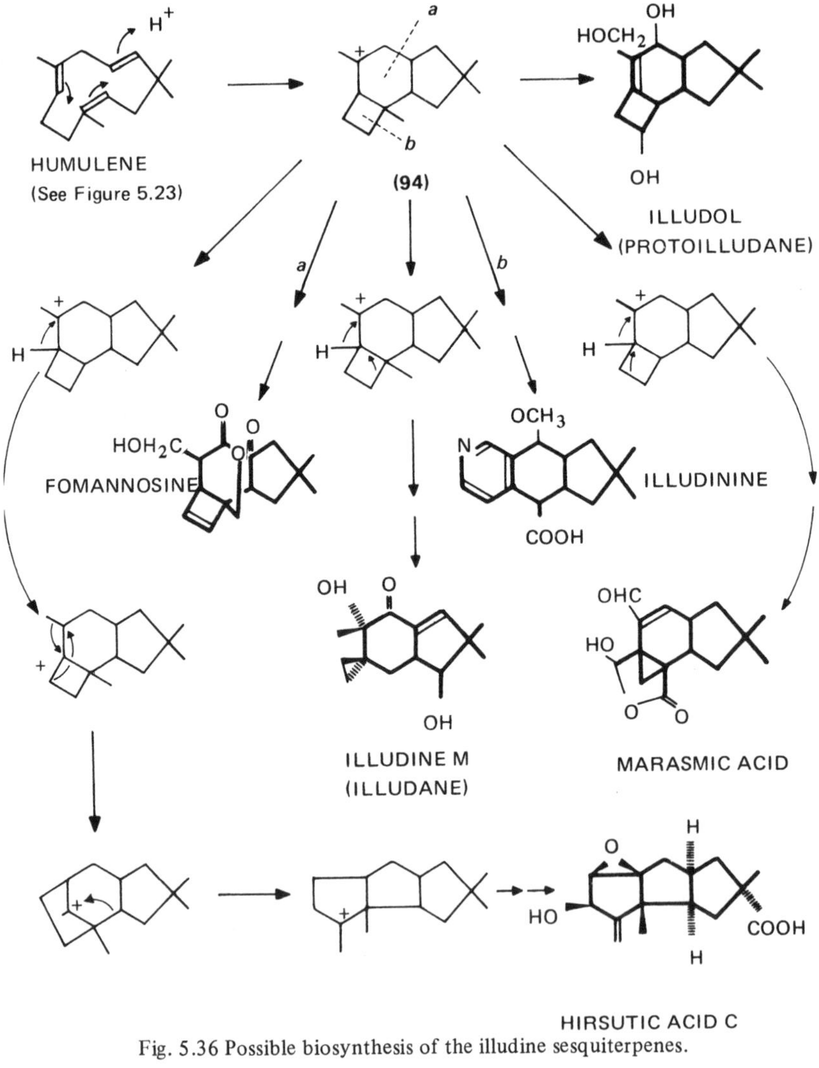

Fig. 5.36 Possible biosynthesis of the illudine sesquiterpenes.

5.4.2 Bicyclofarnesol Sesquiterpenes

A small number of bicyclic sesquiterpenes are derived from the cyclisation shown in Fig. 5.37. The departure of the pyrophosphate anion is the starting point for all the cyclisations shown in the previous sections, an incipient (or actual) carbocation being formed at the tail position of the FPP chain (Fig. 5.23).

In forming the bicyclofarnesyl skeleton (Fig. 5.37, see iresin and polygodial) the cyclisation instead takes place after an electrophilic attack, e.g. by a proton, at the double bond at the head of FPP, or onto the corresponding epoxide. The relative positions of the double bonds (of E-configuration) in the conformation assumed by the farnesyl chain at the moment of electrophilic attack determines the structure and stereochemistry of the final products. This mode of cyclisation, which is uncommon for the sesquiterpenes, is much more prevalent for the diterpenes and is by far the major process for the triterpenes which derive from an unsaturated hydrocarbon (squalene) (compare Figs. 5.57 and 5.59).

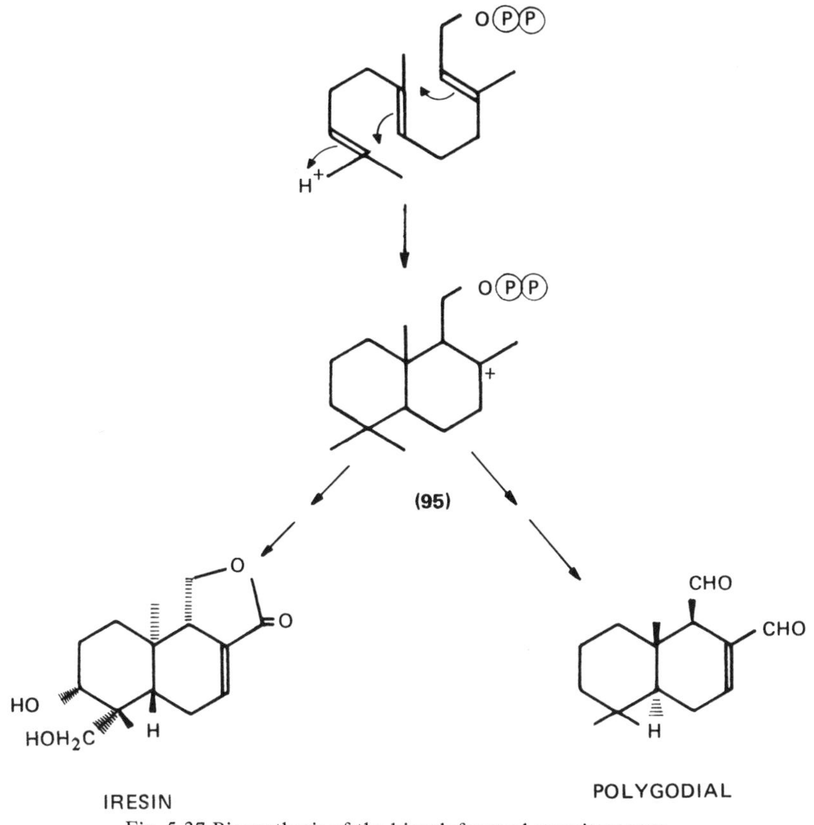

IRESIN   POLYGODIAL

Fig. 5.37 Biosynthesis of the bicyclofarnesol sesquiterpenes.

5.4.3 Sesquiterpenes of Particular Biological Importance

Besides compounds which are highly toxic to mammals (such as picrotoxin and its derivatives) or which have antibiotic or phytotoxic power (such as trichothecin), there are many sesquiterpenes which are extremely interesting from a physiological point of view. Only two groups will be referred to here, the phytohormones of plants and the juvenile hormones of insects.

The hormone regulated growth of the higher plants takes place through the balanced action of stimulating and inhibitory hormones. Such hormones are often terpenoids. The gibberellins, which will be dealt with further on (section 5.5.3) stimulate such growth, whilst abscisic acid (ABA) and its derivatives (typical sesquiterpenoids) are inhibitors. The substitution pattern of the cyclohexyl ring of ABA (Fig. 5.38) closely resembles that of safranal (53), ionone, and vitamin A. Although this pattern suggests an origin from carotenes, recent experimental results show that ABA is biosynthesised from farnesyl pyrophosphate and exclude the degradation of an intermediate carotenoid.

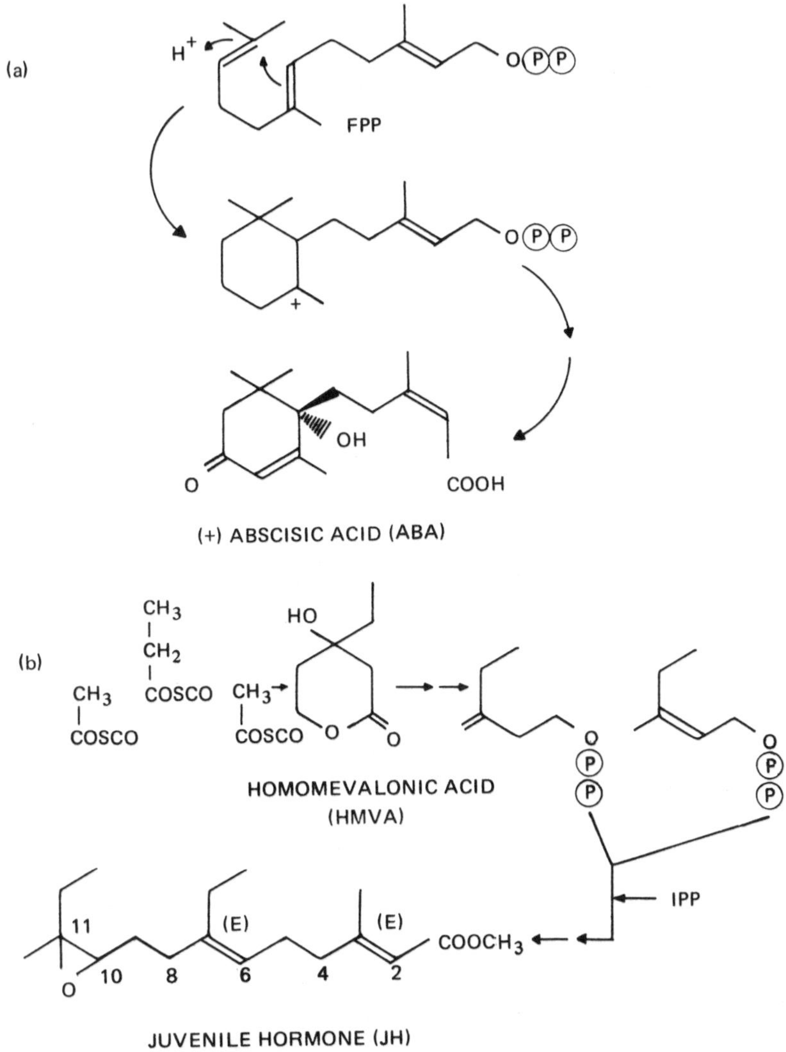

(a)

FPP

(+) ABSCISIC ACID (ABA)

(b)

HOMOMEVALONIC ACID
(HMVA)

IPP

JUVENILE HORMONE (JH)

Fig. 5.38 Derivation of abscisic acid and juvenile hormone.

The physiological role of the insect juvenile hormones is treated on p. 334. The first such hormone to be isolated, which came from tens of thousands of male *Hyalophora cecropia* L. butterflies, was neotenin, which is the methyl ester of 10,11-epoxy-7-ethyl-3,11- dimethyl-10,11-*cis*,2-*trans*,6-*trans*-tridecadienoic acid (also known as JH) (Fig. 5.38). A second juvenile hormone was also extracted from *H. acropia* and this had a methyl group, instead of an ethyl group, at C-7. This discovery, together with the remarkable structural analogies between neotenin and farnesol, leads to the suggestion that JH is derived from FPP through the epoxidation of the terminal double bond, oxidation of the primary alcoholic group followed by esterification and, the rare, addition of a C_1-unit to each methyl group at positions 7 and 11.

5.5 DITERPENES

The main cyclisation process observed in the formation of diterpenes is the same as that operating in the biosynthesis of the bicyclofarnesyl sesquiterpenes (Fig. 5.37). Thus the head part of GGPP is involved and the trans-fused decalin system, typical of bicyclic diterpenes, is formed. In only a few cases do cyclisations induced by the heterolytic cleavage of the CH_2-OPOP bond occur; monocyclic diterpenes are formed in this manner. The CH_2-OPOP group tends to form the allylic carbocation only after the bicyclic diterpene structure has been formed. In this manner the GGPP chain partially cyclises to the decalin system and then new rings can subsequently form, often with skeletal rearrangements leading to the tri-, tetra-, and pentacyclic diterpenes.

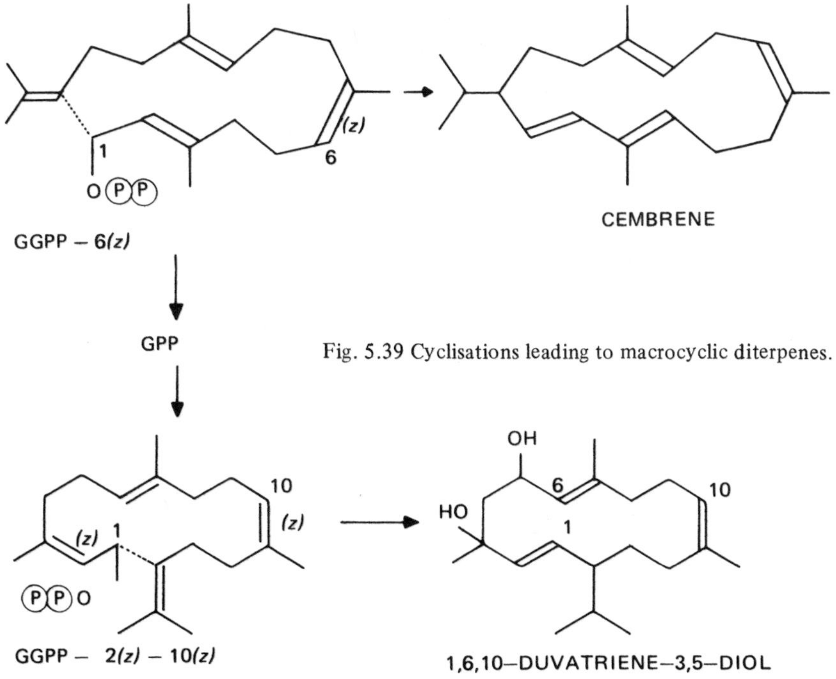

GGPP – 6(z)

CEMBRENE

GPP

Fig. 5.39 Cyclisations leading to macrocyclic diterpenes.

GGPP – 2(z) – 10(z)

1,6,10–DUVATRIENE–3,5–DIOL

5.5.1 Monocyclic Diterpenes

Cembrene and 1,6,10-duvatriene -3,5-diol, are the two most representative
compounds of this series: they are both derived from C_{20} units which differ
from GGPP in the configuration of one, or two, double bonds (Fig. 5.39).
Cembrene is present in the oleo resin of *Pinus albicaulis*. Various duvatrienes
have been isolated from aged tobacco leaves, and so they may be artefacts.

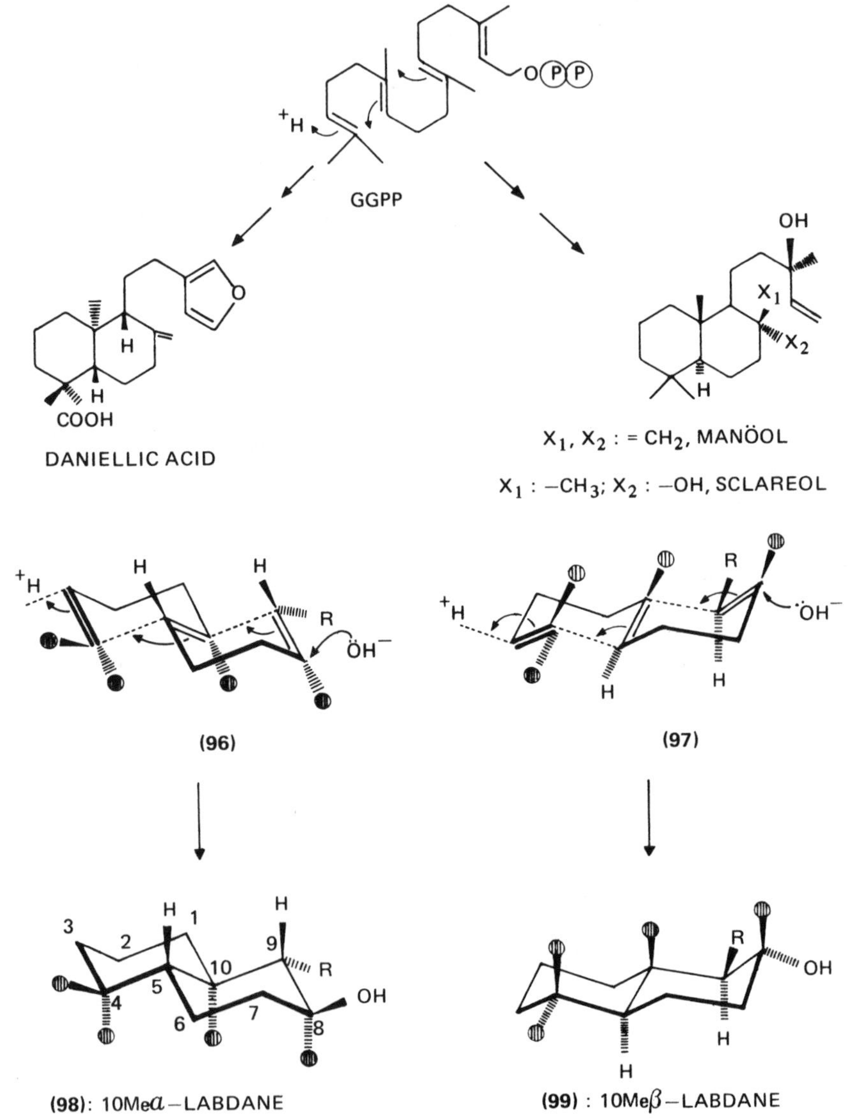

GGPP

DANIELLIC ACID

X_1, X_2 : = CH_2, MANÖOL

X_1 : $-CH_3$; X_2 : $-OH$, SCLAREOL

(96)

(97)

(98): 10Meα –LABDANE

(99) : 10Meβ –LABDANE

Fig. 5.40 Formation of the labdane skeleton.

5.5.2 Bicyclic Diterpenes

The biosynthesis of the labdane group of bicyclic diterpenes is shown in Fig. 5.40. Sclareol and manool are the simplest molecules in this subgroup, and they are directly derived from the cyclisation of GGPP, or of one of its allylic isomers, GLPP or GNPP.

The *trans*-junction between the rings A and B, found in all labdane diterpenes and in compounds with a larger number of rings in which other rearrangements have not taken place, is coherent with the stereoelectronic requirements of a concerted mechanism, in which multiple additions of non-conjugated double bonds take place. The conformation of the polyenic chain during the cyclisation sequence is trivially called 'chair (pre-ring A)-chair (pre-ring B)', using the same terminology largely used in the triterpene series. A priori, such a conformation can exist in two enantiomeric forms, (96) and (97), from which the two enantiomeric labdane skeletons (98) and (99) derive.

These structures have the same relative configurations at the asymmetric centres at C-5, C-10, and C-9 and, as a consequence, determination of the absolute configuration of any one of these asymmetric carbons is sufficient to determine all of them. Thence skeleton (98) is called 10-Meα-labdane and its enantiomer (99) is called 10-Meβ-labdane. Examples of both stereotypes are known in nature, although they are seldom present in the same plant. The enantiospecific formation of these compounds indicates an enzymatically controlled process. Since many labdane compounds are subsequently transformed in plants into tri-, tetra-, and pentacyclic diterpenes through mechanisms involving well-defined stereochemistry, such compounds are also divided into the two, 10-Meα and 10-Meβ series.

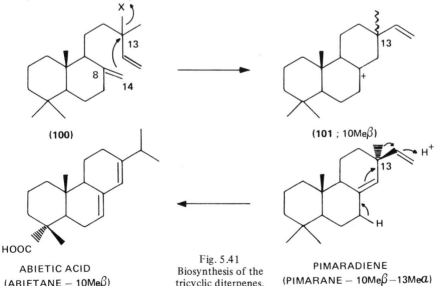

(100)

(101 ; 10Meβ)

ABIETIC ACID
(ABIETANE − 10Meβ)

Fig. 5.41
Biosynthesis of the
tricyclic diterpenes.

PIMARADIENE
(PIMARANE − 10Meβ −13Meα)

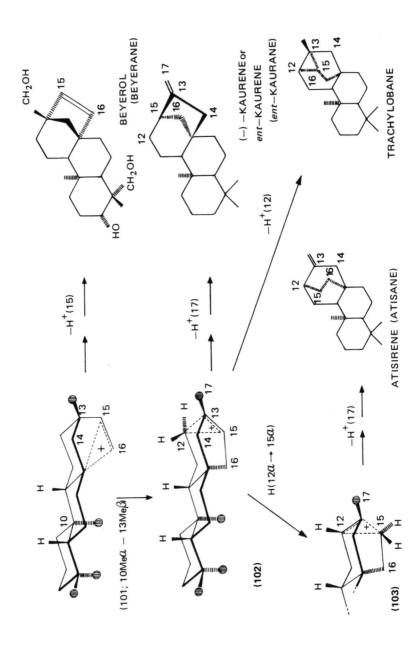

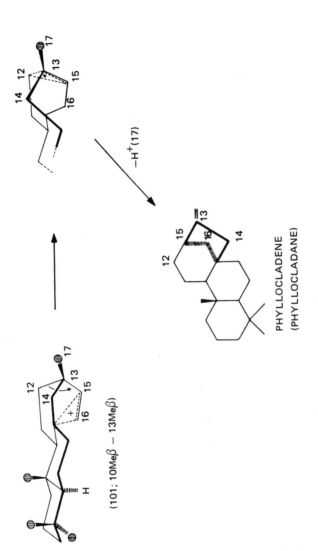

Fig. 5.42 Formation of tetra- and penta-cyclic diterpene structures.

5.5.3 Tri-, Tetra-, and Pentacyclic Diterpenes

The following compounds and mechanisms have been chosen because of their frequent occurrence in nature and examples of the 10-Meα and 10-Meβ series are known.

The tricyclic diterpenes can derive from a labdane intermediate such as (100), by interaction between the incipient C(13) carbocation and the 8,14-double bond (Fig. 5.41). 10 Meα-compounds are called *ent*-diterpenes.

The four possible isomeric forms (10-Meα-, 13-Meα or 13-Meβ; 10-Meβ-, 13Meα or 13-Meβ of the pimarane cation, e.g. (101), produces the various derivatives such as pimanes (methyl groups on C(10) and C(13) anti), sandaacopimaranes (C(10) and C(13) syn), and of abietane. The structural isoprene rule is not represented in this latter compound, since a methyl shift takes place (Fig. 5.41).

The pimarane cation also leads to the tetra- and pentacyclic structures (Fig. 5.42). There is a close relationship amongst these and the illustrated schemes attempt to rationalise their biosynthetic derivations. Non-classical ions are largely invoked as intermediates (Appendix 3).

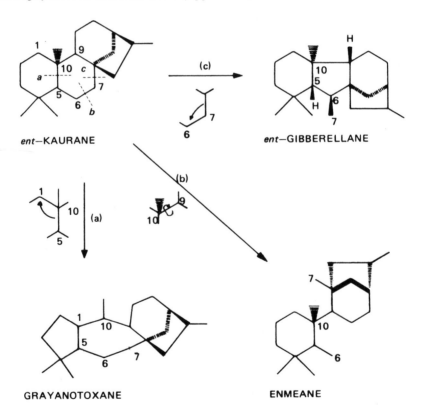

Fig. 5.43 Sesoditerpenes.

The derivatives grayanotoxane, enmeane, and gibbane possibly derive from the kaurane skeleton. During the rearrangement processes the bonds at positions 5,10-, 6,7-, and 7,8- (Fig. 5.43) are respectively broken (see Fig. 1.43). The gibbane skeleton is rather important, as it is characteristic of the gibberellins (Fig. 5.44) which are a group of phytohormones stimulating the growth of plants.

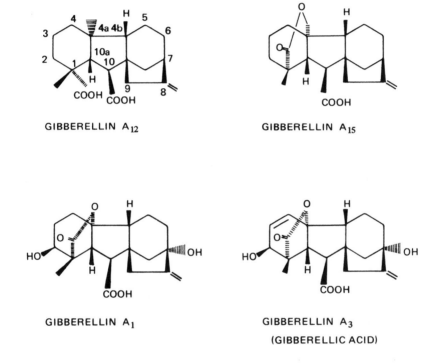

GIBBERELLIN A_{12}

GIBBERELLIN A_{15}

GIBBERELLIN A_1

GIBBERELLIN A_3

(GIBBERELLIC ACID)

Fig. 5.44 Source of the principal gibberelins (C_{20} and C_{19}).

The gibberellins were initially isolated from a fungus, *Gibberella fujikuroi* and subsequently from higher plants. It is now thought that they are synthesised by all higher plants. These substances were discovered by studying a rice plant illness caused by *G. fujikuroi*. It was noticed that the infected plants grew much faster than uninfected ones during the first stages of the disease. A crystalline product, active in the rice plants, was isolated from the filtrate of the fungal culture and proved to be a mixture of gibberellic acid and other gibberellins. The biological importance of the gibberellins and their isolation from fungal cultures, stimulated biogenetic studies. These studies reveal the biosynthetic paths shown in Fig. 5.45 to be highly probable.

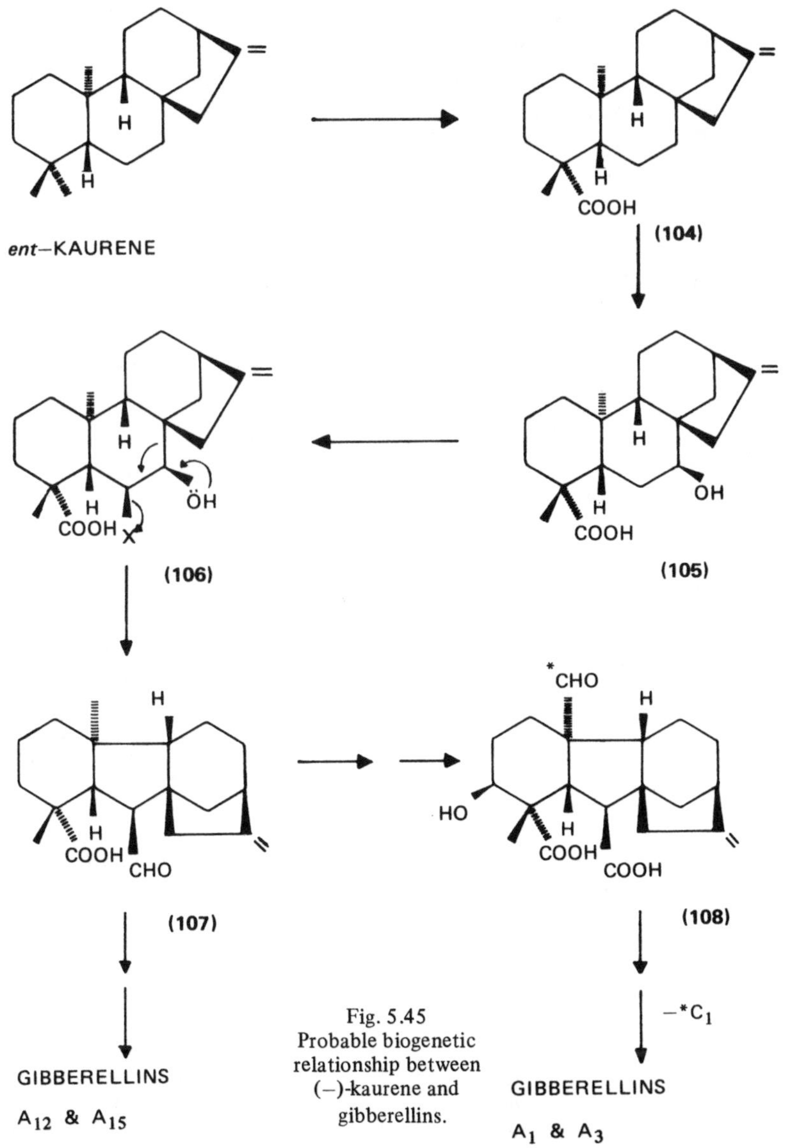

ent–KAURENE

COOH (104)

OH

COOH (105)

ÖH

COOH X (106)

H
COOH
CHO (107)

*CHO

HO

H
COOH
COOH (108)

GIBBERELLINS

A_{12} & A_{15}

Fig. 5.45
Probable biogenetic
relationship between
(−)-kaurene and
gibberellins.

$-*C_1$

GIBBERELLINS

A_1 & A_3

5.5.4 Diterpenes of Fungal Origin

Two other fungal diterpene families are known besides the gibberellins, to which pleuromutilin (113) and rosenonolactone (116) belong. The biosynthesis of pleuromutilin, produced by *Pleurotus* sp. has been studied by incorporating $[2\text{-}^{14}C]$ mevalonic acid (Fig. 5.46); an unusual 1,5-hydride ion shift occurs in (111). For rosenonolactone, the incorporation data of $[1\text{-}^{14}C]$ and $[2\text{-}^{14}C]$ acetic acid agree with the scheme of Fig. 5.47.

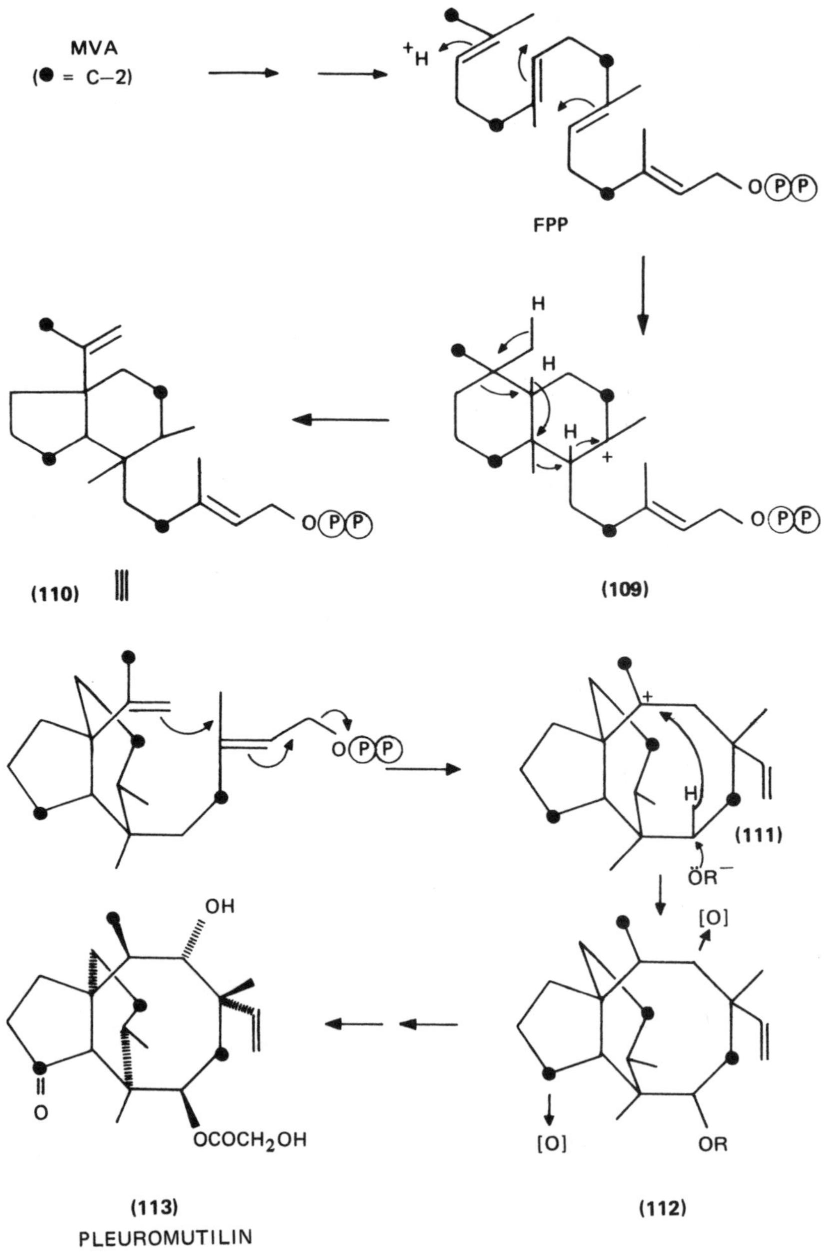

Fig. 5.46 Biosynthesis of pleuromutilin.

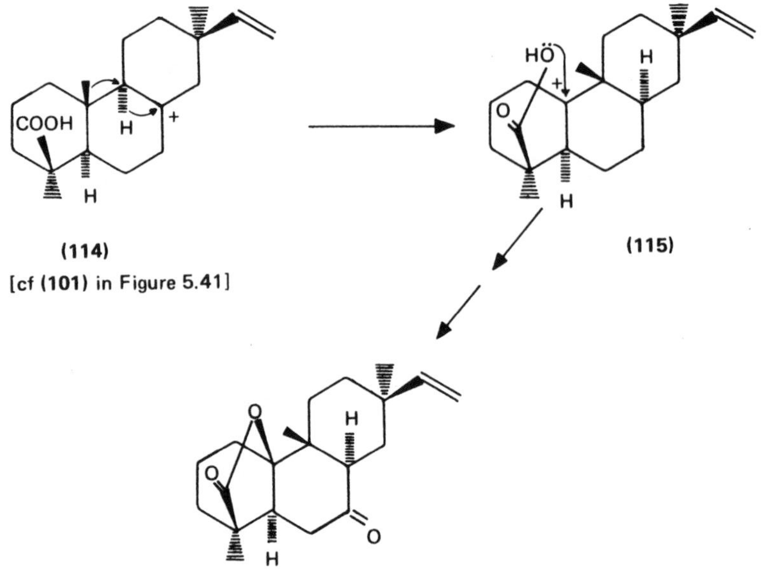

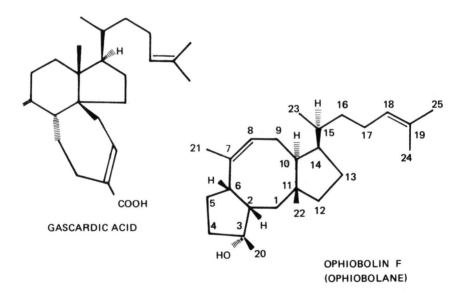

(114)

[cf (101) in Figure 5.41]

(115)

(116)

ROSENONOLACTONE

Fig. 5.47 Biosynthesis of rosenonolactone.

GASCARDIC ACID

OPHIOBOLIN F
(OPHIOBOLANE)

Fig. 5.48 Sesterterpenes.

5.6 SESTERTERPENES

The first compound belonging to this group of terpenes was isolated in 1965 from the secretion of an insect, *Gascardia madagascariensis.* The compound was called gascardic acid and was assigned to the structure indicated in Fig. 5.48. Other sesterterpenes have since been isolated from various species of *Helmintho-sporium* and *Cochliobolus* fungi. These compounds usually show a remarkable phytotoxic activity. The skeleton of ophiobolane, in which a cycloctane ring is fused to cyclopentane rings is peculiar to this second family of sesterterpenes (see ophiobolin F, Fig. 5.48).

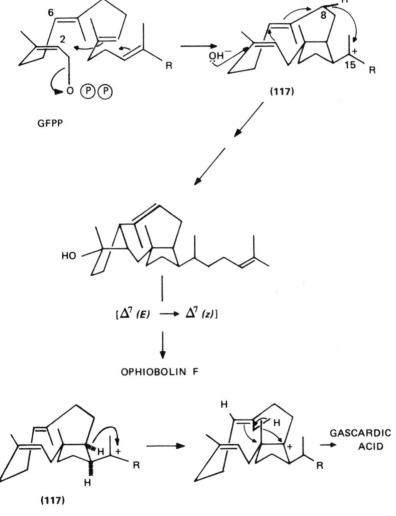

Fig. 5.49 Probable biosynthesis of some ophiobolin skeleta and gascardic acid.

Numerous biogenetic researches have been carried out the ophiobolins, whilst not much is known about the formation of the insect product, gascardic acid. The postulated procursor geranylfarnesyl pyrophosphate has itself been isolated from the protective wax of the insect, *Ceroplastes albolincatus*. For the ophiobolins this has the 2(3) double bond in a (*Z*)-configuration. The first part of the biosynthetic path, up to the cation (117) (Fig. 5.49) is likely to be common to both structural types. An interesting feature of the biosynthetic process for the ophiobolins is the 8,15-hydride ion shift. This process has been proven by incorporation of $[2\text{-}^3H_2]$ MVA. An analogous shift has been invoked by Arigoni in pleuromutilin biosynthesis (Fig. 5.46).

5.7 TRITERPENES

The triterpenes are derived from MVA through squalene (118) and, in most cases, *via* 2,3-epoxysqualene (119). Their various structures, of which there are about 20 skeletal types, depend on the tendency of squalene, with its six double bonds, to undergo multiple cyclisations. These cyclisations are mediated by enzymes (cyclases) capable of exerting rigorous stereochemical controls, such as those described for the formation of bicyclofarnesol sesquiterpenes (Fig. 5.37) and of the labdanes (Fig. 5.40). Cyclisation starts with the formation of an incipient carbocation at the tertiary carbon of the end double bond of the polyisoprene chain. This type of cyclisation is the only one possible, since squalene is a symmetrical, unsaturated hydrocarbon and doesn't possess the ability to form a carbocation by departure of a good leaving group, such as the pyrophosphate group. In this respect squalene is totally different from all the other primary terpene precursors.

The cyclisation of squalene can be promoted either by oxidative or non-oxidative agents. The former lead to cyclic compounds, with ring A bearing a hydroxyl (or ketonic) group, in position 3. By alternatively using either $H_2{}^{18}O$ or $^{18}O_2$ in rat liver homogenates, it was shown that the C-3-oxygen of lanosterol derives from atmospheric oxygen. The hypothesis of a monooxygenase operating in the cyclisation process was thus confirmed. Two routes can be envisaged: *either* a prior conversion of squalene into 2,3-epoxysqualene (119), followed by its transformation into cyclic derivatives via protonation and Markownikoff opening of the epoxide ring (Fig. 5.50, route a); or OH^+ (protonated oxene, p. 90) attack on the 2,3 double bond (route b), in a similar way to and with the same effects of a proton attack (cyclisation, non-oxidative processes; route c). Corey and van Tamelen independently found, in 1966, that the transformation of squalene into lanosterol implied the formation of 2,3-epoxysqualene as an intermediate: $[^{18}O]$ 2,3-epoxysqualene gave $[^{18}O]$ -lanosterol (see Fig. 5.51), when incorporated into a rat liver homogenate. In the microsomes of hepatic cells two enzymes were identified: the insoluble squalene epoxidase NADP- and O_2- dependent and so of the mixed-function type oxidases (p. 82) and the soluble,

epoxysqualene cyclase, which does not depend on cofactors, and is inhibited by
2,3-iminosqualene, which differs from epoxysqualene only in having an aziridine
ring rather than the oxirane one. These results include route (a), (Fig. 5.50) as a
likely process for other oxidative cyclizations.

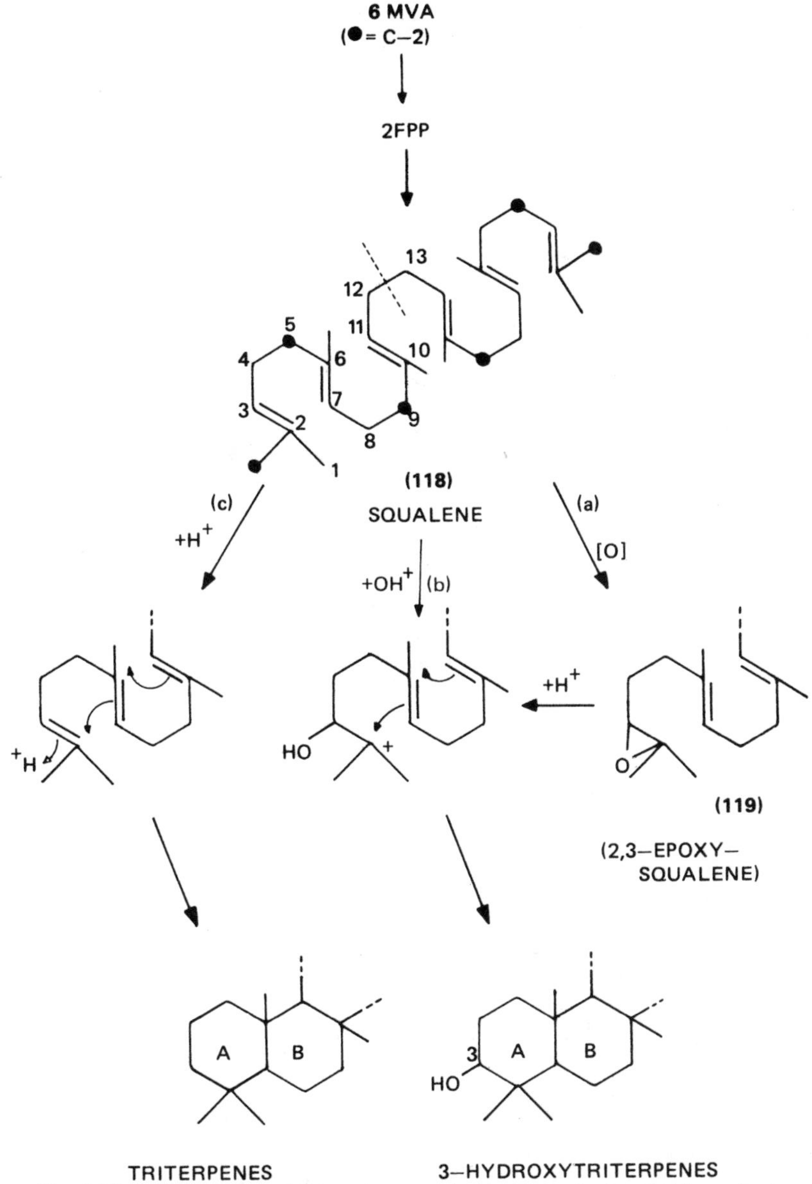

Fig. 5.50 Oxidative, (a) and (b), and non-oxidative, (c), paths for the biosynthesis
of triterpenes.

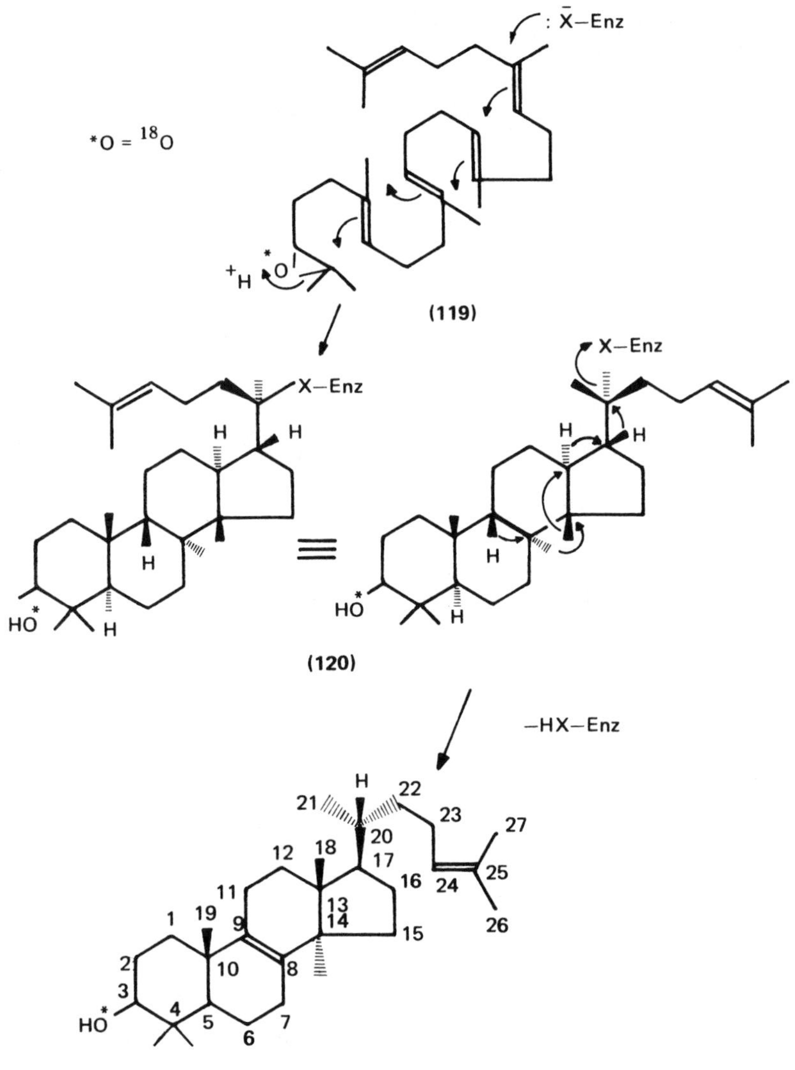

$*O = {}^{18}O$

(119)

(120)

−HX−Enz

LANOSTEROL

Fig. 5.51 Possible rationalisation for the biosynthesis of lanosterol.

As far as the mechanism of formation of all the different triterpene structures are concerned, general schemes can be postulated based on the available experimental evidence.

Three processes can be envisaged for the formation of the *trans*-decalin system of rings A and B (Fig. 5.52): a concerted one (type 1), a multi-step one involving non-classical cations (type 2) and a multistep one involving classical cations (type 3).

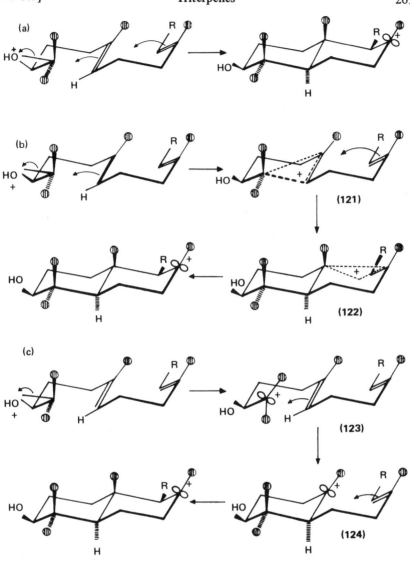

⬢ = METHYL GROUP

Fig. 5.52 Possible mechanism for the formation of the *trans*-decalin system (rings A and B) in triterpenes.

In the first process the enzyme has to stabilise the conformation of the polyisoprene chain, so that the relative positions of the double bonds are properly aligned: the stereoelectronic requirements of the concerted process automatically determine the stereochemistry of the final products (see Appendix 2).

In process 2 the enzyme could stabilise the non-classical cations (121) and (122): the bridged structures of such cations (see Appendix 3) only allows the approach of the incoming nucleophilic agents (e.g. double bonds), from certain directions, thus determining the stereochemistry of the products.

In the third process the enzyme has to stabilise the single, classical cations, (123) and (124), as well as the attack of the nucleophile to one of the carbo-cation faces (*re* or *si*, see Appendix 2): labile bonds could be formed between the enzyme and carbocation in this process. The cyclisation schemes illustrated in the following pages correspond to the biogenetic hypothesis proposed by Ruzicka and Eschenmoser in 1955. Mechanisms as concerted as possible are invoked and the classical or non-classical carbocations are only postulated when necessary in order to explain certain dichotomies of the synthetic routes or the steric formulae of the final products.

Triterpenoids occur widely in both the plant and animal kingdoms. The tetracyclic triterpenoids are more common in animals (especially lanosterol and its steroidal derivatives) while pentacyclic triterpenes are more common in plants. The latter can be found in nature either in their free forms, or conjugated with one or more sugar units through the hydroxyl group at position 3 (the triterpene saponins).

5.7.1 The Biosynthesis of Squalene and its 2,3-epoxide

The formation of squalene from farnesyl pyrophosphate (FPP) can be con-sidered a reductive dimerisation (tail-tail). A net addition of two hydrogen atoms to the molecules of FPP is required in order to obtain one molecule of squalene, liberating two molecules of pyrophosphoric acid (or one of acid and one of the anion) (Fig. 5.53(a)). Cornforth and Popjak demonstrated that the reaction requires NADPH as hydride ion donor and they have clarified the overall stereo-chemistry of the dimerisation process. FPP, stereospecifically tritiated at the tail carbon and specifically tritiated NADPH (at the hydride donor carbon) were used with enzymes from rat liver homogenates. The FPP molecule (see Fig. 5.53(b)) loses one of the hydrogens at C-1, which becomes C-12 or C-13 in squalene; H_F (*pro*-S) is also lost as H^+, corresponding to the 5-*pro*-S of MVA (see Figs. 5.6 and 5.5), and is replaced by H_B of the NADPH coenzyme (β-stereo-specificity, see p. 59). The C-1 of the second FPP unit, meanwhile, undergoes complete inversion of its configuration, although all the hydrogen atoms are retained in this second unit.

Two mechanisms accounting for these stereochemical facts and explaining the formation of the central squalene C–C bond were originally proposed (Fig. 5.54). These are now known to be incorrect since the discovery of a new intermediate has been made which indicates the operation of the more sophisticated reaction sequence shown in Fig. 5.55. Rilling called the intermediate presqualene pyrophosphate, which he isolated from a yeast microsomal preparation provided with FPP, but not with NADPH. The structure (125) which Rilling proposed for presqualene pyrophosphate has now been definitely confirmed by degradation and by synthesis. Only that shown, of the eight possible stereoisomers of (125) (four enantiomeric couples), is biologically active.

Presqualene pyrophosphate accumulates in the absence of NADPH, clearly indicating how the pyridine coenzyme operates *after* the formation of the cyclopropane ring. The cyclopropane unit is formed by interaction of two FPP molecules through a mechanism similar to that proposed by Cornforth for the assembly of C_5-units (S_N2-Ad-*trans* + E2) (compare to Fig. 5.7).

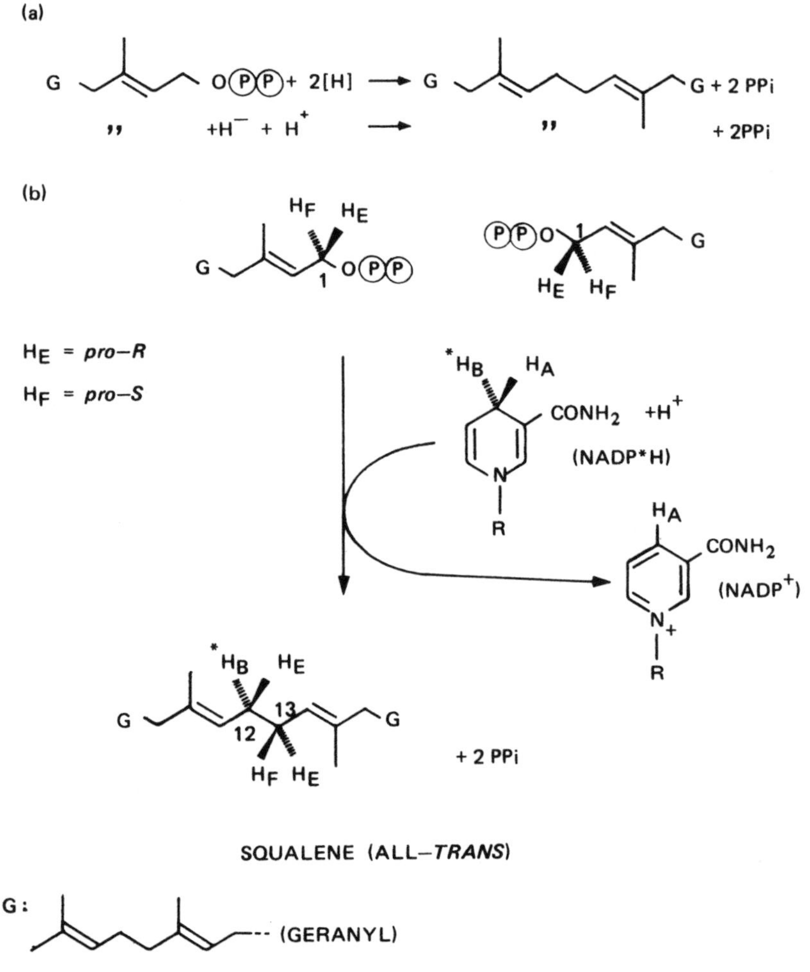

(a)

(b)

H_E = pro—R

H_F = pro—S

SQUALENE (ALL—*TRANS*)

G:

Fig. 5.53 Stoichiometry and stereochemistry of the process leading to squalene.

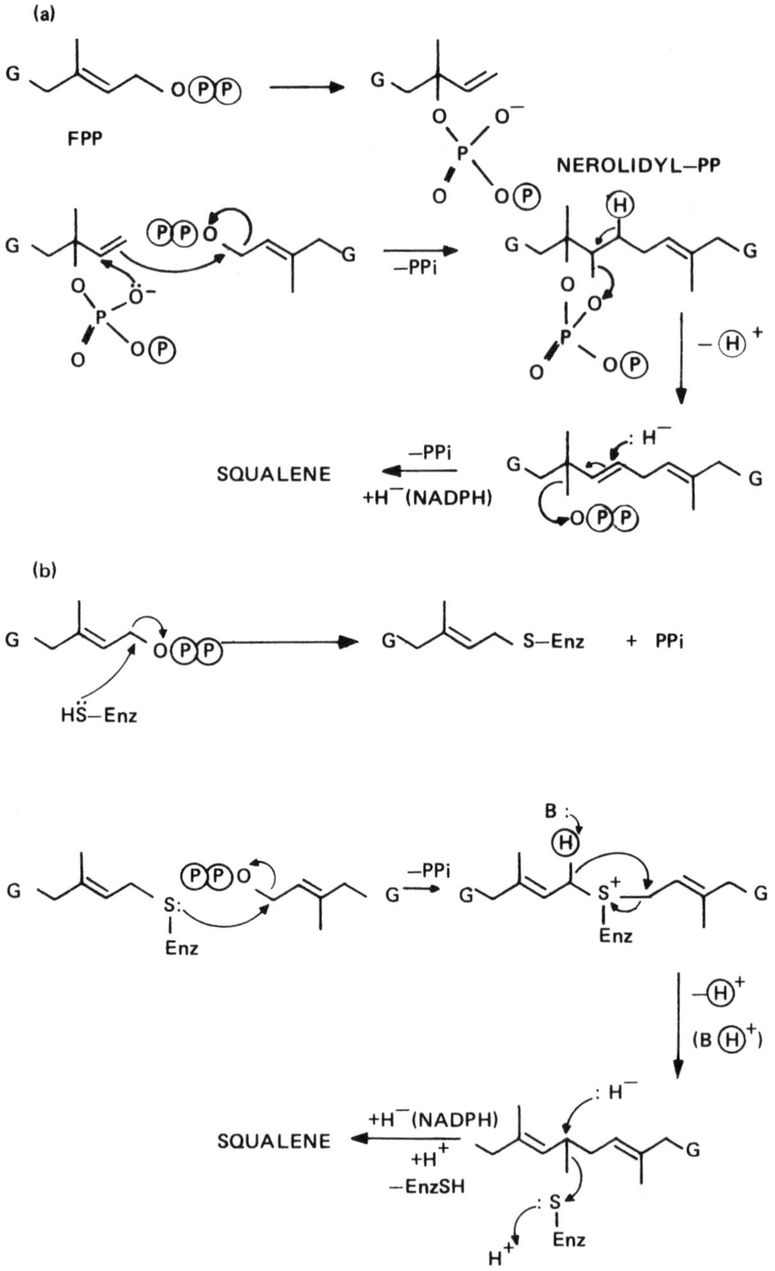

Fig. 5.54 Early hypothetical mechanism for the dimerisation of farnesylpyrophosphate leading to squalene.

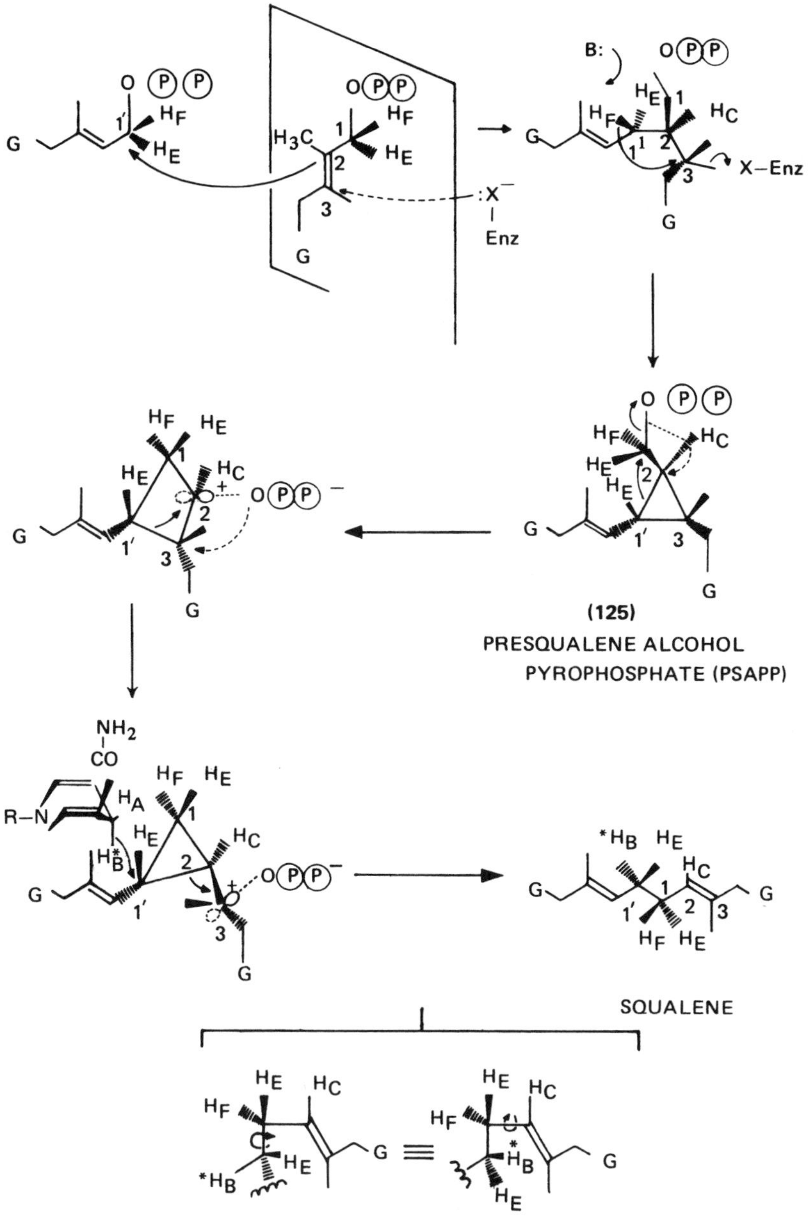

Fig. 5.55 Probable route for the synthesis of squalene from farnesylpyrophosphate.

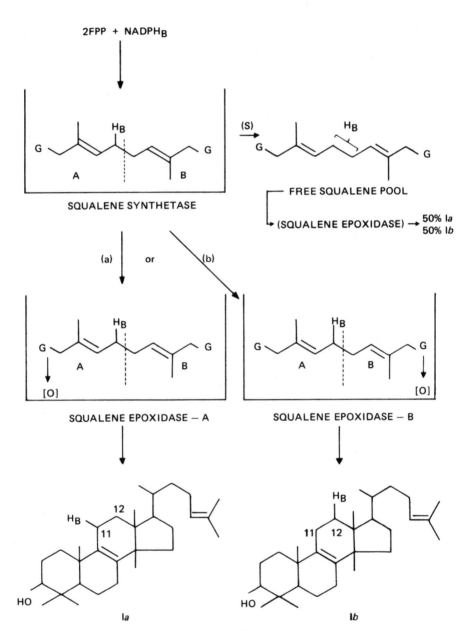

Fig. 5.56 (a), (b): Asymmetric route and (c) symmetric route to lanosterol from FPP.

The conversion of FPP into the symmetrical molecule of squalene is an asymmetric process, because of the asymmetry of the squalene synthetase enzyme. In most organisms squalene is formed and immediately transformed into the 2,3-epoxide. In the microsomal fraction from rat livers both squalene-synthetase and squalene epoxidase, catalysing the epoxidation of a terminal double bond of squalene, have been found. This co-occurrence of the enzymes suggested an orderly array of them in the endoplasmic reticulum. The transfer of squalene, containing the H_B hydrogen of NADPH in only one half of the molecule, to the squalene epoxidase enzyme site should take place according to a well-defined pathway, such that epoxidation might only occur selectively at one half of the symmetrical squalene molecule (Fig. 5.56). Since 2,3-epoxy-squalene is rapidly transformed into lanosterol (Fig. 5.51) one can check this hypothesis, since label should be localised on only one carbon atom (11 or 12) of lanosterol. To date, the experimental results obtained with the incorporation of either $[1\text{-}^3H_2]$ FPP or $[^3H_B]$ NADPH into lanosterol and its derivatives show an almost identical radioactivity distribution in the C-11 and C-12 positions. Such experiments cannot be considered conclusive, as they have not been done on intact cells, but on microsomal fractions from rat livers, in which the orderly array of enzymes has perhaps been lost during the homogenisation and isolation process.

The triterpenes which directly derive from the cyclisation of squalene or of its 2,3-epoxide are called *fundamental* (or *primary*) *triterpenes*. The fundamental triterpenes can be divided into subgroups according to their origin from: (a) oxidative cyclisations arising from only one terminal part of squalene; (b) non-oxidative cyclisations, also arising from only one terminal part of squalene, and (c) cyclisations (oxidative and non-oxidative) which independently involve both the terminal double bonds of squalene.

5.7.2 Oxidative Cyclisation of Squalene; Lanosterol and the Amyrins

The biosynthetic routes to the main fundamental triterpenes derived from 2,3-epoxysqualene are shown in Figs. 5.57–5.66. The tetracycles, dammar-enediols (Fig. 5.57), euphol and tirucallol (Fig. 5.58), lanosterol (Fig. 5.59), cycloartenol and protosterol (Fig. 5.60) differ slightly from each other in their cyclisation mechanisms:

(1) The dammarenediols, euphol and tirucallol derive from the 2,3-epoxysqualene molecule (all-*trans*), held in the active site of the enzyme in a 'chair-chair-chair-boat' conformation, while lanosterol and cycloartenol derive from a 'chair-boat-chair-boat' conformation.

(2) The formation of dammarenediol I instead of dammarenediol II, as well as of tirucallol instead of euphol, depends on the isomerisation of the non-classical cation (127) into the non-classical cation (128). An analogous isomerisation of (132) into (133) produces an *R*-configuration at C-20 in lanosterol and cycloartenol.

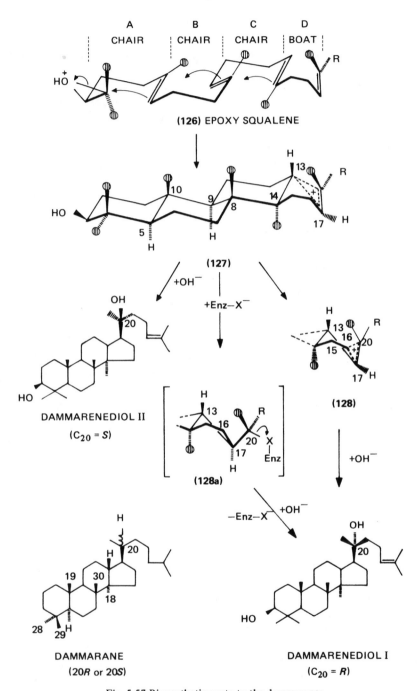

Fig. 5.57 Biosynthetic route to the dammaranes.

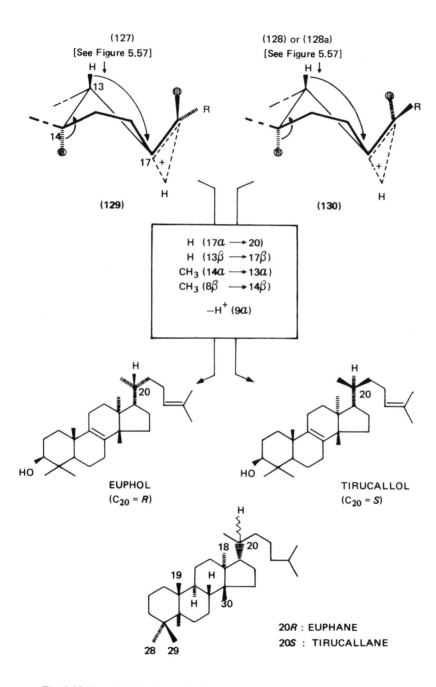

Fig. 5.58 Biosynthetic scheme for the production of euphanes and tirucallanes.

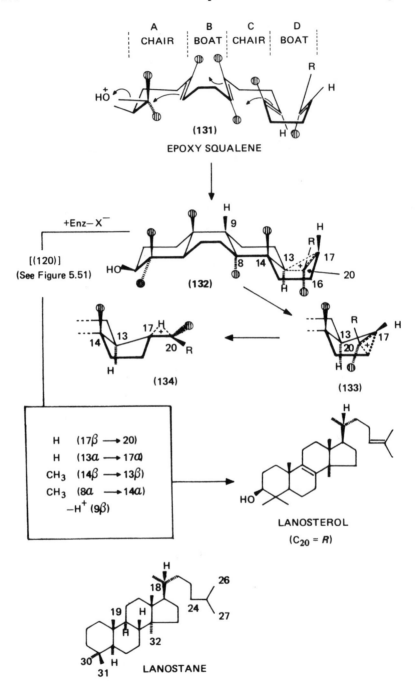

Fig. 5.59 Biosynthetic route to lanostanes.

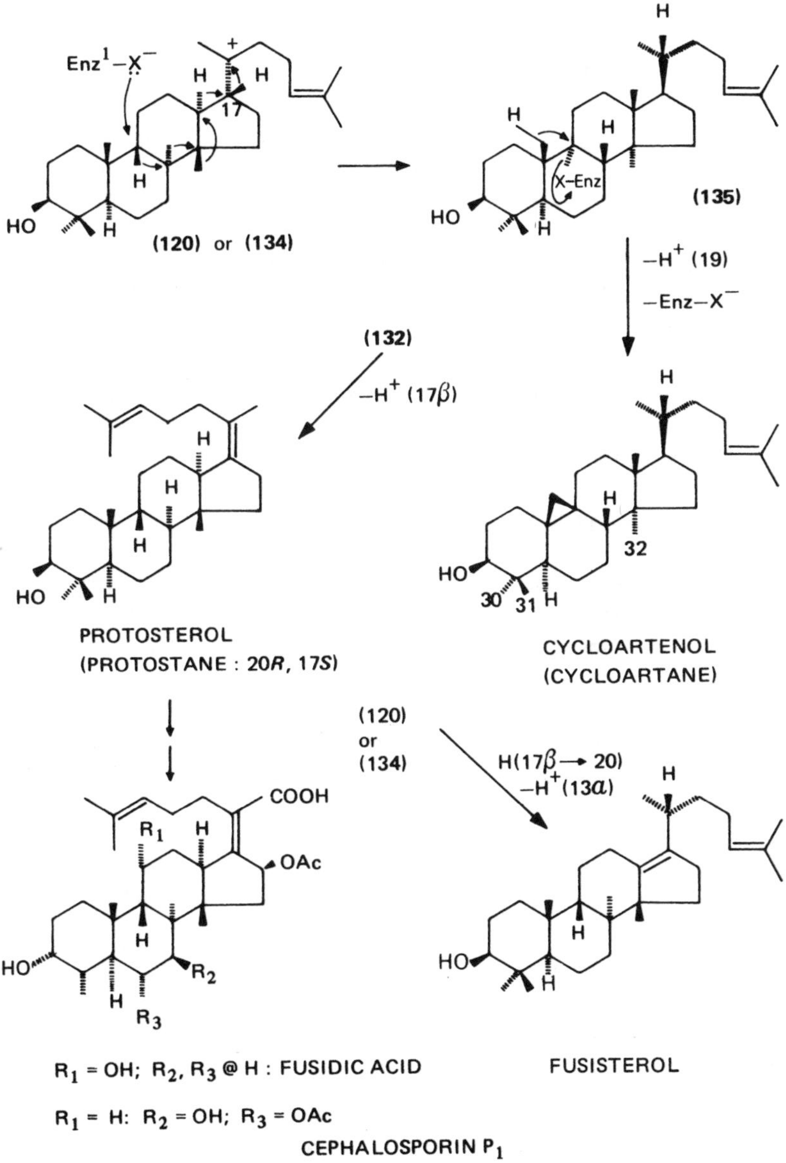

Fig. 5.60 Biosynthetic routes to the cycloartanes and protosterols.

(3) The skeleton of euphane, tircallane, and lanostane are produced by a concerted series of hydride and methyl group shifts; each group is antiparallel to both the preceding and the successive one, along the links C-20, C-17, C-13, C-14, C-8, and C-9 of the corresponding tetracyclic systems. The cyclisation process and the successive series of shifts presumably take place without interruption between the two processes.

The lanosterol-cycloartenol dichotomy is rationalised by admitting the formation of the intermediate (135) bound with a covalent bond at C-9 to the enzyme, for the latter. The formation of the cyclopropane ring is thus explained. The cyclopropane ring could not result from a mechanism entirely concerted with the cyclisation and the transpositions because of the stereoelectronic requirements imposed on E2-type eliminations (see Appendix 2).

In the tetracyclic triterpenes ring D is five-membered, whilst rings A, B and C are six-membered. This result could depend on the tendency of the C-13 electrophilic centre to attack the penultimate double bond of squalene with Markownikoff regiospecificity. An analogous, thermodynamically favoured, regiospecificity should also occur on forming ring C, but this is instead formed under enzymic control, via an anti-Markownikoff attack. The cyclisation of synthetic 2,3-epoxysqualene under non-enzymic conditions takes place mainly in the Markownikoff manner, producing tricyclic skeleta of the type (136). The extremely poor distribution of such structures in nature (e.g. malabaricol, isolated from *Ailantus malabaricum* (Simarubaceae)) attests to the important role played by enzymes in cyclising squalene.

The lanosterol-cycloartenol pair of compounds has a primary importance among tetracyclic triterpenes: all of the steroid derivatives, such as cholesterol, derive from them (Chapter 6). Lanosterol is especially widespread in vertebrate animals and its biosynthesis has been largely investigated by using liver homogenates; cycloartenol is distributed in higher plants. Derivatives of protosterol, for example fucisterol, and of non-protosterol, such as fusidic acid (skeleton of fusidane) (see Fig. 5.60), are often found in fungi. The protostane derivatives could derive from the protosterol cation (132) as an alternative to its isomerising to (133) or stabilisation as (120). The fundamental pentacyclic triterpenes originating from 2,3-epoxysqualene (all-*trans*) can be derived by interacting the non-classical cation (128) (the cation which also leads to dammarenediol I and tirucallol) with the last double bond still present in the molecule, so that the new non-classical cation (137) is formed; the latter may undergo many further transformations, mostly shifting hydride ions and methyl groups, before stabilising by expulsion of a proton (Figs. 5.62, 5.63, 5.64).

The possibility of a concerted cyclisation mechanism involving the whole 2,3-epoxysqualene molecule exists in nature, and molecules, such as hydroxy-hopanone, arborinol and moretenol are formed. These arise from different conformations of the isoprene chain; a 'chair-chair-chair-chair-chair' conformation gives the first molecule; a 'chair-boat-chair-chair-chair-boat' gives the second; and a 'chair-chair-chair-chair-boat' conformation gives the third (Fig. 5.64).

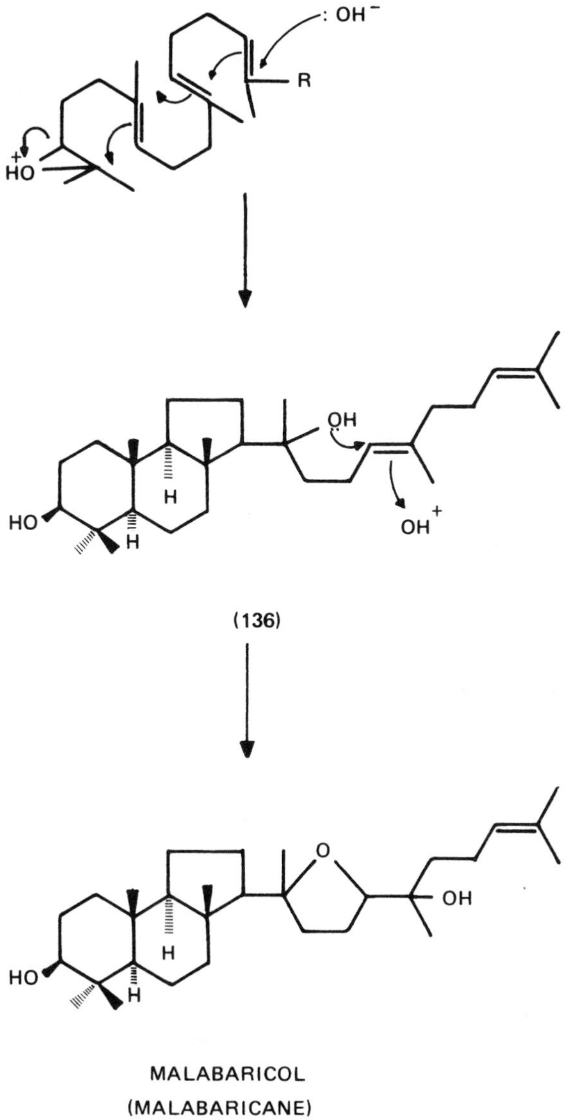

(136)

MALABARICOL
(MALABARICANE)

Fig. 5.61 Biosynthesis of the malabaricanes.

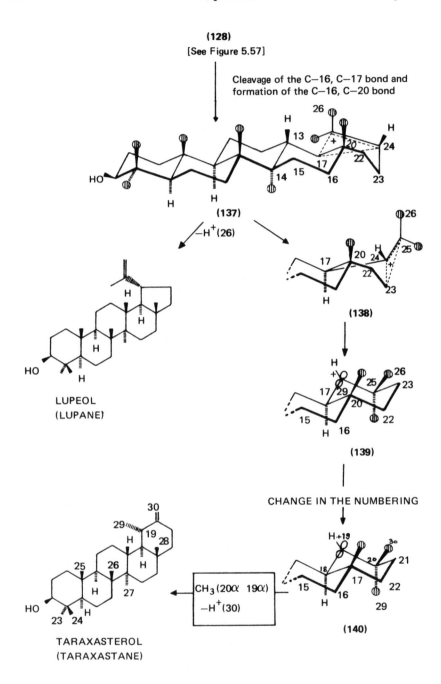

Fig. 5.62 Biosynthetic scheme for the lupane and taraxastane structures.

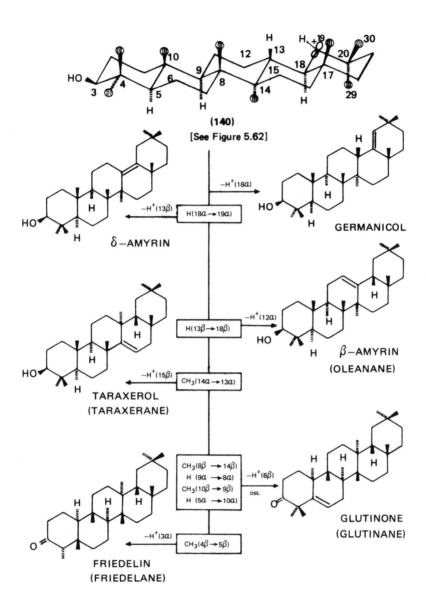

Fig. 5.63 Biosynthesis of some pentacyclic triterpenes.

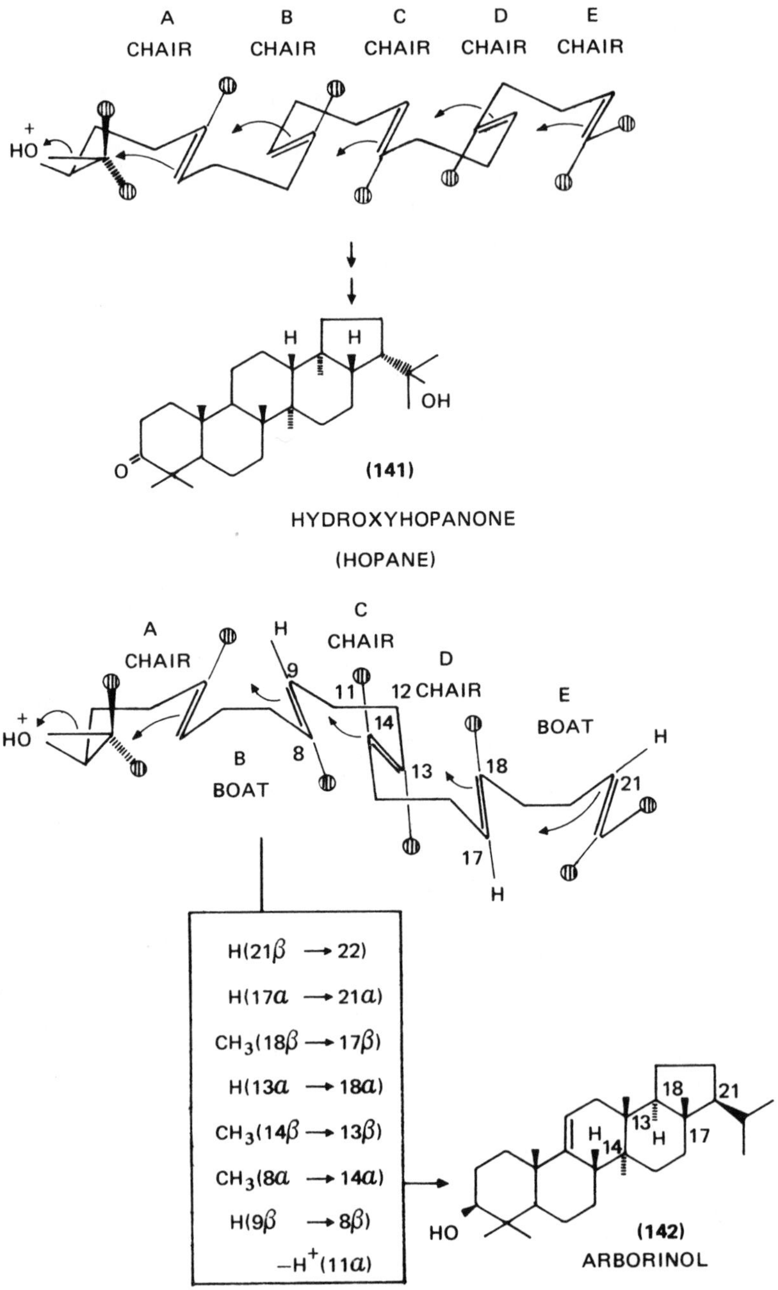

(141)

HYDROXYHOPANONE

(HOPANE)

$$H(21\beta \longrightarrow 22)$$

$$H(17\alpha \longrightarrow 21\alpha)$$

$$CH_3(18\beta \longrightarrow 17\beta)$$

$$H(13\alpha \longrightarrow 18\alpha)$$

$$CH_3(14\beta \longrightarrow 13\beta)$$

$$CH_3(8\alpha \longrightarrow 14\alpha)$$

$$H(9\beta \longrightarrow 8\beta)$$

$$-H^+(11\alpha)$$

(142)

ARBORINOL

Fig. 5.64 (contd. next page)

Fig. 5.64 (contd.)

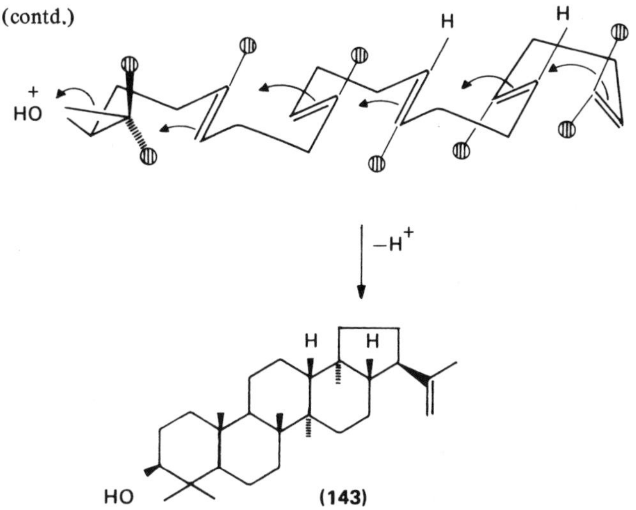

(143)

MORETENOL

Fig. 5.64 Cyclisations of 2,3-epoxysqualene with formation of a pentacyclic skeleton.

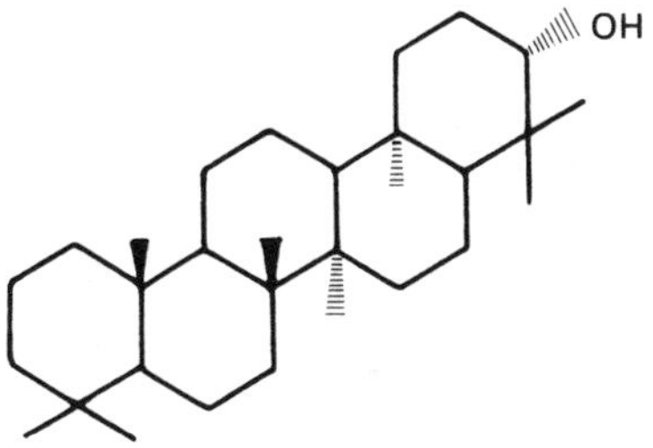

(144)

TETRAHYMANOL

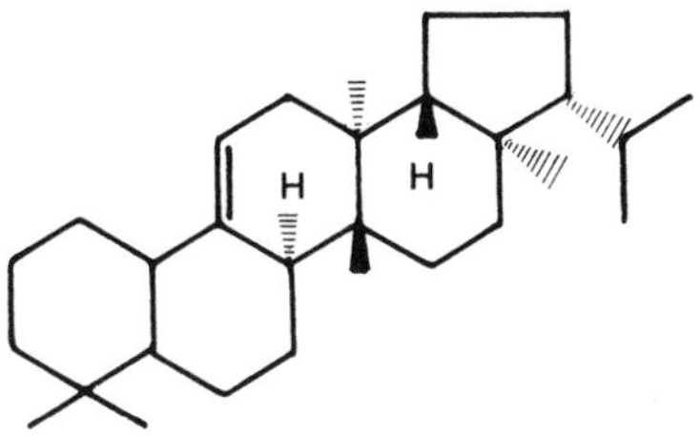

(145)

FERN—9—ENE

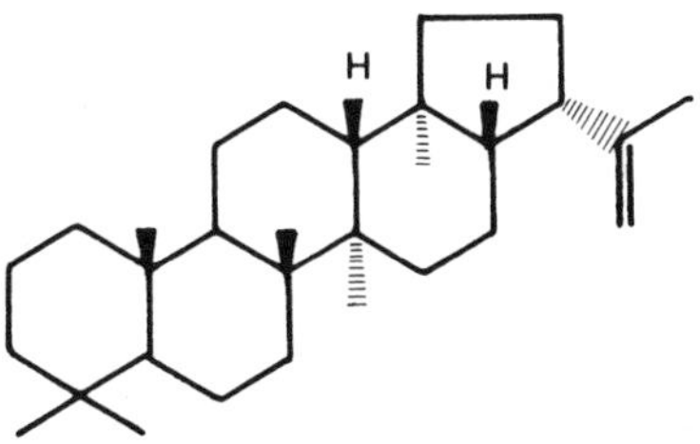

(146)

DIPLOPTENE

Fig. 5.65 Non-oxidative cyclisation products from squalene.

5.7.3 Non-Oxidative Cyclisations

Such cyclisations are rarer than the oxidative ones. All *trans*-squalene, not its 2,3-epoxide, is the actual intermediate and in some cases this has been proved. If [³H]-2,3-epoxidosqualene and [¹⁴C]-squalene are given to the *Tetrahymena pyriformis* protozoan, or to an enzyme preparation from such cells, tetrahymanol (144) labelled only with ¹⁴C was obtained; the anaerobic cyclisation of squalene in media containing D_2O or $H_2^{18}O$ gave tetrahymanol containing deuterium at C-3, or ¹⁸O at the hydroxyl alcohol group. From these results the tetrahymanol should be formed after a proton attack at the 2,3-double bond of squalene (Fig. 5.65).

A similar cyclisation process can be extended to other triterpenes non-hydroxylated at position C-3, for instance to fern 9-ene (145) and to diploptene (146). For the formation of the various sketeta in the non-oxidative cyclisations, the mechanisms and the reasons already explained for oxidative cyclisations can be invoked. Thus diploptene could derive from the 'chair-chair-chair-chair-chair-' squalene conformation (*cf* hydroxyhopanone, Fig. 5.64).

5.7.4 Cyclisations at both ends of the Squalene molecule

The biosyntheses of terpenes such as onocerin (147) and ambrein (148) are explained by admitting two independent electrophilic attacks at both ends of the squalene molecule (Fig. 5.66). The two attacks are both oxidative in the onocerin molecule, but not in ambrein; it is not known whether they are simultaneous or not.

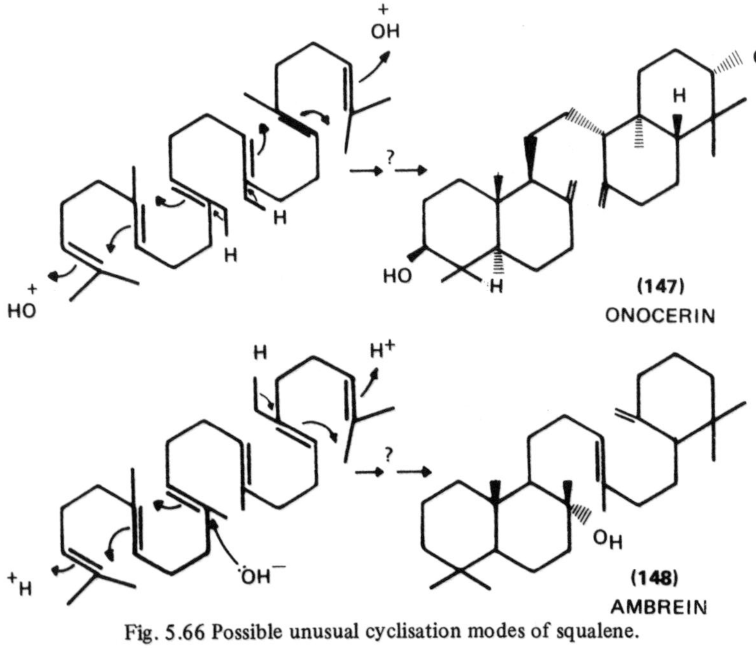

Fig. 5.66 Possible unusual cyclisation modes of squalene.

5.8 CAROTENOIDS

The carotenoids are yellow and red pigments, usually associated with chlorophyll in the chloroplasts of green plants; they are also present in algae, fungi and lichens. Animals cannot synthesise carotenoids and the presence of such pigments in animal fats arises from ingestion. The first substance called 'carotene' was isolated in 1831 by H. W. F. Wackenroder from carrots, and thereafter from green leaves. Kuhn showed, in 1931–33, that these red crystals were actually a mixture of three compounds, which he named α-, β,- and γ-carotene. Kuhn used elution chromatography for their separation. This technique had already been suggested by Tswett in 1906, but most chemists had ignored the technique. Karrer elucidated the structure of β-carotene, the most abundant of the three (85%) and of lycopene, a fourth carotenoid isolated from tomatoes, at about the same.

The structural features of carotenes are shown in Fig. 5.67: (a) A 40 carbon atom skeleton (the tetraterpenes) constituted of a unique polyisoprene chain, symmetrical with respect to the central $(15-15')$ bond. The symmetry of these compounds refers to the arrangement of the isoprene units, as the chain can be cyclised at either one or both of its ends, forming trimethyl-cyclohexyl rings. A few carotenes containing 45 or 50 carbon atoms are also known. These differ from the normal carotenes by the presence of one or two additional isoprene groups at the 2- and 2'- positions.
(b) numerous double bonds of E-configuration (the Z ones are quite rare) often conjugated into very long chromophores. These are responsible for their absorption of light in the visible region and, hence, for the yellow-red colour of such substances.

About 350 carotenoids are known. They are usually divided into three groups: the *carotenes,* which are hydrocarbons (e.g. Fig. 5.67); the *xanthophylls,* which the oxygenated (oxo or hydroxyl group) derivatives of carotenes (e.g. **(153)** and **(154)**, Fig. 5.68); the carotenic acids, obtained via a degradative oxidation of the C_{40} carotenes, such that the number of carbons is reduced by some units and one or two carboxylic functions are formed (e.g. bixin and crocetin, Fig. 5.68).

Some probable biosynthetic routes to the carotenes and xanthophylls based on known experimental data are shown in Figs. 5.69 and 5.70.

The dimerisation mechanism of the two GGPP units, giving the *cis-* and the *trans-* natural phytoenes (geometrical isomers with respect to the central double bond) is schematised in Fig. 5.71. It has close analogies with the reductive synthesis of squalene from two farnesyl units (Fig. 5.55).

The losses of the 1-*pro-S*-H (H_F) from two GGPP molecules in *cis*-phytoene, and of 1-*pro-R*-H (H_E) and 1-*pro-S*-H(H_F), each from different C_{20} unit in *trans*-phytoene have been shown experimentally. The prephytoene alcohol pyrospospate (PPAPP) has been obtained synthetically, and recognised as an actual intermediate in the biosynthesis of phytoenes. In Fig. 5.71 the same stereochemistry of PPAP is assumed as for presqualene alcohol pyrophosphate (PSAPP) (Fig. 5.55).

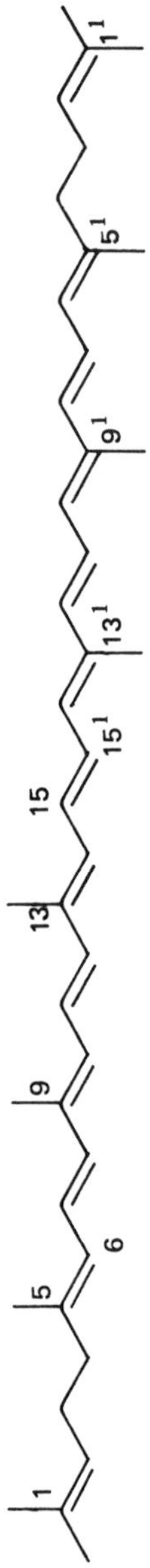

(149)
LYCOPENE
[λmax = 475.5 nm (hexane)]

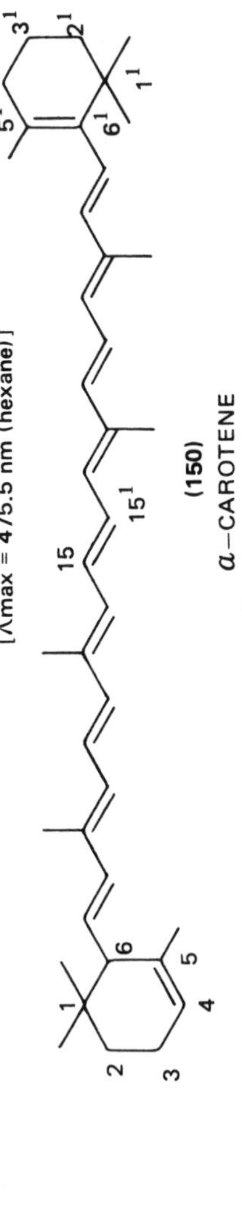

(150)
α–CAROTENE
[λmax = 447.5 nm (hexane)]

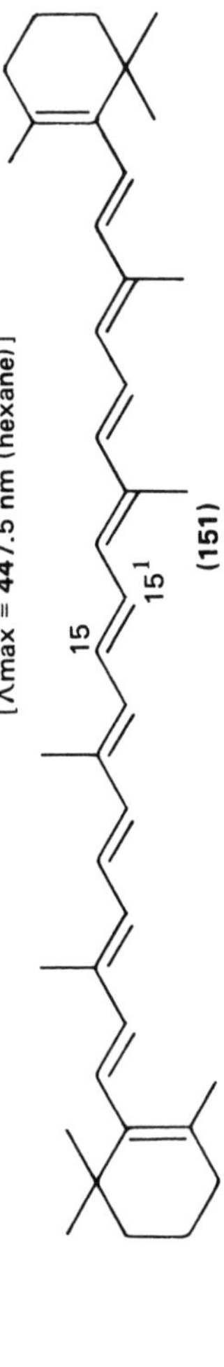

(151)
β–CAROTENE
[λmax = 451 nm (hexane)]

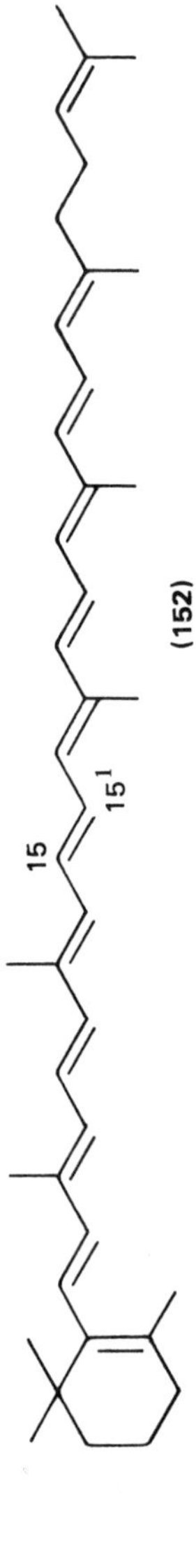

(152)
γ–CAROTENE
[λmax = 462 nm (hexane)]

Fig. 5.67 Carotenes.

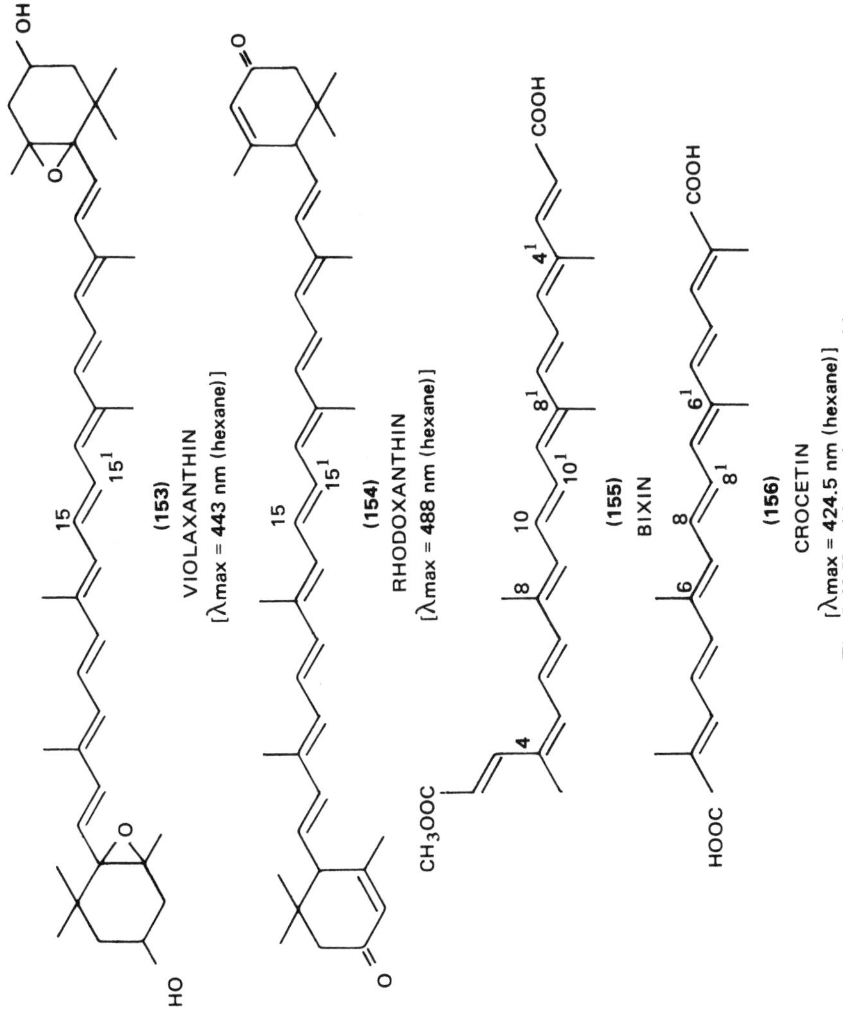

VIOLAXANTHIN
(153)
[λmax = 443 nm (hexane)]

RHODOXANTHIN
(154)
[λmax = 488 nm (hexane)]

BIXIN
(155)

CROCETIN
(156)
[λmax = 424.5 nm (hexane)]

Fig. 5.68 Xanthins and carotene acids.

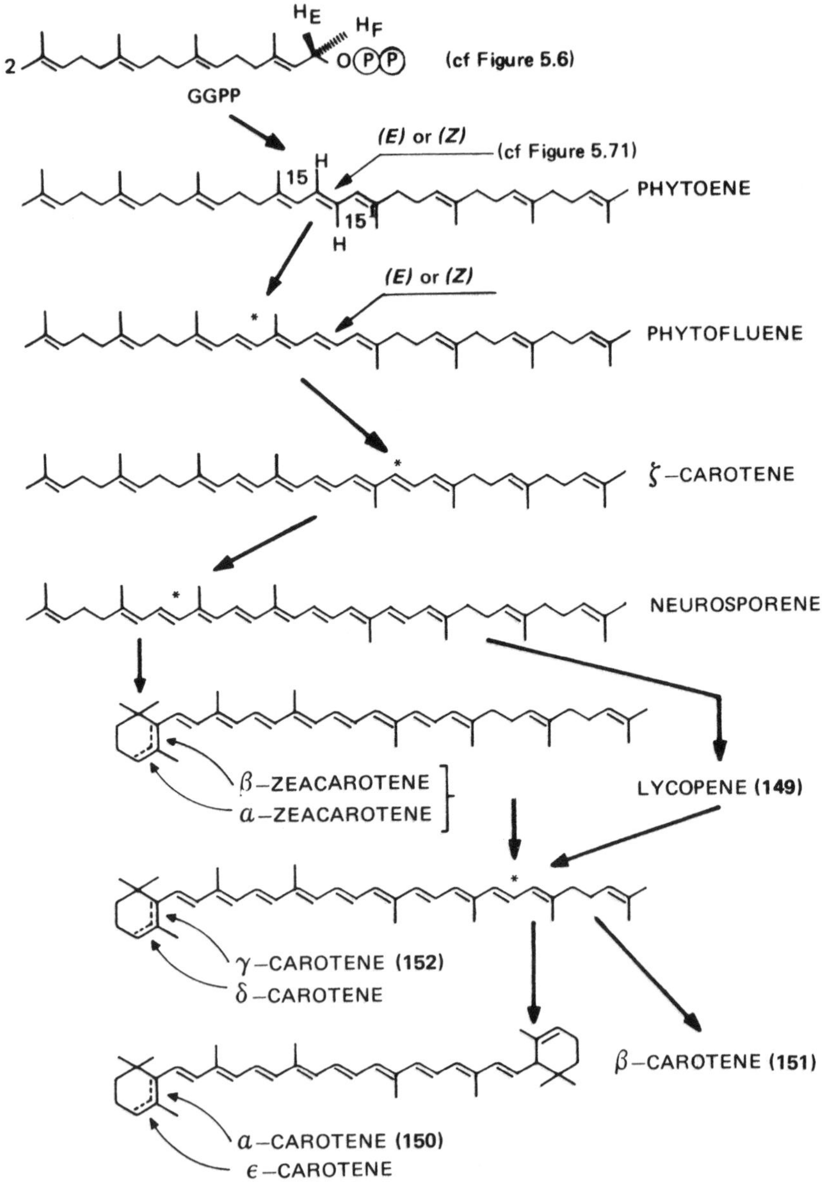

Fig. 5.69 Hypothetical scheme for the biosynthesis of carotenes in higher plants, fungi and non-photosynthetic bacteria. (The asterisk indicates the position of the newest-formed double bond).

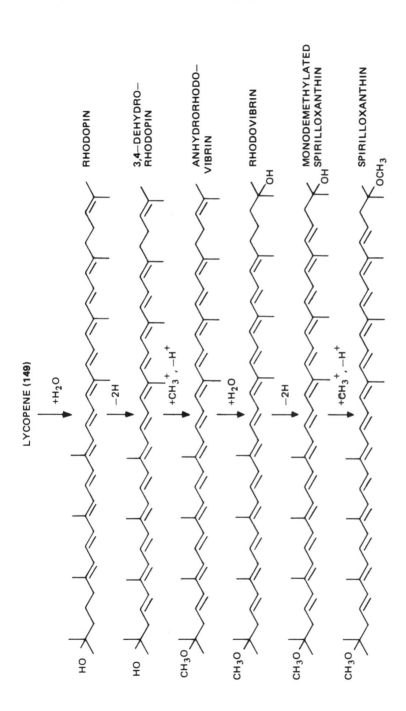

Fig. 5.70 Hypothetical scheme for the biosynthesis of some xanthins occurring in photosynthetic bacteria.

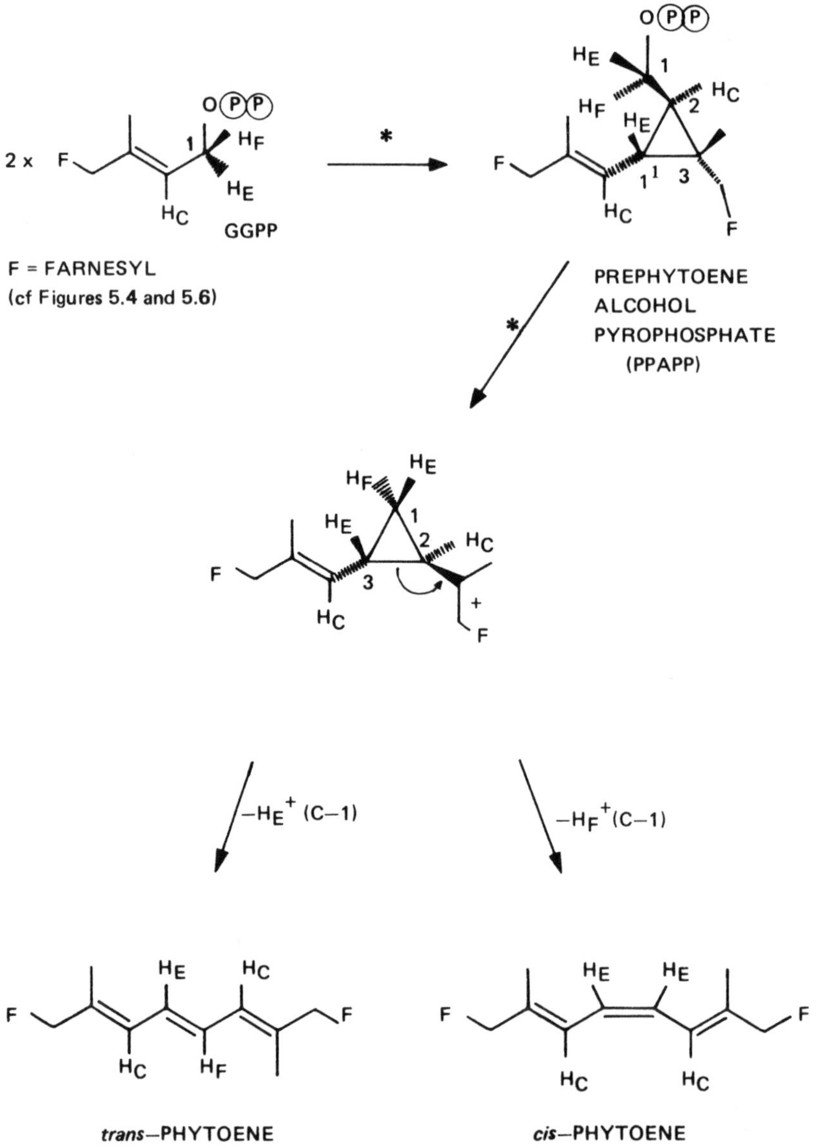

*See Figure 5.55

Fig. 5.71 Probable routes for the formation of the geometrical isomers of phytoene.

The mechanism of formation of the terminal cyclohexyl rings, found in many carotenes, possibly involves cyclisation induced by a Markownikoff proton attack to the last double bond of the polyisoprene chain (Fig. 5.72).

As previously mentioned, the carotenes are precursors in the biosynthesis of some other natural substances with the same number of carbons of other groups of terpenes. Safranal (C_{10}) has already been mentioned. The α and β-ionones (Fig. 5.72) can also be considered as apparent sesquiterpenes, since they originate from the oxidative cleavage of (cyclohexane) carotenes at the 9(10)-double bond. The presence of one carbon more in the irones, isolated from *Iris florentina*, presumably derives by a methylation process of a carotene molecule, or of a degradation product from it.

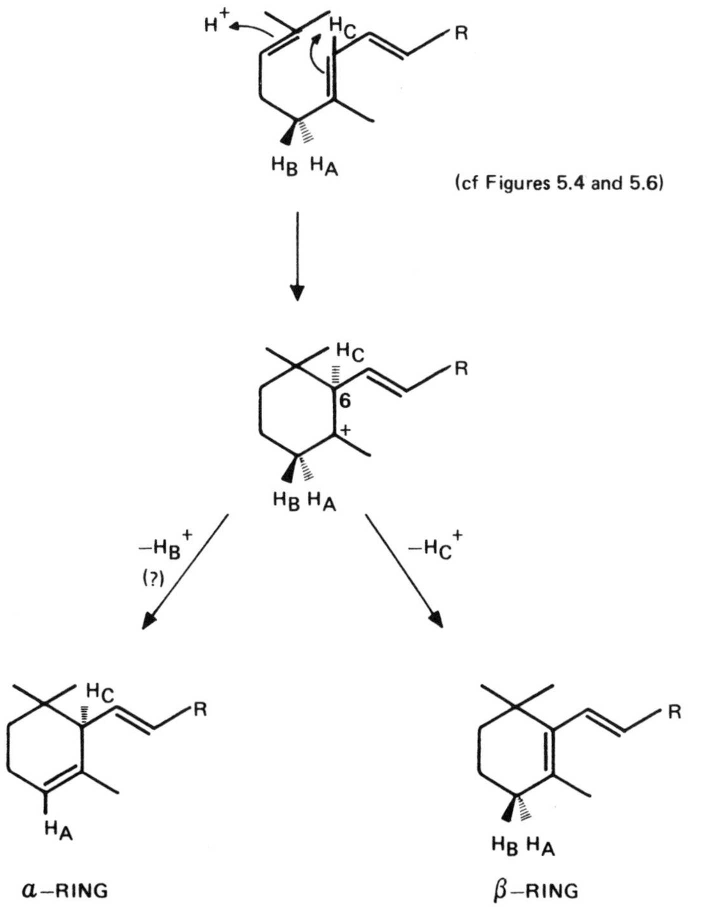

Fig. 5.72 Cyclisations of carotenoids; probable formation of the α- and β-rings.

A carotene (C_{20}) compound of considerable biological interest is all-*trans*-retinol (vitamin *A*) (Fig. 5.73). Vitamin A interferes with the processes of hormonal regulations and influences the biosynthesis of steroids; it plays an especially important role in the visual perception process. The visual pigment present in the human eye retina is an enzymic protein, opsin, to which 11-*cis*-retinal is bound rhodopsin.

The rhodopsin is decolourised by light. In this process, the retinal isomerises into its all-*trans* form, while the protein changes its secondary structure. This sequence of events produces the various stimuli determining vision. The process by which the initial rhodopsin-(11-*cis*)-retinal complex is reformed is not completely understood.

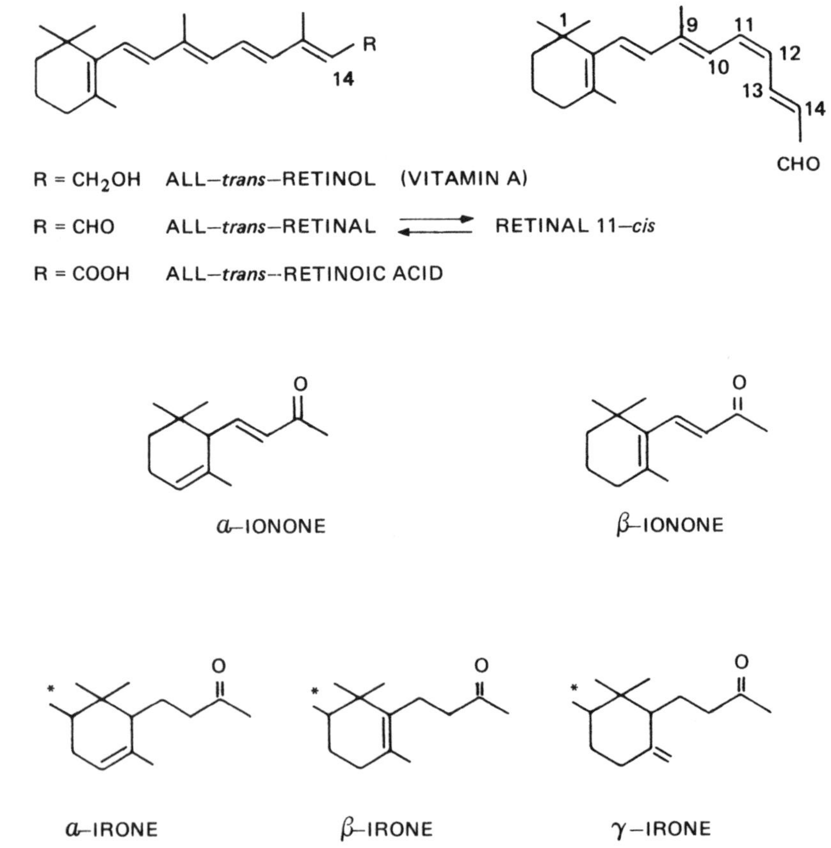

R = CH$_2$OH   ALL–*trans*–RETINOL   (VITAMIN A)

R = CHO   ALL–*trans*–RETINAL   ⇌   RETINAL 11–*cis*

R = COOH   ALL–*trans*–RETINOIC ACID

α–IONONE   β–IONONE

α–IRONE   β–IRONE   γ–IRONE

Fig. 5.73 Degradation products of carotenoids.

5.9 POLYISOPRENES

Polyisoprene chains made up with more than 10 C_5 units, bound tail to head, frequently occur in nature. They can exist in a free form, as for the alcohols called polyprenols and in the natural rubber polymers, or bound through C-C bonds to structures of a non-terpene origin.

(a)

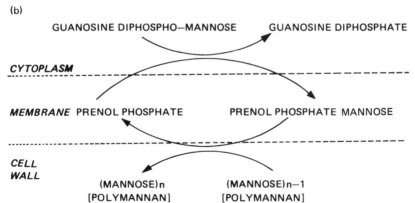

(157)
CASTAPRENOLS

(158)
DOLICHOLS

(159)
HEXAHYDROPOLYPRENOLS

(b)

GUANOSINE DIPHOSPHO—MANNOSE   GUANOSINE DIPHOSPHATE

CYTOPLASM

MEMBRANE PRENOL PHOSPHATE   PRENOL PHOSPHATE MANNOSE

CELL WALL
    (MANNOSE)n   (MANNOSE)n−1
    [POLYMANNAN]   [POLYMANNAN]

Fig. 5.74 (a) Natural polyprenes and (b) possible function of polyprenes in *Micrococcus lysodeikticus.*

Such polyisoprene chains arise by the combination of C_5 units, affording pyrophosphate esters of polyunsaturated alcohols, e.g. (157), having their double bonds *E* or *Z*, or partially *E* and partially *Z*, according to the stereochemistry of the chain lengthening mechanism. On using MVA stereospecifically tritiated at C_4, it could be established that all the *E*-bonds were *biogenetically-trans* (loss of the 4-*pro-S*-H), and the *Z*-ones *biogenteically-cis* (loss of the 4-*pro-R*-H) (see Fig. 5.7). The polyisoprenylphosphates can then undergo molecular modifications, especially hydrolyses and hydrogenations of double bonds, see (158) and (159), or interact with nucleophilic agents, giving alkylated products.

Amongst the most important families of polyprenols one could mention the *castaprenols* (157), present in the horse-chestnut tree (*Aesculus hippocastanum*) leaves, the *dolichols* (158), isolated from human, rabbit, ox, and sheep tissues, from yeast (*Saccharomyces cerevisiae*), and some hexhydro-polyprenols (159), produced by *Aspergillus fumigatus*.

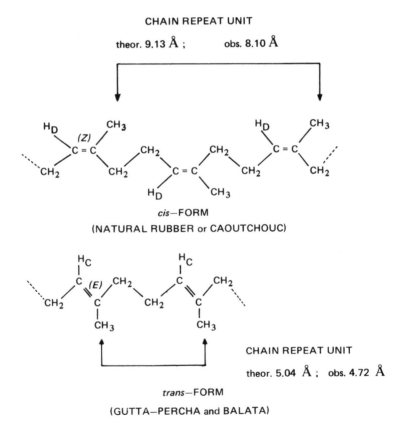

CHAIN REPEAT UNIT

theor. 9.13 Å ;   obs. 8.10 Å

cis—FORM

(NATURAL RUBBER or CAOUTCHOUC)

CHAIN REPEAT UNIT

theor. 5.04 Å ; obs. 4.72 Å

trans—FORM

(GUTTA—PERCHA and BALATA)

Fig. 5.75 Hydrocarbon rubbers (for the significance of H_C and H_D see Fig. 5.7).

Polyprenols are commonly associated with plasma membranes and, in bacteria, they possibly carry carbohydrate components from the cytoplasm into the external cellular walls (Fig. 5.74(b)).

The common natural rubber is obtained by coagulating the latex produced by some tropical or subtropical trees (for instance *Hevea-brasiliensis*) with acetic acid. The latex is extruded from the tree by cutting its bark. The latex consists of an emulsion of tiny droplets of what can be called the 'rubber hydrocarbon'. The hydrocarbon is composed of macromolecules with an average weight of 200 000–400 000. The monomeric isoprene unit is regularly bound head to tail, according to the schemes of Fig. 5.75. The difference between rubber and 'gutta-percha' or 'balata', is in the configuration and in the latter ones it is E. Some incorporation experiments of $(4S)$-$[4$-$^3H]$ and $(4R)$-$[4$-$^3H]$ MVA have been carried out in the *Hevea brasiliensis* latex; the 4-*pro-S* hydrogen of MVA is retained in rubber (Z-double bonds), thus differing from what usually happens during the biosynthesis of terpenes (E-double bonds), in which the 4-*pro-S* hydrogen is lost.

5.10 MEROTERPENES AND PRENYLATION REACTIONS

Besides normal terpenes, whose skeleta arise from MVA, many substances exist in nature which result from the union of terpene units with substances of a different biosynthetic origin. Cornforth named such substances *meroterpenes*. Lysergic acid, emetine and lupulone (Fig. 5.76) are typical examples of meroterpenes.

The terpene chain can be introduced into a molecule *via* a Mannich reaction, as for emetine, or through alkylation processes, by DMAPP or, more often, with one of its higher prenylogues (GPP, FPP).

These alkylation processes, known as *prenylation reactions* are very common in fungi and in plants (see for instance lupulone).

The prenyl group may be further transformed, especially *via* oxidative processes, after it has been bound to the different basic structures, such as a phenolic, aromatic ring of either polyketide (see Chapter 4) or shikimic (see Chapter 7 and 8) origin. Examples of such modified prenyl groups are shown in Fig. 5.77. The 2,3-benzofuran system is often formed, the 2'- and 3'- carbon atoms of the furan ring respectively deriving from C-4 and C-5 of MVA.

The prenyl group is clearly recognisable in certain naphthoquinones and anthraquinones biosynthesised by certain plants.

There are at least four different ways of synthesising the naphthoquinone system *in vivo*. The first one is the polyketide route outlined in Chapter 4. The second route utilises the whole molecule of shikimic acid, which gives seven out of ten carbon atoms of the naphthaquinone molecules (p. 000). The third and fourth routes are illustrated in Fig. 5.78(a) and (b) for chimaphilin, produced by *Pyrola media* (Pyrolaceae) and for alkannin produced by many Boraginaceae and Euphorbiaceae; one and two prenyl units are respectively involved (only one C_5 unit contributes to the naphthalenic skeleton).

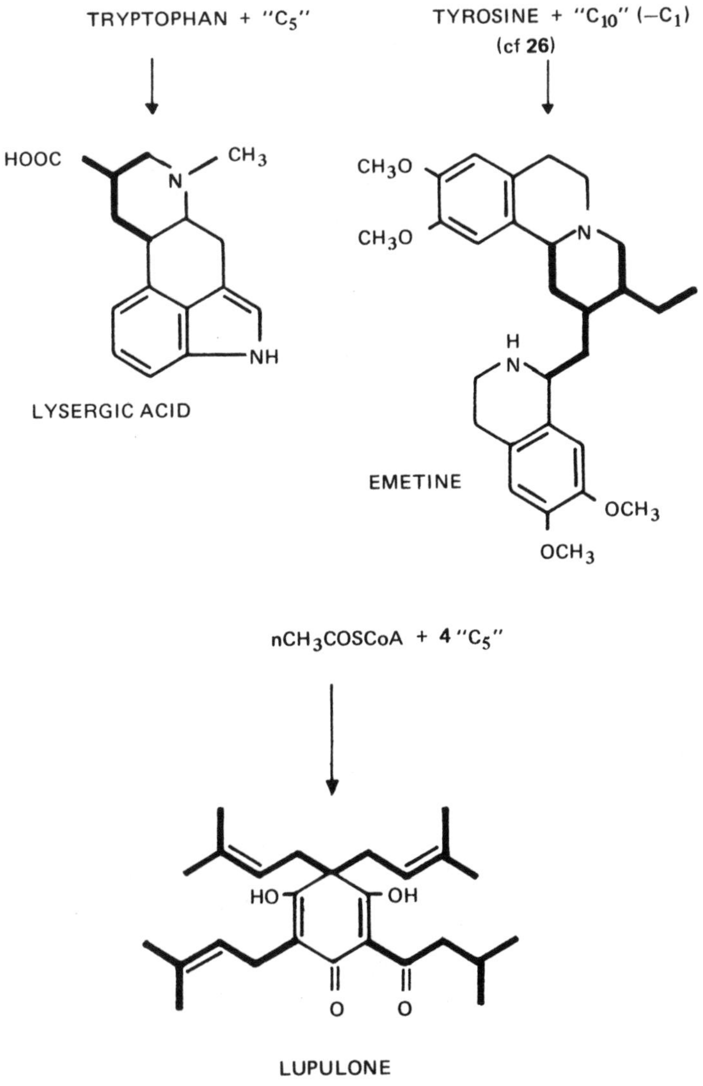

Fig. 5.76 Meroterpenes.

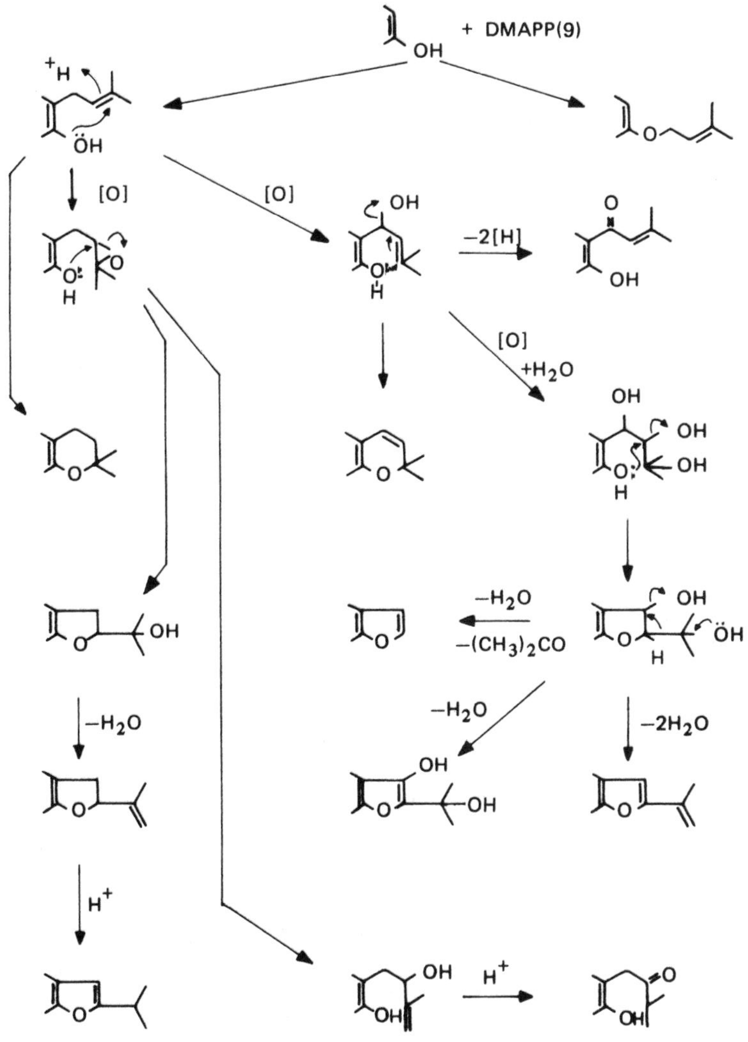

Fig. 5.77 Some of the more common transformations of the prenyl group.

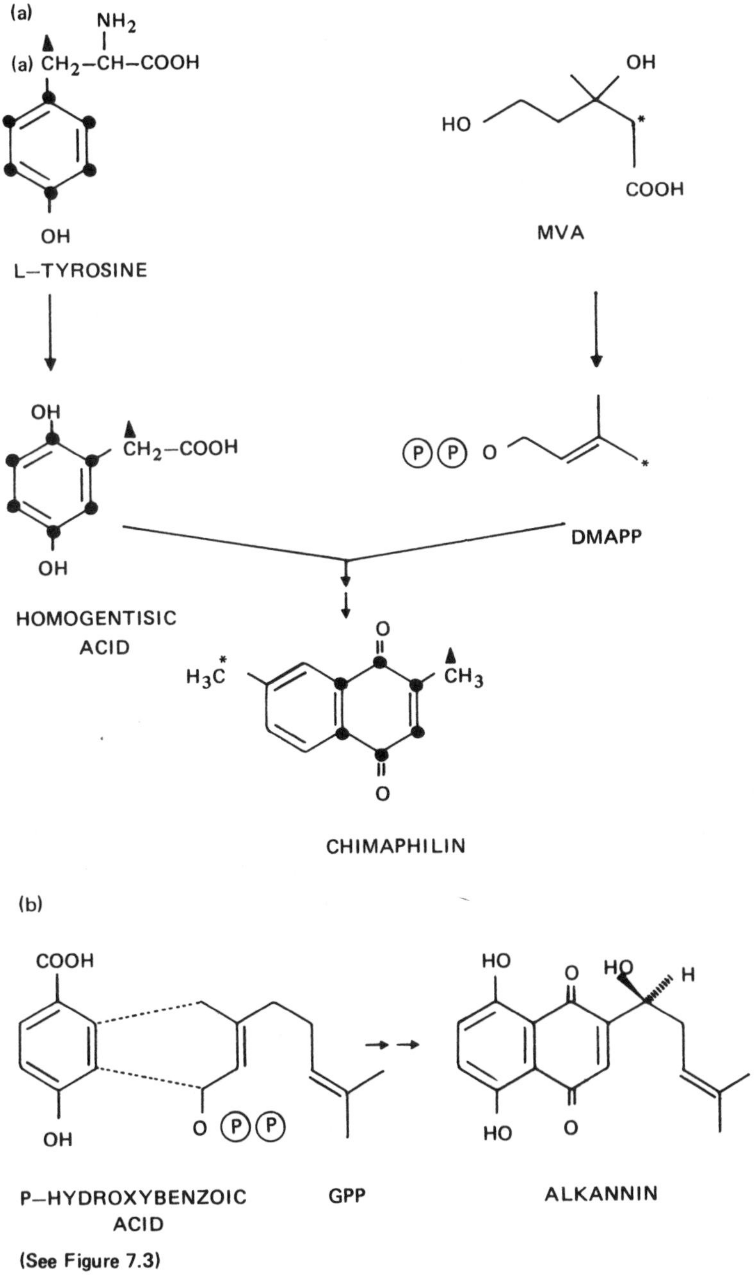

Fig. 5.78 Biosynthesis of the naphthoquinone meroterpenes.

The anthraquinones produced by plants can be distinguished into two categories, depending on how only one, or both A and C rings bear particular substituents (OH, OCH_3, CH_3 etc.). The compounds substituted in both rings A and C, usually with a C_{15} skeleton, have a polyketide origin (e.g. emodin) whilst the others are formed (Fig. 5.79) by prenylation of a naphthalene precursor, e.g. the acid (160).

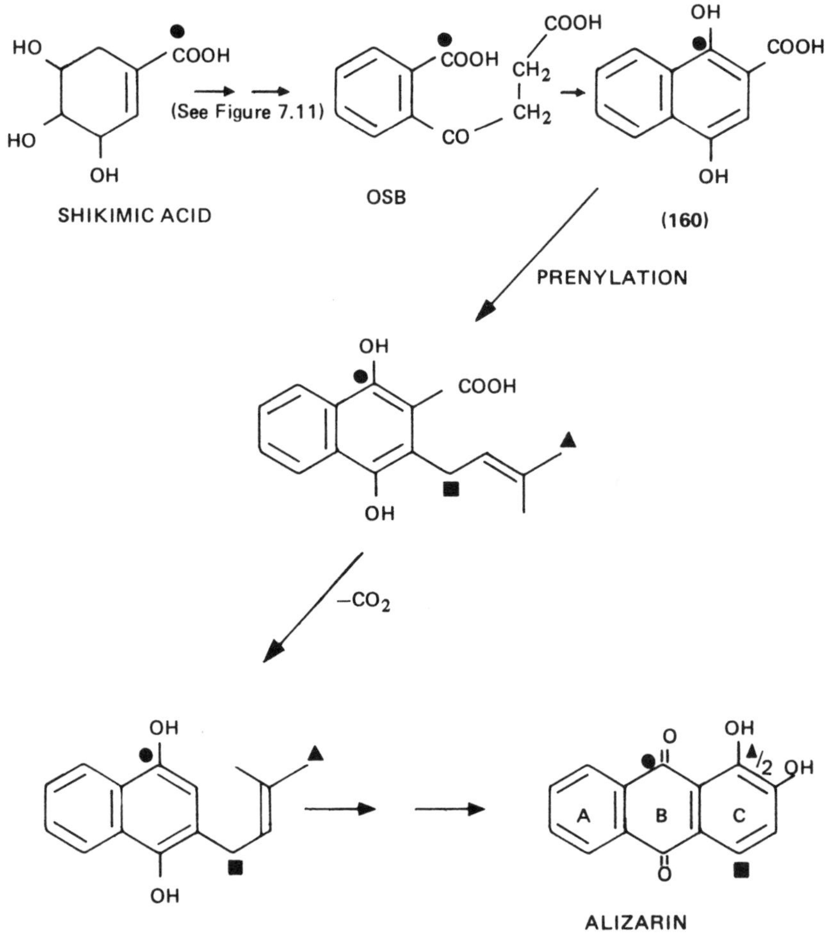

Fig. 5.79 Biosynthesis of alizarin.

By separately feeding shikimic acid (^{14}C at the carboxyl), and (R, S)-[5-^{14}C] mevalonic acid to *Rubia tinctorum* (Rubiaceae), radioactive alizarins have been isolated. Their chemical degradation indicated the following. Ring C derives from MVA, the final hydroxylation taking place on the C-3 and C-6 MVA carbon atoms; DMAPP links at the *meta*-position to the shikimic acid carboxyl carbon.

Three families of meroterpenoids have a particular biological importance: (a) the ubiquinones (161); (b) the plastoquinones (165)-(168) and the tocopherols (169); (c) the menaquinones (170) and the phylloquinones (171). All these substances possess a quinone system (p-benzo- or 1,4-naphtho-), and they transport electrons in living organisms, playing a fundamental role in oxidation-reduction chains (p. 000).

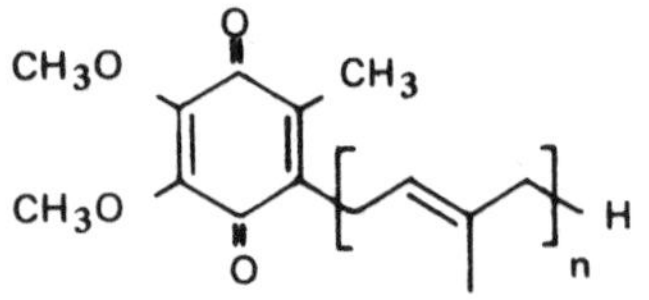

(161, n = 1 – 12)
UBIQUINONES

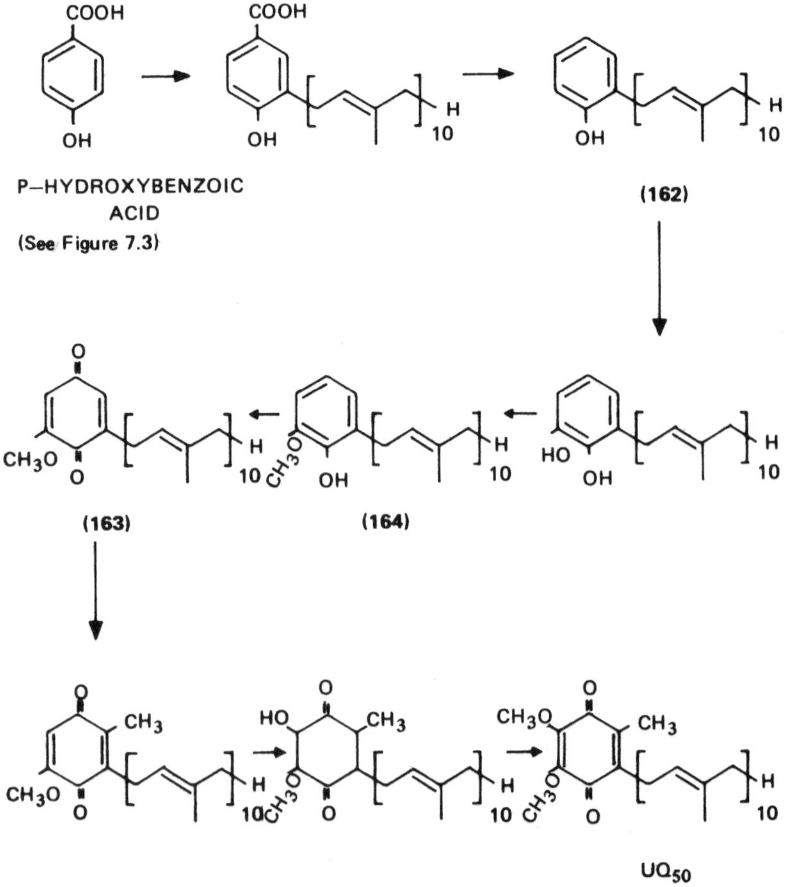

P–HYDROXYBENZOIC
ACID
(See Figure 7.3)

(162)

(163)   (164)

UQ_{50}

Fig. 5.80 Probable biosynthesis of ubiquinone-50 in *Rhodospirillium rubrum*.

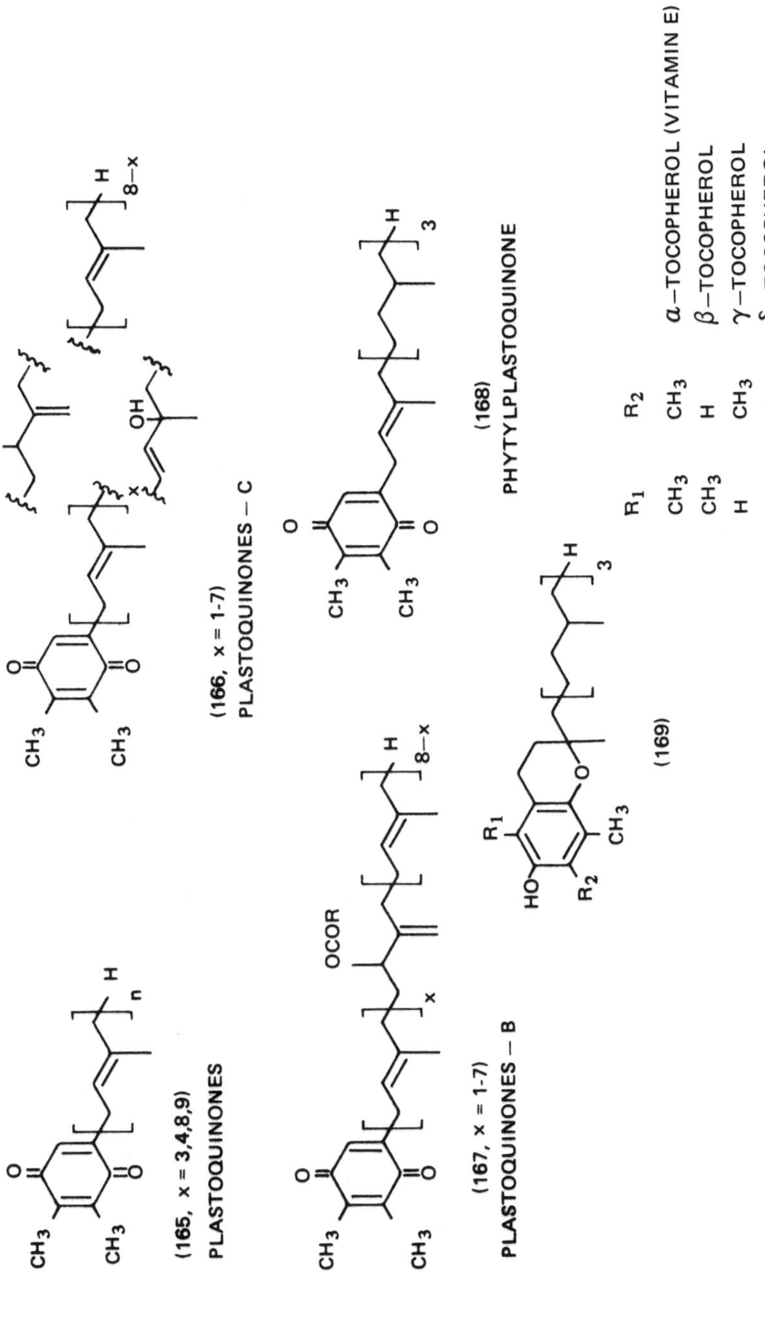

Fig. 5.81 Plastoquinones and tocopherols.

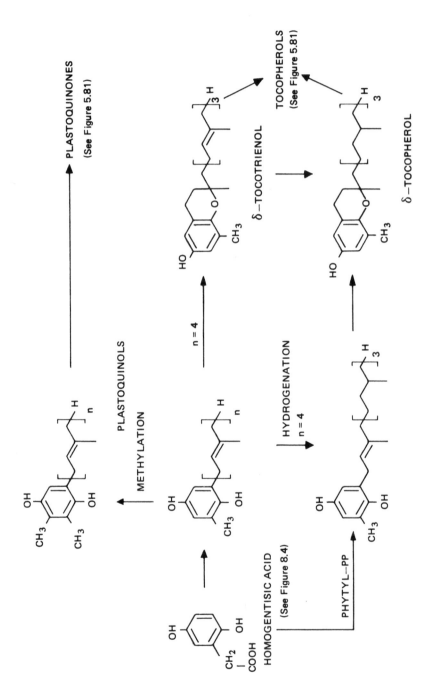

Fig. 5.82 Biosynthesis of plastoquinones and tocopherols.

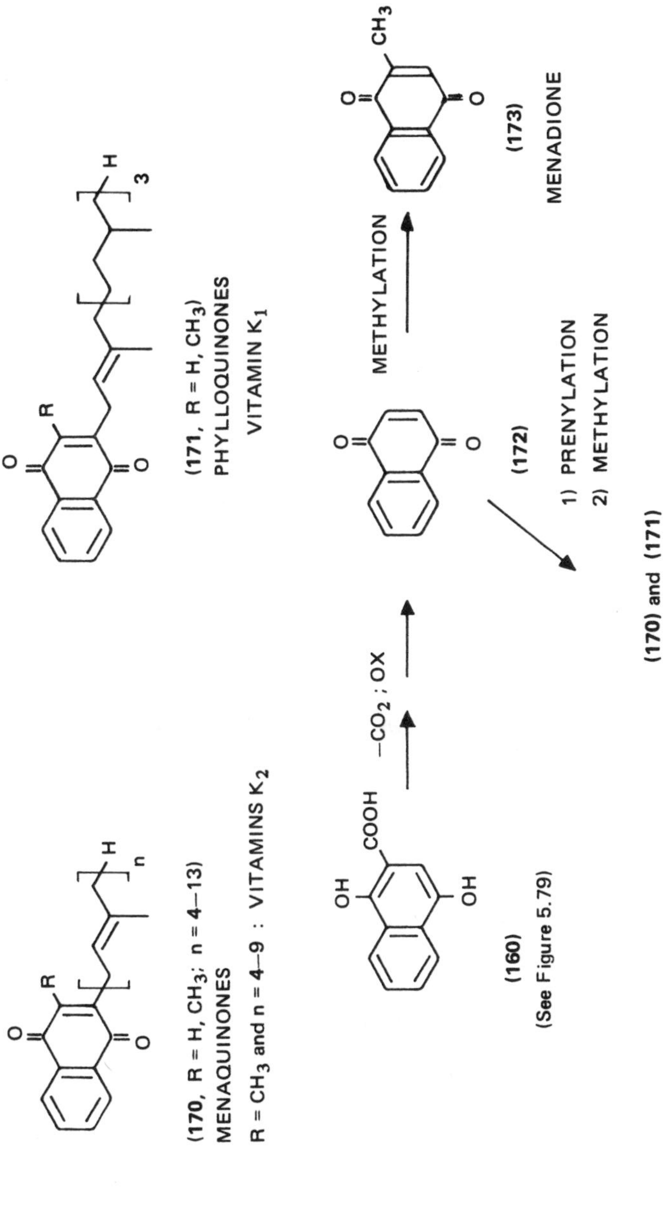

Fig. 5.83 Naphthoquinones with polyisoprenyl chains and their biosynthesis.

The ubiquinones (UQ), also called coenzymes-Q (CoQ), are implied in the primary metabolism of the cell, as constituents of the respiratory chain (Fig. A1.2). The number of the isoprene units (n) varies from 1 to 12, but the most abundant members in nature have n = 7 to 10.

The ubiquinones take their name from being present in virtually all living organisms, exceptions being the Gram-negative bacteria and the blue algae or cyanophytes. The ubiquinone molecule is very likely attached to the internal mitochondrial membrane, being able to penetrate into the lipid layer with its polyisoprene chain, leaving the quinone ring free and available for oxidation-reduction reactions.

The biosynthesis of ubiquinones is uncertain as far as the exact sequence of reactions is concerned (Fig. 5.80). In the cases examined the prenylation reaction, with a polyprenyl phosphate of appropriate length, takes place on p-hydroxy-benzoic acid, which is then decarboxylated. Both 2-decaprenylphenol (162) and 2-decaprenyl-6-methoxyphenol (164) have been intimated as intermediates in the UQ_{50} biosynthesis; the two O-methyl groups and the C-methyl function all come from methionine.

The plastoquinones (Fig. 5.81) differ from ubiquinones in the substitution pattern of the quinone ring and in the modifications of the isoprene chain, as in (166) (plastoquinones-C) and (167) (plastoquinones-B). The isoprene chain can also be partially saturated, as in (168) (phytylplastoquinones). Plastoquinones have been detected in all higher plants and algae so far examined. In the photosynthetic tissues the plastoquinones are concentrated within the chloroplasts, where they are associated with the lamellae.

The structural differences between the ubiquinones and plastoquinones are also caused by the different origins of the quinone ring: the plastoquinones derive from homogentisic acid, which is derived from p-hydroxyphenylpyruvic acid; the methyl group *meta* to the prenyl chain corresponds to C-3 of tyrosine, whilst the other methyl group comes from methionine (Fig. 5.82).

The tocopherols (169) are closely related to the plastoquinones in their structural and genetic aspects (Fig. 5.82). Their hydroquinone character is blocked by the formation of a chromene ring by interaction of the hydroxyl group *ortho* to the polyprenyl or phytyl chain with the first double bond of the latter. Another feature of the tocopherols is the completely saturated nature of their isoprene chains.

α-Tocopherol is undoubtedly the most important tocopherol because of its well-known vitamin action (Vitamin E). The remarkable antioxidising power of such vitamins is particularly important for protecting unsaturated lipids such as Vitamin A.

Menaquinones and phylloquinones are widely found in plants and in bacteria; they are distinguished from the former quinones in possessing a napththoquinone or methylnaphthoquinone system instead of the benzoquinone unit. Certain menaquinones have a specific anti-haemorrhagic activity: they are taken up with food, and participate in the hepatic synthesis of some plasmatic factors necessary to the blood coagulation process. The term Vitamin K is a generic one, and it refers to every quinone having anti-haemorrhagic activity.

The biosynthesis of menaquinones and of phylloquinones has not yet been completely investigated.

The alkylation of the naphthoquinone nucleus is likely to take place on both 1,4-napthoquinone (172) of shikimic origin and menadione (Vitamin K_3) (173), which derives from (172) via methylation with methionine.

SOURCE MATERIALS AND SUGGESTED READING

[1] L. Ruzicka, (1953), The Isoprene Rule and the Biogenesis of Terpenic Compounds, *Experientia*, 10, 357.

[2] A. Eschenmoser, L. Ruzicka, O. Jeger, and D. Arigoni, (1955), Eine stereochemische Interpretation der biogenetischen Isoprenregel bei den Triterpenen, *Helv. Chim. Acta*, 38, 1890.

[3] L. Ruzicka, (1959), History of the Isoprene Rule, *Proc. Chem. Soc.*, 341.

[4] R. B. Bates, Terpenoid Antibiotics in *Antibiotics*, ref. 13 of Chap. 4.

[5] J. Bonner, Rubber Biogenesis in *Biogenesis of Natural Compounds*, ref. 24 of Chap. 1.

[6] C. O. Chichester and T. O. M. Nakayama, The Biosynthesis of Carotenoids and Vitamin A in *Biogenesis of Natural Compounds*, ref. 24 of Chap. 1.

[7] T. W. Goodwin, The Biological Significance of Terpenes in Plants in ref. 16.

[8] F. W. Hemming, Polyisoprenoid Alcohols (Prenols) in ref. 16.

[9] W. D. Loomis, Biosynthesis and Metabolism of Monoterpenes in ref. 16.

[10] R. Mechoulam and Y. Gaoni, (1967), Recent Advances in the Chemistry of Hashish in *Progress in the Chemistry of Organic Natural Products*, ref. 1 of Chap. 1, 25.

[11] H. J. Nicholas, The Biogenesis of Terpenes in Plants in *Biogenesis of Natural Compounds*, ref. 24 of Chap. 1.

[12] W. Parker, J. S. Roberts and R. Ramage, (1967), Sesquiterpene Biogenesis, *Quarterly Rev.*, 21, 331 (1967).

[13] J. F. Pennock, The Chemistry of Isoprenoid Quinones in ref. 16.

[14] L. A. Porter, (1967), Picrotoxinin and Related Substances, *Chem. Rev.*, 67, 441.

[15] J. W. Porter and D. G. Anderson, (1967), Biosynthesis of Carotenes, *Ann. Rev. Plant Physiol.*, 18, 197.

[16] J. B. Pridham (Ed.), (1967), *Terpenoids in Plants* (Proceedings of the Phytochemical Group Symposium, Aberystwyth, April 1966), Academic Press.

[17] D. R. Threlfall, Biosynthesis of Terpenoid Quinones in ref. 16.

[18] B. C. L. Weedon, Structural studies in the Carotenoid Field in ref. 16.

[19] R. Stevens, (1967), The Chemistry of Hop Constituents, *Chem. Rev.*, 67, 19.

[20] G. Berti and F. Bottari, (1968), Constituents of Ferns in *Progress in Phyto-chemistry*, ref. 56 of Chap. 1, 1.

[21] J. W. Cornforth, (1968), Terpenoid Biosynthesis, *Chem. in Britain*, 4, 102.

[22] J. W. Cornforth, (1968), Olefin Alkylation in Biosynthesis, *Angew. Chem. Int. Ed. Engl.*, 7, 903.

[23] B. E. Cross, (1968), Biosynthesis of Gibberellins in *Progress in Phyto-chemistry*, ref. 56 of Chap. 1, 1.

[24] J. R. Hanson, (1968), *The Chemistry of the Tetracyclic Diterpenes*, Pergamon Press.

[25] J. R. Hanson, (1968), Recent Advances in the Chemistry of Tetracyclic Diterpenes in *Progress in Phytochemistry*, ref. 56 of Chap. 1, 1.

[26] J. R. Hanson and Achilladelis, (1968), The Biosynthesis of the Diterpenes, *Perfumery and Essential Oil Record*, 59, 802.

[27] J. W. Rowe, (1968), *The Common and Systematic Nomenclature of Cyclic Diterpenes*, U. S. Department of Agriculture, Forest Products Laboratory, Madison, Wisconsin.

[28] W. S. Johnson, (1968), Nonenzymic Biogenetic-like Olefin Cyclization, *Accounts Chem. Res.*, 1, 1.

[29] R. L. Stedman, (1968), The Chemical Composition of Tobacco and Tobacco Smoke, *Chem. Rev.*, 68, 153.

[30] E. E. Van Tamelen, (1968), Bioorganic Chemistry: Sterols and Acyclic Terpene Terminal Epoxides, *Accounts Chem. Res.*, 1, 111.

[31] R. Bryant, (1969), *Mono- and Sesqui-terpenoids*, Wiley-Interscience.

[32] G. M. Sanders, J. Pot, and E. Havinga, (1969), Some Recent Results in the Chemistry and Stereochemistry of Vitamin D and Its Isomers in *Progress in the Chemistry of Organic Natural Products*, ref. 1 of Chap. 1, 27.

[33] W. I. Taylor and A. R. Battersby (Eds.), (1969), *Cyclopentanoid Terpene Derivatives*, Marcel Dekker.

[34] A. R. Battersby, Biosynthesis of Terpenoid Alkaloids in ref. 40.

[35] A. F. Brodie *et al.*, 'Biological Function of Terpenoid Quinones in ref. 40.

[36] L. Canonica and A. Fiecchi, (1970), Structure and Biosynthesis of Ophiobolins *Rec. Progr. Org. Biol. Med. Chem.*, 2, 51.

[37] J. D. Connolly, K. H. Overton, and J. Polonsky, (1970), The Chemistry and Biochemistry of the Limonoids and Quassinoids in *Progress in Phyto-chemistry*, ref. 56 of Chap. 1, 2.

[38] J. W. Cornforth, Chemistry of Mevalonic Acid in ref. 40.

[39] L. J. Goad, Sterol Biosynthesis in ref. 40.

[40] T. W. Goodwin (Ed.), 1970, Natural Substances Formed Biologically from Mevalonic Acid *(Biochemical Society Symposia, No 29)*, Academic Press.

[41] F. W. Hemming, Polyprenols in ref. 40.

[42] V. Herout, (1970), Some Relations between Plants, Insects and Their Isoprenoids, in *Progress in Phytochemistry*, ref. 56 of Chap. 1, 2.

[43] P. Karlson, Terpenoids in Insects, in ref. 40.

[44] A. Lang, (1970), Gibberellins: Structure and Metabolism, *Ann. Rev. Plant Physiol.*, **21**, 537.

[45] S. Liaaen-Jensen, (1970), Developments in the Carotenoid Field, *Experentia*, **26**, 697.

[46] A. R. Pinder, (1970), *The Chemistry of Terpenes*, Wiley.

[47] G. Popjak, Conversion of Mevalonic Acid into Prenyl Hydrocarbons as Exemplified by the Synthesis of Squalene in ref. 40.

[48] H. Rudney, The Biosynthesis of Terpenoid Quinones in ref. 40.

[49] B. M. Trost, (1970), The Juvenile Hormone of Hyalophora cecropia, *Accounts Chem. Res.*, **3**, 120.

[50] J. C. Wallwork and F. F. Crane, (1970), The Nature, Distribution, Function and Biosynthesis of Prenyl Phytoquinones and Related Compounds in *Progress in Phytochemistry*, ref. 56 of Chap. 1, 2.

[51] O. Wiss and V. Gloor, Nature and Distribution of Terpene Quinones in ref. 40.

[52] S. Yamamoto and K. Bloch, Enzymatic Studies on the Oxidative Cyclization of Squalene in ref. 40.

[53] A. J. Birch, (1971), Terpenoid Compounds of Mixed Biogenetic Origin, *J. Agr. Food Chem.*, **19**, 1088.

[54] G. Britton, General Aspects of Carotenoid Biosynthesis in ref. 59.

[55] R. B. Clayton, The Biological Significance of the Terpenoid Pathway of Biosynthesis in ref. 59.

[56] M. J. O. Francis, Monoterpene Biosynthesis in ref. 59.

[57] D. Goldsmith, 1971, Biogenetic-type Synthesis of Terpenoid Systems in *Progress in the Chemistry of Organic Natural Products*, ref. 1 of Chap. 1, **29**.

[58] T. W. Goodwin, (1971); The Biogenesis of Terpenes and Sterols in *Rodd's Chemistry of Carbon Compounds*, 2nd ed. (S. Coffey, Ed.), Elsevier, **II/E**, supplement to, (1974), **II/E** (M. F. Ansell, Ed.).

[59] T. W. Goodwin (Ed.), (1971), *Aspects of Terpenoid Chemistry and Biochemistry* (Proceedings of the Phytochemical Society Symposium, Liverpool, April 1970), Academic Press.

[60] T. W. Goodwin, Algal Carotenoids in ref. 59.

[61] J. R. Hanson, 1971, The Biosynthesis of Diterpenes in *Progress in the Chemistry of Organic Natural Products*, ref. 1 of Chap. 1, **29**.

[62] V. Herout, 'Biochemistry of Sesquiterpenoids in ref. 59.

[63] O. Isler, (1971), *Carotenoids*, Birkhäuser Verlag, Basel.

[64] J. H. Law, (1971), Biosynthesis of Cyclopropane Rings, *Accounts Chem. Res.*, **4**, 199.

[65] S. Liaaen-Jensen, Recent Progress in Carotenoid Chemistry in ref. 59.

[66] J. Macmillan, Diterpenes − The Gibberellins in ref. 59.

[67] B. V. Milborrow, Abscisic Acid in ref. 59.

[68] G. P. Moss, (1971), The Biogenesis of Terpenoid Essential Oils, *J. Soc.*

Cosmet. Chem., 22, 231.

[69] K. H. Overton (Ed.), *Terpenoids and Steroids,* A Specialist Periodical Report, 1, 1971; 2, 1972; 3, 1973; 4, 1974; 5, 1975; 6, 1976; J. R. Hanson (Ed.), 7, 1977; 8, 1978; 9, 1979; The Chemical Society, London, (Biosynthesis of Terpenoids and Steroids by G. P. Moss in Vols. 1-3, and by D. V. Banthorpe and B. V. Charlwood in Vols. 4-7).

[70] A. Pfiffner, Juvenile Hormones in ref. 59.

[71] V. Plouvier and J. Favre-Bonvin, (1971), Les Iridoides et Séco-iridoides: Répartition, Structure, Propriétés, Biosynthèse, *Phytochemistry,* 10, 1697.

[72] H. H. Rees, Ecdysones in ref. 59.

[73] D. R. Threlfall and G. R. Whistance, Biosynthesis of Isoprenoid Quinones and Chromanols in ref. 59.

[74] O. B. Weeks, Biosynthesis of C_{50} Carotenoids in ref. 59.

[75] D. W. Banthorpe, B. W. Charwood, and M. J. O. Francis, (1972), The Biosynthesis of Monoterpenes, *Chem. Rev.,* 72, 115.

[76] J. R. Hanson, (1972-1977), Biosynthesis of Terpenoid Compounds: C_5-C_{20} Compounds in *Biosynthesis,* ref. 30 of Chap. 1, 1-5.

[77] J. R. Hanson, (1972), The Bicyclic Diterpenes in *Progress in Phytochemistry,* ref. 56 of Chap. 1, 3.

[78] L. J. Mulheirn and P. G. Ramm, (1972), The Biosynthesis of Sterols, *Chem. Soc. Rev.,* 1, 259.

[79] A. A. Newman (Ed.), (1972), *Chemistry of Terpenes and Terpenoids,* Academic Press.

[80] H. H. Rees and T. W. Goodwin, (1972-1975), Biosynthesis of Triterpenes, Steroids and Carotenoids in *Biosynthesis,* ref. 30 of Chap. 1, 1-3.

[81] D. Whittaker and D. V. Banthorpe, (1972), The Chemistry of Thujane Derivatives, *Chem. Rev.,* 72, 305.

[82] W. W. Epstein and C. D. Poulter, (1973), A Survey of Some Irregular Monoterpenes and Their Biogenetic Analogies to Presqualene Alcohol, *Phytochemistry,* 12, 737.

[83] D. A. Evans and C. L. Green, (1973), Insect Attractants of Natural Origin, *Chem. Soc. Rev.,* 2, 75.

[84] T. A. Geissman, (1973), The Biogenesis of Sesquiterpene Lactones in *Recent Advances in Phytochemistry,* ref. 57 of Chap. 1, 6.

[85] T. W. Goodwin, (1973), Recent Developments in the Biosynthesis of Plant Triterpenes in *Recent Advances in Phytochemistry,* ref. 57 of Chap. 1, 6.

[86] K. E. Harding, (1973), On the Stereochemistry of Biogenetic-like Olefin Cyclizations, *Bioorganic Chem.,* 2, 248.

[87] E. Havinga, (1973), Vitamin D, Example and Challenge, *Experentia,* 29, 1181.

[88] R. S. Irving and R. P. Adams, (1973), Genetic and Biogenetic Relationships of Monoterpenes in *Recent Advances in Phytochemistry,* ref. 57 of Chap. 1, 6.

[89] D. Loomis and R. Croteau, (1973), Biochemistry and Physiology of Lower Terpenoids in *Recent Advances in Phytochemistry*, ref. 57 of Chap. 1, **6**.

[90] J. G. MacConnel and R. M. Silverstein, (1973), Recent Results in Insect Pheromone Chemistry, *Angew. Chem. Int. Ed. Engl.*, **12**, 644.

[91] J. Polonsky, (1973), Chemistry and Biogenesis of the Quassinoids (Simaroubolides) in *Recent Advances in Phytochemistry*, ref. 57 of Chap. 1, **6**.

[92] J. Polonsky, (1973), Quassinoid Bitter Principles in *Progress in the Chemistry of Organic Natural Products*, ref. 1 of Chap. 1, **30**.

[93] G. Rucker, Sesquiterpenes, *Angew. Chem. Int. Ed. Engl.*, **12**, 793 (1973).

[94] S. K. Agarwal and R. P. Rastogi, (1974), Triterpenoid Saponins and Their Genins, *Phytochemistry*, **13**, 2623.

[95] G. A. Cordell, (1974), The Occurrence, Structure Elucidation and Biosynthesis of the Sesterterpenes, *Phytochemistry*, **13**, 2343.

[96] D. E. Gregonis and H. C. Rilling, (1974), The Stereochemistry of *trans*-Phytoene Synthesis. Some Observations on Lycopersene as Carotene Precursor and a Mechanism for the Synthesis of *cis*- and *trans*-Phytoene, *Biochemistry*, 13, 1538.

[97] J. MacMillan, (1974), Recent Aspects of the Chemistry and Biosynthesis of the Gibberellins in *Recent Advances in Phytochemistry*, ref. 57 of Chap. 1, **7**.

[98] J. A. Marshall, S. F. Brady and N. H. Andersen, (1974), The Chemistry of Spiro-4,5-Decane Sesquiterpenes in *Progress in the Chemistry of Organic Natural Products*, ref. 1 of Chap. 1, **31**.

[99] B. V. Milborrow, (1974), Chemistry and Biochemistry of Abscisic Acid in *Recent Advances in Phytochemistry*, ref. 57 of Chap. 1, **7**.

[100] Nomenclature of Carotenoids (Rules approved 1974), in *Pure and Applied Chemistry* – The Journal of IUPAC, **41**, p. 405, Butterworths.

[101] C. D. Poulter, O. J. Muscio and R. J. Goodfellow, (1974), Biosynthesis of Head-to-Head Terpenes. Carbonium Ion Rearrangements Which Lead to Head-to-Head Terpenes, *Biochemistry*, 13, 1530.

[102] R. B. Boar and C. R. Romer, (1975), Cycloartane Triterpenoids, *Phytochemistry*, **14**, 1143.

[103] D. Gross, (1975), Growth Regulating Substances of Plant Origin, *Phytochemistry*, **14**, 2105.

[104] G. Popjak, H. Ngan, and W. Agnew, (1975), Stereochemistry of the Biosynthesis of Presqualene Alcohol, *Bioorganic Chem.*, **4**, 279.

[105] E. D. Beytia and J. W. Porter, (1976), Biochemistry of Polyisoprenoid Biosynthesis, *Ann. Rev. Biochem.*, **45**, 113.

[106] G. Britton, (1976), Later reactions of Carotenoid Biosynthesis in *Pure and Applied Chemistry* – The Official Journal of IUPAC, **43**, 223.

[107] G. Britton, (1976), Biosynthesis of Carotenoids in *Chemistry and Biochemistry of Plant Pigments*, ref. 39 of Chap. 1, **1**.

[108] B. H. Davies and R. F. Taylor, (1976), Carotenoid Biosynthesis – The Early Steps in *Pure and Applied Chemistry* – Official Journal of IUPAC, 43, 211, Pergamon Press.

[109] R. M. Coates, (1976), Biogenetic-type Rearrangements of Terpenes in *Progress in the Chemistry of Organic Natural Products*, ref. 1 of Chap. 1, 33.

[110] G. A. Cordell, (1976), Biosynthesis of Sesquiterpenes, *Chem. Rev.*, 76, 425.

[111] T. W. Goodwin, (1976), Distribution of Carotenoids in *Chemistry and Biochemistry of Plant Pigments*, ref. 39 of Chap. 1, 1.

[112] W. S. Johnson, (1976), Biomimetic Polyene Cyclizations, *Agnew. Chem. Int. Ed. Engl.*, 15, 9.

[113] G. P. Moss and B. C. Weedon, (1976), Chemistry of the Carotenoids in *Chemistry and Biochemistry of Plant Pigments*, ref. 39 of Chap. 1, 1.

[114] L. J. Mulheirn, (1976-1977), Triterpenoids, Steroids and Carotenoids in *Biosynthesis*, ref. 30 of Chap. 1, 4-5.

[115] E. Rodriguez, G. H. N. Towers and J. C. Mitchell, (1976), Biological Activities of Sesquiterpene Lactones, *Phytochemistry*, 15, 1573.

[116] A. Stoessl, J. B. Stothers and E. W. B. Ward, (1976), Sesquiterpenoid Stress Compounds of the Solanaceae, *Phytochemistry*, 15, 855.

[117] M. Sumper, H. Reitmeier and D. Oesterhelt, (1976), Biosynthesis of the Purple Membrane of Halobacteria, *Angew. Chem. Int. Ed. Engl.*, 15, 187.

[118] G. A. Cordell, (1977), The Sesterterpenes in *Progress in Phytochemistry*, ref. 56 of Chap. 1, 4.

[119] P. E. Georghiou, (1977), The Chemistry of Vitamin D: The Hormonal Calciferols, *Chem. Soc. Rev.*, 6, 83.

[120] A. R. Pinder, (1977), The Chemistry of the Eremophilane and Related Sesquiterpenes in *Progress in the Chemistry of Organic Natural Products*, ref. 1 of Chap. 1, 34.

[121] R. L. Wain, (1977), Chemicals Which Control Plant Growth, *Chem. Soc. Rev.*, 6, 261.

[122] D. V. Banthorpe, B. M. Modawi, I. Poots and M. Rowan, (1978), Redox Interconversion of Geraniol and Nerol in Higher Plants, *Phytochemistry*, 17, 1115.

[123] B. V. Charlwood and D. V. Banthorpe, (1978), The Biosynthesis of Monoterpenes in *Progress in Phytochemistry* ref. 56 of Chap. 1, 5.

[124] M. F. Grundon, The Biosynthesis of Aromatic Hemiterpenes, *Tetrahedron*, 34, 143 (1978).

[125] C. D. Poulter and H. C. Rilling, (1978), The Prenyl Transfer Reaction. Enzymatic and Mechanistic Studies of the 1'-4 Coupling Reaction in the Terpene Biosynthetic Pathway, *Accounts Chem. Res.*, 11, 307.

[126] The Biochemical Functions of Terpenoids in Plants, (1978), A discussion organized by T. W. Goodwin, *Phil. Trans. Roy. Soc.*, London, B284, 439.

[127] I. Uritani, (1978), The Biosynthesis of Monoterpenes in *Progress in Phytochemistry*, ref. 56 of Chap. 1, **5**.

[128] R. F. Chandler and S. N. Hooper, (1979), Friedelin and Associated Triterpenoids, *Phytochemistry*, **18**, 711.

[129] Fifth International Symposium on Carotenoids, 1979, held at Madison, USA, 23-28 July 1978, *Pure and Applied Chemistry*, **51**, No 3.

[130] N. H. Fischer, E. J. Olivier and H. D. Fischer, (1979), The Biogenesis and Chemistry of Sesquiterpene Lactones, in *Progress in the Chemistry of Organic Natural Products*, ref. 1 of Chap. 1, **38**.

[131] R. G. Kelsey and F. Shafizadeh, (1979), Sesquiterpene Lactones and Systematics of the Genus Artemisia, *Phytochemistry*, **18**, 1591.

[132] P. Pant and R. P. Rastogi, (1979), The Triterpenoids, *Phytochemistry*, **18**, 1095.

[133] A. J. Weinheimer, C. W. J. Chang and J. A. Matson, (1979), Naturally Occurring Cembranes, in *Progress in the Chemistry of Organic Natural Products*, ref. 1 of Chap. 1, **36**.

[134] C. A. West, M. W. Dudley and M. J. Dueber, (1979), Regulation of Terpenoid Biosynthesis in Higher Plants, in *Recent Advances in Phytochemistry* ref. 57 of Chap. 1, **13**.

[135] D. E. Cane, (1980), The Stereochemistry of Allylic Pyrophosphate Metabolism, *Tetrahedron*, **36**, 1109.

[136] E. Caspi, (1980), Biosynthesis of Tetrahymanol by *Tetrahymena pyriformis:* Mechanistic and Evolutionary Implications, *Accounts Chem. Res.*, **13**, 98.

[137] S. Liaaen-Jensen, (1980), Stereochemistry of Naturally Occurring Carotenoids in *Progress in the Chemistry of Organic Natural Products* ref. 1 of Chap. 1, **39**.

Chapter 6

Biosynthesis of Steroids

6.1 INTRODUCTION

The steroid nucleus is shown in formula (1), where R^1, R^2, and R^3 differentiate various steroid families. Besides some tetracyclic triterpenes, called the trimethyl steroids, many substances of both animal and vegetable origin possess the skeleton (1).

These natural steroids are derived, by a series of chemical transformations, from the two parent triterpenes lanosterol and cycloartenol (Fig. 6.1).

The earlier steps of biosynthesis are common to all the natural steroids and proceed from acetic acid to lanosterol (or to cycloartenol) through mevalonic acid and squalene.

These steps have been discussed in the preceding chapter. The reactions through which the organisms elaborate the parent tetracyclic triterpenes will be described here. In many cases compounds possessing important physiological and hormone properties are produced.

It is generally recognised that all steroids in animals originate from lanosterol, whilst cycloartenol is the precursor of the steroids in plants. The basis of this deduction rests on three facts:

(a) the excellent incorporations of cycloartenol into the phytosteroids;

(b) the large amount of cycloartenol in plants and the rare occurrence of lanosterol in them;

(c) the inability of liver homogenates to utilise cycloartenol, rather than lanosterol, for synthesising cholesterol and other steroidal derivatives.

Plants probably contain an isomerase which transforms the 9,19-cyclopropyl ring of cycloartenol into the 10-methyl-Δ^8 system; it is located at an undefined point in the reaction sequence, which transforms the original triterpene structure into a steroidal one (loss of methyls C-4, C-14, etc.)

The fundamental problems involved in the metabolism of steroids include:

(a) The actual sequence of the reactions transforming a given steroid into another one.

(b) The stereochemistry and mechanisms of individual reactions.

(c) The catalytic actions of the enzymes involved and the structure of their active sites.

(d) The mechanisms for the control of the various transformations.

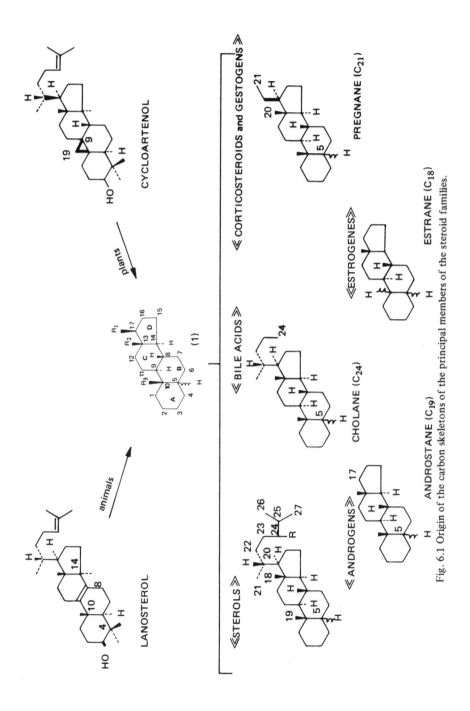

Fig. 6.1 Origin of the carbon skeletons of the principal members of the steroid families.

The structural differences in the natural steroids depends on many factors, including the length of the side chain [R^1 in (1)]; the functions in the side chain and at position 18 (R^2) and 19 (R^3); the number and position of oxygen functions and double bonds; the configuration of the asymmetric centres.

The traditional division of these diverse natural steroids is into families based on their different kinds of physiological action, for example, the sex hormomes (gestogens, androgens, estrogens), and the bile acids, etc.

In every family the structures, as well as their biogenesis, are remarkably homogenous (Fig. 6.1). Each skeleton offers two series of compounds, corresponding to a *trans* or *cis* junction between the rings A and B ring: 5α (or *allo*) and 5β (or *normal*) respectively. The 5β-cholestane isomer is also called coprostane.

6.2 STEROLS

The natural sterols possess the cholestane, ergostane, or stigmastane frameworks (Fig. 6.1). They usually bear a hydroxyl group at position 3 and a double bond at position 5(6) (see Figs. 6.2, 6.5, 6.7 and 6.8).

Depending on their origin, they are called *zoosterols* (from animals, especially vertebrates), *phytosterols* (from plants), *mycosterols,* (from fungi) and *marine sterols* (from invertebrate, marine organisms, particularly sponges).

Cholesterol is a typical zoosterol and is present in animal tissues in amounts between 0.05 to 5%. It is particularly abundant in the human brain, comprising 17% of the solid matter; it is also the main constituent of many biliary calculi.

The biosynthesis of cholesterol has been studied with rat liver homogenates, including subcellular fractions, for instance hepatic microsomes, and with enzymic preparations. Cholesterol arises from lanosterol differing in the absence of the C-4, and C-14 methyl groups, the double bond at position 5(6), and not at 8(9), and the saturated isooctyl side chain.

The *in vivo* manner in which these modifications are incorporated is not unique. The different metabolic changes occur at cellular sites separated from one another and are not necessarily interdependent. The enzymes promoting the various structural transformations are often not absolutely specific and several enzymes each catalysing the same reaction on a different substrate can. Thus, formation of cholesterol follows a metabolic grid (see p. 15) rather than a unique metabolic pathway.

In this respect every species differs slightly from other species and so too, do different tissues in the same species. The scheme (Fig. 6.3) indicates the mechanisms of each single step rather than their actual sequence.

Of the transformations in the conversion of lanosterol to cholesterol, the hydrogenation of the side chain double bond is most variable. In some cases it occurs early on in the sequence whereas in the human brain, for example, it appears to be the last step (see Fig. 6.2). The hydrogen adding to C-24 arises from the medium, whilst the one reacting at C-25 comes from NADPH. The

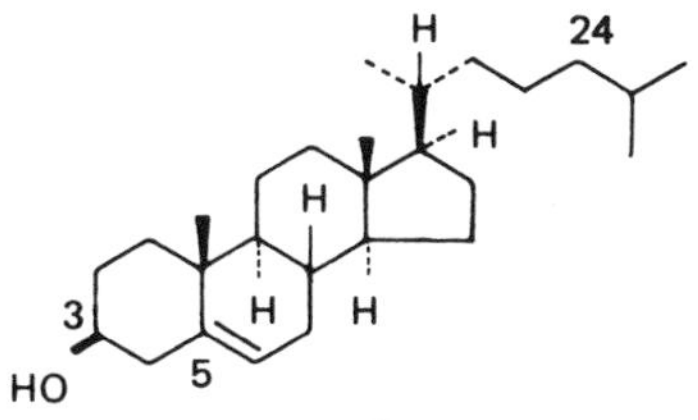

CHOLESTANE

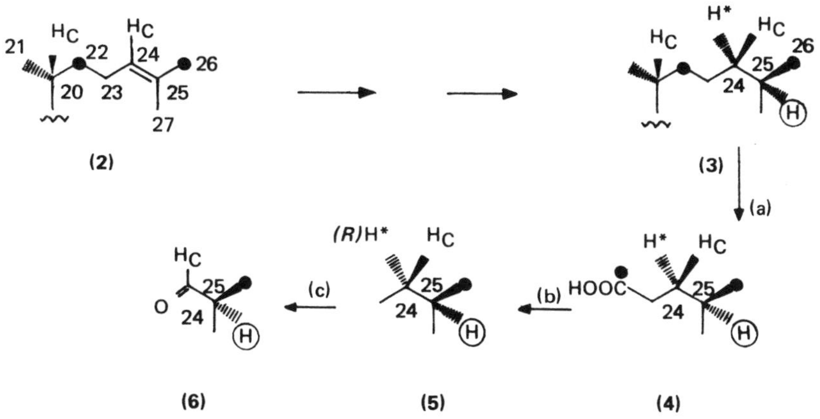

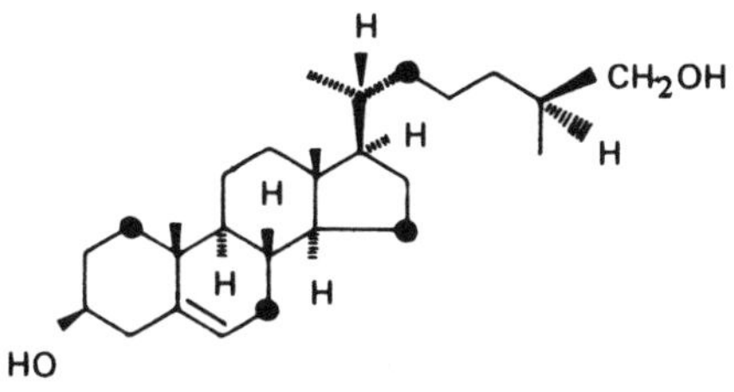

(2)   (3)

(6)   (5)   (4)

(25S)–26–HYDROXYCHOLESTEROL

● = C(2) of MVA

H_C = 4–*pro*–R of MVA

Ⓗ = H_B of NADPH

H* = H of the medium

(a) = BOVINE ADRENAL ENZYME PREPARATION

(c) = YEAST ALCOHOL DEHYDROGENASE

Fig. 6.2
Reduction of the
side-chain during
cholesterol biosynthesis.

stereochemistry of the addition is *cis*, with a *si,si*-stereospecificity (see Appendix 2). Attack on the *si* face of the trigonal carbon at C(24) has been demonstrated through reactions (a)-(c) of Fig. 6.2. Degradation of the side chain of cholesterol, biosynthesised from $(4R)$-$[2$-$^{14}C, 4$-$^3H]$ mevalonic acid was studied. The key step, from (5) to (6), was carried out with a yeast alcohol dehydrogenase, which is known to remove only the 1-*pro-R* proton [H* in (5)] of primary aliphatic alcohols. The stereochemistry of H^- addition to C-25 has been deduced from two results, the microorganism *Mycobacterium smegmatis* oxidises cholesterol to (25-*S*)-26-hydroxycholesterol (the absolute configuration of C-25 being obtained by X-ray studies) and this oxidation only involves one of the two methyls at the C-25 pro-chiral centre and which is the one corresponding to C-2 in mevalonic acid. Since this methyl group is *trans* with respect to the side chain in (2) (see Fig. 5.4) the attack of the NADPH hydride must consequently take place on the *si*-face of the C-25 carbon.

The loss of the methyl groups from lanosterol generally appears to take place in the following sequence: the 14α-methyl is first eliminated, (7) to (9), followed by 4α-methyl, (10) to (12), and finally the 4β-methyl, (12) to (14) (Fig. 6.3).

The fates of the various hydrogens, from Figs. 6.3 and 5.4 can be summarised as follows:

Mevalonic acid (Figure 5.4)		Lanosterol (Figure 6.3)		Cholesterol (Figure 6.3)
H_A (2-*pro-R*)	→	H_L (15β)	→	H_L (15α)
H_B (2-*pro-S*)	→	H_M (15α)	→	lost; taken up from the medium (at 15β)
H_B (2-*pro-S*)	→	H_N (7β)	→	lost; taken up from NADPH (at 7α)
H_A (2-*pro-R*)	→	H_O (7α)	→	H_O (7β)
H_E (5-*pro-R*)	→	H_P (6β)	→	H_P
H_F (5-*pro-S*)	→	H_Q (6α)	→	lost
H_C (4-*pro-R*)	→	H_T (5α)	→	lost

The 4β-methyl group, after the loss of the 4α-methyl, adopts the α-orientation on equilibration of C-4, that is having the more stable equatorial substituent. The methyl groups are eliminated after oxidation up to the aldehyde level at 14α, and to the carboxyl level for the 4-methyls, thus being eliminated respectively as formic acid and carbon dioxide. The oxidations proceed stepwise (methyl to primary alcohol to aldehyde to acid) and are catalysed throughout by mixed function microsomal oxygenases. The extended role of these oxygenases is unusual since alcohols are generally converted into aldehydes and carboxylic acids anaerobically, by dehydrogenases. The decarboxylations are favoured by the presence of a neighbouring ketone group at a position β to the point of attack. Thus C-3 is alternatively oxidised to carbonyl and reduced to alcohol during the demethylation processes.

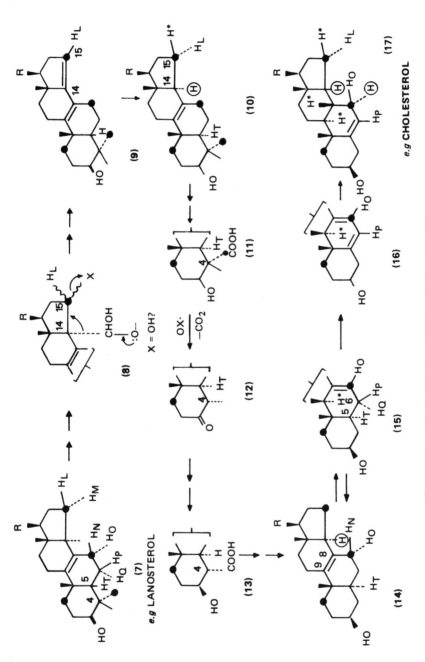

Fig. 6.3 Biosynthesis of cholesterol (for symbols see Fig. 6.2).

The intermediate (8) had not been isolated, suggesting that the deformylation is simultaneous to hydride ion abstraction from position 15. Reduction of the $\Delta^{8,14}$-diene (9) to the Δ^8-intermediate (10) takes place by attack of a proton from the medium to the 15β-position, and of hydride from NADPH on to the 14α-position; the addition is *trans*. The transposition of the double bond from the 8,9- to the 5,6-position implies the loss of the 7β, 6α and 5α hydrogens of lanosterol. The first hydrogen atom is lost in the (reversible) isomerisation of (14) to (15); the reaction is prototropic and takes place on exchanging the proton with the medium. Formation of the $\Delta^{5,7}$-diene (16) causes the loss of two extra hydrogens through a process requiring molecular oxygen.

Attempts to demonstrate a hydroxylation-dehydration process have been unsuccessful. It has therefore been hypothesized that a direct dehydrogenation by molecular oxygen, enzymatically activated, acts in a six-centre process (Fig. 6.4). This hypothesis accounts for the observed stereochemistry of the reaction involving *cis*-elimination.

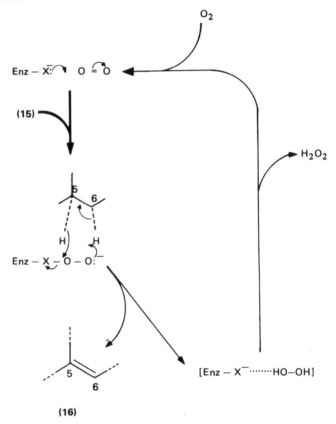

Fig. 6.4 Hypothetical mechanism for the introduction of the double bond into position-5(6) during cholesterol biosynthesis.

Final reduction of the 7(8)-double bond in the diene (16) is effected by NADPH, which transfers a hydride ion into the 7α-position, whilst the β hydrogen arises from the medium.

A protein, capable of specifically binding to squalene and to the successive intermediates leading to cholesterol, has been isolated. This protein has a dramatic accelerating effect on cholesterol biosynthesis through the above described reactions which are catalysed by microsomal enzymes. It is therefore thought that this protein, called 'sterol carrier protein' (SCP), originates in the endoplasmic reticulum of the hepatocytes, where it binds the various substrates and enables the activated forms to reach the various enzymes.

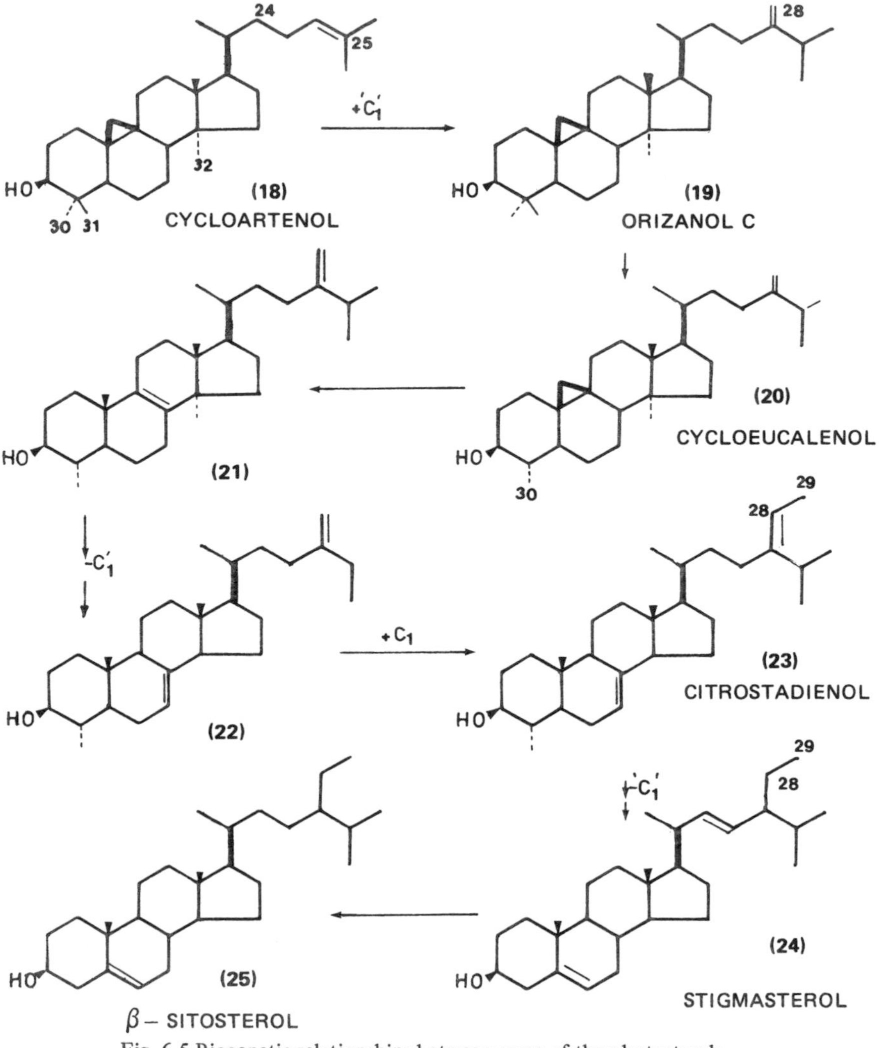

Fig. 6.5 Biogenetic relationships between some of the phytosterols.

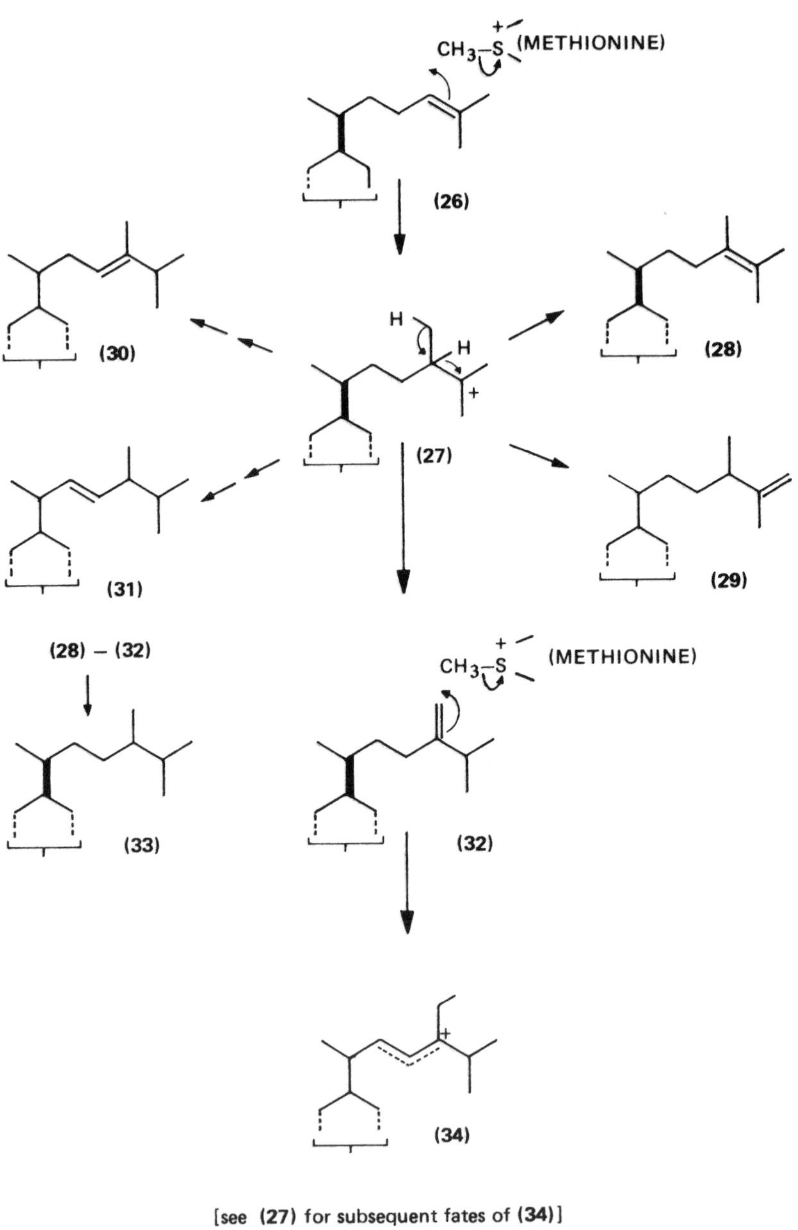

Fig. 6.6 Modifications to the side-chain of the phytosterols.

The commonest phytosterol is stigmasterol **(24)**, which is particularly abundant in soya-beans and in calabar seeds. A plausible biosynthetic scheme is shown in Fig. 6.5. As previously described, the phytosterols derive from cyclo-artenol. The opening of the cyclopropane ring probably occurs after loss of the 4α-methyl group and before loss of the 14α-methyl group. The 4α-methyl group is eliminated before the 4β-methyl group, in the same order found for the biosynthesis of cholesterol. The phytosterols can also be synthesised in plants through sequences varying in timing from the ones shown in Fig. 6.5. Thus compounds such as stigmasta-8,14,(Z)-24(28)-trien-3β-ol, and 31-norcycloartenol have been isolated from various tissues. Furthermore, in some plants, such as *Camelia sinensis,* the process of the double bond shift from the 8(9)- into the 7(8)-position involves the loss of the 7β-hydrogen, as occurs in rat livers and which contrasts with the results obtained in the *Aspergillus* and *Saccharomyces* fungi.

The phytosterol skeleta have the peculiarity of possessing one or two carbon atoms more than the cholestane series. These extra carbons are bound to the 24-position as a methyl or methylene group, or as an ethyl or ethylidene group, to produce ergostane (C_{28}) and the stigmastane (C_{29}) structures (Fig. 6.1). Both of the additional carbons arise from methionine. The mechanisms for the single and double methylation probably follow the scheme shown (Fig. 6.6), in which the first carbon adds to the 24(25)-double bond and the second one to the C-28-methylene group. The presence of the double bond in the side chain at positions distant from the cationic site is more difficult to explain (see, for instance, **(30)** and **(31)**).

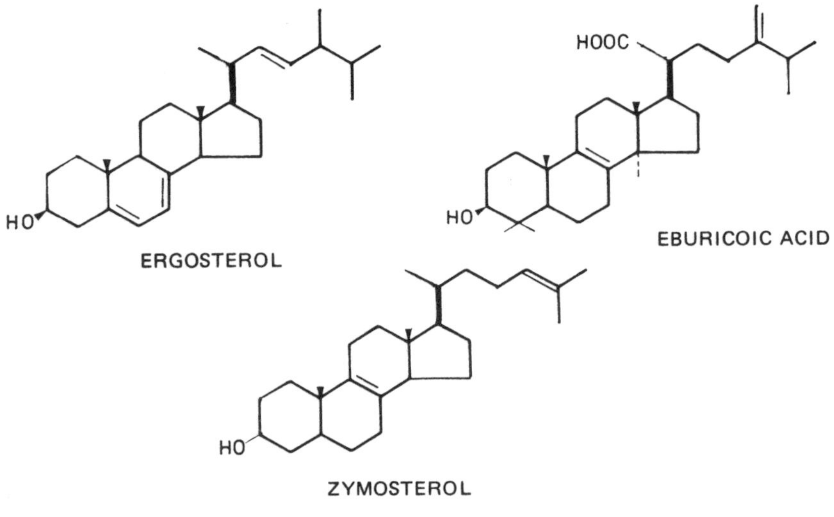

Fig. 6.7 Some common mycosterols.

Ergosterol is the most typical mycosterol, and is synthesised in large amounts by yeasts. Eburicoci acid, possessing a trimethylergostane skeleton (Fig. 6.7) and zymosterol are other mycosterols. The known features of the biosynthesis of these sterols still include some obscure points. They are very likely derived from lanosterol, in a similar manner to cholesterol; the yeasts also produce lanosterol. Some interesting sterols have been isolated from marine invertebrates including cholesterol, and poriferasterol from sponges; chalinasterol; (or ostreasterol) from sponges, sea anemones (coelenterates), and oysters (bivalve molluscs); stellasterol from star-fish (echinoderms); desmosterol from crustaceae; etc. (Fig. 6.8).

It is not known whether such organisms, as well as other invertebrates such as insects, always synthesis the steroid nucleus or simply modify sterols procured by feeding. Recent data indicate that some invertebrates are able to synthesise their own steroid structures. Steroid biosynthesis also takes place in procaryotes such as the blue algae and bacteria.

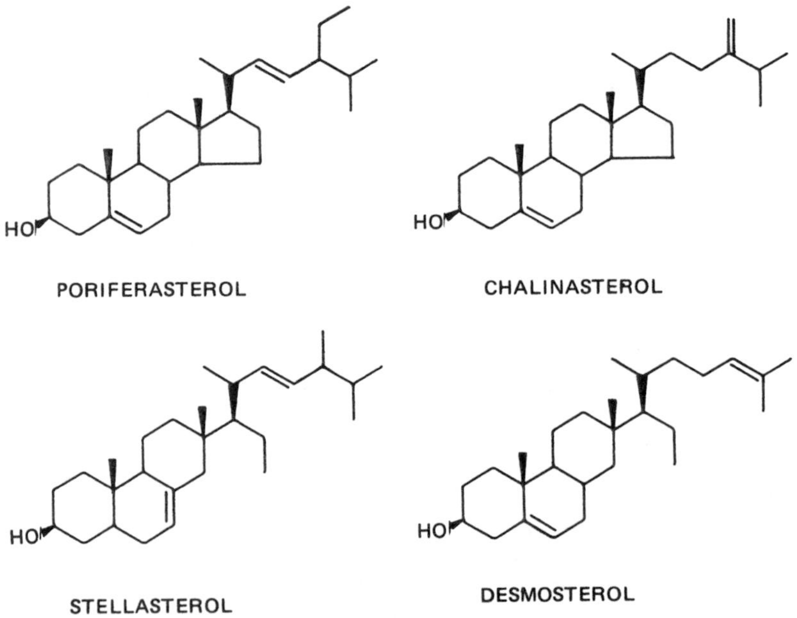

PORIFERASTEROL

CHALINASTEROL

STELLASTEROL

DESMOSTEROL

Fig. 6.8 Some sterols from marine invertebrates.

6.3 VITAMIN D

The vitamin D group of compounds are derived from the chemical and photochemical transformations of a few steroid nuclei (Fig. 6.9). Ultraviolet light promotes the fragmentation of the cyclohexadiene ring. Vitamin D_3 (cholecal-

ciferol) is formed from 7-dehydrocholesterol, present in skin tissue, by exposure
to sunlight. The most active compound in humans is Vitamin D_2, (ergocalciferol),
which is derived from ergosterol and which is produced in large amounts by
irradiated yeast cells. The D-group of vitamins play an important role in the
metabolism of calcium and phosphorus in bones; a deficiency of the vitamin in
children causes bone deformations, called rickets.

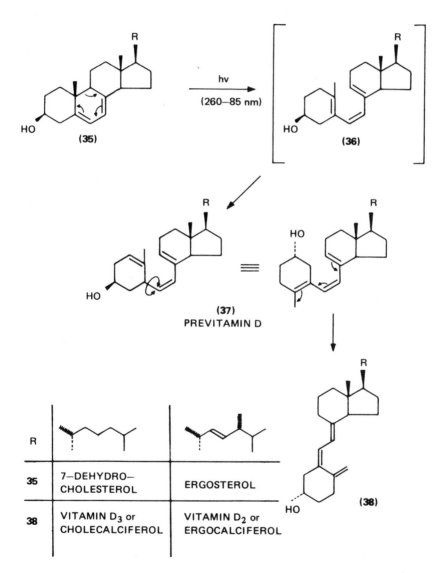

Fig. 6.9 Formation of vitamin D compounds.

6.4 TRANSFORMATION OF STEROLS IN PLANTS

Cholesterol and the phytosterols can undergo a variety of structural transformations in plants, especially involving the side chains. The resulting products are classified into two families: the *cardiotonic glycosides*, and the *steroidal saponins*.

6.4.1 Cardiotonic Glycosides

These compounds are so called because of their powerful heart stimulating activity. On acidic or enzymatic hydrolysis they afford aglycones, sometimes called genins. These genins possess the usual steroidal nucleus and a characteristic side chain, in the form of a butenolide ring for the *cardenolide aglycones*, such as digoxigenin a 2-pyrone ring for the *scilladienolide*, or *bufadienolide*, *aglycones*, such as scillarenin, two carbon atoms only as in the *digitenolide aglycones*, such as diginigenin (Fig. 6.10).

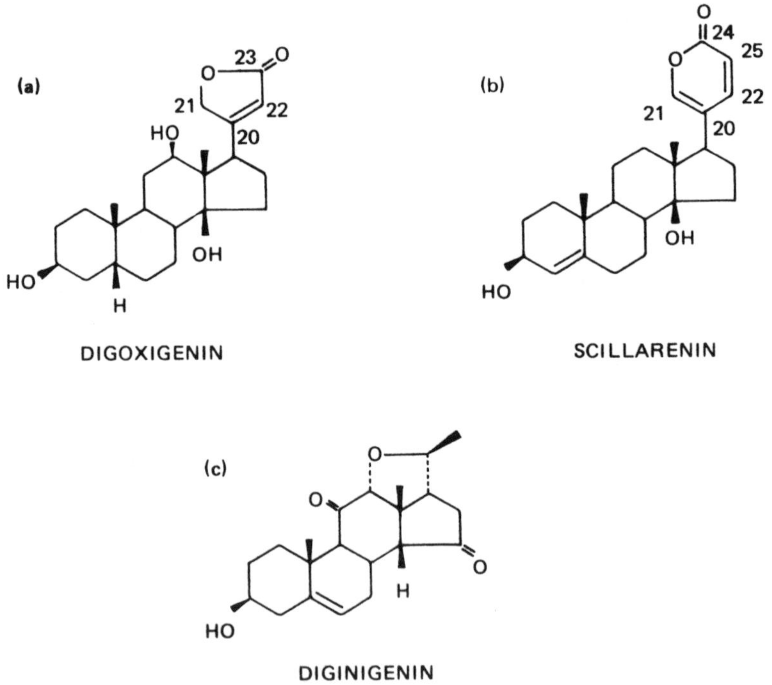

Fig. 6.10 Some aglycones of cardiotoxic sterols, (a) cardenolide type; (b) bufadienolide type; (c) digitenolide type.

The molecule k-strophenthoside (Fig. 6.11), a constituent of *Strophanthus kombé*, contains the structural elements usually found in cardenolides besides the αβ-unsaturated-γ-lactone ring; a tertiary hydroxyl group in the 14β-position, a *cis* junction between the A/B and C/D rings, and a *trans*-8β-H, and 9α-H connection between the B, C rings. Sometimes anomalous structures are also found, such as a *trans*-junction between the A, B rings (5 α-H) or the α-orientation of the butenolide ring at C-17 and of the hydroxyl at C-3. Oxidation of the C(10) methyl substituent of the steroidal skeleton to an alcohol or aldehyde group is typical of the *Strophanthus* group of *cardenolides*, whilst the *Digitalis* family frequently have alcohol groups at positions 12 and 16. The glycoside residue bound to position 3 varies, being constituted of 2 to 5 sugar units, such as D-glucose, L-rhamnose, and other rare monosaccharides including digitoxose (39), digitalose (40) and cymarose (41).

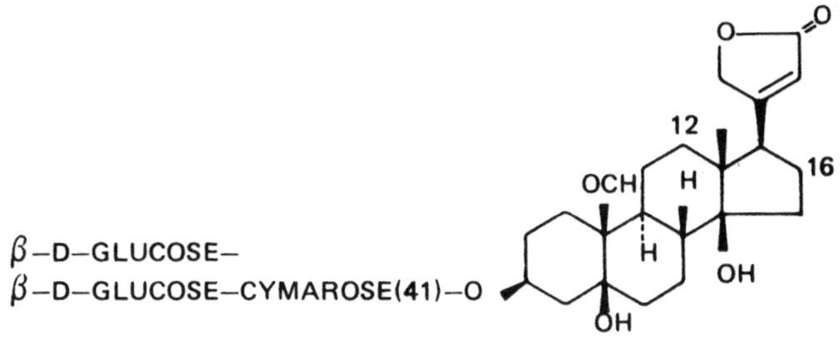

β−D−GLUCOSE−
β−D−GLUCOSE−CYMAROSE(41)−O

k−STROPHANTHOSIDE

CHO	CHO	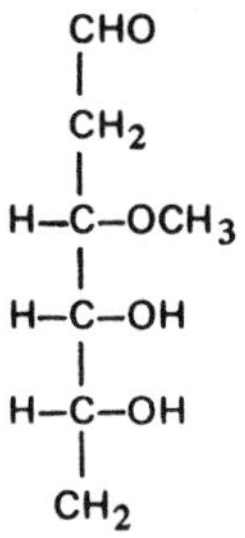 CHO
CH₂	H— C-OH	CH₂
H–C-OCH₃	CH₃O–C–H	H–C-OCH₃
H–C–OH	HO–C–H	H–C–OH
H–C–OH	H–C–OH	H–C–OH
CH₂	CH₃	CH₃
(39)	(40)	(41)

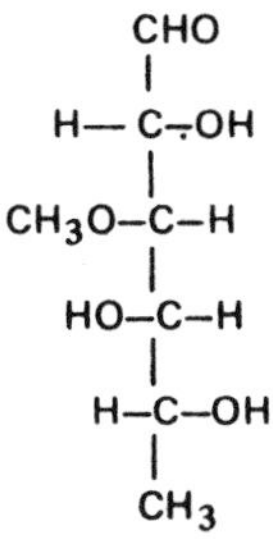

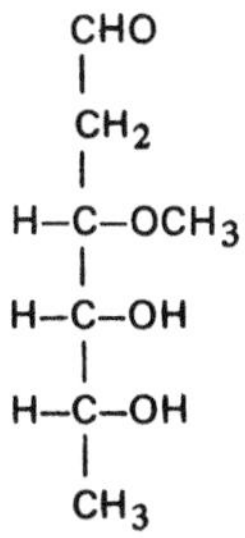

Fig. 6.11 A typical cardenolide glycoside and some monosaccharides frequently found in such glycosides.

The cardenolide glycosides are commonly found in the Apocynaceae, Liliaceae, Ranunculaceae and Scrophulariaceae plant families. They probably originate from a pregnane skeleton (C_{21}) intermediates, itself derived from cholesterol or phytosterols by processes similar to those occurring in vertebrates (see Fig. 6.20). Progesterone (43) is specifically incorporated into cardenolides, the two carbon atoms necessary to complete the butenolide ring coming from an acetate unit. A possible biosynthetic scheme of cardenolides is shown in Fig. 6.12 for digitoxigenin.

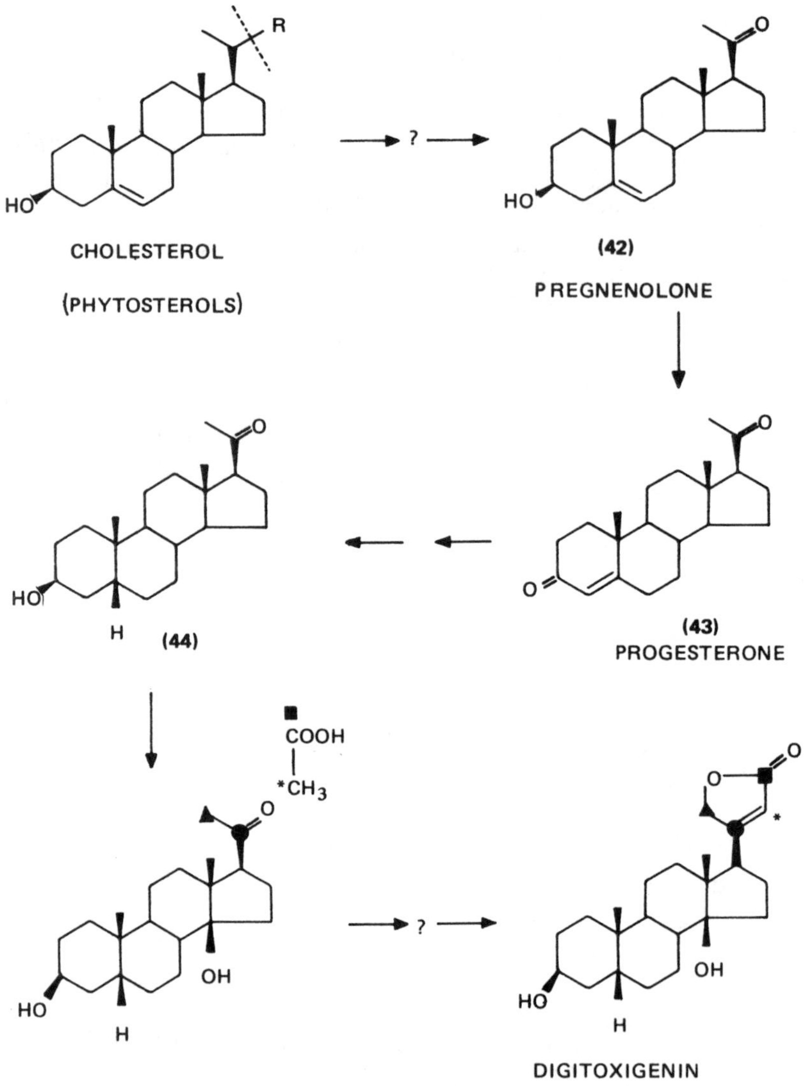

Fig. 6.12 Probable biosynthesis of cardenolides.

The presence of a 14β-hydroxyl group in cardenolides is of biosynthetic interest. At first sight it appears to indicate the hydroxylation of a saturated carbon with inversion of configuration, which fact would be in disagreement with the general mechanism of hydroxylations promoted by monooxygenases.

Scilladienolides, particularly widespread in the *Scilla* genus of plants, are also called bufadienolides (or bufatenolides), as some of them have been found, in their free forms or conjugated with suberylarginine, in the poison secreted by the skin and parotid glands of certain toads. Suberylarginine is bound to the 14β-hydroxyl group, as in bufotoxin isolated from the poison of the common European toad (*Bufa vulgaris*) (Fig. 6.14). Apart from the lactone ring, the bufadienolides show the same structural features as the cardenolides.

The mechanism of formation of the 2-pyrone ring has still not been established biogenetically although cholesterol appears to be an obligatory intermediate of the bufadienolides, in both plants and toads. As in the cardenolides, the lactone ring should be formed in plants following the degradation of the isooctyl side-chain to give pregnane derivatives. The three carbon atoms necessary to complete the pyrone ring do not arise from acetic acid but from an as yet unknown C_3 unit. Certain derivatives of cholanic acid (56) have been found in toads which act as better precursors for the bufadienolides than the pregnane derivatives. In animals, rather than plants, all the δ-lactone ring carbons could arise from the cholesterol side chain.

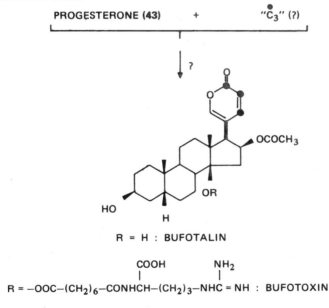

PROGESTERONE (43)   +   "$\overset{\bullet}{C}_3$" (?)

R = H : BUFOTALIN

$$R = -OOC-(CH_2)_6-CONHCH-(CH_2)_3-NHC=NH \; : \; BUFOTOXIN$$
$$\qquad\qquad\quad\;\; \overset{|}{COOH} \qquad\;\; \overset{|}{NH_2}$$

(SUBERYLARGININE)

Fig. 6.13 Origin of the bufadienolide.

The digitenolides have no cardiokinetic activity, both as their aglycones or in conjugated forms with sugars typical of the cardenolide glycosides; they are nevertheless described together with the cardenolides and scilladienolides because of their structural similarity. The digitenolides do not possess a lactone ring. Since they do have a pregnane skeleton the digitenolides are in an uncertain biogenetic position with respect to the cardiotonic aglycones; they can be either intermediates of the latter or catabolites.

6.4.2 Steroidal Saponins

Like the cardiotonic glycosides, steroidal saponins are comprised of a steroidal aglycone, the sapogenin, conjugated to an oligosaccharide bound through the hydroxyl group in the 3β-position (Fig. 6.15). The oligosaccharides generally involve common hexoses and pentoses, such as D-glucose, D-xylose, etc. numbering between 2 and 6. Saponins behave as soaps and aqueous solutions readily form foams. Because of this property and the consequent hemolytic properties, they are extremely toxic to animals if introduced directly into their blood, whilst being non-poisonous to ingestion.

Economic interest in the steroidal saponins arises from the use of these substances as starting material for the industrial production of the sex hormones, corticosteroids, and steroid derivatives in general. The main sources of the saponins have been from higher plants, especially of the in the Liliaceae, Solanaceae and Scrophulariaceae families.

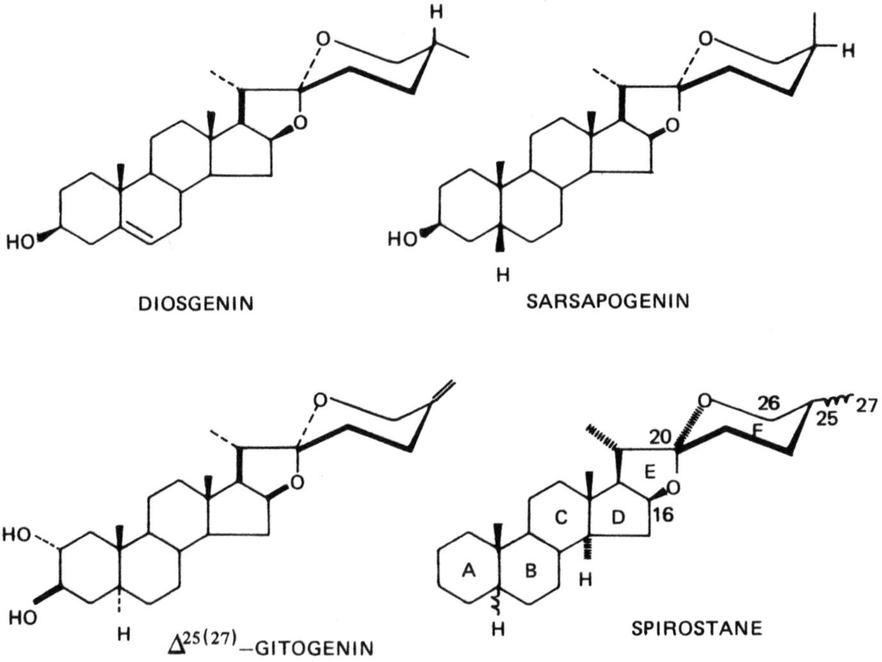

Fig. 6.14 Sapogenins and the spirostane skeleton.

The carbon skeleton of the sapogenins is as in the 5α or 5β cholestane system. The aglycones, obtained from the corresponding saponins *via* mild hydrolysis, also generally possess a characteristic *spiro*-junction formed from two heterocyclic rings: a tetrahydrofuran, ring E, and a tetrahydropyran, ring F.

These rings include carbon atoms 16 and 17 from the nucleus, and carbons from the isooctyl side-chains. The combination of this *spiro*-ring system and the cholestane nucleus is called the spirostane structure, which represents the basic skeleton of all the sapogenins (Fig. 6.14). The 25-methyl substituent can adopt either the equatorial configuration, as in the 25α-sapogenins, or *isosapogenins*, such as diosgenin, or the axial one as in the 25β-sapogenins, or *neosapogenins*, for example sarsasapogenin. A methylene group is sometimes present at C-25 ($\Delta^{25(27)}$-sapogenins, e.g. $\Delta^{25(27)}$-gitogenin).

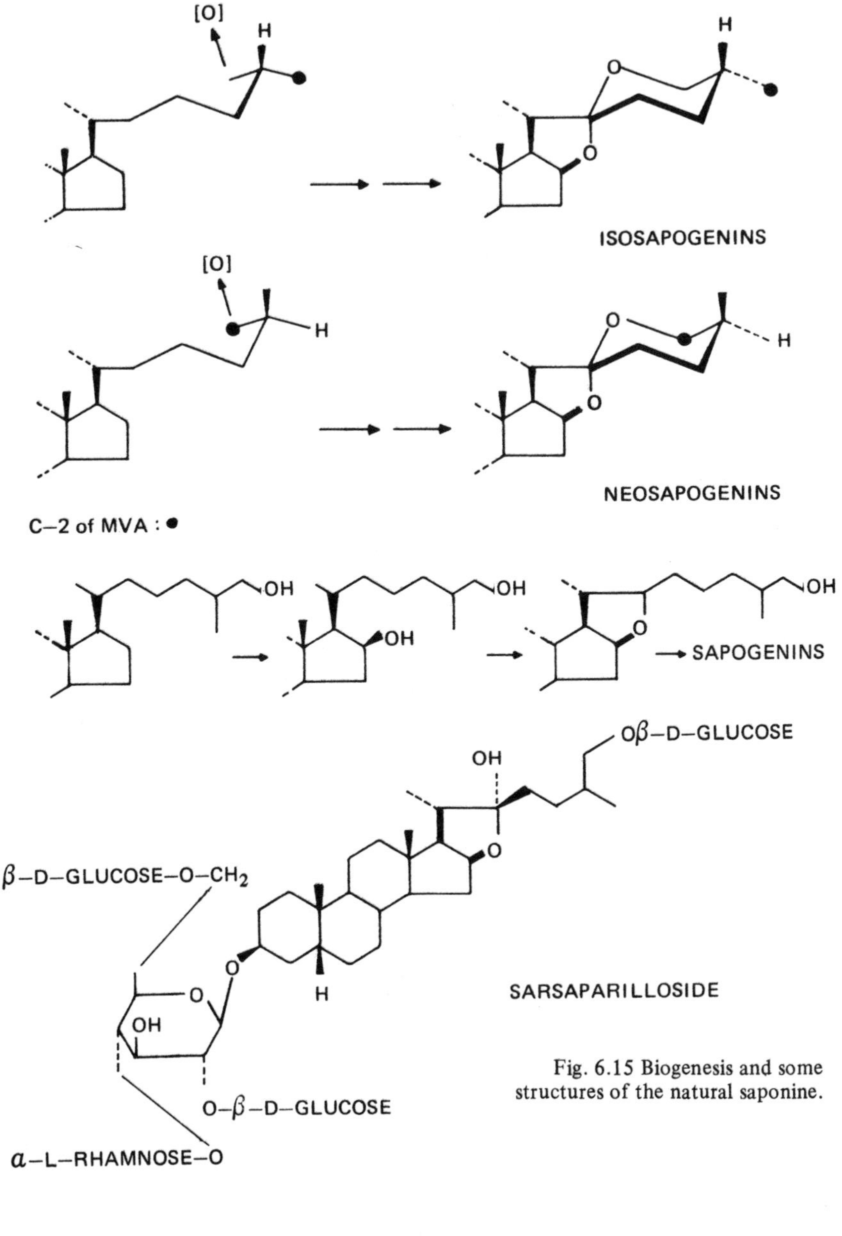

Fig. 6.15 Biogenesis and some structures of the natural saponine.

Biogenetically, the interesting feature of these compounds is formation of the spirostane system, or of the correspondong open form, as in sarsaparilloside (Fig. 6.15).

The origin of the sapogenins from cholesterol, with the retention of all the carbon atoms, has been experimentally confirmed by incorporation experiments. The biosynthesis of the iso- and neosapogenins is shown in Fig. 6.15. Slight variations in different plants are possible, as is observed for formation of the phytosterols. The neosapogenins form through hydroxylation of the methyl derived from C-2 of MVA and the isosapogenins through hydroxylation of that from C-6 (see Fig. 6.15).

A question, still not fully answered, concerns the form in which the saponins are present *in vivo*. Sarsaparilloside has been isolated without the F ring and with the primary alcoholic group of its chain blocked through conjugation to a molecule of D-glucose. This observation has been accepted as an indication that most saponins occur in plants in their open form, the spirostane system forming spontaneously after hydrolysis of the glucosidic bond at C-26, for instance, by the action of emulsin or of analogous β-glucosidases liberated from the cells during the extraction process.

6.5 ECDYSONES

The name, ecdysone, was given to the first insect moulting hormone isolated in a crystalline form. The name is now collectively applied to a group of about thirty compounds; their common features are their stimulating effect on the moulting of arthropods and structural resemblances. These similarities include a steroid nucleus with a *cis* junction between the A and B rings, an $\alpha\beta$-unsaturated ketone system in ring B (6-keto-Δ^7) and numerous hydroxyl groups, especially at positions 2β, 14α, 20, 22, 25 and 26 (Fig. 6.16). 'Ecdysis' is the entomological term for the moulting of larvae. Ecdysone was first isolated from pupae of the silk worm, *Bombyx mori* (25 mg from 500 Kg!). The structure of ecdysone (45) was obtained via X-ray analysis. Only six compounds of the ecdysone type have been isolated from arthropods whilst all of the others are obtained from plants. The ecological function of the ecdysones in plants is still obscure.

In order to understand the physiological role of ecdysones, some of the post-embryonic developments of insects have to be understood. Two periods can be distinguished in the life of such invertebrates: the larval period, and the imago, or adult period, which are separated by a metamorphosis stage. In some insects (e.g. Lepidoptera) a third phase occurs between the two larval and perfect insect phases. This is the pupal phase (the pupa is called a chrysalis in the Lepidoptera). During the larval period the insect grows up, and several moults follow each other; they are controlled by complex hormonal mechanisms. Once formed the adult does not undergo further changes; its only task is the reproduction of its species. The moulting process is necessary for growth of the larva, as they are enclosed within an exoskeleton which must be replaced when it becomes too small.

ECDYSONES ISOLATED FROM ARTHROPODS (AND PLANTS)

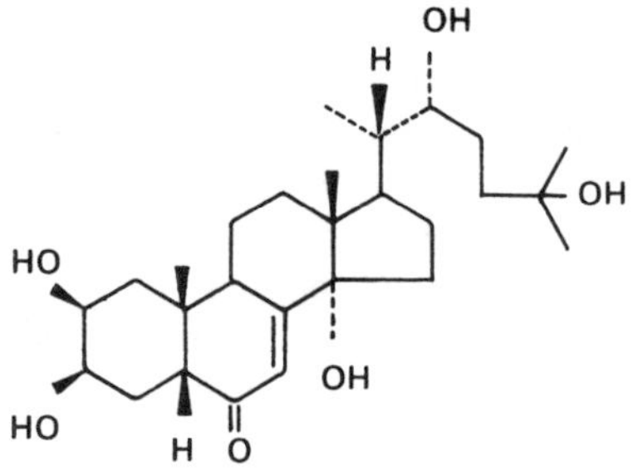

(45)
ECDYSONE

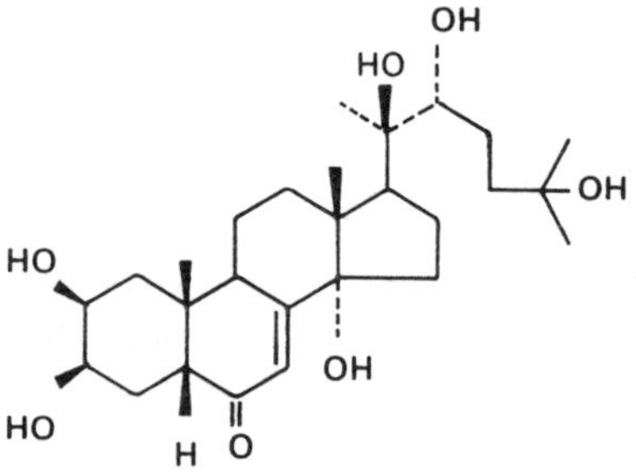

(46)
ECDYSTERONE

PLANT ECDYSONES

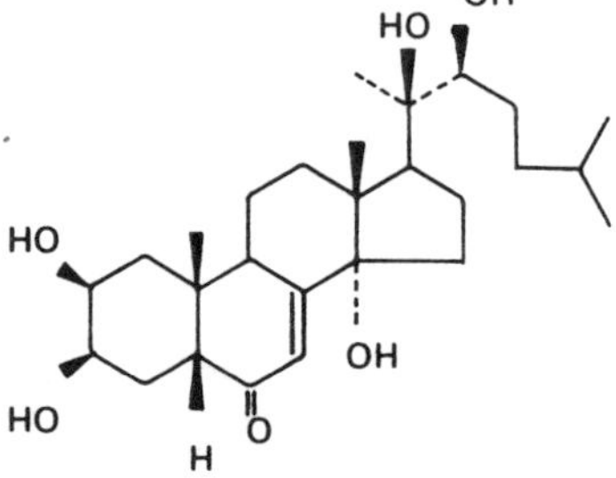

(47)
PONASTERONE A

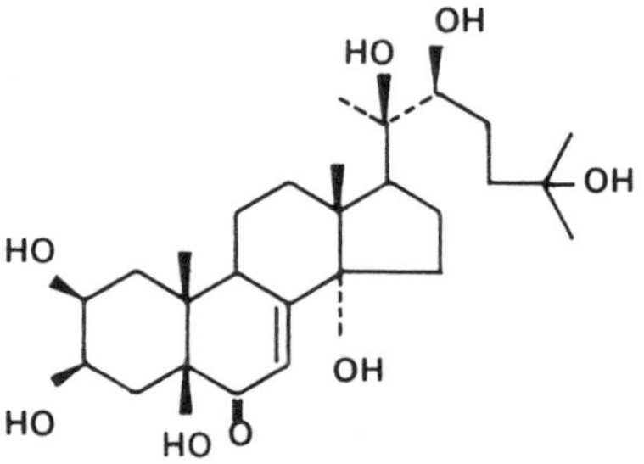

(48)
AJUGASTERONE

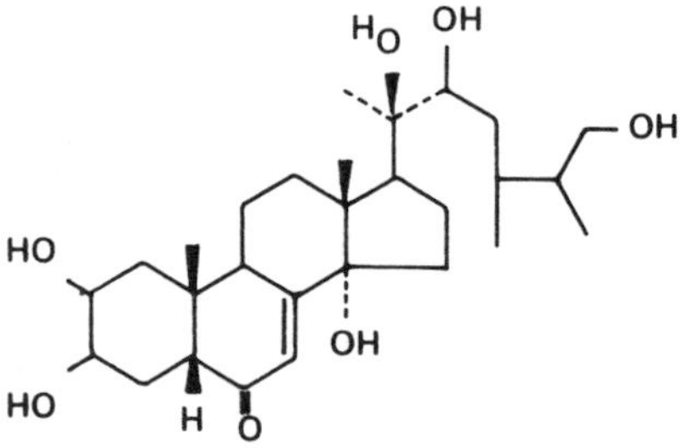

(49)
MAKISTERONE B

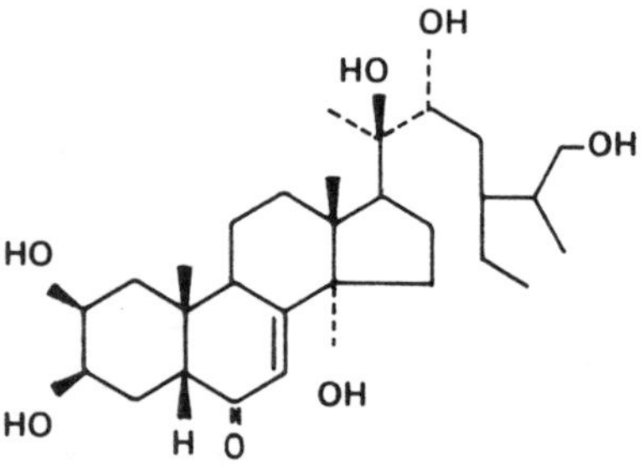

(50)
AMARASTERONE A

Fig. 6.16 Some ecdysones.

The endocrinology of insects is schematised in Fig. 6.17. The neurosecretory cells of the brain of the immature insect secrete a polypeptide hormone, 'brain hormone', (B.H.) which reaches the corpora cardiaca and successively diffuses into the hemolymph. This hormone activates certain endocrine glands situated in the thorax (thoracic glands), which in turn produce a 'moulting hormone., (M.H.) or ecdysone. The ecdysones are thought to activate chromosomal genes which synthesise particular enzymes responsible for the renewal of the epidermal cells. Thus the moulting hormone stimulates the growing and removal of these epidermal cells. Thus the moulting hormone stimulates the growing and removal of these epidermal cells. The structure of the secreted cuticle is also controlled by a third hormone, called neotenin, or 'juvenile hormone', (J.H.).

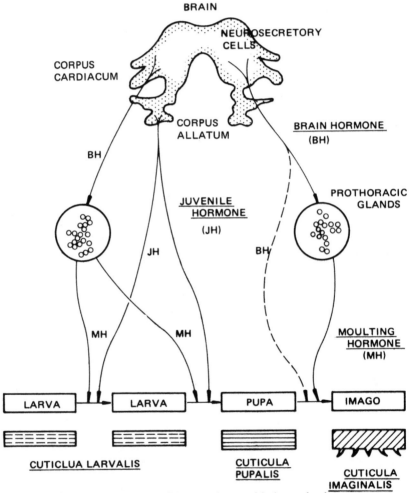

Fig. 6.17 Schematic representation of hormonal control in insect development. BH = brain hormone; MH = moulting hormone; JH = juvenile hormone.

Juvenile hormone is produced by the corpora allata, which are endocrine glands situated near the brain. If its concentration in the hemolymph is high, the newly secreted cuticle is of the larval type; if its concentration is low, a pupal cuticle is formed; if juvenile hormone is absent the adult cuticle is produced under the synergistic action of M.H. and B.H. and the insect is transformed into the perfect insect.

The study of ecdysones and of their physiological role has rapidly developed in recent years as there is the possibility of obtaining them easily from vegetable sources and for using them in the chemical struggle against insect pests. Certain insecticides are being developed, which rely on the manufacture of hormonal analogues.

Little information is available on the biosynthesis of ecdysone and of its relatives in insects. This is partly because of the extremely small quantities of these hormones present in insects. It is still not clear whether these steroidal derivatives derive from exogenous or endogenous sources, although it appears that insects are not capable of synthesising steroids from terpene precursors such as squalene.

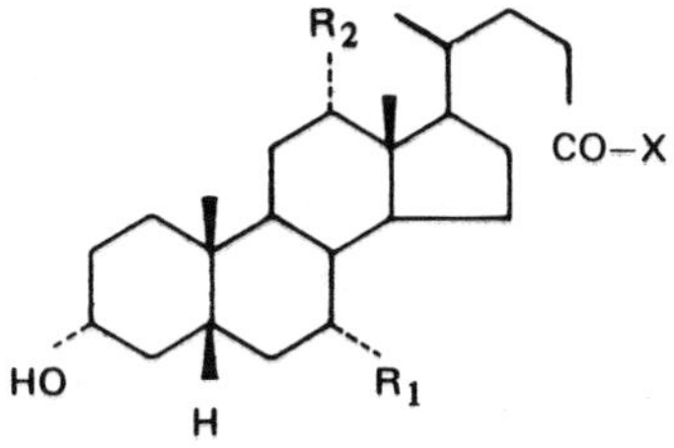

(51 : $R_1, R_2, X = OH$)
CHOLIC ACID

(52 : $R_1 = H; R_2, X = OH$)
DEOXYCHOLIC ACID

(53 : $R_1, R_2 = H; X = OH$)
LITHOCHOLIC ACID

(54 : $R_2 = H: R_1, X = OH$)
CHENODEOXYCHOLIC ACID

BILE SALTS

$X = -NH-CH_2-COONa$ (GLYCINE)

$X = -NH-CH_2CH_2-SO_3Na$ (TAURINE)

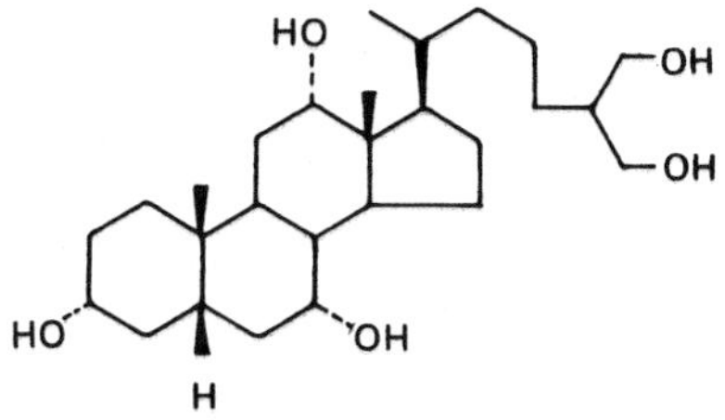

(55)
β-CYPRINOL

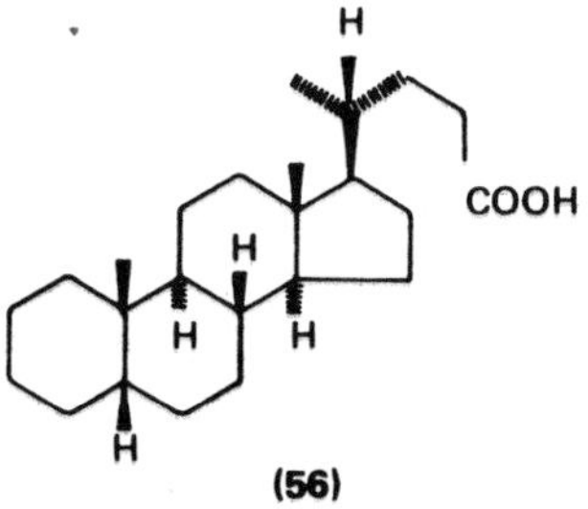

(56)
5β-CHOLANIC ACID

Fig. 6.18 Bile acids and alcohols.

6.6 BILE ACIDS AND ALCOHOLS

In all animals studied, the bile acids and alcohols are formed from cholesterol in the liver, where they are also conjugated with taurine and glycine (Fig. 6.18). The bile salts are transported into the intestine where microbial flora operate many further chemical transformations on the steroidal nucleus, such as reduction, hydroxylation, etc. The secondary acids and alcohols, so produced, are partly eliminated from the gut and partly recovered through the enterohepatic circulation. In studying the biosynthesis of the bile acids, two parts can therefore be distinguished. The first part is on cholesterol metabolites before any contact with the intestinal flora and the second part is on the transformations carried out by microbial flora such as occur in the intestine.

From a structural point of view the bile acids should be considered derivatives of 5β-cholestane (or coprostane, C_{27}, see Fig. 6.1). It is current practice to use 5β-cholanic acid (56) as the basic structure for naming the bile acids. Frequently a hydroxyl group in the 7α-position is found, the C-3-hydroxyl is also frequently α-oriented, compare cholic acid (51) and 5β-cyprinol (55). These structural features can be explained from the metabolic sequences leading to cholic acid and most primary bile acids and alcohols (Fig. 6.19). The first step is almost always the hydroxylation of cholesterol into 7α-hydroxycholesterol (57) and the corresponding enzyme has been observed in rat liver microsomes: it is a very active 7α-hydroxylase, requiring NADPH and molecular oxygen. The successive steps are saturation of the 5(6)-double bond and inversion of the 3-hydroxyl orientation, via oxidation to a carbonyl group and successive reduction by hydride ion attack); (b) further oxidation of the molecule at C-12 and, eventually, C-6, C-22, and C-23; (c) the oxidative cleavage of the side chain; (d) conjugation, for instance with taurine and glycine. The above reaction sequence is probably the most common but certainly not the only possible one. The variation in timing of these metabolic changes leads to a metabolic grid for formation of the bile acids.

Degradation of the side chain of cholesterol takes place through a process analogous to the β-oxidation of saturated fatty acids: the first step is the oxidation to a carboxylic group of one of the C-25 geminal methyls. If this oxidation stops at a primary alcohol stage, for example because of absence of a dehydrogenase, the bile alcohols are obtained. The isooctyl chain of these compounds generally bear other hydroxyl groups as well.

The bile acids without a hydroxyl group at C-7 result from early degradation of the cholesterol side chain. An example is lithocholic acid (53), which, at least partially derives from the sequence: 26-hydroxycholesterol to 3β-hydroxycholest-5-en-26-oic acid to lithocholic acid. They also originate from microbial dehydroxylation of primary 7α-hydroxy acids, as happens in the formation of desoxycholic acid (52), which is therefore a bile acid of the secondary type, and, probably, for some of the lithocholic acid deriving from chenodesoxycholic acid (54).

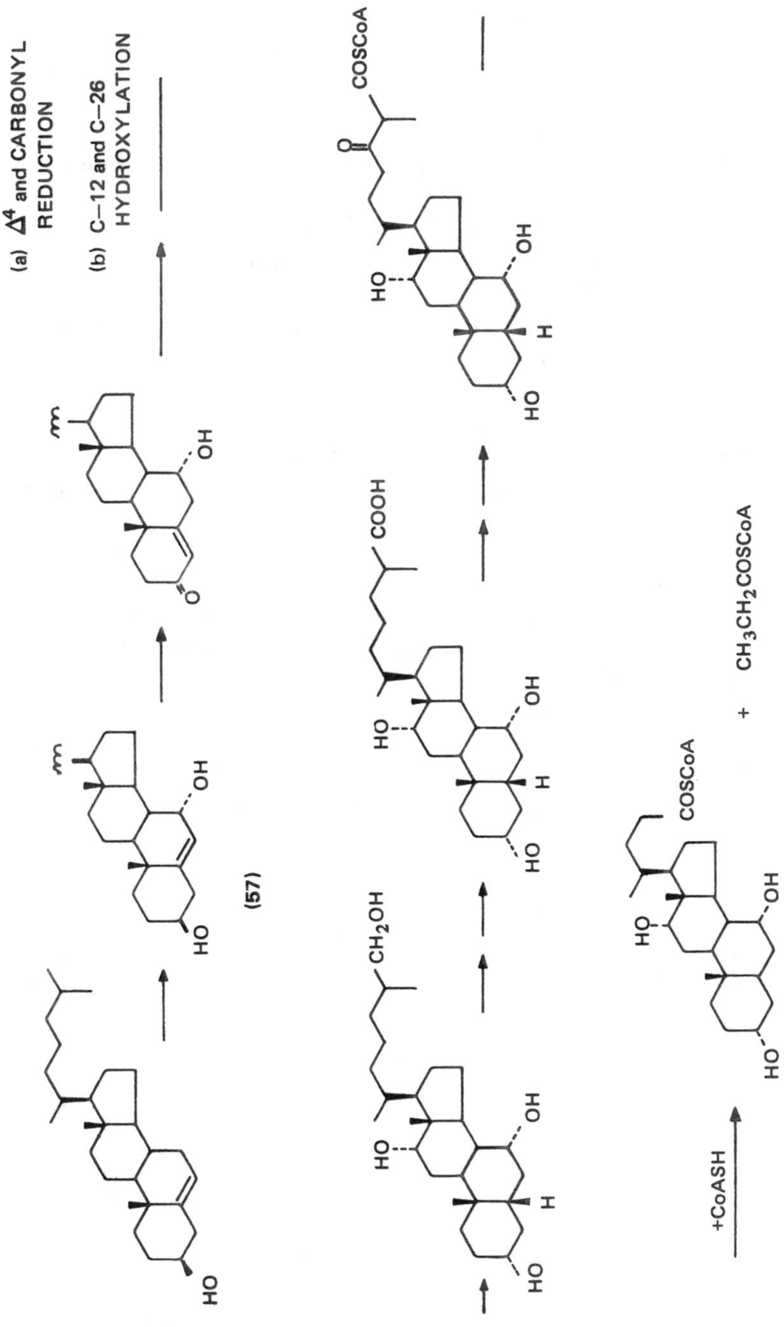

Fig. 6.19 Biosynthesis of cholic acid.

The sodium salts of the bile acid conjugates are water soluble and have emulsive properties: they favour the absorption of fats and of water insoluble substances, such as the carotenes and vitamin K compounds in the intestine.

6.7 STEROIDAL HORMONES

The steroidal hormones are usually divided into four distinct families and each family is structurally and physiologically homogenous. The *corticosteroids* and the *gestogens* (Fig. 6.21) all possess a pregnane skeleton. These, e.g. (60), (63), (67) and (68), possess the Δ^4-3-keto- and 20-keto groups typical of progesterone and also have a primary alcohol function at C(21), and, occasionally, a hydroxyl group in the 17α- and/or 11β-positions.

The *androgens*, or male sex hormones (Fig. 6.22), have the androstane (C_{19}) skeleton and carry an oxygen function at C-17, in the place of the side chain. A further reduction in the number of carbon atoms is found in the estrane skeleton (C_{18}), typical of *estrogens*, or female sex hormones. The key structural features of these molecules are the aromaticity of ring A, necessarily requiring the expulsion of the C-19 methyl group, and the presence of a phenolic hydroxyl function at C-3.

All the steroid hormones derive in animals from cholesterol, through biosynthetic paths having a common first step. This is the oxidative cleavage of the isooctyl chain at the C-20/C-22 bond (Fig. 6.20) and this removal of the side chain takes place in various anatomical parts: in the adrenal cortex, the corpus luteum, and the gonads, etc. In every case pregnenolone (42) and isocaproic acid are formed. The pregnenolone is successively utilised for synthesising the various steroidal hormones.

The process of pregnenolone formation is strongly stimulated by the adrenocorticotropic hormone (ACTH), produced by the anterior hypophysis.

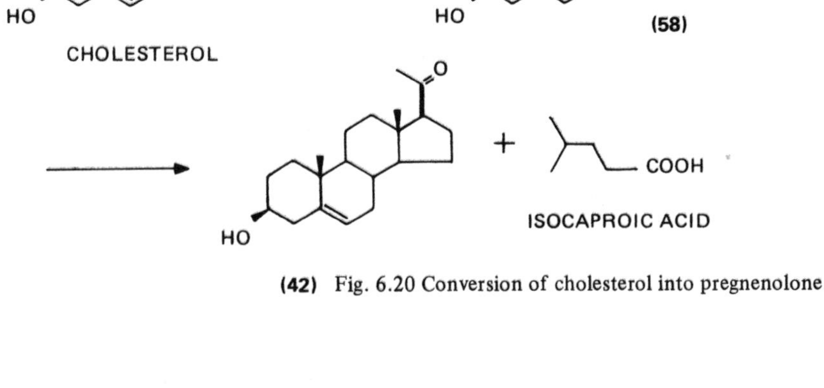

CHOLESTEROL

(58)

ISOCAPROIC ACID

(42)   Fig. 6.20 Conversion of cholesterol into pregnenolone.

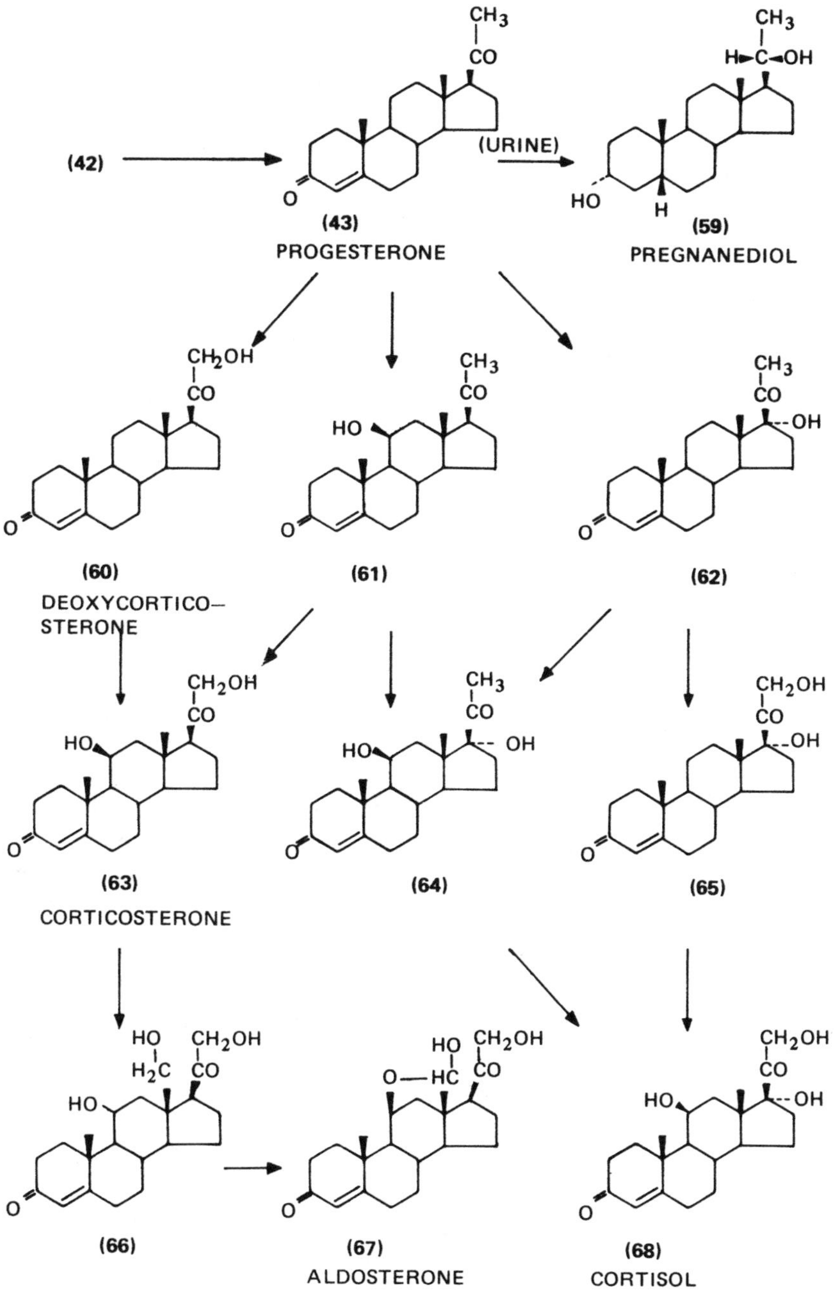

Fig. 6.21 Biosynthesis of the corticosteroids and gestogenins.

6.7.1 Corticosteroids

These hormones, elaborated by the adrenal cortex, can be separated into: (a) *glycocorticoids*, such as cortisol (68) and corticosterone (63), acting on the metabolism of carbohydrates and maintaining normal glycemia and (b) *mineralocorticoids*, such as desoxycorticosterone (60) and aldosterone (57), which help to control the metabolism of cations.

A biogenetic picture of the various corticosteroids, is shown in Fig. 6.21. It consists of a metabolic grid. The transformations in Fig. 6.21 do not take place equally throughout the suprarenal cortex; aldosterone is exclusively synthesised in the more external part of the cortex, whilst cortisol is produced in the interior zone: corticosterone is synthesised in both zones. This fact can be ascribed to a reduced 17α-hydroxylase activity in the external part of the cortex. The synthesis of aldosterone differs from that of the glycocorticoids in that it is not stimulated by ACTH.

6.7.2 Gestogens

The typical gestogen is progesterone (43), which is formed from pregnenolone by oxidation of the C-3-secondary alcoholic function and by migration of the double bond from the 5(6)- to the 4(5)-position.

The corpus luteum, which develops from the burst follicle, is an endocrine gland which produces progesterone. Two gonadotropic hormones are known and which are produced by the anterior hypophysis. They are follicle stimulating hormone, FSH, which promotes maturation of the follicles in the ovary and of spermatogenesis in the testicles; interstitial cell stimulating hormone (ICSH), called luteinising hormone in women (LH), determines the secretion of estrogens in the follicle and transforms the burst follicle into the corpus luteum; ICSH stimulates the secretion of testosterone from the testicles. Progesterone induces uterine modifications favourable to implantation of the ovum and, if pregnancy ensues, it is essential for completion of the term. Progesterone has numerous catabolites: the principal one being pregnanediol, or 5β-pregnane-3α,20α-diol (59), isolated from urine, and its 3α-conjugate with D-glucuronic acid.

Pregnanediol is not excreted in the follicular phase of the menstrual cycle, but about 10 mg per day are excreted in the luteal phase. During pregnancy the amount of pregnanediol eliminated keeps constant for the first two months, then increases up to between 60 to 100 mg per day near delivery and thereafter rapidly decreasing to very low values.

6.7.3 Androgens

The androgens, testosterone (73), dehydroepiandrosterone (DHA) (70), and androstenedione (72), are produced by the intestitial tissue of the testicles and by the adrenal cortex (Fig. 6.22).

The C-17 and C-20 bond is broken from progesterone precursors with introduction of a 17α-hydroxyl group.

The cleavage can be imagined as either a Bayer-Villiger type oxidation or as a process involving TPP.

The most powerful male steroidal hormone is testosterone (73). It is formed in the testicles and its most common urinary metabolite is androsterone (74).

The synthesis of testosterone is stimulated by gonadotropin ICSH. In an adult man about 20 mg testosterone are secreted daily.

Testosterone has many physiological functions primarily causing development of male sexual features, the increase of protein anabolism, and the stimulation of spermotogenesis.

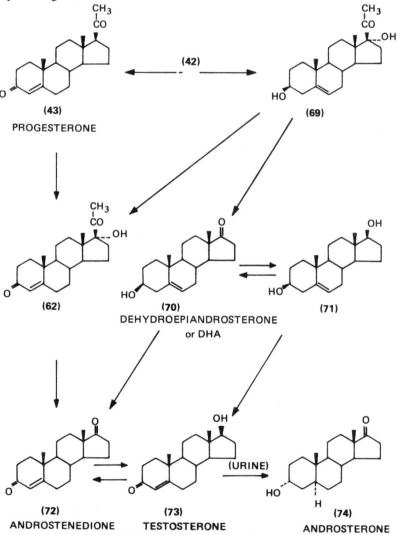

Fig. 6.22 Biosynthesis of the androgens.

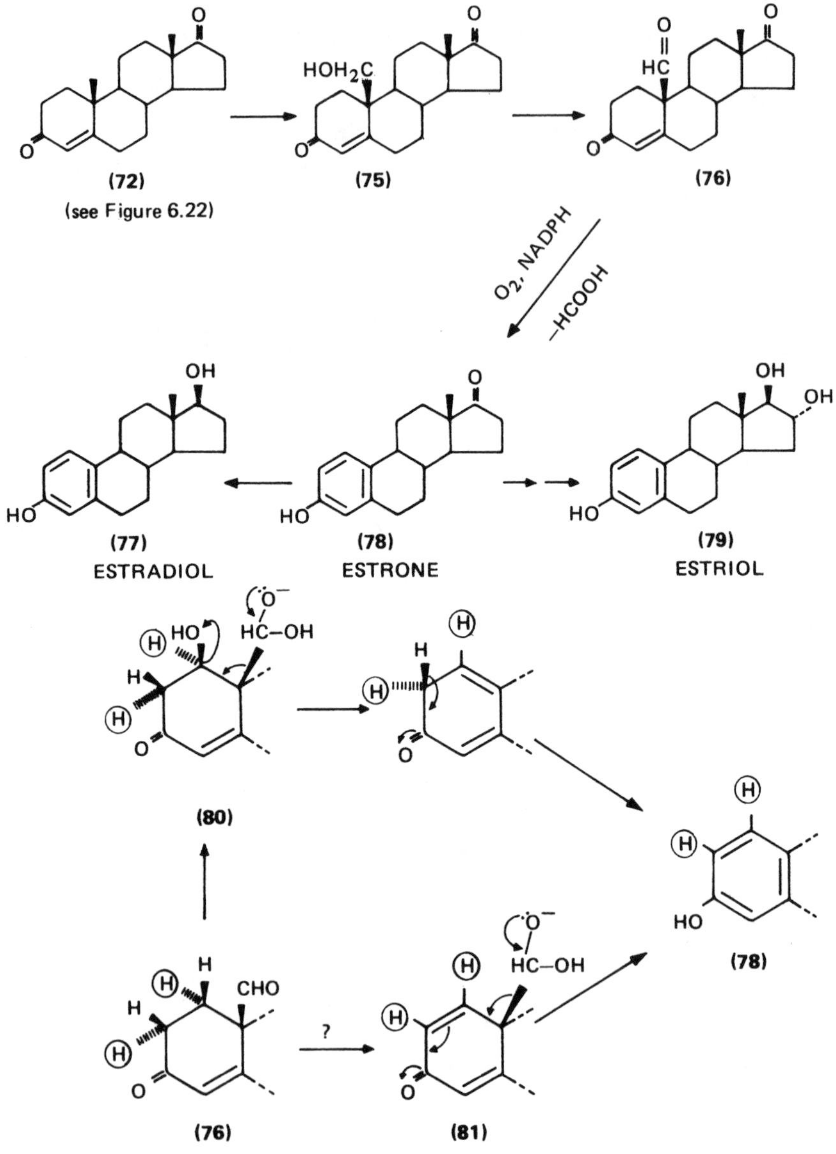

Fig. 6.23 Biosynthesis of the estrogens.

6.7.4 Estrogens

The estrogens, estrone (78), estradiol (77) and estriol (79), are biosynthesised by follicles and, in small amounts, by the adrenal cortex, under synergistic action of the gonadotropins FSH and LH.

The biosynthesis of the three fundamental estrogens is illustrated in Fig. 6.23. The multienzymic complex which catalyses the conversion of androstendione (72) into estrone (78), is usually called aromatase: the name emphasises the transformation of ring A of the steroid nucleus into an aromatic ring. The expulsion of the 19-methyl group takes place at the aldehyde stage, as in (80) or (81), formic acid being liberated. As far as the stereochemical mechanism of aromatisation is concerned, the 1β- and 2β-hydrogen atoms are lost.

Neither of the two possible processes, schematised in Fig. 6.25 have yet been proven. Each of them implies a different intermediate: the 1β-hydroxy-derivative (80), deformylating via an unlikely *cis*-elimination and the dienone (81), arising from a *cis*- dehydrogenation of (76). The last reaction is analogous to the introduction of a double bond into the 5,6-position of.cholesterol (see Fig. 6.4). Whatever the process the deformylation step requires O_2 and NADPH.

The estrogens are transformed into glycoside sulphates and ethers during catabolism and as they undergo further hydroxylations.

The specific function of estrogen in mammals is to stimulate the oestrus.

In women the tissues involved in reproduction are stimulated to grow, such as the endometrium, vaginal mucoses and the mammary glands. Other important physiological actions of the estrogens are the development of secondary sexual features in females and their behaviour as antagonists towards androgens.

SOURCE MATERIALS AND SUGGESTED READING

[1] A. Wettstein, (1961), Biosynthese des Hormones Steroides, *Experentia,* **17,** 329.

[2] C. W. Shoppee, (1964), *Chemistry of the Steroids,* 2nd ed., Butterworths.

[3] R. B. Clayton, (1965), Biosynthesis of Sterols, Steroids and Terpenoids. Part I. Biogenesis of Cholesterol and the Fundamental Steps in Terpenoids Biosynthesis, *Quart. Rev.,* **19,** 168; (1965), Part II. Phytosterols, Terpenes and the Physiologically Active Steroids, *ibidem,* **19,** 201.

[4] I. D. Frantz and G. J. Schroepfer, (1967), Sterol Biosynthesis, *Ann. Rev. Biochem.,* **36,** 691.

[5] L. J. Goad, Aspects of Phytosterol Biosynthesis in *Terpenoids in Plants,* ref. 16 of Chap. 5.

[6] W. F. Loomis, (1967), Skin Pigment Regulation of Vitamin D Biosynthesis in Man, *Science,* **157,** 501.

[7] E. Staple, The Biosynthesis of Sterols in *Biogenesis of Natural Compounds,* ref. 24 of Chap. 1.

[8]   R. Tschesche, Biosynthesis of Cardenolides, Bufadienolides, and Steroid
      Sapogenins in *Terpenoids in Plants,* ref. 16 of Chap. 5.

[9]   J. R. Hanson, (1968), Introduction to Steroids Chemistry, Pergamon
      Press.

[10] P. Morand and J. Lyall, (1968), The Steroidal Estrogens, *Chem. Rev.,* **68,**
      85.

[11] C. E. Berkoff, (1969), The Chemistry and Biochemistry of Insect Hormones,
      Quart. Rev., **23,** 372.

[12] G. M. Sanders, J. Pot and E. Havinga, (1969), Some Recent Results in the
      Chemistry and Stereochemistry of Vitamin D and Its Isomers in *Progress
      in the Chemistry of Organic Natural Products,* ref. 1 of Chap. 1, **27.**

[13] C. T. Sawin, (1969), *The Hormone Endocrine Physiology,* Churchill.

[14] L. G. Goad, Sterol Biosynthesis in *Natural Substances Formed Biologically
      from Mevalonic Acid,* ref. 40 of Chap. 5.

[15] P. Karlson, Terpenoids in Insects in *Natural Substances Formed Biologically
      from Mevalonic Acid,* ref. 40 of Chap. 5.

[16] E. Heftmann, (1970), Insect Molting Hormones in Plants in *Recent
      Advances in Phytochemistry,* ref. 57 of Chap. 1, **3.**

[17] E. Heftmann, (1970), *Steroid Biochemistry,* Academic Press.

[18] H. Hikino and Y. Hikino, (1970), Arthropod Moulting Hormones in *Progress
      in the Chemistry of Organic Natural Products,* ref. 1 of Chap. 1.

[19] C. E. Berkoff, (1971), Insect Hormones and Insect Control, *J. Chem.
      Educ.,* **48,** 577.

[20] T. W. Goodwin, (1971), The Biogenesis of Terpenes and Sterols in *Rodd's
      Chemistry of Carbon Compounds,* 2nd ed. (S. Coffey, Ed.), Elsevier, **II/E,**
      supplement to, (1974), **II/E** (M. F. Ansell, ed.).

[21] K. H. Overton (Ed.), *Terpenoids and Steroids,* A Specialist Periodical
      Report, The Chemical Society, London, **1,** 1971; **2,** 1972; **3,** 1973; **4,**
      1974; **5,** 1975; **6,** 1976 (Biosynthesis of Terpenoids and Steroids by G. P.
      Moss in **1-3,** and by D. V. Banthorpe and B. V. Charlwood in **4-6.**

[22] H. H. Rees, *Ecdysones* in Aspects of Terpenoid Chemistry and Biochemistry,
      ref. 59 of Chap. 5.

[23] L. J. Mulheirn and P. J. Ramm, (1972), The Biosynthesis of Sterols, *Chem.
      Soc. Rev.,* **1,** 259.

[24] H. H. Rees and T. W. Goodwin, (1972-1975), Biosynthesis of Triterpenes,
      Steroids and Carotenoids in *Biosynthesis,* ref. 30 of Chap. 1, **1-3.**

[25] B. A. Knights, (1973), Sterol Metabolism in Plants, *Chem. in Britain,* **9,**
      106.

[26] R. Tschesche and G. Wulff, (1973), Chemie und Biologie der Saponine in
      Progress in the Chemistry of Organic Natural Products, ref. 1 of Chap. 1,
      30.

[27] J. D. Weete, (1973), Sterols of the Fungi: Distribution and Biosynthesis,
      Phytochemistry, **12,** 1843.

[28] K. Slama, M. Romanuk and F. Sorm, (1974), *Insect Hormones and Bio-analogues,* Springer-Verlag.

[29] E. Heftmann, (1975), Functions of Steroids in Plants, *Phytochemistry,* **14,** 891.

[30] A. M. Malkinson, (1975). *Hormone Action,* Chapman and Hall.

[31] V. J. A. Novak, (1975), *Insect Hormones,* 2nd ed., Chapman and Hall.

[32] K. Lubke, E. Scillinger and M. Topert, (1976), Hormone Receptors, Angew. Chem. Int. Ed. Engl.

[32] K. Lubke, E. Scillinger and M. Topert, (1976), Hormone Receptors, *Angew. Chem. Int. Ed. Engl.* **15,** 741.

[33] L. J. Mulheirn, (1976-1977), Triterpenoids, Steroids and Carotenoids in *Biosynthesis,* ref. 30 of Chap. 1, **4-5.**

[34] E. Heftmann, (1977), Function of Steroids in Plants in *Progress in Phytochemistry,* ref. 56 of Chap. 1, **4.**

[35] W. R. Nes and M. L. McKean, (1977), *Biochemistry of Steroids and Other Isoprenoids,* University Park, Baltimore.

[36] H. H. Rees, (1977), *Insect Biochemistry,* Chapman and Hall.

[37] J. M. C. Geuns, (1978), Steroids Hormones and Plant Growth and Development, *Phytochemistry,* **17,** 1.

[38] G. Ourisson, M. Rohmer and R. Anton, (1979), From Terpenes to Sterols: Macroevolution and Microevolution, in *Recent Advances in Phytochemistry,* ref. 57 of Chap. 1, **13.**

[39] H. Jones and G. H. Rasmusson, (1980), Recent Advances in the Biology and Chemistry of Vitamin D, in *Progress in the Chemistry of Organic Natural Products,* ref. 1 of Chap. 1, **39.**

Chapter 7

The Shikimate Pathway

Fungi produce most of their secondary metabolites starting from acetate, through polyketide type intermediates (Chapter 4). Plants of higher order than these thallophytes differ in using intermediates of the so-called shikimate pathway for their synthetic purposes (see Fig. 7.2).

The shikimate pathway was initially observed in bacteria (*Escherichia coli*), in yeasts (*Saccharomyces cerevisiae*) and in moulds (*Neurospora crassa*) thanks to the brilliant work of Davis, who used the *mutant method* for such investigations (see p. 000). A similar biogenetic route has subsequently been shown to exist in all the cormophytes.

Shikimic acid (Fig. 7.1) was first obtained in 1885 from the plant *Illicium religiosum*, whose Japanese name is shikimi-no-ki and has since been identified in many plants. It can be easily isolated in good yields from the young twigs of conipherae, for instance from the vegative apices of *Gingkyo biloba*. Quinic acid is also found in plants; it is formed by (reversible) reduction of the carbonyl group of dehydroquinic acid, an obligatory intermediate of shikimic acid.

When (−)-[G-^{14}C] quinic and (−)-[G-^{14}C]shikimic acids were isolated from young rose plants, kept under a $^{14}CO_2$ atmosphere, the specific radioactivity of the former was higher than that of the latter. Furthermore, when the labelled quinic acid was given to other rose plants, ^{14}C-labelled-shikimic acid was obtained. Analogous results obtained with other vascular plants demonstrate the general validity of the scheme represented in Fig. 7.1.

Quinic acid is often found combined as esters of aromatic carboxylic acids, e.g. gallic and caffeic acids. The quinic esters of cinnamic acids, e.g. chlorogenic acid (3), can accumulate in large amounts in plants. These esters also undergo chemical modifications, especially hydroxylation of the aromatic rings, but these derivatives do *not* appear to be utilised as the precursors of the more common phenylpropanoids (see Chapter 8).

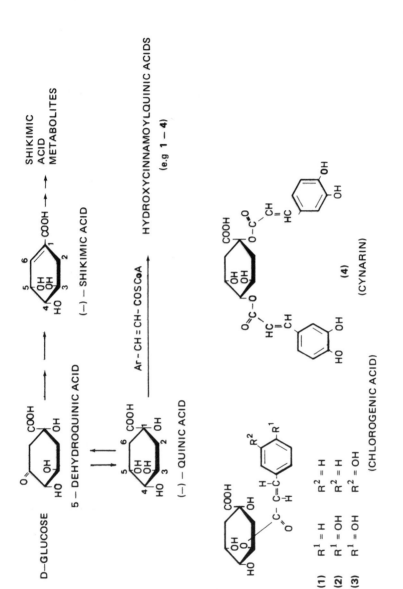

Fig. 7.1 Origin of some quinic acids and cinnamate derivatives.

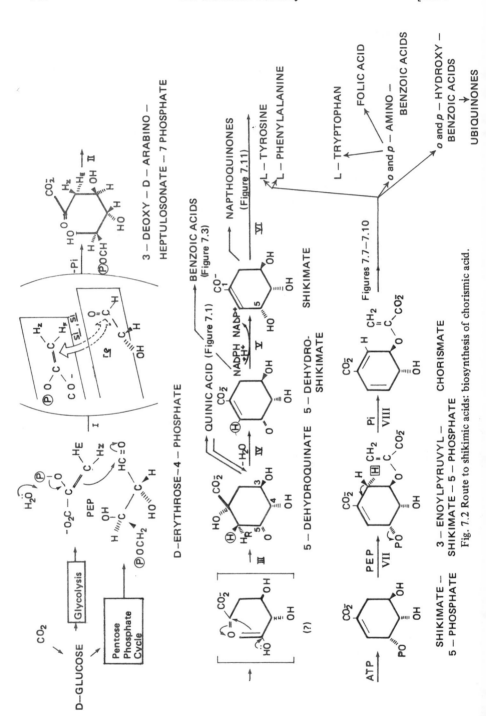

Fig. 7.2 Route to shikimic acids: biosynthesis of chorismic acid.

7.1 BIOSYNTHESIS OF CHORISMIC ACID*

The biogenetic route (Fig. 7.2) illustrates the central role of shikimic acid, which is comparable to that of mevalonic acid in the biosynthesis of terpenes. Two main stages are evident: the first proceeds from phosphoenolpyruvate and D-erythrose-4-phosphate to chorismate and is composed of a single pathway with few deviations. The second stage involves many ramifications, all departing from chorismate and which is thus the last common intermediate of a large range of biologically important molecules, including the three aromatic amino acids, nicotinic acid, tetrahydrofolic acid, and the ubiquinones.

The biosynthesis of shikimic and chorismic acids begins with the condensation between a molecule of D-erythrose-4-phosphate, coming from the pentosephosphate cycle (p. 141), and one of phosphoenolpyruvic acid (PEP), arising from glycolysis (p. 139). A typical nucleophilic addition to the tetrosephosphate carbonyl group takes place: the PEP methylene group acting as the nucleophile in concert with the hydrolytic cleavage of the bond between the enolic oxygen and the phosphoric acid. An analogous process (steps II, III) results in formation of the cyclohexane ring of 5-dehydroquinic acid. The C-1 and C-3 asymmetric centres of this intermediate have specific, absolute configurations, which result from stereospecific enzymatic control of the steps I and III, involving reaction of sp^2, prochiral carbon atoms; the configuration at position 4 corresponds exactly to that of the starting D-erythrose at position 2.

The asymmetric centre at position 1 of dehydroquinic acid is lost in shikimic acid, although a new chiral centre is formed at C-5 by stereospecific reduction of the corresponding carbonyl group. It is noteworthy that eliminations IV and VIII occur with overall *cis* and *trans* geometry respectively, i.e. in contrast with the stereoelectronic requirements of conceted mechanisms.

Some of the enzymes catalysing the processes shown in Fig. 7.2 have been detected in higher plants. For example, 5-dehydroquinate dehydratase (E.C. 4.1.2.10), promoting reaction IV, has been found in spinach.

7.2 PROTOCATECHUIC AND GALLIC ACIDS-HYDROLYSABLE TANNINS

Gallic acid is one of the more important natural hydroxybenzoic acids; large amounts are found in nature, particularly as a constituent of tannins. It can be formed in both the higher plants and in microorganisms together with protocatechuic acid, through three different metabolic paths: (a), (b) and (c) of Fig. 7.3. The shortest route (a) affords the two carboxylic acids by direct transformation (dehydration or dehydrogenation) of 5-dehydro-shikimic acid; the longest (c) passes through the synthesis of L-phenylalanine (or L-tyrosine), the hydroxylation of cinnamic acid, and the degradative oxidation of caffeic

---

*Chorismic acid is so-called because of its particular position in the metabolic grid.
($\chi \omega \rho \iota \sigma \mu o$ S) = separation).

acid via 3,4,5-trihydroxycinnamic acid. Route (b) involves the conversion of chorismic acid into *p*-hydroxybenzoic acid, followed by its hydroxylation into protocatechuic and gallic acid. The latter route is the least important in plants, and has been proven in only a few species such as in *Pelargonium hortorum* (Geraniaceae).

The three biosynthetic routes to the benzene ring illustrated in Fig. 7.3 all start from D-glucose and differ by the stage at which aromatisation occurs. This can occur (i) at the 5-dehydroshikimic acid level (Fig. 7.3, route (a)) (ii) at the chorismic acid level (Fig. 7.7) and (iii) from prephenic acid (Fig. 7.10).

A fourth route for synthesising the benzene nucleus, starting from acetyl CoA, has been widely discussed in Chapter 4.

In plants gallic acid is usually found esterified with various alcohols, for instance with quinic acid; it is also the predominant aryl system in the hydrolysable tannins (Fig. 7.4). Most gallic acid in plants is derived from the direct aromatisation of 5-dehydroshikimate.

Incorporation experiments showed that both ^{14}C-labelled D-glucose and L-phenylalanine are incorporated into gallic acid in *Geranium pyrenaicum*.

The tannins are a group of vegetable-derived substances, which have been used since ancient times to transform animal skins into waterproof, long-lasting leather. The term tannin was introduced by Seguil in 1796, before it was realised the tannins are comprised of a diverse mixture and not a narrow range of substances. The ferric chloride test, formerly used to test for tannin-like compounds, is also inadequate. Ferric chloride can also develop a green or black colour with many polyphenols, for instance with chlorogenic acid (3) and with the simple catechins (see p. 403): such compounds have few structural analogies with common tannins.

The tannins are nowadays defined as those natural substances with a molecular weight between 500 and 3,000, and possessing a number of free phenolic hydroxyls (1-2 per 100 molecular weight units) allowing the formation of stable crosslinks with proteins and other biopolymers such as cellulose and pectins. Polyphenolic compounds of lower molecular weights can also bind with proteins but such complexes are usually labile. Likewise, phenolic polymers of very high molecular weight do not show tanning properties, since their large dimensions prevent penetration into the fibrils of collagen, a fundamental, proteinaceous component of mammal skins.

The tannins act as potent enzyme inhibitors when bound to proteins. Upon tasting they produce a typical astringent feeling in the mouth, similar to that felt when eating sour fruit. This reaction is due to the precipitation from the saliva of certain glycoproteins which normally possess lubricating properties. In fruit the degree of polymerisation of the tannins increases with ripening and thus the tannin binding capacity to proteins proportionately decreases.

Tannins can be structurally and chemically divided into two groups: *Hydrolysable tannins*, and *non-hydrolysable*, or *condensed tannins*. The former are

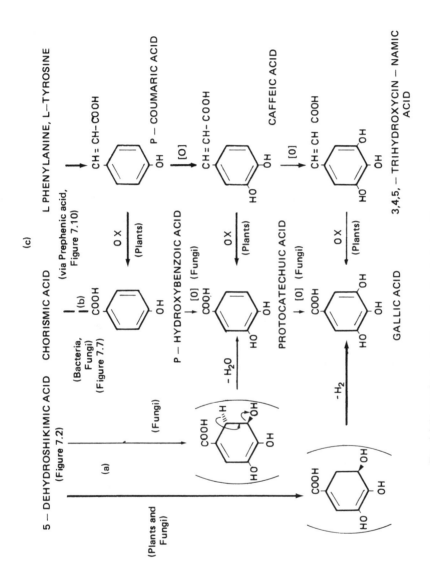

Fig. 7.3 Biosynthesis of gallic and protocatechuic acids.

conjugates of a monosaccharide, usually D-glucose, with the hydroxyl groups (all or part) esterified with gallic, digallic, trigallic and hexahydroxydiphenic acids (Figs. 7.4 and 7.5). These tannins are easily hydrolysed by acids and alkali (and by specific enzymes), to liberate the individual aromatic acids and the carbohydrate unit. In contrast, the second group of tannins do not give low molecular weight compounds on hydrolysis and instead increase their degree of polymerisation, producing amorphous insoluble substances, often coloured red, called phlobaphenes.

(a) GALLOTANNINS, e.g :

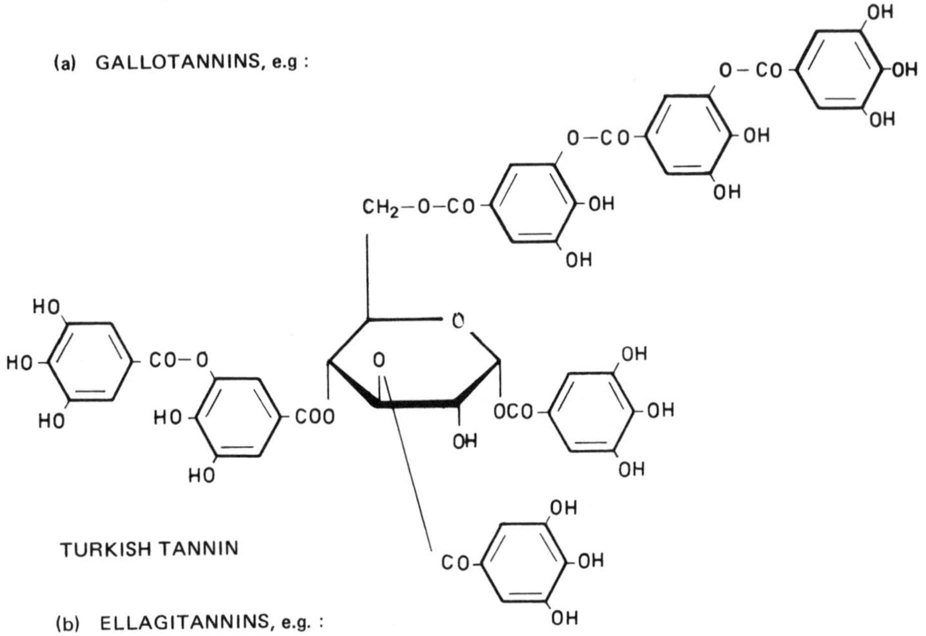

TURKISH TANNIN

(b) ELLAGITANNINS, e.g. :

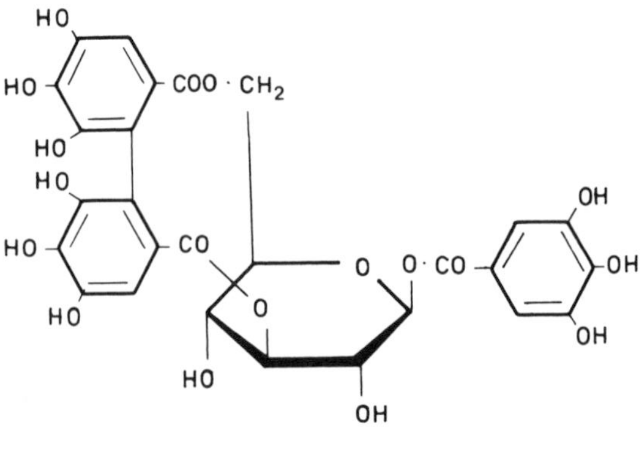

CORILAGIN   Fig. 7.4 Some hydrolysable tannins.

The non-hydrolysable tannins are catechin polymers, and are described on p. 000. The hydrolysable tannins are classified according to the type of acidic residues present: the *gallotannins* yield only gallic acid under strong hydrolysis conditions and the *ellagitannins* also liberate ellagic acid (Fig. 7.4). Ellagic acid is really an artefact, originating by a double, and spontaneous, lactonisation of the natural hexahydroxydiphenic acid (Fig. 7.5).

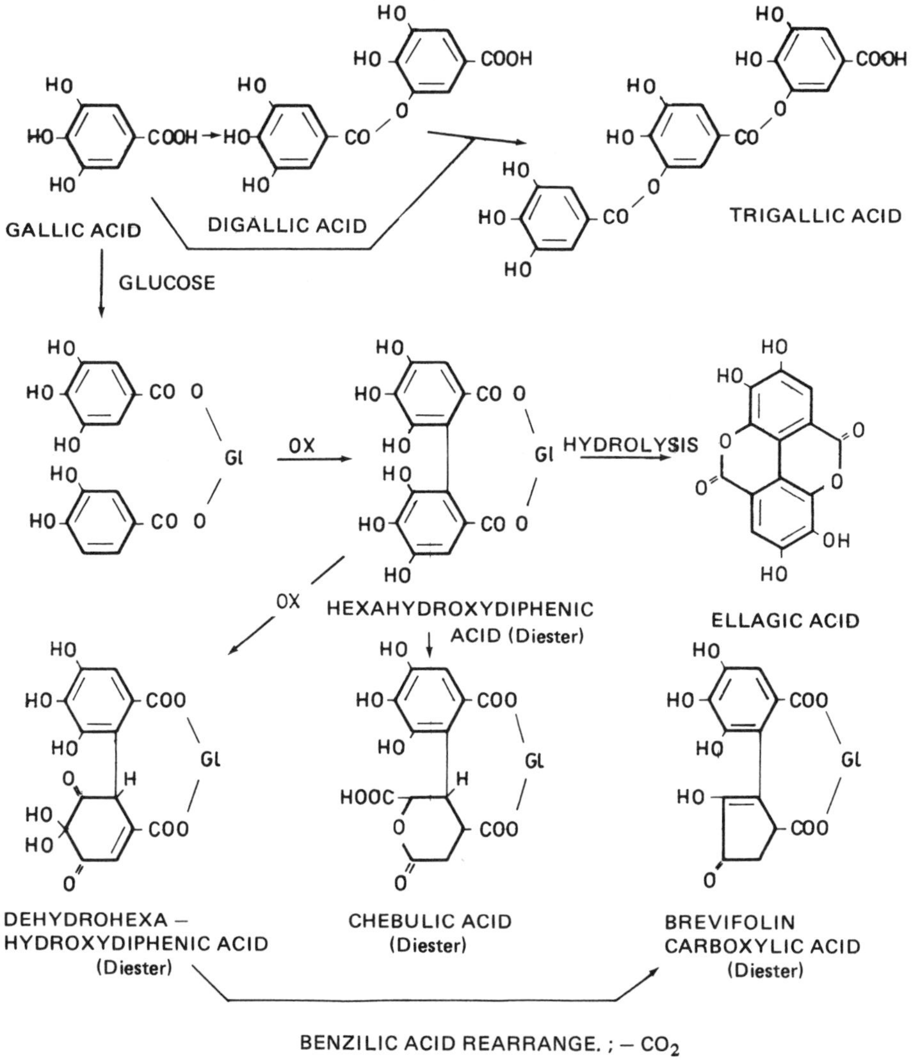

Fig. 7.5 Formation of gallic acid derivatives.

All the hydrolysable tannins are derived from the progressive esterification of sugar hydroxyl groups with either gallic or polygallic acids. The gallic acid residues can subsequently be further modified. Oxidations are particularly common (Fig. 7.5) to give, for example, hexahydroxydiphenic acid (by oxidative coupling of two units of gallic acid) dehydrohexahydroxydiphenic acid (by further oxidation), brevifolincarboxylic acid (by benzylic rearrangement of dehydrohexahydroxydiphenic acid and decarboxylation) and chebulic acid (by oxidative cleavage of the hexahydroxydiphenic acid aromatic ring and hydrogenation of the residual double bond).

The hexahydroxydiphenic acid residues present in the various ellagitannins tend to show optical activity* through diphenyl stereoisomerism (atropoisomerism; Appendix 2) and this is in agreement with the hypotheses schematised in Fig. 7.5, the oxidative coupling of gallic acid residues occurring in the tannin macromolecules.

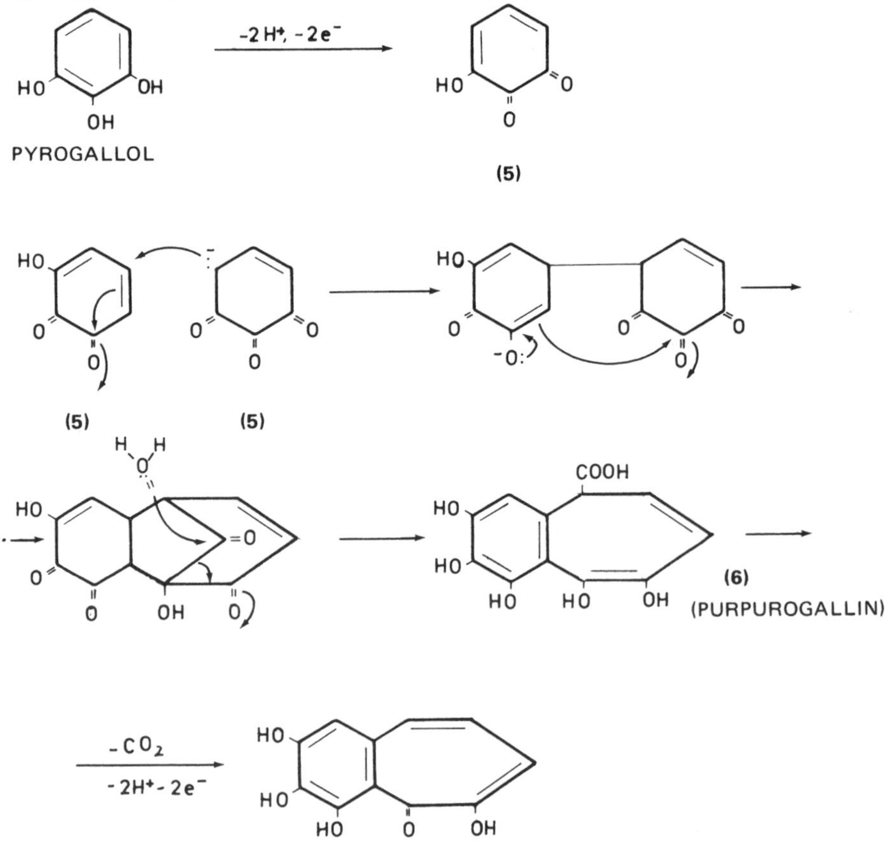

Fig. 7.6 Probable mechanism for the formation of purpurogallin.

*If corilagin is first methylated with diazomethane and then hydrolysed, a hexamethoxydiphenic acid is isolated in an optically active form.

The most probable biogenetic sequence seems to be: gallic acids + glucose → gallotannins → ellagitannins. The latter also contain acids derived from further transformations of the hexahydroxydiphenic acid.

The hydrolysable tannins are widely distributed in the tissues of the higher plants, particularly in galls, which are abnormal development of cells or vegetable tissues, appearing as simple hypertrophies or actual neoplasms. The agents provoking galls include bacteria, fungi, nematodes and insects.

Large amounts of Turkish tannin is produced in the galls of *Quercus infectoria* (Fagaceae) and are instigated by the insect *Cynips tinctoria*. The galls of *Quercus sessiliflora*, which are infested by the insect *Dryophanta taschenbergii*, also contain a red glucoside, having purpurogallin (6) as its aglycone. This pigment can be easily obtained *in vitro* by oxidation of pyrogallol with potassium ferricyanide, or with oxidases, such as peroxidase (see p. 000). It is probable that purpurogallin is also produced *in vivo* by an oxidase acting on pyrogallol, itself formed by a non-oxidative decarboxylation of gallic acid. A possible mechanism is shown in Fig. 7.6. Purpurogallin represents a rare example of a tropolone system of shikimic acid origin.

7.3 *p*-HYDROXY AND *p*-AMINOBENZOIC ACIDS; SALICYLIC AND ANTHRANILIC ACIDS; FOLIC ACID

All four natural benzoic acids substituted with either one amino or hydroxyl group in the *ortho-* or *para-* position to the carboxyl group can be derived directly from chorismic acid.

Some mechanistic routes that have been suggested are shown in Fig. 7.7. Both hydroxy acids (salicylic acid and *p*-hydroxybenzoic acid) can also be found in plants by oxidative cleavage of the side chain of the corresponding cinnamic acids.

In microorganisms, such as *Mycobacterium fortuitum*, which produce both salicylic acid and 6-methylsalicylic acid, it has been shown that the former acid arises exclusively via a shikimate-chorismate pathway, whilst the latter originates, as expected, from acetate (cf. Fig. 4.8). The enzyme, salicylate synthetase, has been isolated from various Mycobacterium species; it catalyses the conversion of isochorismic acid into salicylic acid.

p-Hydroxybenzoic acid is an intermediate in the synthesis of important metabolites, amongst which are the ubiquinones (Fig. 5.80).

Anthranilic acid is the starting point for the synthesis of tryptophan which can also be catabolised back to it. The catabolic degradation of tryptophan is the only way animals can produce anthranilic acid.

p-Aminobenzoic acid (PABA) is an important growth factor for many microorganisms and antagonises the bacteriostatic action of the sulphonamides. The principal biological function of PABA is in the biosynthesis of folic acid; it acts as a bridge between the pteridine nucleus and glutamic acid (Fig. 7.8).

The condensation between the hydroxymethylene group of 2-amino-4-hydroxy-6-hydroxymethylpteridine (7) and the amino group of PABA affords dihydropteroic acid (8), so that folic acid is also called pteroylglutamic acid (PGA). Other synonyms of folic acid are vitamin M, vitamin B_c, and the *Lactobacillus casei* factor. These synonyms respectively indicated an antianaemic factor in monkeys (present in the yeasts and in liver extracts), an antianaemic factor for baby-chickens (present in liver extracts), and a growth factor for *Lactobacillus casei.* All of these factors, as well as a fourth growth factor for *Streptococcus lactis,* called folic acid, were shown to be the same substance in 1940. Folic acid has subsequently been synthesised.

In vivo, folate (F) is in an oxidation-reduction equilibrium with 7,8-dihydrofolate (DHF) and 5,6,7,8-tetrahydrofolate (THF). The last acts as a one-carbon carrier undergoing methylation, hydroxymethylation and formylation at the 5- and 10-nitrogen atoms (Fig. 3.29).

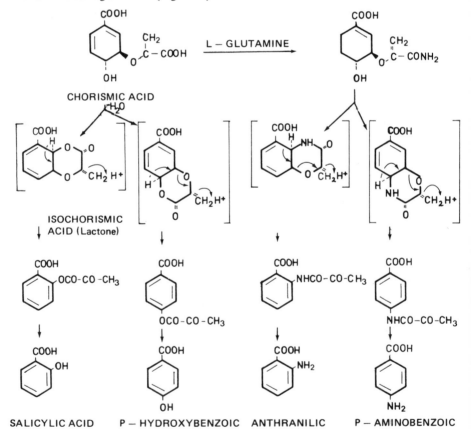

Fig. 7.7 Possible mechanisms for the formation of hydroxy- and amino-benzoic acids.

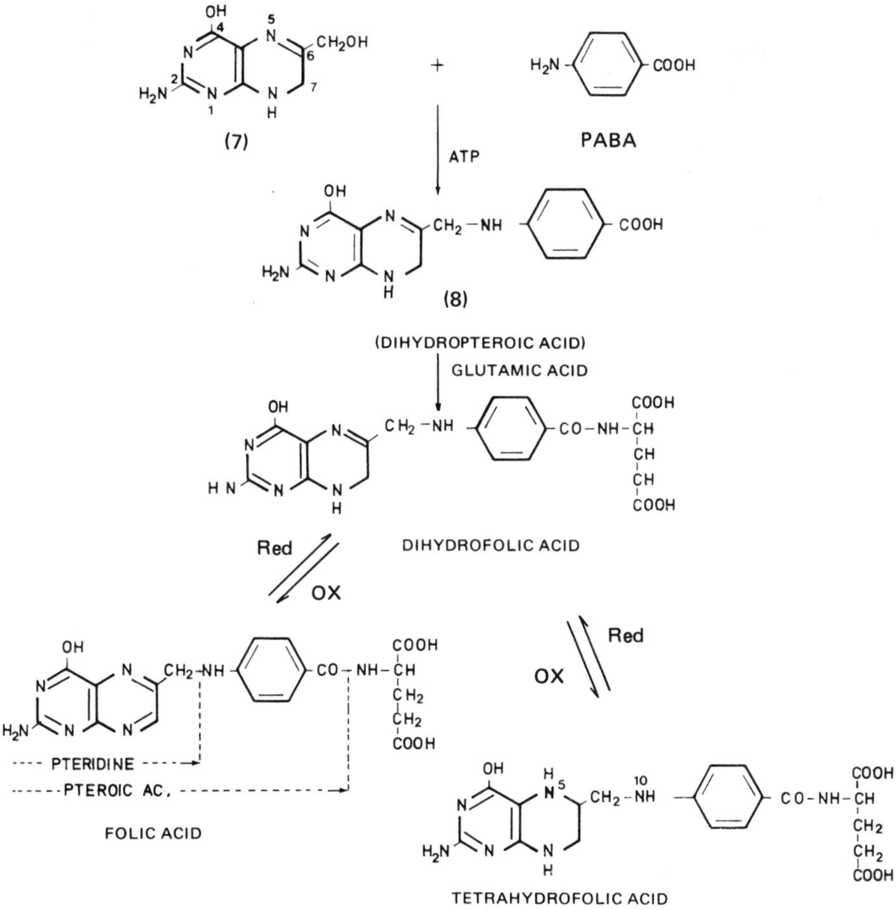

Fig. 7.8 Biosynthesis of folic acid.

7.4 BIOSYNTHESIS OF TRYPTOPHAN

As mentioned above, the first stage in the biosynthesis of tryptophan, starting from chorismic acid, is the formation of anthranilic acid. The α- and β- carbons of the tryptophan indole nucleus arises from D-ribose-1-pyrophosphate-5-phosphate (9) (see the pentosephosphate cycle, Fig. 3.13), and correspond to the carbon atoms 1 and 2 of the pentose molecule (Fig. 7.9). The compounds, 10, 11, 12, and 13 are obligatory biosynthetic intermediates in various microorganisms (*Aerobacter aerogenes, Saccharomyces cerevisiae, Escherichia coli,* etc.); compound 11 derives from an Amadori rearrangement, and accumulates in mutants needing tryptophan, which are characterised by a genetic enzymic block in the reactions from 11 to 12. The last step in the biosynthesis of tryptophan is the exchange of the side chain of the indole-3-glycerolphosphate (12) with the C_3-chain of L-serine. Such exchange takes place by the elimination of the glyceric chain and the successive reaction of the indole (13) with the enzyme-bound Schiff's base (14). The whole process is ruled by a multienzyme complex with two components A and B, each catalysing one of the two consecutive reactions. If part B is missing, due to a mutation for instance, an accumulation of indole is observed.

Tryptophan is also a precursor of nicotinic acid (vitamin PP), and many other secondary metabolites, amongst which are the indolic alkaloids of both chemical and pharmacological importance.

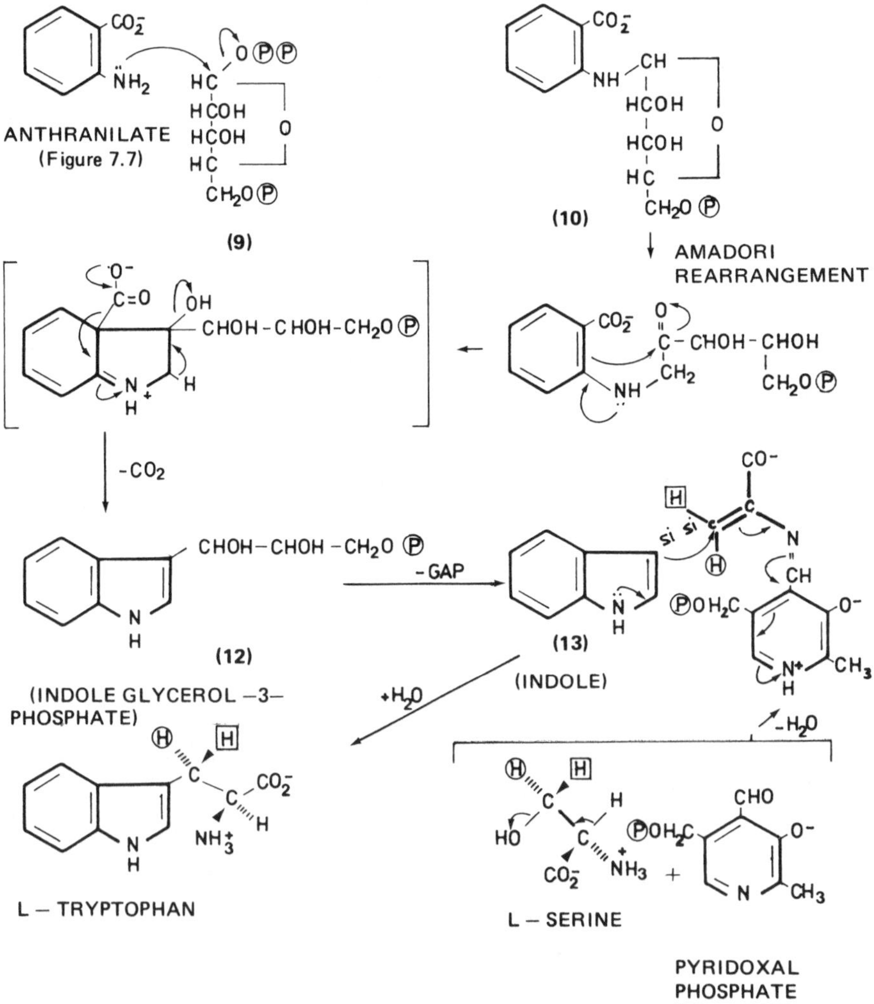

Fig. 7.9 Biosynthesis of tryptophan.

7.5 BIOSYNTHESIS OF PHENYLALANINE AND TYROSINE

Phenylalanine, like tryptophan, is an essential amino acid for animals, which only obtain it from foodstuffs. It is biosynthesised by microorganisms and plants, as schematised in Fig. 7.10.

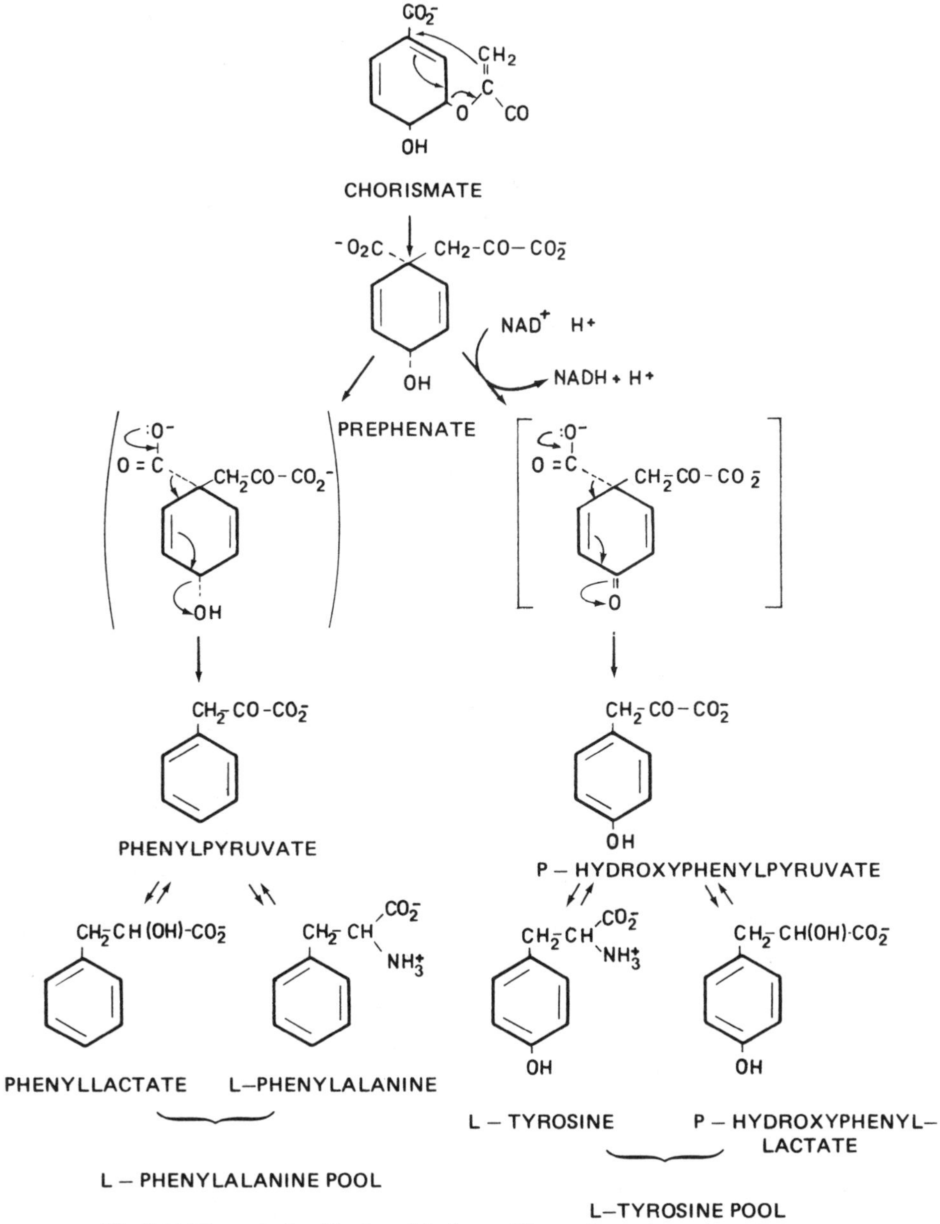

Fig. 7.10 Biosynthesis of L-phenylalanine and L-tyrosine and some reactions.

The hydroxylation of L-phenylalanine to L-tyrosine is indicated in animals by the enzyme phenylalanine hydroxylase, which also occurs in some micro-organisms. Such an enzyme is usually absent in higher plants. In these organisms L-tyrosine is formed from chorismate without passing through L-phenylalanine. The dividing of the biosynthetic routes to the two aromatic amino acids occurs at the prephenate level; it can decarboxylate either oxidatively or non-oxidatively.

In plants, the conversion of compounds in equilibrium with L-phenylalanine (e.g. phenylpyruvate) into those in equilibrium with L-tyrosine is also restricted or nil. Thus two independent metabolic pools exist (Fig. 7.10). The secondary metabolites from these pools form two distinct series, which can be distinguished, experimentally, by separate incorporation experiments involving labelled phenyl-alanine or tyrosine (see Fig. 8.1).

7.6 NAPHTHOQUINONES

Shikimic acid is an obligatory intermediate in the synthesis of the napthoquinone nucleus of the menaquinones found in certain Gram-positive bacteria (e.g. *Bacillus subtilis*), of the phylloquinones distributed in higher plants (e.g. *Zea mays*) (Fig.5.83) and of the anthraquinone nucleus of alizarin (in *Rubia tinctorum*) (see Fig. 5.79). Shikimic acid is also the precursor of two other naphthoquinones of vegetable origin, lawsone (15) present in *Impatiens balsamina,* and juglone (16), present in *Juglans regia.*

Many biosynthetic studies have illustrated how the seven carbon atoms of shikimic acid are incorporated into the napthoquinone nucleus, together with the three central carbon atoms of α-ketoglutaric acid, an intermediate of the tricarboxylic acid cycle (p. 150) or of glutamic acid. *o*-Succinylbenzoic acid (OSB) is a necessary intermediate of all the above quinones. Although this has been proved by incorporation experiments, OSB itself has not so far been isolated from nature. Figure 7.11 depicts the biogenesis of various naphthoquinones and shows three current hypotheses for the formation of OSB, starting from shikimate, chorismate, or prephenate respectively.

Definitive experiments are still awaited to decide which pathway operates in the various organisms.

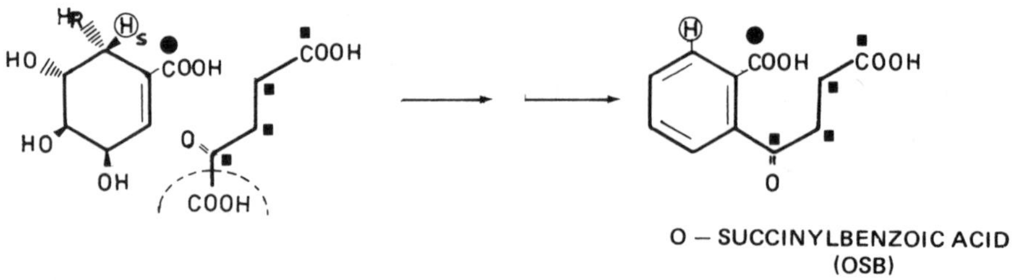

O – SUCCINYLBENZOIC ACID
(OSB)

Fig. 7.11 (contd. overleaf)

Fig. 7.11 (contd.)

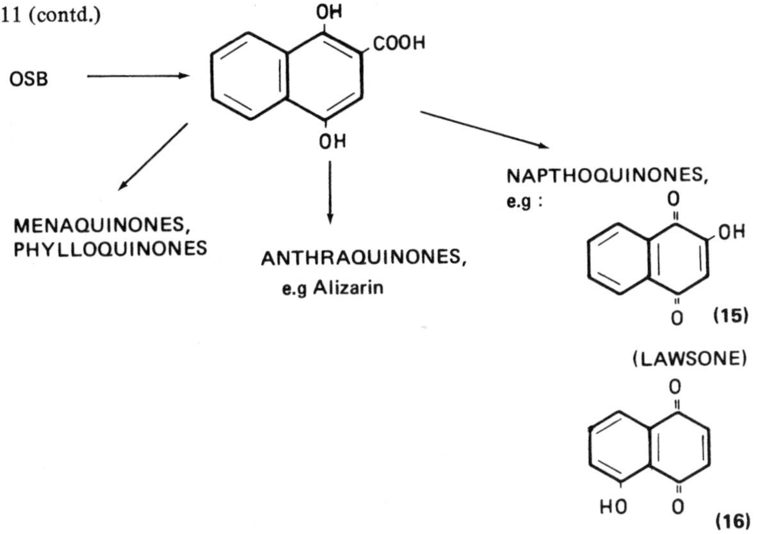

OSB

MENAQUINONES, PHYLLOQUINONES

ANTHRAQUINONES, e.g Alizarin

NAPTHOQUINONES, e.g :

(15)

(LAWSONE)

(16)

(JUGLONE)

POSSIBLE PATHWAYS TO OSB :

(i) from SHIKIMATE

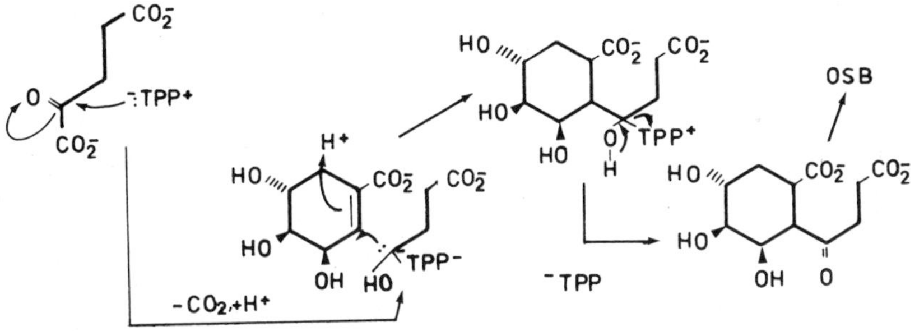

(ii) from CHORISMATE

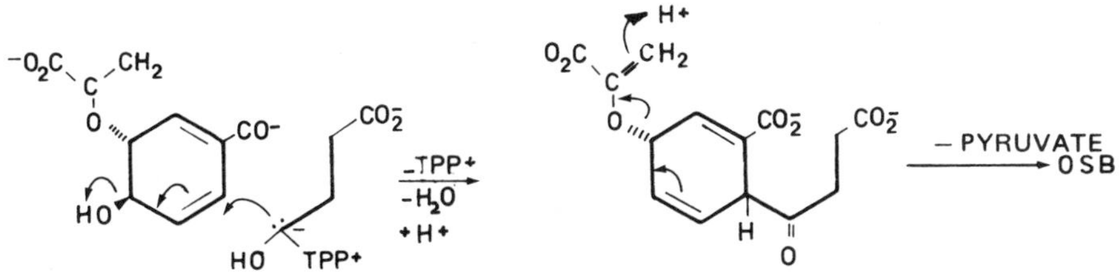

Fig. 7.11 (contd. overleaf)

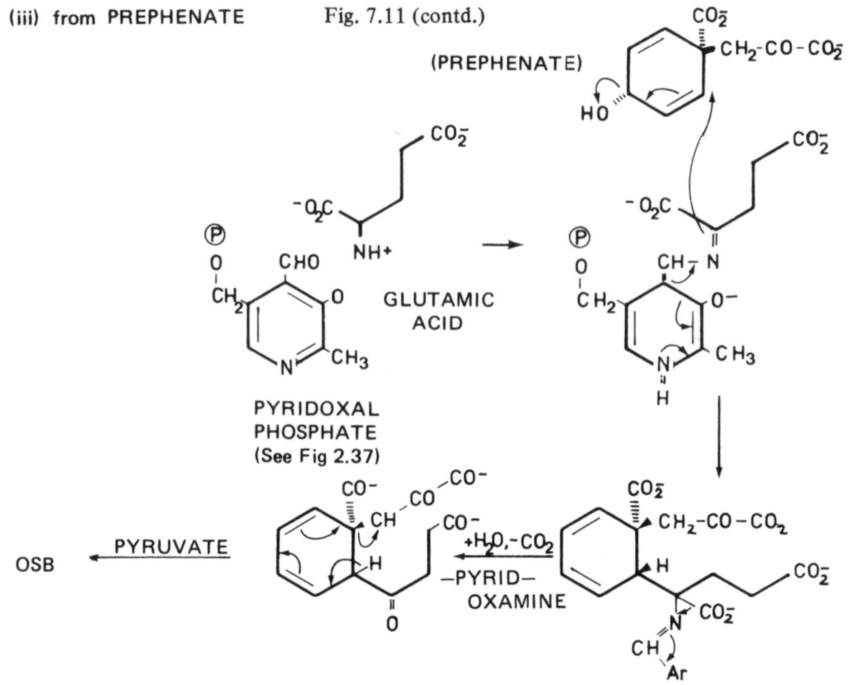

(iii) from PREPHENATE   Fig. 7.11 (contd.)

Fig. 7.11 Biosynthesis of some naphthoquinones.

SOURCE MATERIALS AND SUGGESTED READING

[1] B. D. Davis, (1955), Intermediates in Amino Acid Biosynthesis, *Advances in Enzymol.*, **16**, 247.

[2] D. B. Sprinson, (1961), Biosynthesis of Aromatic Compounds from D-Glucose, *Advances in Carbohydrate Chem.*, **15**, 235.

[3] B. A. Bohm, (1965), Shikimic Acid (3, 4, 5-trihydroxy-1-cyclohexene-1-carboxylic acid). *Chem. Rev.*, **65**, 435.

[4] R. B. Morton (Ed.), (1965), *Biochemistry of Quinones*, Academic Press.

[5] S. G. Humphries, (1967), The Biosynthesis of Tannins in *Biogenesis of Natural Compounds*, ref. 24 of Chap. 1.

[6] M. H. Zenk and E. Leister, (1968), Biosynthesis of Quinones, *Lloydia*, **31**, 275.

[7] F. Lingens, (1968), The Biosynthesis of Aromatic Amino Acids and Its Regulation, *Angew. Chem. Int. Ed. Engl.*, **7**, 350.

[8] S. M. Hopkinson, (1969), The Chemistry and Biochemistry of Phenolic Glycosides, *Quart. Rev.*, **23**, 98.

[9] J. B. Harborne, (1973–1977), Biosynthesis of Phenolic Compounds Derived from Shikimate in *Biosynthesis*, ref. 30 of Chap. 1, **2–5**.

[10] E. Haslam, (1974), *The Shikimate Pathway*, Butterworths.

[11] R. Bentley, (1975), Biosynthesis of Quinones in *Biosynthesis*, ref. 30 of Chap. 1, **3**.

[12] B. Ganem, (1978), From Glucose to Aromatics: Recent Developments in Natural Products of the Shikimic Acid Pathway, *Tetrahedron*, **34**, 3353.

[13] J. Weiss and J. M. Edwards, (1979), *The Biosynthesis of Aromatic Compounds*, Wiley.

Chapter 8

Phenyl-Propanoids

The main transformations of L-phenylalanine (2), R=H and of L-tyrosine (2), R=OH, in plants are summarised in Fig. 8.1. There is an equilibrium, by the usual transamination reaction, between these two amino-acids and phenylpyruvic acid (1), R=H and p-hydroxyphenylpyruvic acid (1), R=OH, respectively, which are produced via two independent paths from shikimic acid (Chapter 7).

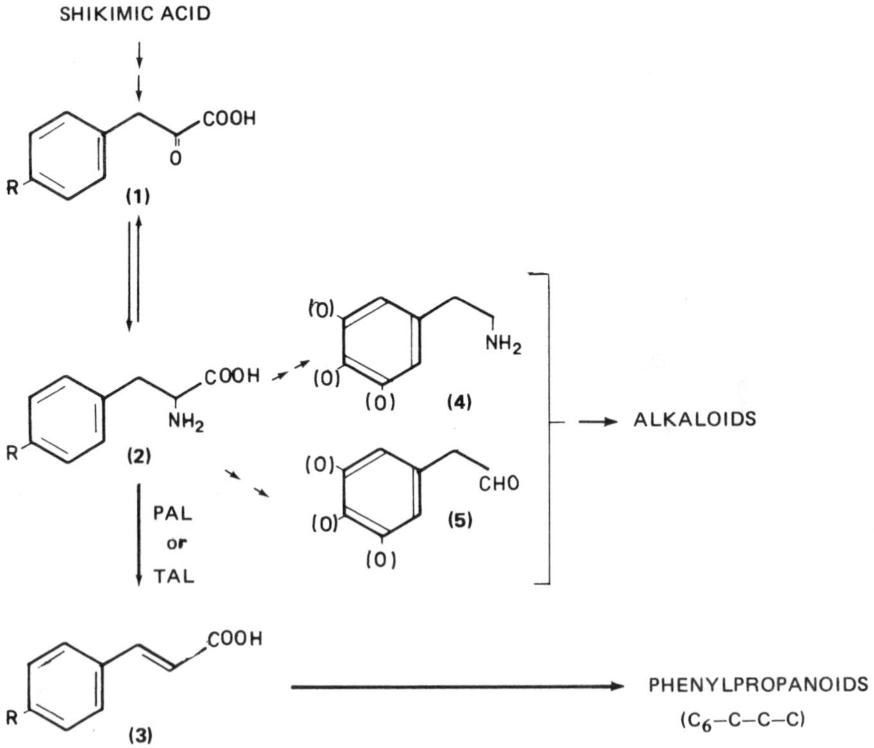

Fig. 8.1 Principal metabolic routes of L-phenylalanine (R=H) and of L-tyrosine (R=OH) in plants.

Two important metabolic routes originate from L-phenylalanine and from L-tyrosine. Phenylalanine is involved in the transformation into *trans*-cinnamic acid (3), R=H, whilst tyrosine is particularly involved with the other process. In this the side chain carboxyl group is expelled and β-phenylethylamine (4) and β-phenylacetaldehyde (5) are formed, generally with extensive hydroxylation of their aromatic rings.

Phenylalanine (and tyrosine) metabolites, with the starting aminoacid skeleton (C_6–C_3) intact, are commonly called the *phenylpropanoids*.

8.1 DERIVATIVES OF PHENYLPYRUVIC ACID

L-Tropic, atropic, and L-α-phenylglyceric acids (Fig. 8.2) are usually found in nature esterified to alkaloid bases, for instance to tropine. Incorporation experiments with [^{14}C] phenylalanine, labelled in each of its three side chain positions showed that the carboxylic carbon of tropic acid corresponds to the carboxylic carbon of the precursor amino acid. Tritium present in the 2-position of phenylalanine is lost during conversion into tropic acid. According to these results, the biosynthetic route schematised in Fig. 8.2(a) appears likely. The transposition of the carboxylic group is an unusual process in organic chemistry and needs to be confirmed. An *in vitro* rearrangement which could be analogous to the *in vivo* mechanism, is the conversion of glycidic acid thiolesters into the corresponding α-formylthiolesters under the influence of boron trifluoride, as indicated in scheme (b) of Fig. 8.2.

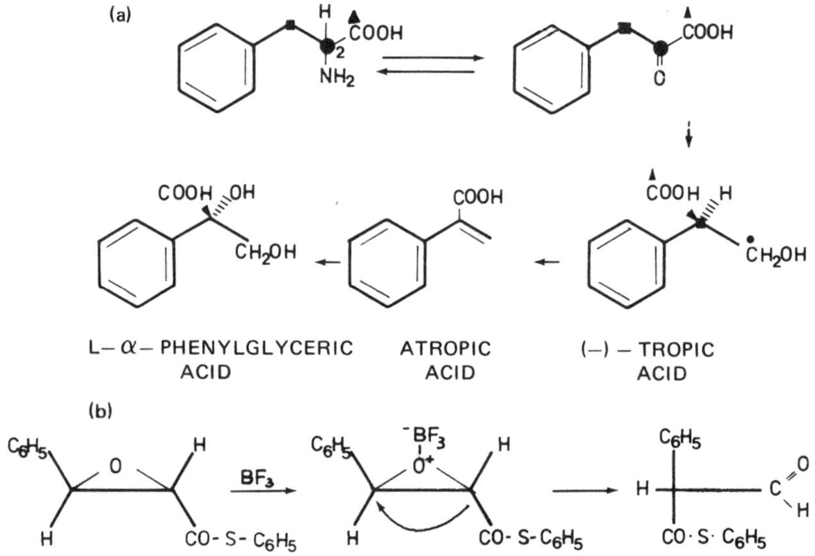

Fig. 8.2 Probable biosynthesis of tropic acid and its derivatives (Scheme *a*) and transposition reactions of glycidic and thiol esters.

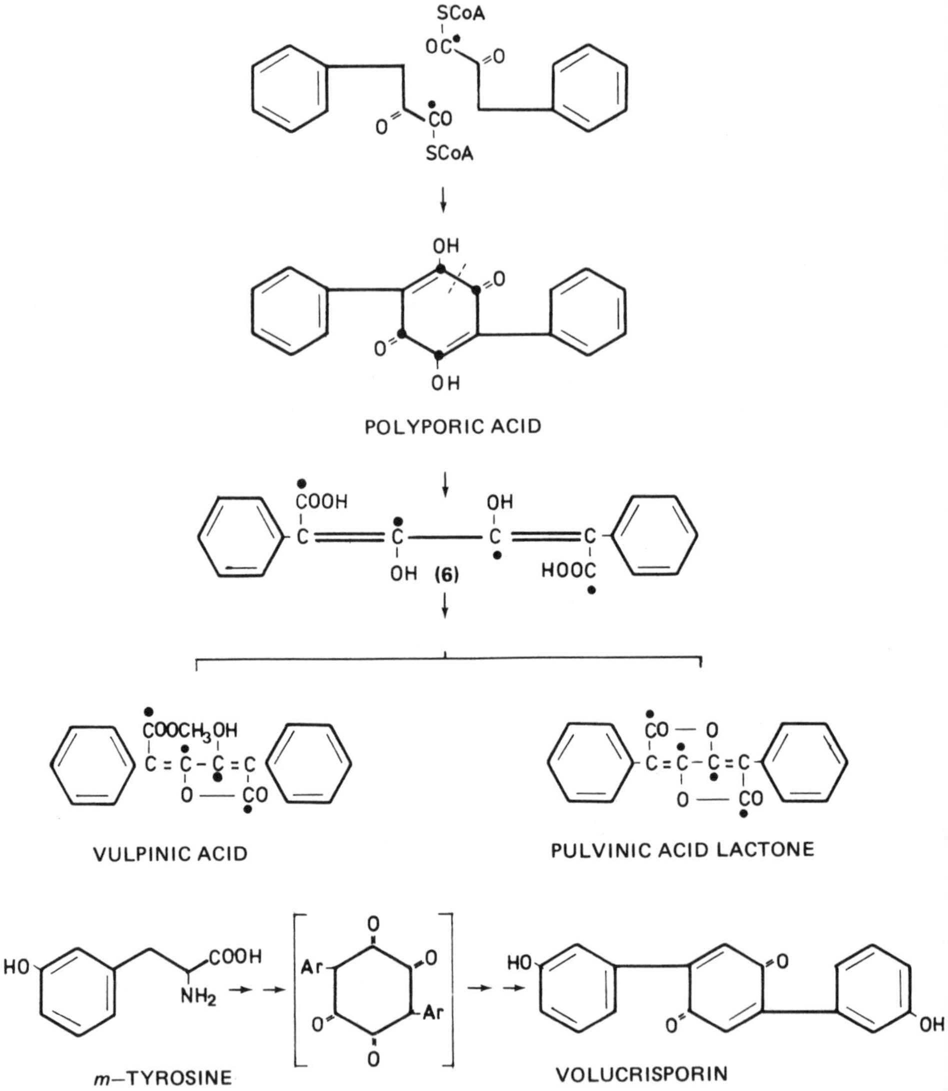

Fig. 8.3 Biosynthesis of some terphenyl derivatives and their catabolism.

In some species of fungi and lichens phenylpyruvic acid and its ring-oxygenated derivatives are further transformed by side chain rearrangements. Some examples are given in Fig. 8.3; the biogenetic routes indicated have been confirmed by much experimental work. After feeding [1-^{14}C] phenylalanine to *Evernia vulpina*, a vulpinic acid was obtained in which the radioactive labels were equally distributed amongst the four carbon atoms of the chain not bound to the aromatic rings.

This result could be explained by the existence of a symmetrical terphenyl intermediate such as polyporic acid. The oxidative cleavage of the dihydroxy-*p*-benzoquinone ring, affording the carboxylic acid (**6**), can be easily carried out in *in vitro* experiments. *m*-Tyrosine is an efficient precursor of volucrisporin in *Volucrispora aurantiaca* Haskins. In this latter case two carbonyl groups of the tetraketo cyclohexane ring are presumably reduced. This type of ring, as occurs in polyporic acid, originates from a Claisen type of condensation between two phenylpyruvic units.

One catabolic route of L-tyrosine, common to both plants and animals, is the conversion of *p*-hydroxyphenylpyruvic acid into homogentisic acid (Fig. 8.4). The known mechanistic details of this process suggest that an oxidative decarboxylation, an oxidation of the aromatic ring and a migration of the side chain taking place. The enzyme, or multienzymatic complex, catalysing the formation of homogentisic acid is known as *p*-hydroxyphenylpyruvate oxidase. It contains copper and needs ascorbic acid as a coenzyme. A mechanism (Fig. 8.4) has been proposed to account for the migration of the side chain.

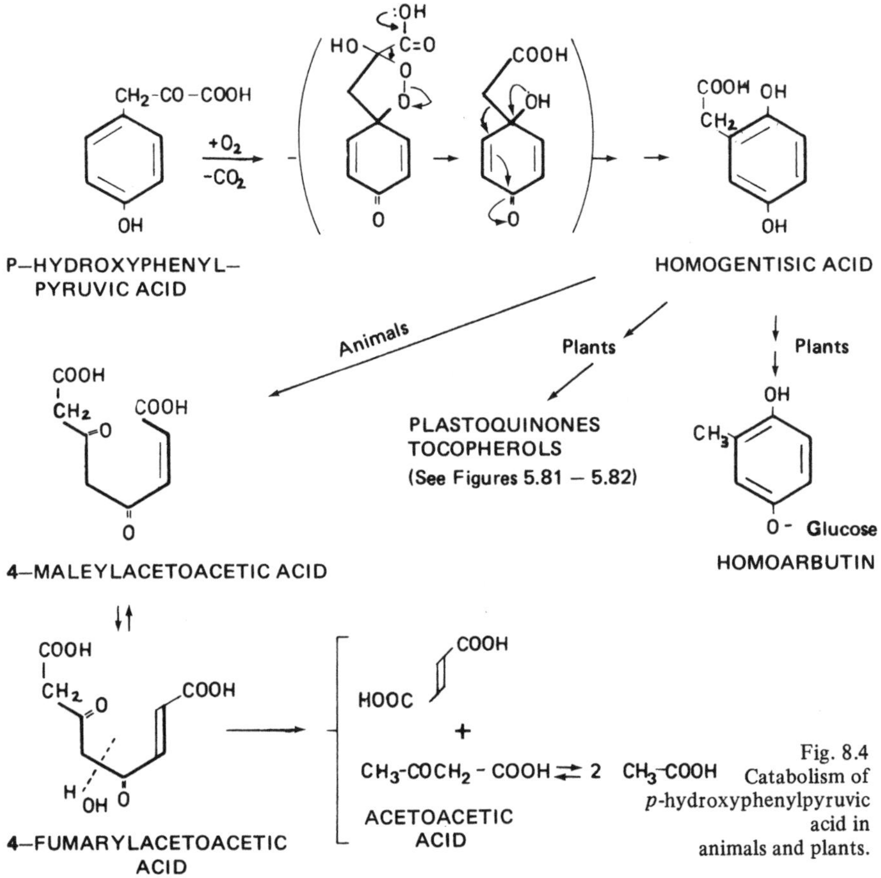

Fig. 8.4 Catabolism of *p*-hydroxyphenylpyruvic acid in animals and plants.

The homogentisic acid in animals is subsequently degraded to 4-maleylaceto-acetic acid, by the enzyme homogentisic acid oxidase. The 4-maleyacetoacetic acid is isomerised into 4-fumarylacetoacetic acid, which is then split into fumaric and acetic acid. The former can enter into the tricarboxylic acid cycle and the latter into the acetic acid pool.

The absence of the homogentisic acid-oxidase enzyme in new-born children sometimes occurs as a hereditary metabolic error and can be discovered by the abnormally large excretion of homogentisic acid into the urine (alkaptonuria). After the addition of alkali such urine quickly turns brown, because of oxidative transformations of the homogentisic acid into quinones, which subsequently polymerise into melanine-type compounds. Although alkaptonuric individuals do not show other abnormalities during their early years, subsequent effects do occur. Their connective tissues become markedly pigmented and, usually, an early onset of arthritis occurs.

Other inborn metabolic errors connected with the catabolism of phenyl-alanine and tyrosine include phenylketonuria and albinism. In the first case an abnormally high elimination of phenylpyruvic and phenyllactic acid is observed in the urine, since L-phenylalanine cannot be converted into L-tyrosine because of an absence of L-phenylalanine hydroxylase. A child affected by this congenital metabolic illness has his psychological development impaired. Lack of pigmenta-tion of the skin, hair and eyes is a sign of albinism. This condition is due to a block on melanogenesis, which is the oxidation of L-tyrosine into melanin pigments.

8.2 CINNAMIC ACIDS

The conversion of L-phenylalanine (2), R=H, into *trans*-cinnamic acid (3), R=H, is catalysed by an enzyme in all vascular plants. This enzyme is called phenyl-alanine ammonia-lyase (PAL, E.C. 4.3.1.5.; see Appendix 5) and it plays an important role in the metabolism of the higher plants, since it controls an essential biosynthetic step on the route to a very large group of important natural products, such as the lignins and flavanoids.

The stereochemical mechanism of the L-phenylalanine deamination (Fig. 2.6) is of the Hofmann elimination type (E2; Appendix 1), which specifically involves only one of the two diastereotopic hydrogens at position 3 (H_S). The discrimination between the two hydrogens, which could also be due to the enzyme, can be explained on the stereoelectronic requirements for an E2 type reaction, in which the leaving groups are antiperiplanar in the transition state, together with the (S) configuration of the asymmetric centre at C-2 and on the more stable conformation (steric effects) of the transition state. The transition state is indicated in the Newman formula (A) (Fig. 2.6) where the NH_2 group and the *pro-S* hydrogen are antiperiplanar. The enzyme should further stabilise this, already stable, transition state. The geometry of the transition state also determines the *trans*-configuration of the resulting cinnamic acid. The catalytic action of the enzyme depends on the presence, in the active site, of a basic centre and of a dehydroalanine residue (Fig. 8.5).

The dehydroalanine binds the nucleophilic amino group of phenylalanine whilst the base operates at the benzylic proton to help liberate the cinnamic acid residue.

The controlling mechanism of PAL activity appears to be of the type shown in Fig. 2.16(b); the PAL enzyme appears to be composed of two isoenzymes: one of them is inhibited by various natural cinnamic acids and the other by benzoic acids.

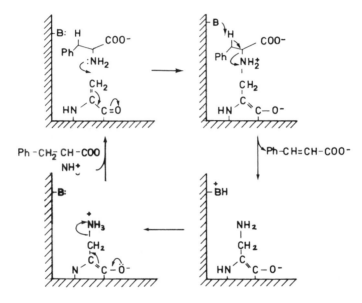

Fig. 8.5 Enzymic conversion of L-phenylalanine into *trans*-cinnamic acid.

Tyrosine can undergo an enzyme-catalysed deamination, affording *p*-coumaric acid (3), R=OH. This reaction is catalysed by tyrosine ammonia-lyase (TAL) and it has a similar mode of action and stereochemical requirements to PAL. Its importance in the overall metabolism of higher plants appears to be quite limited (TAL has been identified with certainty in only a few monocotyledon plants).

Most *p*-coumaric acid has been shown to arise in nature by hydroxylation of cinnamic acid (Fig. 8.6). This latter fact, as well as the total inability of plant tissue to transform phenylalanine into tyrosine and the more typical metabolites of tyrosine, such as (4) and (5) (Fig. 8.1), is in agreement with the almost complete lack of reciprocal interference between the metabolic paths of the two aromatic amino acids.

Correlations between the various natural cinnamic acids is indicated in Fig. 8.6. 3,4,5-Trihydroxycinnamic acid (7) and 3,4-dihydroxy-5-methoxycinnamic acid (8) are possible intermediate metabolites, although they have not been isolated in nature up to now. The acid (7) gives gallic acid and (8) is converted to sinapic acid. A similar situation exists for the *o*-hydroxycinnamic acids. *o*-Coumaric and 2,4-dihydroxycinnamic acids, derived by *o*-hydroxylation of the corresponding (*E*)-cinnamic acids, are probably the intermediates in the biosynthesis of coumarins although neither has been isolated. An explanation for the failure to isolate some of these polyhydroxylated acids is that they are bound in a stable way to multienzymatic complexes, or to carrier proteins, and that they are released from them only after conversion into other products (for instance after methylation). Alternatively, they might exist *in vivo* only as very transient intermediates.

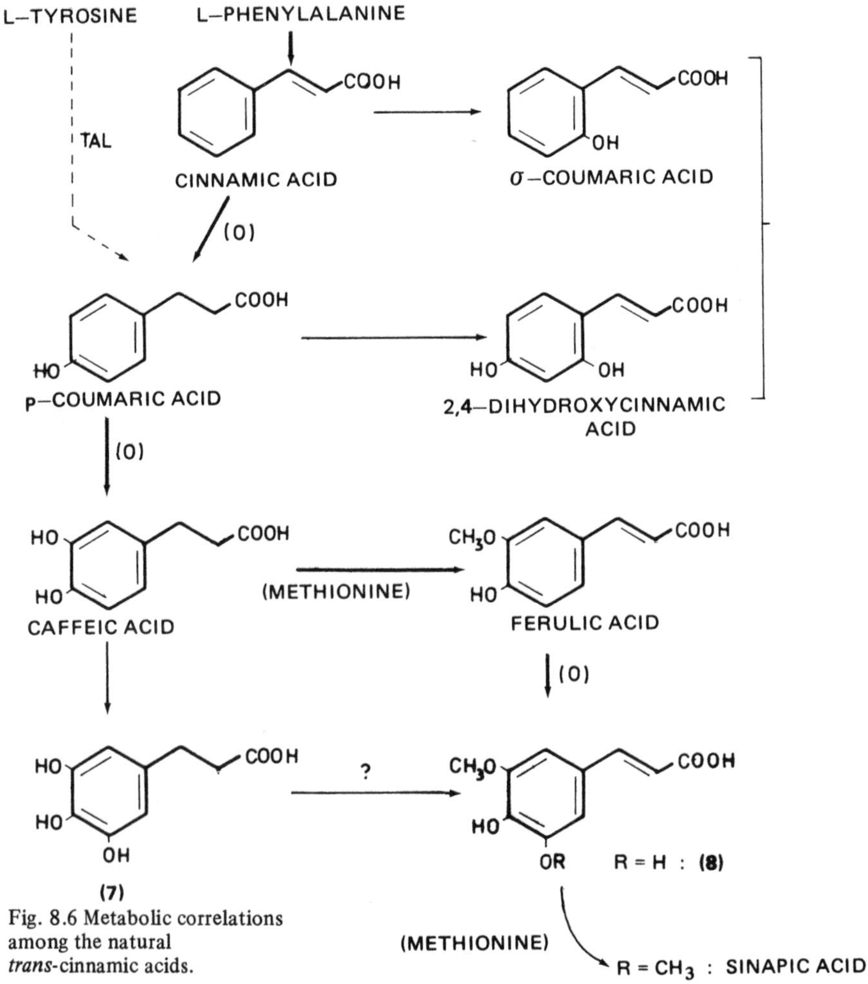

Fig. 8.6 Metabolic correlations among the natural *trans*-cinnamic acids.

Hydroxylations of the aromatic ring of the cinnamic acids are mediated by mixed-function oxygenases and are characterised by the typical NIH-shifts. The introduction of the hydroxyls into a benzene ring with a side chain according to the sequence p; m, C-3, m, C-5 with o-hydroxyl groups being introduced somewhere along this sequence, is customary in higher plants. The resulting oxygenation patterns have a considerable diagnostic value for the biogenesis of the aromatic ring itself, since they indicate a shikimate origin, whilst oxygenation in alternate positions in the aromatic ring suggest a polyketide origin. O-Methylation of cinnamic acids are catalysed by O-methyltransferases, using S-adenosylmethionine as methyl donor.

Cinnamic acids in vegetable tissues are usually combined with polyhydroxylated molecules, which impart high water solubility to the products. The conjugation of hydroxylated cinnamic acids can involve the carboxyl group or a phenolic hydroxyl. The commonest conjugating compounds are quinic acid and D-glucopyranose. Combination products with tartaric acid are also known, such as monocaffeyltartaric and cicoric acids (Fig. 8.7).

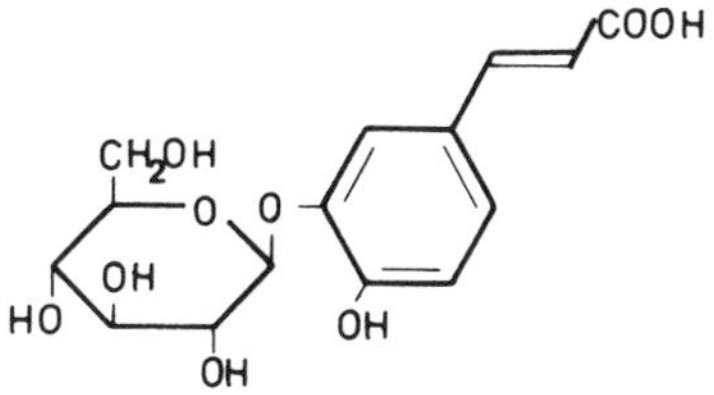

3–(β–D–GLUCOSYL) CAFFEIC ACID

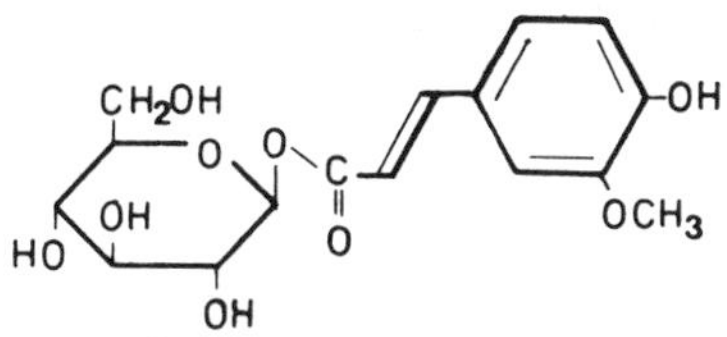

1β– FERULYL – D–GLUCOSE

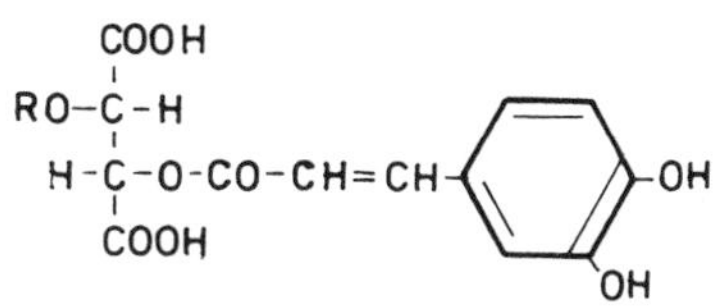

R = H,   MONOCAFFEYLTARTARIC ACID
R=CAFFEYL, CICORIC ACID

Fig. 8.7 Conjugated forms of some cinnamic acids.

These conjugations probably have a physiological explanation since certain transformations of cinnamic acid, for instance hydroxylations, take place on the acids esterified to quinic acid. Such esters are more soluble in nature and possibly reach the hydroxylation sites more readily.

Rosmarinic acid (Fig. 8.8) is a particularly interesting biogenetic example. Through incorporation experiments in mint plants it has been demonstrated that the caffeic acid residue arises from L-phenylalanine, whilst the 3,4-dihydroxyphenyllactic acid one comes from the L-tyrosine pool via DOPA (3,4-dihydroxyphenylalanine) and 3,4-dihydroxyphenylpyruvic acid.

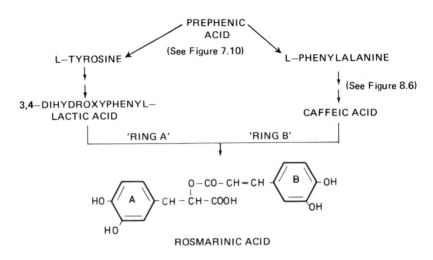

Fig. 8.8 Biosynthesis of rosmarinic acid.

8.3 COUMARINS AND NEOFLAVONOIDS

The coumarins (9) are typical metabolites of higher plants, and have only seldom been isolated from microorganisms. The benzo-2-pyrone nucleus of the simple coumarins (Fig. 8.9) derives from the phenylacrylic skeleton of cinnamic acids via *ortho*-hydroxylation, *trans-cis* isomerisation of the side chain double bond, and lactonisation. The sequences and mechanisms of such processes are still uncertain in most cases. The biosynthesis of the simplest molecule, coumarin, has been investigated in two plant species, *Hierochloe odorata* (Graminaceae) and *Melilotus alba* (Leguminosae) and the results confirm the reaction sequence (a)–(d) of Fig. 8.9. In particular, it has been found that *ortho*-hydroxylation of *trans*-cinnamic acid is under the control of a specific gene in *M. alba*. The *ortho*-glucoside (melilotoside) and the *o*-coumarinyl glucoside are probable inter-mediates, as is confirmed by a high incorporation of the former (^{14}C-labelled) into coumarin, and by the presence in *M. alba* cells of a β-glucosidase which specifically hydrolyses the *cis*-glucoside. This enzyme is liberated on crushing the cells, affording the aglycone, *o*-coumarinic acid, which spontaneously cyclises to coumarin and which led some people to consider coumarin as purely an artefact. There are no doubts, however, about the existence of free coumarin, in small amounts, in plants, which may indicate that the β-glucosidase also operates in intact cells.

Coumarin is sometimes formed through the very simple path shown by reactions (e) and (f) (Fig. 8.9). The *trans-cis* isomerisation of the double bond of *o*-coumarinylglucoside, or of cinnamic acid itself, is still not clear. It could occur under enzymatic catalysis, through a photochemical process, or through other mechanisms, such as a reduction-dehydrogenation sequence.

The path (a)-(d) (Fig. 8.9) should be followed in the biosynthesis of all coumarins oxygenated at position 7: umbelliferone, esculetin and scopoletin are the most widespread coumarins in nature. During the synthesis of these compounds *o*-hydroxylation should respectively take place on *p*-coumaric, caffeic and ferulic acid; umbelliferone cannot be ruled out as an intermediate to coumarins oxygenated at positions 6 or 8, besides position 7.

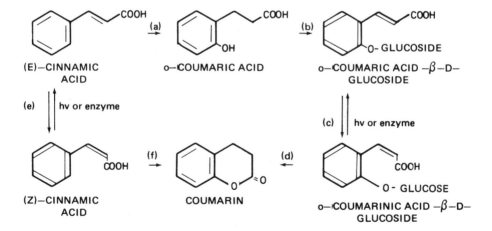

(9)

	C-6	C-7	C-8	
R =	H	H	H	COUMARIN
	H	OH	H	UMBELLIFERONE
	H	OCH$_3$	H	HERNIARIN
	OH	OH	H	AESCULETIN
	H	OH	OH	DAPHNETIN
	OCH$_3$	OH	H	SCOPOLETIN
	OCH$_3$	O-Gl	H	SCOPOLIN
	OCH$_3$	OH	OH	FRAXETIN

Fig. 8.9 Principal simple coumarins and their biosynthetic pathways..

A scheme of simultaneous *o*-hydroxylation and lactonisation has been proposed for the 7-hydroxycoumarins: phenolic oxidation of the corresponding 4-hydroxycinnamic acids should take place (Fig. 8.10).

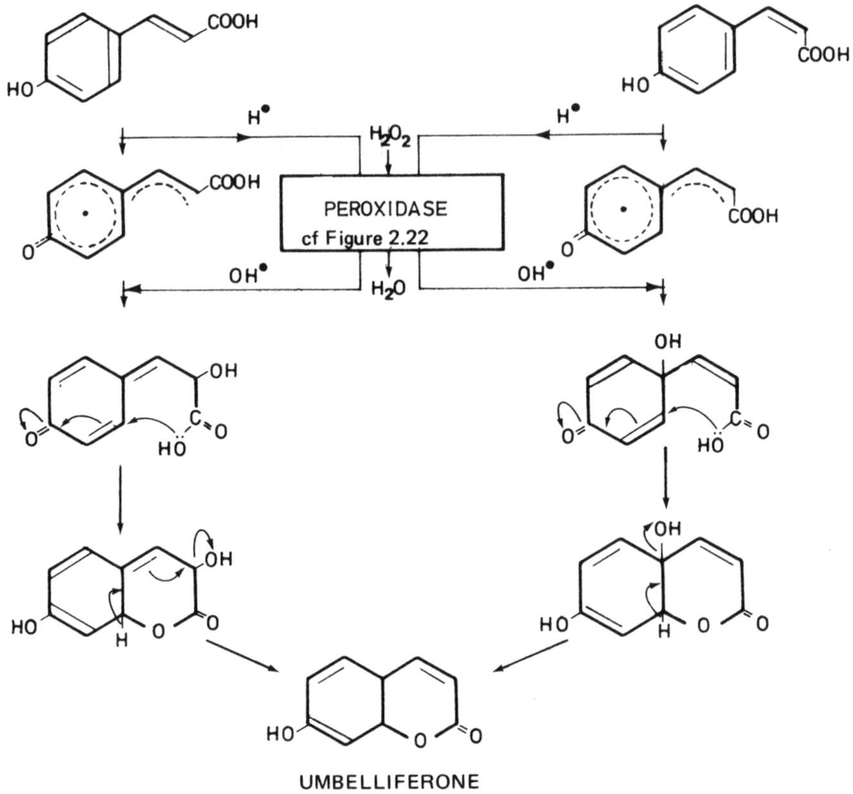

Fig. 8.10 Possible formation schemes of 7-hydroxycoumarins via phenolic oxidation.

During fermentation of hay some microorganisms (such as *Aspergillus fumigatus*), transform coumarin into dicoumarol (Fig. 8.11). Such spoiling of hay can be dangerous to animals as dicoumarol is a powerful blood anticoagulant and can cause fatal haemorrhages in cattle which eat it.

The immediate precursor of dicoumarol is 4-hydroxycoumarin (12), which is derived from coumarin through melilotic acid, *o*-coumaric acid, β-hydroxy-melilotic acid (10) and β-ketomelilotic acid (11). The central methylene group of dicoumarol arises from formaldehyde, also produced during the fermentation, and not from methionine or serine.

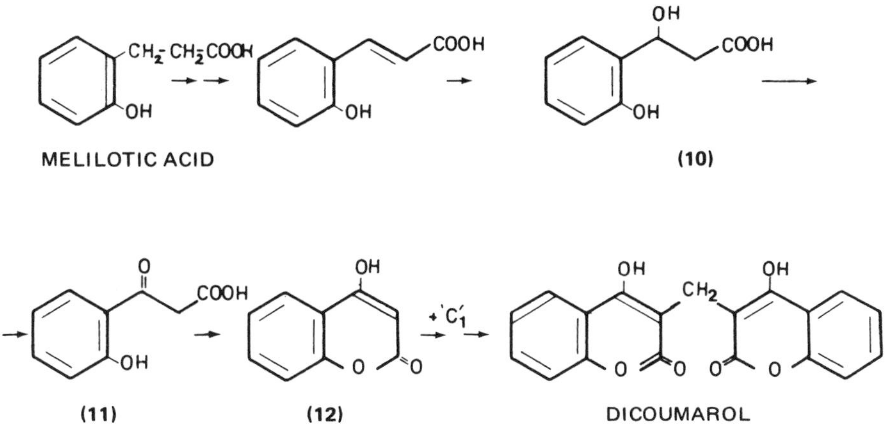

MELILOTIC ACID   (10)

(11)   (12)   DICOUMAROL

Fig. 8.11 Biosynthesis of dicoumarol.

Many complex coumarins are known in nature besides the simple coumarins (Fig. 8.12). The benzopyrone nucleus often bears alkyl substituents, such as the prenyl group in suberosin (13), or pendant aryl rings, such as in dalbergin (18); they can also occur condensed with other heterocyclic rings, such as in compounds (14)-(16). The furocoumarins, such as bergapten, are common coumarin derivatives. These compounds are derived from prenylcoumarins through mechanisms similar to those schematised in Fig. 5.77. Apart from a few exceptions, a constant structural element of simple coumarins, as well as of prenyl- and furocoumarins, is the presence of an oxygen atom at position 7. This fact suggested a polyketide biogenesis as an alternative to the shikimate pathway. All evidence to date, however, supports the latter pathway with a common intermediate, umbelliferone (Fig. 8.9), which is a catabolite of p-coumaric acid.

The coumarins with an aryl group at C-3 or C-4 or, much more rarely, with a non-prenylic alkyl group in C-4 are biogenetically very different from the above systems. In the 3-aryl coumarins, of which coumestrol (16) is a typical example, the C-3 benzene ring and the three carbon atoms of the pyrone ring derive from the same molecule of phenylalanine, whilst the benzene ring of the coumarin nucleus is formed by the condensation of three acetate units.

The 3-arylcoumarins belong to the isoflavonoid group, which will be discussed later on (p. 424). The biosynthesis of the 4-aryl and 4-alkylcoumarins has still not been elucidated. The former such as dalbergin (18) are also called neoflavonoids because of their structural similarities to the flavanoids (cf. Fig. 8.39) and the isoflavanoids (cf. Fig. 8.60).

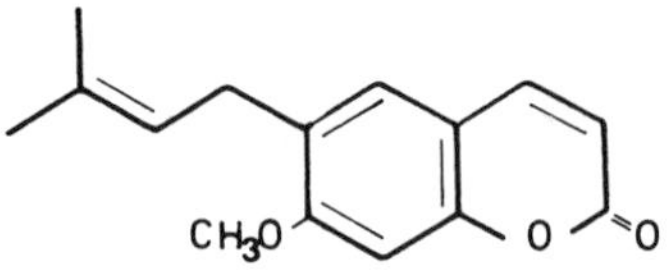

(13)
SUBEROSIN

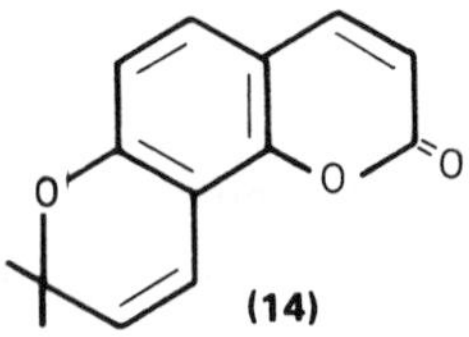

(14)

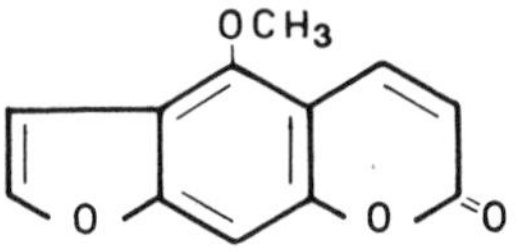

(15)
BERGAPTEN

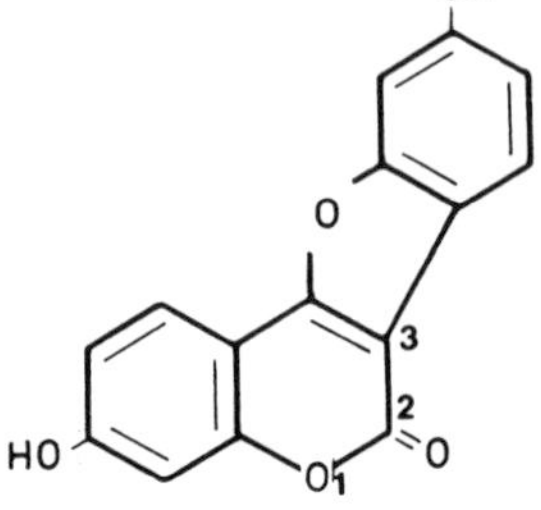

(16)
COUMESTROL

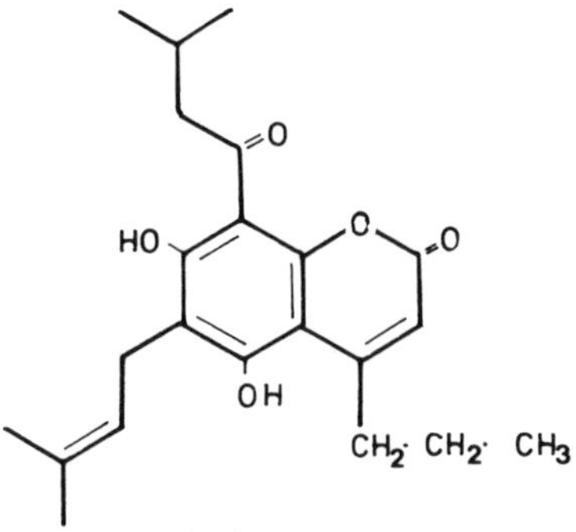

(17)
MAMMEIN

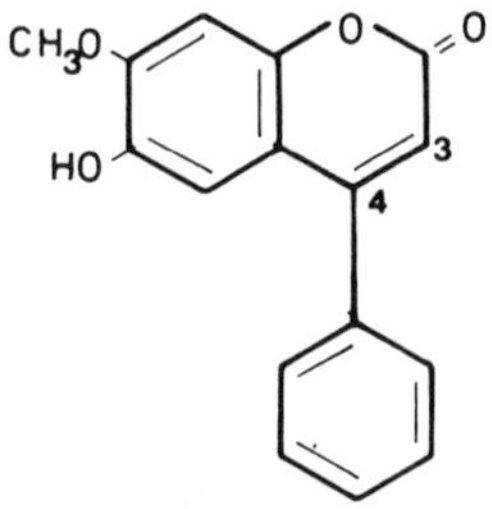

(18)
DALBERGIN

Fig. 8.12 Complex coumarins.

Difficulties are encountered in studying the biosynthesis of the neoflavanoids, as they are almost exclusively produced by large exotic trees (genus *Pterocarpus* and *Dalbergin* of the Leguminosae family). An early hypothesis suggested the 4-arylcoumarins were derived from isoflavone intermediates through a migration of the aryl group from position 3 to position 4. This hypothesis was shown to be wrong by a unique incorporation experiment (Fig. 8.13). After feeding [3-^{14}C] phenylalanine to *Calophyllum inophyllum* (Guttiferae), Polonsky isolated calophyllolide specifically labelled in position 4. According to the previous hypothesis the label should have been at C-2, in agreement with the biosynthesis of iso-flavones (see Fig. 8.61).

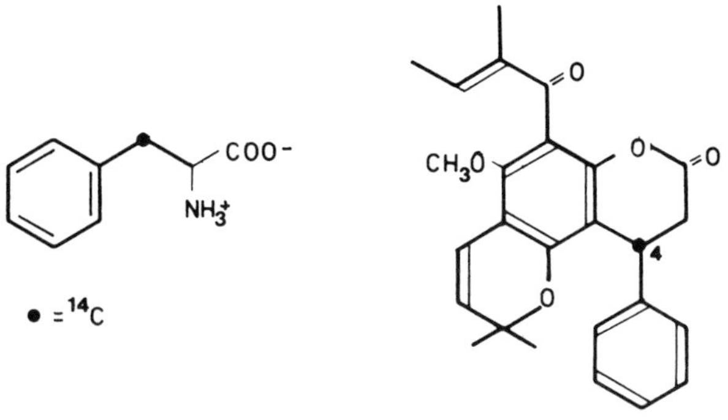

CALOPHYLLOLID

Fig. 8.13 Incorporation of phenylalanine into calophyllolid.

The three hypotheses schematised in Fig. 8.14 and 8.15 are consistent with Polonsky's experimental data. Of these, the most likely route is that of Ollis and Gottlieb, in which either resorcinol or phloroglucinol (both of which are of polyketide origin, see Fig. 4.3) are precursors of the condensed aromatic ring and one cinnamyl unit, for instance the pyrophosphate ester of cinnamyl alcohol (which is derived from phenylalanine), acts as the precursor of the other carbon atoms of the C_{15}-skeleton.

The origin of a wide range of compounds having very different structures and yet occurring in the same plant can often be rationalised. Dalbergiquinols, dalbergiones (also called dalbergenones or dalbergiquinones), neoflavenes (or dalbergichromenes), 4-arylcoumarins (or dalbergins) and brazilins can be accommodated into the same biosynthetic scheme (Fig. 8.15). The three modes of nucleophilic attack of the polyphenolic residue onto the C_6-C_3 unit show close analogies to the mechanisms hypothesised by Birch for the biosynthesis of allyl and propenylphenols and of the various prenylation reaction (Fig. 5.77).

4-Alkylcoumarins, such as mammein are formed through mechanisms analogous to those illustrated in Figs. 8.14 and 8.15. The precursor differs in having an $\alpha\beta$-unsaturated acidic or allylic structure, being aliphatic rather than aromatic.

Although speculative, a biogenetic relationship has been suggested from the 4-arylcoumarins to some benzophenones (19), and natural xanthones (20) and (21) (Fig. 8.16). This route to benzophenones and xanthones should be included with those already described but operates only in plants producing neoflavonoids (e.g. Guttiferae).

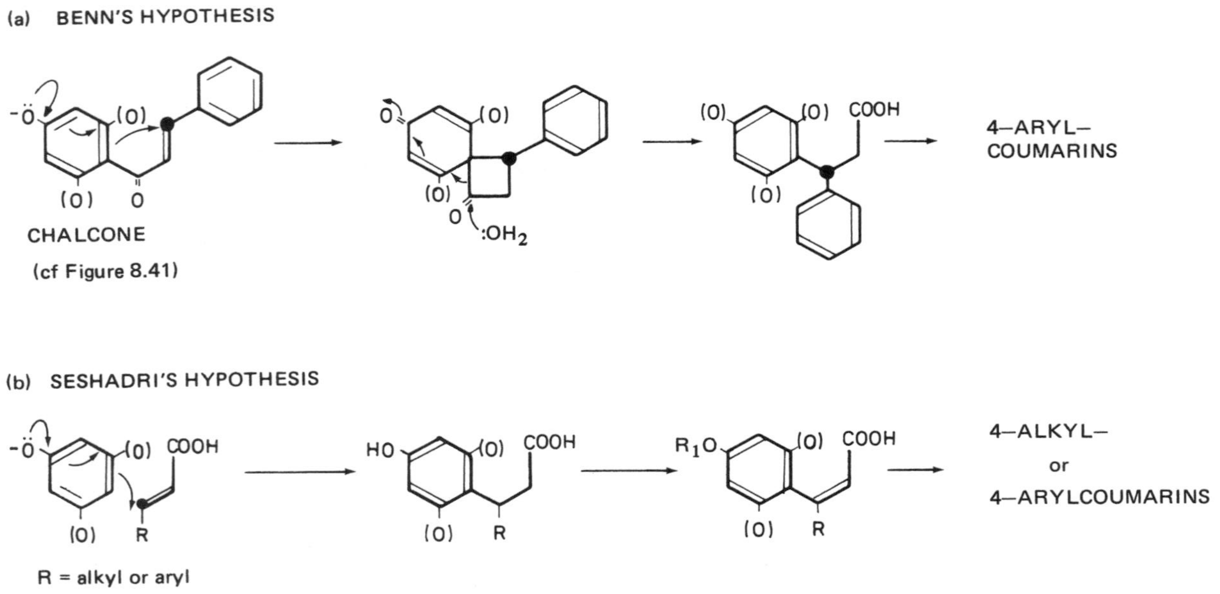

Fig. 8.14 The Benn and Seshadri hypotheses on the biosynthesis of neoflavanoids (● corresponds to the 3 carbon atom of phenylalanine).

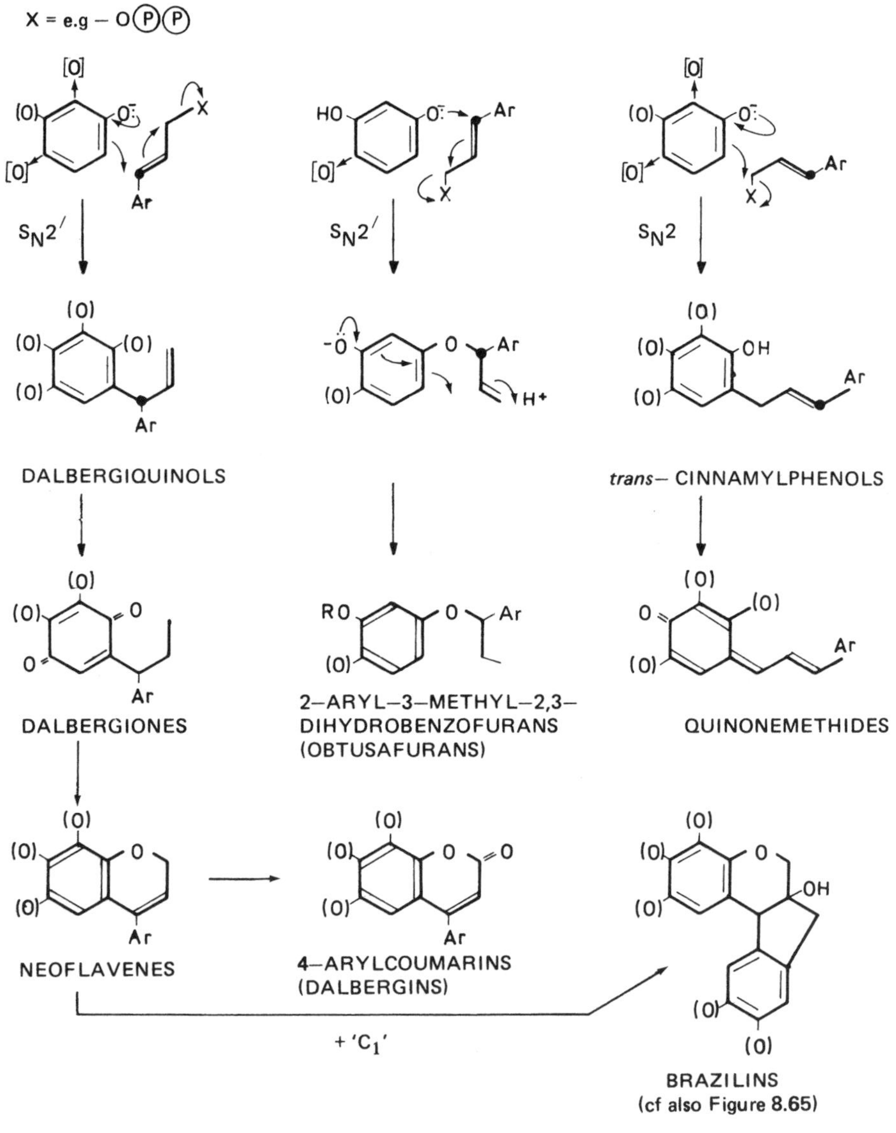

Fig. 8.15 The Ollis and Gottlieb hypothesis on the biosynthesis of neoflavonoids (● corresponds to the 3-carbon atom of phenylalanine).

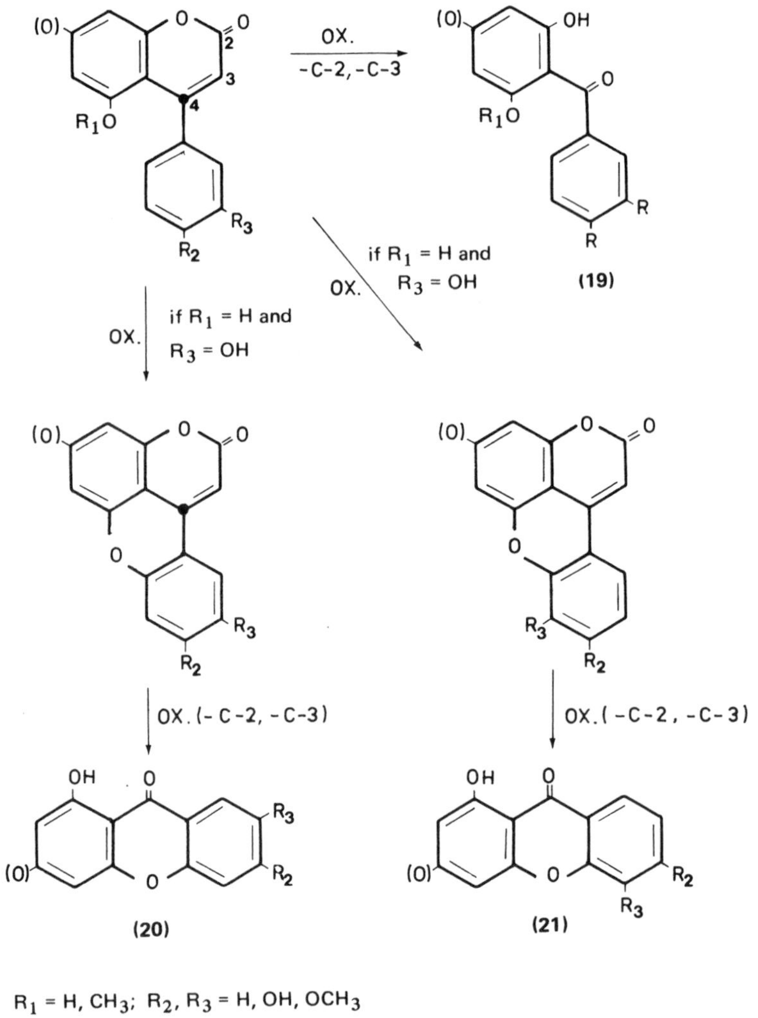

Fig. 8.16 Possible formation of benzophenones and xanthones from
4-arylcoumarins.

R_1 = H, CH_3; R_2, R_3 = H, OH, OCH_3

8.4 TRANSFORMATIONS OF THE SIDE CHAIN OF CINNAMIC ACIDS

The acrylic residue of the cinnamic acids can undergo various modifications of
which there are four main types:

(1) Change of oxidation level, without any alteration in the length of the
carbon chain (cinnamyl alcohols, allyl and propenyl phenols).

(2) New C–C bond formation between different phenylpropane units (bis-
arylpropanoids: lignans and neolignans; lignins).

(3) Reduction in size of the carbon chain (styrenes; acetophenones; benzoic acids).

(4) Increase in the size of the carbon chain (stilbenes; flavonoids; isoflavanoids; rotenoids).

These groups are discussed below.

8.4.1 C_6–C_3 Compounds. Allyl and Propenyl Phenols

Some phenylpropanoids are derived from cinnamic acids by the reduction of the carboxyl group, (22)-(27), or of the double bond, (28), or of both, (29). The *cinnamyl alcohols* play a primary metabolic role as they are obligatory intermediates in the formation of lignins (section 8.4.2). The reduction of the carboxylic group probably proceeds via the esters with CoASH. Like most carbonyl reductions it is catalysed by NAD(P)-dependent dehydrogenases.

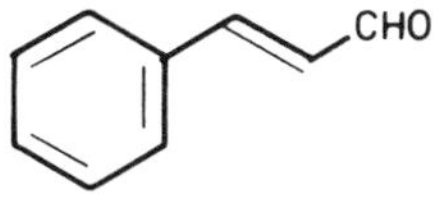

(22)
CINNAMALDEHYDE

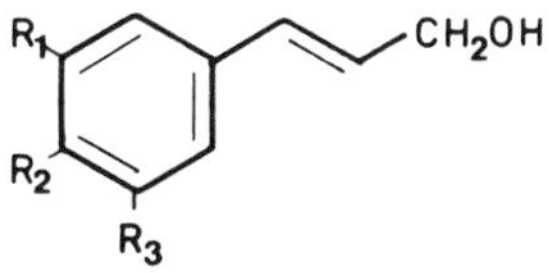

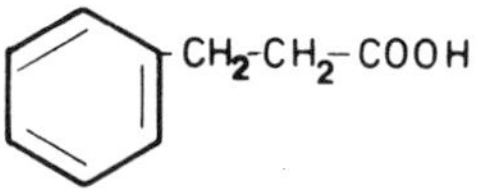

(28)
DIHYDROCINNAMIC ACID

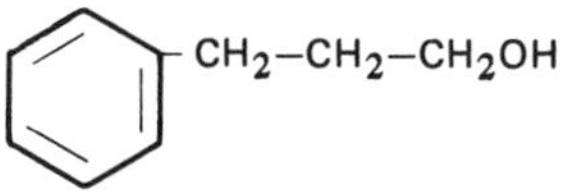

(29)
DIHYDROCINNAMYL
ALCOHOL

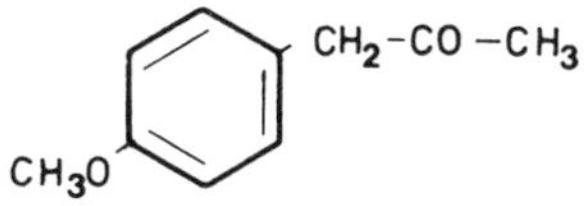

(23: R_1, R_2, R_3 = H)
CINNAMYL ALCOHOL

(24: R_1, R_3 = H, R_2 = OH)
P–COUMARYL ALCOHOL

(25: R_1 = H, R_2 = OH, R_3 = OCH$_3$)
CONIFERYL ALCOHOL

(26: R_1 = H, R_2 = OGl, R_3 = OCH$_3$)
CONIFERIN

(27: R_1 = OCH$_3$, R_2 = OH, R_3 = OCH$_3$)
SINAPYL ALCOHOL

(30)
ANISE KETONE

Fig. 8.17 Some C_6–C_3 compounds.

The allyl and propenyl phenols (Fig. 8.18) are two related types of C_6–C_3 compounds. They all have a phenolic hydroxyl, or an ether function at C-4, occasionally accompanied by other methoxyl or methylenedioxy groups.

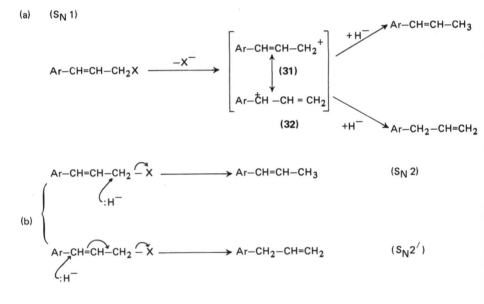

ALLYLPHENOLS	RING SUBSTITUENTS				(E)—PROPENYLPHENOLS
	2	3	4	5	
CHAVICOL			OH		—
ESTRAGOLE			OCH_3		ANETHOLE
EUGENOL		OCH_3	OH		ISOEUGENOL
SAFROLE		$O - CH_2 - O$			ISOSAFROLE
ELEMICIN		OCH_3	OCH_3	OCH_3	ISOELEMICIN
MYRISTICIN		$O - CH_2 - O$		OCH_3	ISOMYRISTICIN
APIOLE	OCH_3	$O - CH_2 - O$		OCH_3	ISOAPIOLE

Fig. 8.18 Principle natural allyl- and propenyl-phenols.

As discussed in earlier chapters, the conversion of a carboxyl group ($-CO_2H$) into a primary alcohol ($-CH_2OH$) frequently occurs *in vivo*. Oxidation of a methyl group into a primary alcohol function is also known. The direct reduction of a hydroxyl-substituted carbon to a saturated carbon (methyl, methylene, or methine) is, however, rather unusual owing to the difficulty of substituting a hydroxyl group (even an activated one) with a hydride ion arising from NAD(P)H. This difficulty should be less for allylic hydroxyls, as the corresponding carbocation (discrete or incipient) is more resonance stabilised (see Appendix 1). A biogenetic hypothesis accounting for formation of the allyl and propenylphenols has been proposed by Birch and is schematised in Fig. 8.19: the two mechanisms (a) and (b) correspond to a monomolecular nucleophilic substitution and to a bimolecular one, H^- being the nucleophile. An *a priori* dichotomy exists for these two mechanisms, in trying to explain the formation, under enzymic control of both propenylphenols (carbocation **(31)**, or S_N2') and allylphenols (benzylic carbocation **(32)**, or S_N2).

The most recent experimental data on the biosynthesis of anethole (propenyl group) in *Pimpinella anisum* (Umbelliferae) confirms Birch's hypothesis. Further work is necessary to fully elucidate the mechanisms for the *in vivo* formation of the allyl and propenyl side chains.

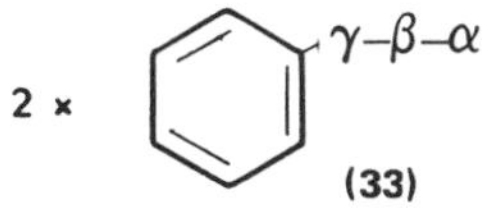

(33)

PHENYLPROPANOID UNIT

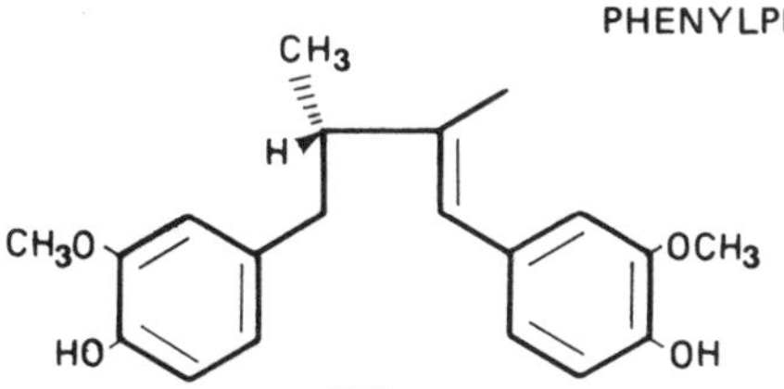

(34)

(–)–GUAIARETIC ACID

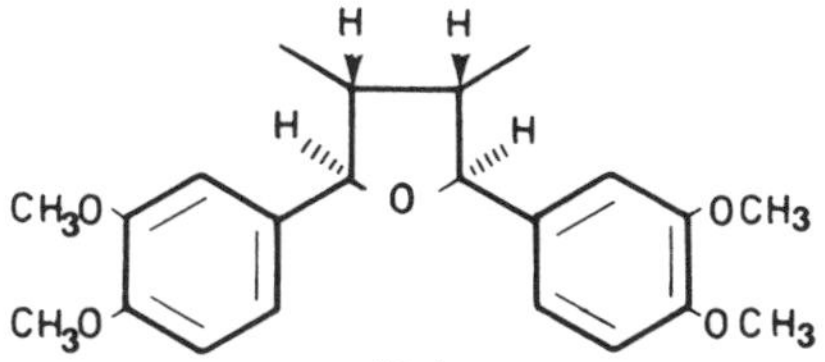

(35)

GALGRAVIN

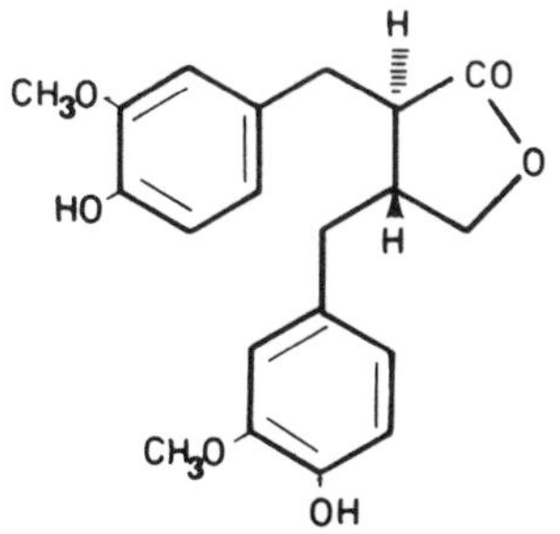

(36)

(–)–MATAIRESINOL

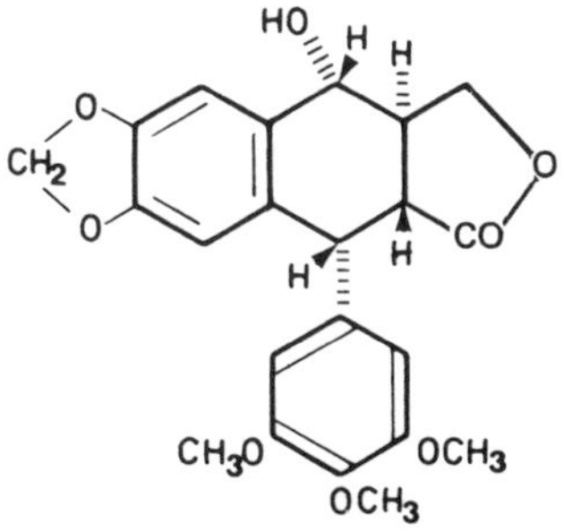

(37)

PODOPHYLLOTOXIN

Fig. 8.20 Typical lignan structures.

8.4.2 Lignans and Neolignans

The *lignans* comprise the group of natural compounds with their carbon skeletons derived from two phenylpropane units joined together by at least one C-C bond between the two central positions (β and β') of the two C_3 chains (Fig. 8.20). The lignans are widely found in higher plants. Both phenylpropane units are almost certainly derived from shikimic acid.

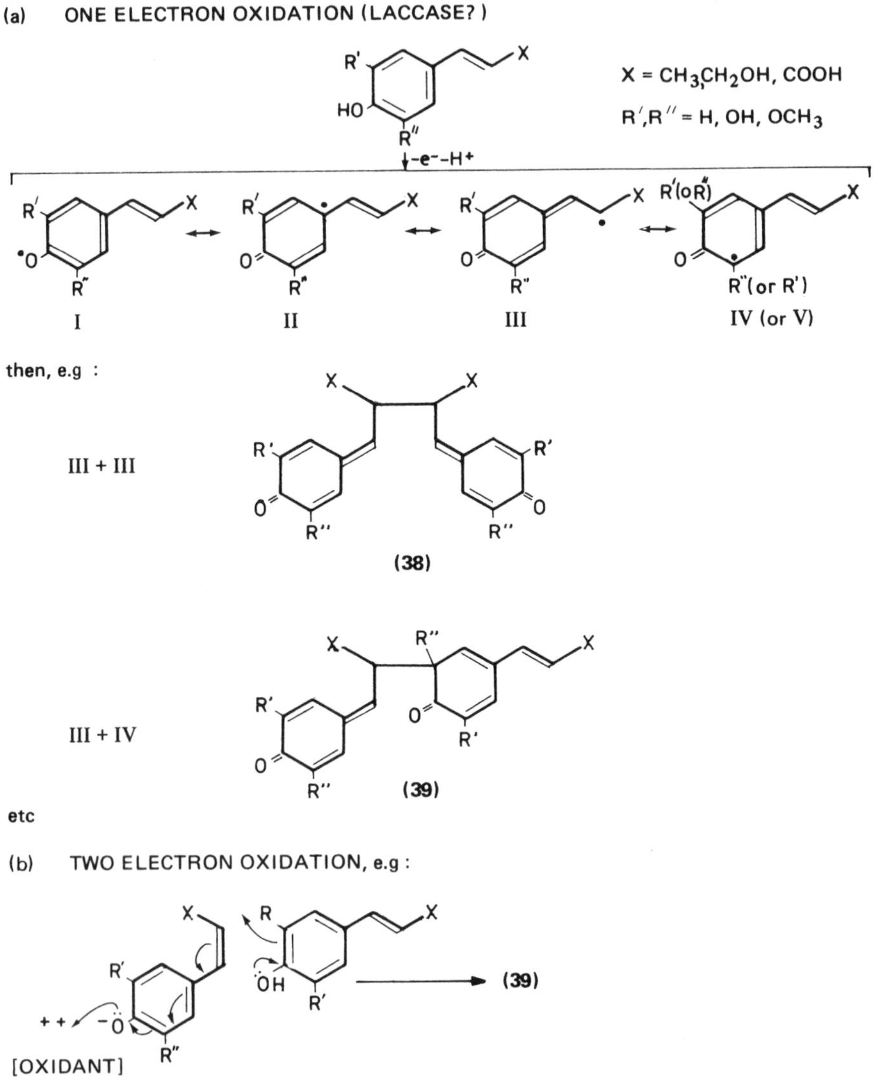

(a) ONE ELECTRON OXIDATION (LACCASE?)

X = CH₃CH₂OH, COOH

R',R'' = H, OH, OCH₃

then, e.g :

III + III

(38)

III + IV

(39)

etc

(b) TWO ELECTRON OXIDATION, e.g :

[OXIDANT]

Fig. 8.21 Dehydrodimerisation of β-hydroxystyrenes by either 1- or 2-electron mechanisms. The quinomethide structures, such as (38) react further as shown in Figs. 8.22, 8.24, 8.27.

The most popular explanation for the origin of the β,β-bond is an oxidative coupling between two C_6-C_3 units. For this explanation the p-hydroxystyrene radicals (Fig. 8.21), form quinomethide (*quinone methide*) dimers (38) which, either directly or after rearrangement to more stable structures, can be used by plants for synthesising lignin as well as lignens (Fig. 8.22). This hypothesis is in contrast with the following arguments:

(a) Some lignans are formed in nature with the oxidation level of the β and β' carbon atoms in a lower state than that expected from such styrene precursors and from the oxidative dimerisation and polymerisation processes which usually give lignins, as will be seen below [cf (34) and (35)].

(b) All the known lignans show optical activity (sometimes this optical activity is masked by internal compensation) so differing from the lignins and their degradation products.

(c) Trimeric or tetrameric lignans have not been found in nature; such polymers should be expected as intermediates if conversion of dimeric lignans into polymeric lignins occurs.

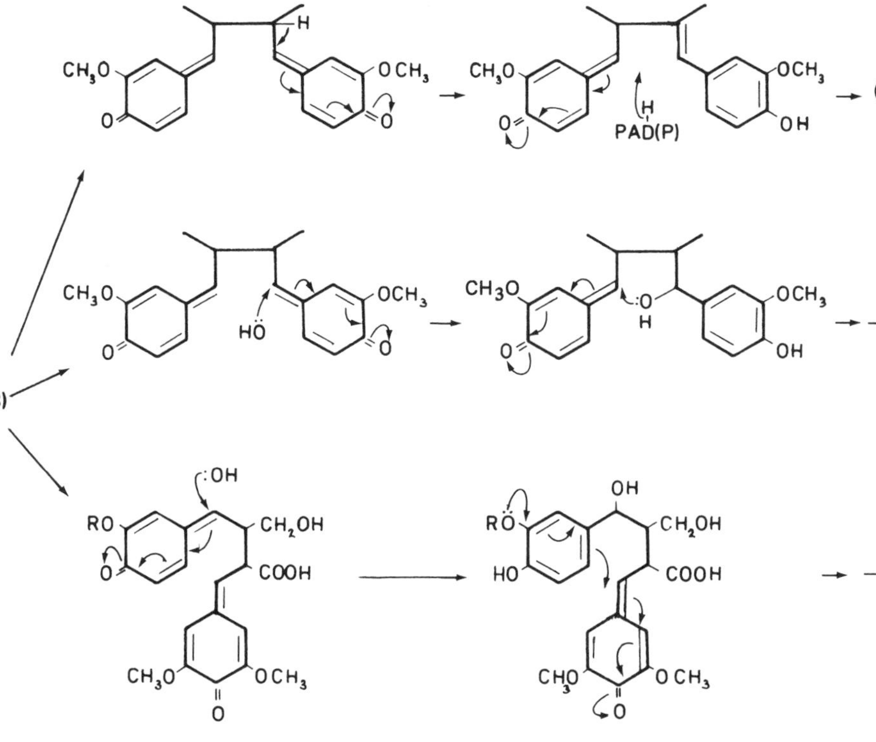

Fig. 8.22 Probable biosynthesis of some lignans (*cf.* Figs. 8.20 and 8.21).

For these reasons Neish's hypothesis appears to be more likely, namely, that the lignans are formed through initial *stereospecific* oxidation processes of *p*-hydroxystyrene units by different enzymes to those active in the biosynthesis of lignins; after such dehydrodimerisations stereospecific reductive processes can occur. The sites for lignan synthesis (oxidation-reduction sites) are probably separate from those involved in lignification.

Podophyllotoxin (Fig. 8.20) isolated from *Podophyllum emodi,* is a lignan of therapeutic importance, for its anti-tumour activity. Most lignans have been isolated from plants of the Magnoliaceae and Piperaceae families.

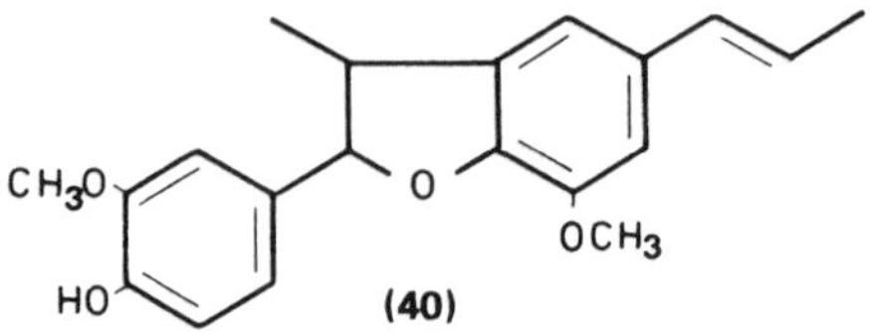

(40)

DEHYDRODIISOEUGENOL

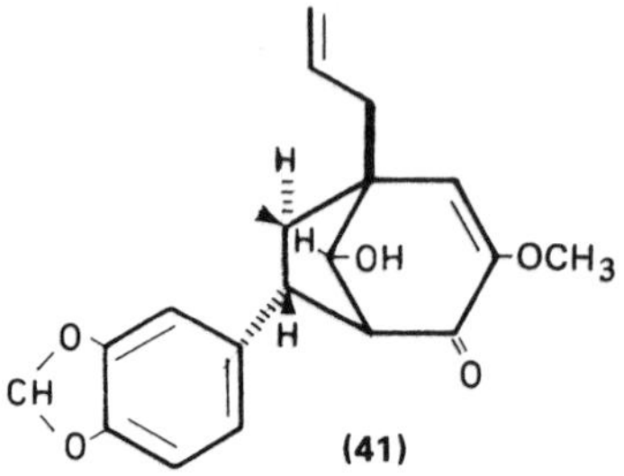

(41)

GUAIANIN

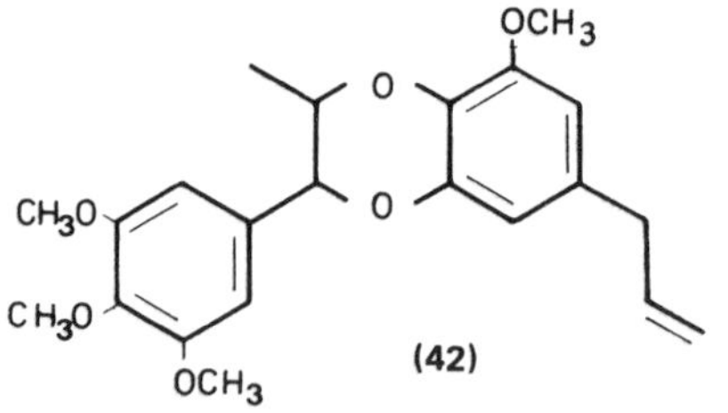

(42)

EUSIDERIN

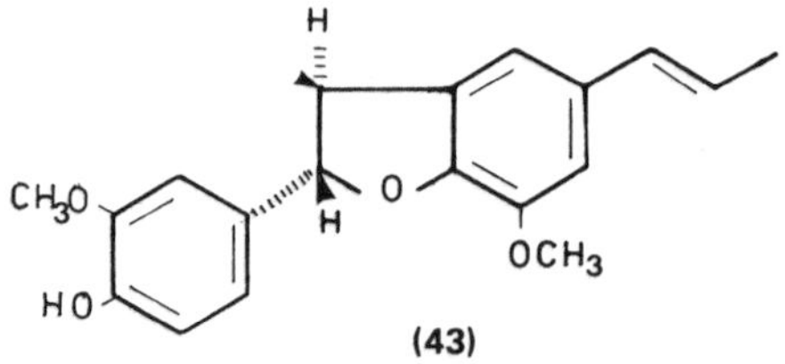

(43)

LICARIN A

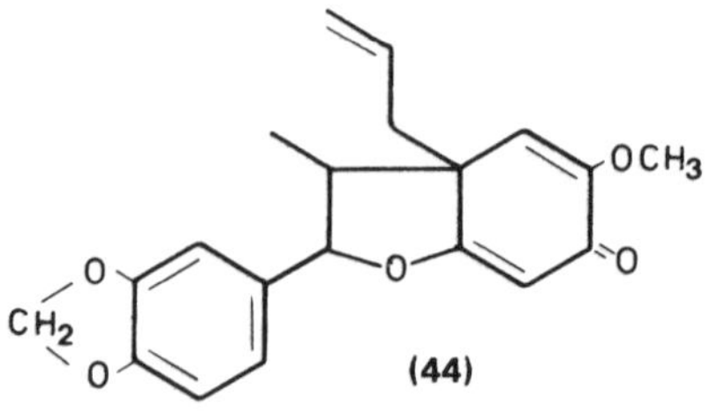

(44)

BURCHELLIN

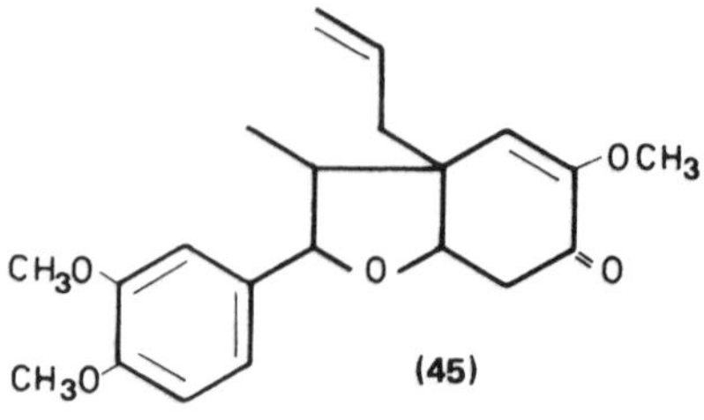

(45)

POROSIN

Fig. 8.23 Some natural neolignans.

Gottlieb proposed the name *neolignans* for those bis-arylpropanoids in which two C_6-C_3 units of 4-allyl- or 4-propenylphenols are joined together by bonds other than the $\beta\beta'$-carbon-carbon bond typical of lignans (Fig. 8.23). According to Gottlieb the neolignans are formed through oxidative coupling of the type shown in Fig. 8.21, followed by typical degradative processes of the intermediate quinomethides (Michael-type additions, aromatisation, etc.), as observed for the lignans. In the synthesis of neolignans at least one position of coupling of the two C_6-C_3 units is different from the β one (see Fig. 8.21). The units involved need not be of the styrene type, i.e. allyl phenols can be incorporated.

The neolignans probably arise at sites similar to those producing lignins. Enzymes of the same kind (laccases and peroxidases) act either on pairs of molecules with a 4-propenyl and/or 4-allylphenol structure (to give the neolignans, Fig. 8.24) or on *p*-hydroxycinnamyl units (to give products called lignols, Fig. 8.27).

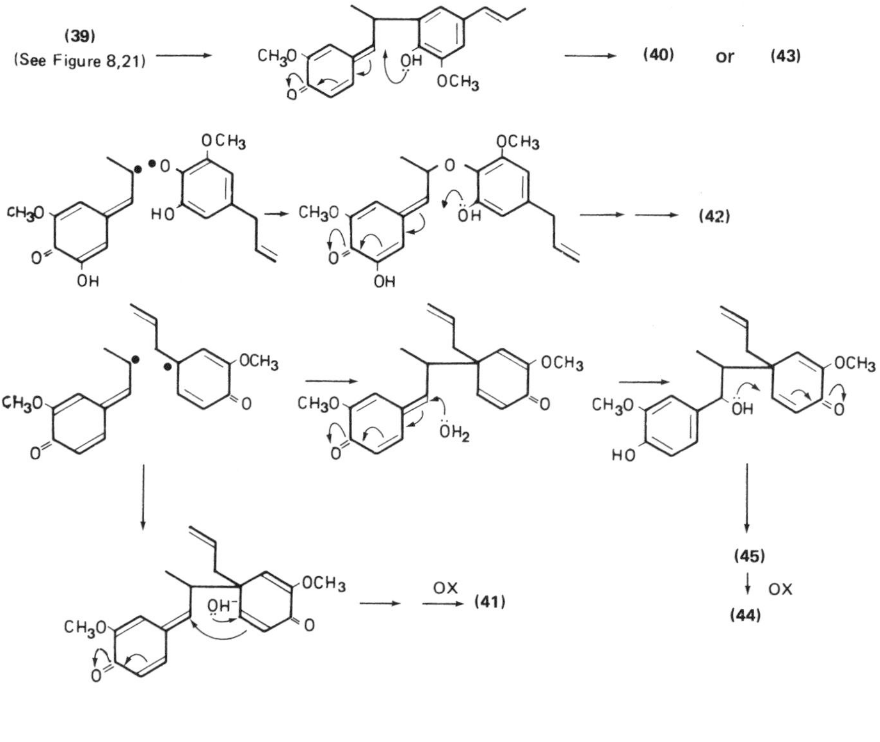

Fig. 8.24 Probable formation mechanisms for the principal neolignan skeletons.

Erdtman assigned structure (40) to a sample of dehydrodiisoeugenol be prepared by an *in vitro* method in 1933; Cousin and Herissey had already obtained its racemic form in 1909, by treating isoeugenol with ferric chloride. Erdtman saw in this molecule and in the *in vitro* preparation a plausible model for the structure and biogenesis of the lignins. This view has been largely substantiated by subsequent work on lignification (section 8.4.3) and recently dehydrodiisoeugenol has been isolated from a natural source, *Licaria aritu*. The extent to which the neolignans are incorporated into the macromolecular lignins is difficult to assess. Alcoholic groups are generally absent in the neolignans thus limiting their polymerisation. According to Erdtman the polymerisation process leading to lignins is largely owing to formation of oxido bridges, by a Michael type addition of the hydroxyl group (alcoholic or phenolic) of one oligomeric unit to the quinomethidic group of another unit. Thus the lignols more aptly fit the structural requirements as intermediates to the lignins.

8.4.3 Lignins

On our planet Earth the vascular plants, particularly the Gymnospermae and the Angiospermae, play a fundamental and well known role. Their characteristic vegetative body, made up of roots, stem and leaves, and their often large dimensions are only possibly as woody tissues develop during plant growth. Such tissues give structural support to the plant and act as channels for the transport of water and nutrients in the various parts of the plant.

A typical lignified tissue is the *xylem,* particularly abundant in arboreal plant stems. The xylem cells (vessels, tracheides and fibres) are elongated and their walls are thickened and coated by a characteristic material called lignin. Lignin is generally absent in fungi and in thallophytes.

During the growth of woody tissues, the cells are initially enriched with carbohydrates some of which are successively transformed into lignin through a long series of reactions called the 'lignification process'. As a result, the spaces in the polysaccharide fibres of the cell walls are gradually filled with lignin. The cellulose fibres become cemented together and protected against physical and chemical damage.

Although lignin was discovered about a century ago, many structural and biogenetic aspects are still not clear. Knowledge of its fundamental structural elements, such as the aromatic rings and their substituents, which allow us to write down an approximate formula of lignin (Fig. 8.25), has come mainly from the work of K. Freudenberg and his Heidelberg school. Lignin is a high molecular weight polymer, prevalently, if not totally, made up of phenylpropane units joined together in a wide variety of ways (C–C bonds between propyl chains, or between chains and aromatic rings, oxido bridges, etc). It contains the highest percentage of methoxy groups present in wood; it is easily oxidisable (Fig. 8.26); it is soluble in hot alkali, and condenses easily with phenols and thiols.

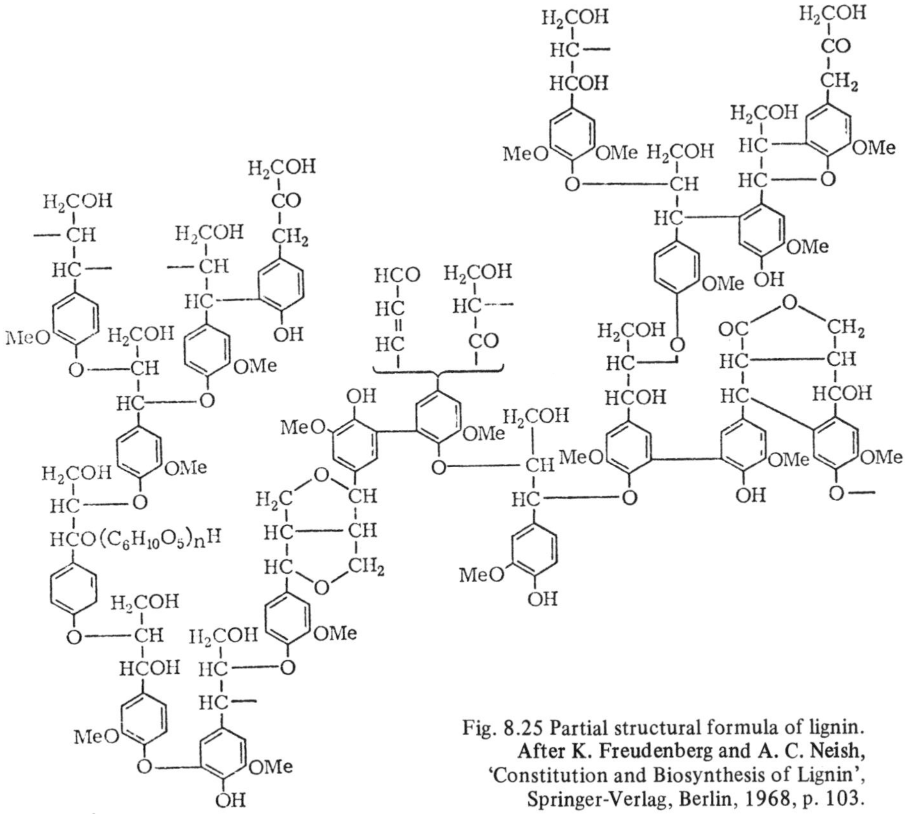

Fig. 8.25 Partial structural formula of lignin.
After K. Freudenberg and A. C. Neish,
'Constitution and Biosynthesis of Lignin',
Springer-Verlag, Berlin, 1968, p. 103.

The above properties, together with its occurrence only in plants, constitute the best definition of lignin up to now. The term, lignin does not imply a single, well defined compound. It is, rather, a collection of similar macromolecules, having molecular weights mostly between 6,000 and 10,000. Lignin is a much more complex polymer than other natural materials, such as cellulose, starch, pectins, etc., because of its particular biosynthesis, which consists of the combination of medium and high molecular weight polymer units (oligo- and polylignols). These polylignols arise from the polymerisation of three fundamental monomeric units: the coniferyl (25), sinapyl (27) and p-coumaryl (24) alcohols; these three units differ from each other and each has many oxidative polymerisation centres (Fig. 8.21). Other units (cinnamic acids, propenylphenols, etc.) can add and stable covalent bonds to polysaccharide structures can also be formed. Up to 1950 the biosynthesis of lignin was exclusively studied by the systematic analysis of chemical degradation products (oxidative, reductive, or simply solvolytic) (Fig. 8.26). With the introduction of radioisotopic techniques biosynthetic studies on lignin received a new impetus and new conclusive data could be collected. Today the lignification process can be schematised as follows:

(a) Biosynthesis of the monomeric units (the cinnamyl alcohols, section 8.4.1).
(b) Dehydropolymerisation of monomers (DHP), by phenoloxidase enzymes (Fig. 8.21). The monomers are oxidised to lignols (monomeric radicals, aldehydes, etc.), which then polymerise via dilignols, trilignols, tetralignols, oligolignols, polylignols, and, finally, high molecular weight lignin (Fig. 8.27). The quinomethide intermediate of lignols can also react intermolecularly with hydroxyl groups (Fig. 8.22), forming bonds which increase the structural complexity, e.g. crosslinks to carbohydrates. This biosynthetic scheme has been recently confirmed by studies on the *in vitro* dehydropolymerisation of the fundamental monomeric alcohols, taken either separately or mixed together, by laccase or peroxidase enzymes. Operating with dilute solutions of monomers (0.2 M) in dioxane-water the average life of the various monolignol radicals have been measured – the coniferyl radical, for example, had a mean life time of 45 seconds – as well as of some quinomethide intermediates; the diquinomethide intermediate leading to *d,l*-pinoresinol has a mean life time of about one hour. Interrupting these enzyme-

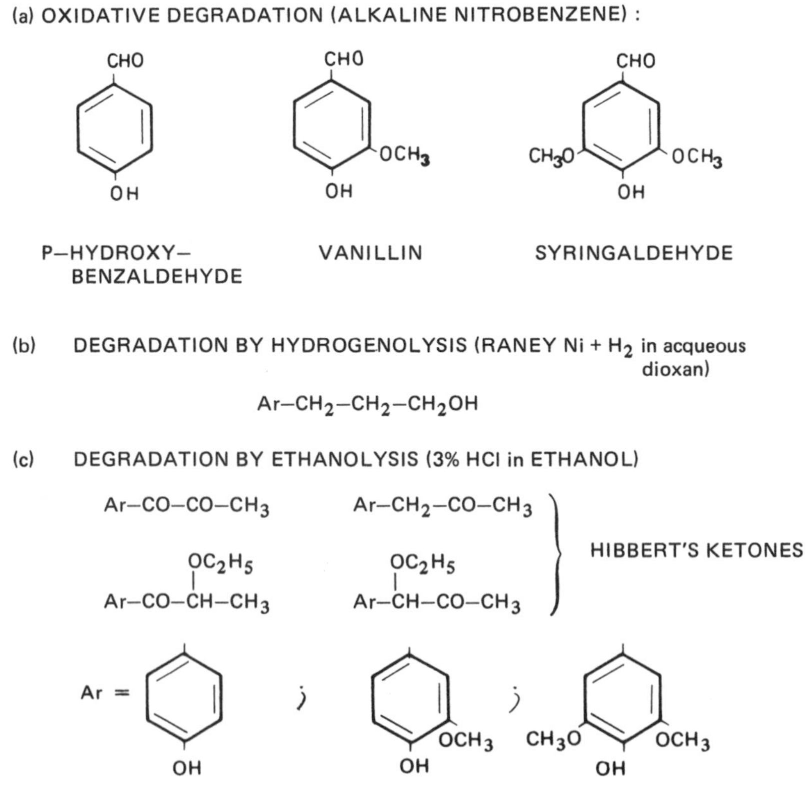
(a) OXIDATIVE DEGRADATION (ALKALINE NITROBENZENE) :
CHO
OH
OCH3
CH3O
P–HYDROXY–BENZALDEHYDE
VANILLIN
SYRINGALDEHYDE
(b) DEGRADATION BY HYDROGENOLYSIS (RANEY Ni + H2 in acqueous dioxan)
Ar–CH2–CH2–CH2OH
(c) DEGRADATION BY ETHANOLYSIS (3% HCl in ETHANOL)
Ar–CO–CO–CH3
Ar–CH2–CO–CH3
OC2H5
Ar–CO–CH–CH3
Ar–CH–CO–CH3
HIBBERT'S KETONES
Ar =

Fig. 8.26 Chemical degradation products of lignins.

catalysed oxidations after increasing intervals of time, afforded lignols of increasing molecular weights. The structures of many of these oligolignols have been elucidated (Fig. 8.27). After long times, and starting with appropriate mixtures of coniferyl, sinapyl and p-coumaryl alcohols, artificial lignins very similar to the natural ones were formed.

DILIGNOLS : e.g

I + III →

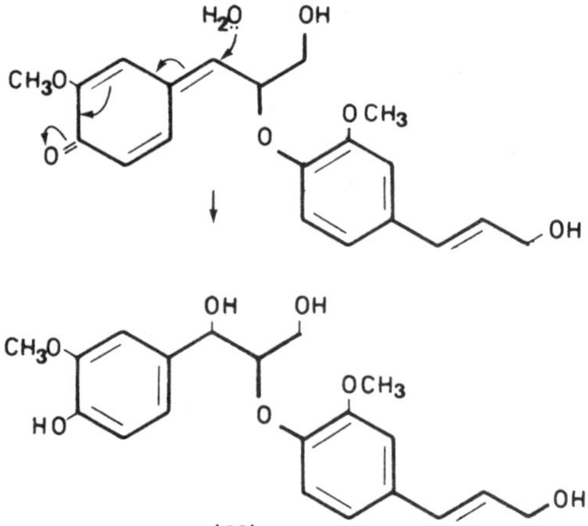

(See Figure 8.21;
$R' = OCH_3$; $R'' = H$
$X = CH_2OH$)

(46)

GUAIACYLGLYCEROL–β–CONIFERYL ETHER

III + III →

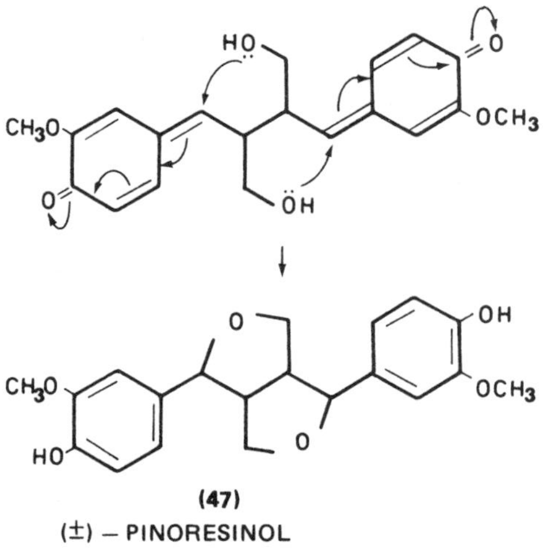

(47)
(±) – PINORESINOL

Fig. 8.27 (contd. overleaf)

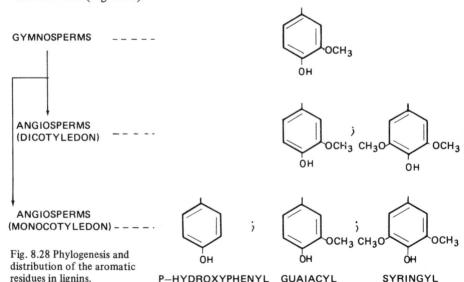

TRILIGNOLS : e.g   Fig. 8.27 (contd.)

(47) + (25) $\xrightarrow[+H_2O]{-2H^+ - 2e^-}$

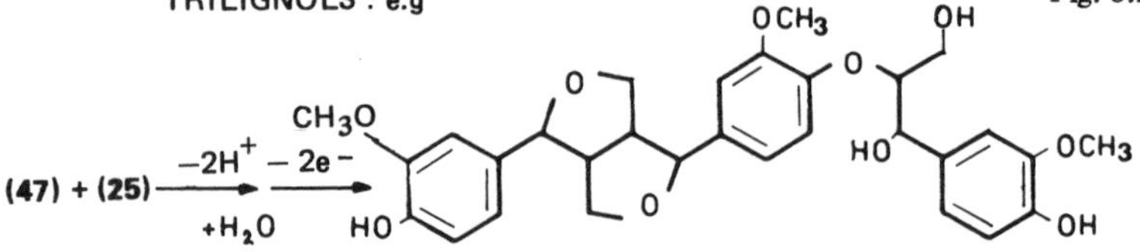

(48)

GUAIACYLGLYCEROL−PINORESINOL ETHER

TETRALIGNOLS : e.g

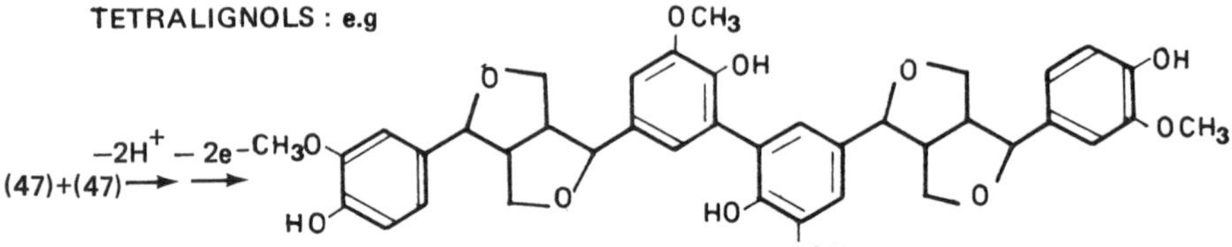

(47)+(47) $\xrightarrow[]{-2H^+ - 2e -CH_3O}$

(49)

DEHYDROPINORESINOL

Fig. 8.27 Some examples of simple lignols produced by dehydrogenation of coniferyl alcohol.

The relative amounts of the three fundamental monomers used by the plants in the lignification process vary from species to species and a remarkable structural heterogenity amongst lignins of different origin is observed. If the ratios of the aromatic residues guaiacyl, syringyl and p-hydroxyphenyl are examined along the series of the phylogenetically ordered plants, an important regularity can be found (Fig. 8.28).

GYMNOSPERMS   – – – – –

ANGIOSPERMS
(DICOTYLEDON)   – – – – –·

ANGIOSPERMS
(MONOCOTYLEDON) – – – – –·

Fig. 8.28 Phylogenesis and distribution of the aromatic residues in lignins.

P−HYDROXYPHENYL   GUAIACYL   SYRINGYL

The lignin of Gymnosperms, e.g. Coniferae, is essentially composed of guaiacyl residues (biogenetically, coniferyl units), the dicotyledon Angiosperms are made of guaicyl and syringyl residues (coniferyl and sinapyl units), and lignin from the more evolved monocotyledon Angiosperms, e.g. Graminaceae, is made of guaiacyl, syringyl and p-hydroxyphenyl residues (coniferyl, sinapyl, and p-coumaryl units).

8.4.4 C_6-C_2 Compounds, Styrenes

The C_6-C_2 compounds, derived by the loss of a carbon atom from the phenylpropane skeleton can be divided into three biogenetic groups.

The first group comprises the β-arylethylamines (4), directly produced from the aromatic amino acids (phenylalanines, tyrosine, and DOPA) by the action of PLP-dependent decarboxylases (Fig. 2.37); the β-arylacetaldehydes (5), produced from the corresponding amines by the action of monoamineoxidases (MAO); the acids and alcohols produced by the oxidation and reduction of the latter aldehydes. The β-arylethylamines and the β-arylacetaldehydes (or their biogenetic equivalents) are mostly used by plants for the biosynthesis of alkaloids (Fig. 8.1).

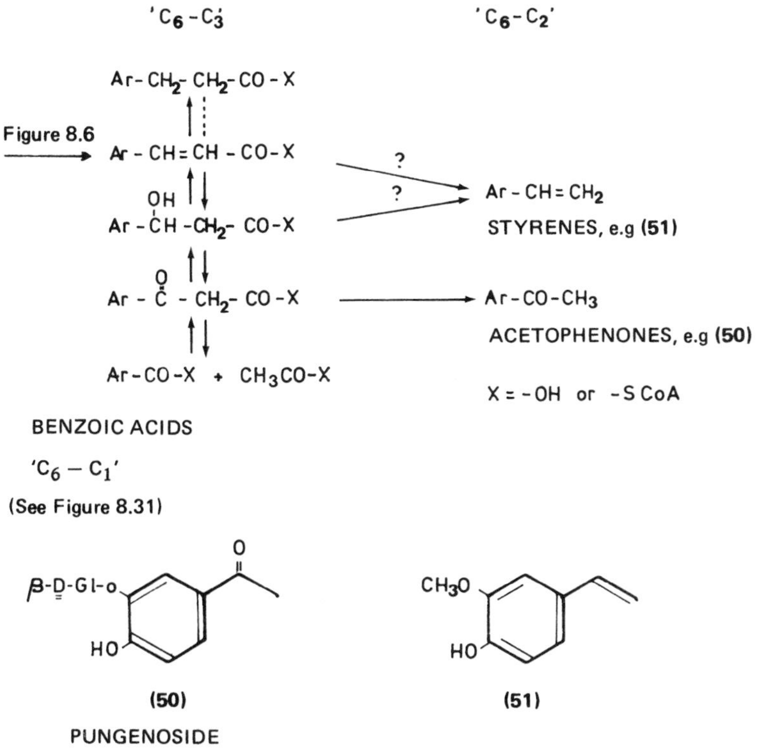

Fig. 8.29 Biodegradation of cinnamic acids and biogenetic relationships amongst C_6-C_3, C_6-C_2, and C_6-C_1 compounds.

The second group includes those compounds formed by shortening of the side chain of phenylpyruvic acids. In this way homogentisic acid (Fig. 8.4) is produced and, especially in yeasts, the β-arylethyl alcohols, such as β-phenyl-ethanol and β-(p-hydroxyphenyl)ethanol (also called tyrosol). The production of such alcohols from the corresponding phenylpyruvic acids, with loss of the carboxylic carbon, can be considered as an extension of the fermentation process leading from pyruvic acid to ethyl alcohol (see Fig. 3.20).

The third group contains catabolites of cinnamic acids (Fig. 8.29). The acetophenones, such as pungenoside (50) present in *Picea pungens,* most likely derives from the decarboxylation of β-ketoacids (see Fig. 2.35), which are formed by addition of water to cinnamic acids and further oxidation, along a route analogous to fatty acid catabolism. Acetophenones of polyketide origin are seldom encountered in higher plants but can be recognised by the hydroxy-lation pattern of the aromatic ring.

Styrenes are specifically produced by fungi and bacteria.

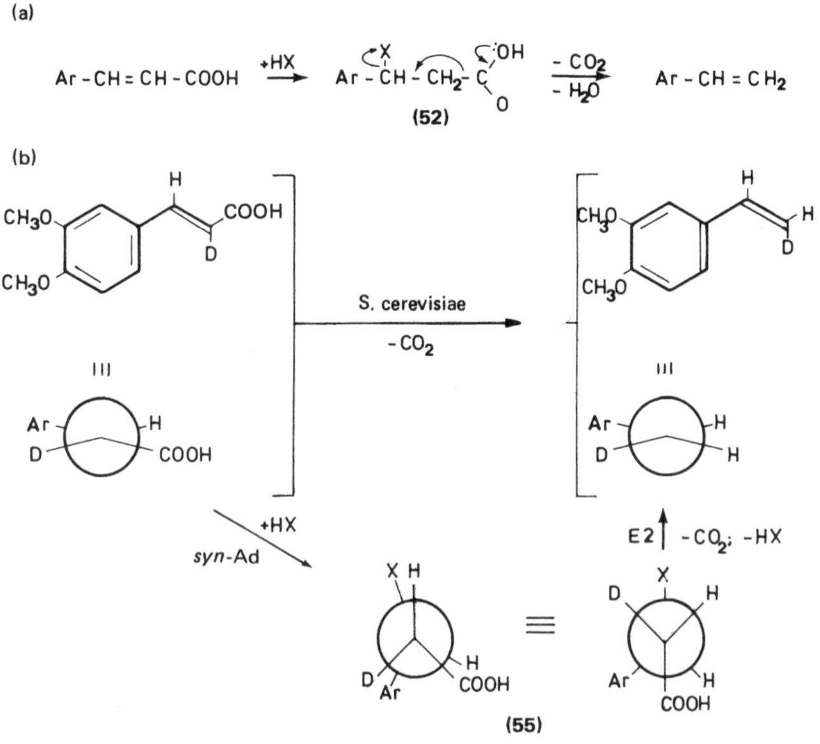

X^- = Nucleophile, e.g OH^-, RS^- (R could be the active site of the enzyme)

Fig. 8.30 Probable mechanisms of microbiological decarboxylations of cinnamic acids.

The mechanism of styrene formation has not yet been fully elucidated. *A priori* they could derive by an E2 type decarboxylation of 3-arylhydracrylic acids, such as (52) either free or bound to an enzyme, or they could be produced by direct decarboxylation of the corresponding protonated cinnamic acids. In recent work of the author and his coworkers (E)-3,4-dimethoxy-[α-^{2}H] cinnamic acid (53) was fed to a culture of *Saccharomyces cerevisiae* and showed how the overall stereochemistry of the decarboxylation process takes place in this microorganism (Fig. 8.30(b)). There is retention of the double bond configuration in proceeding from the acid (53) to the styrene (54). This means that if the mechanism is a two step process, as in scheme (a) of Fig. 8.30, and if eliminative decarboxylation takes place in the usual antiperiplanar manner (compare Fig. 2.35), an initial *cis*-addition of HX to the double bond of cinnamic acid must be postulated. A *trans*-addition of HX to (53), followed by an E2 type decarboxylation, would in fact give a monodeuterated styrene of (E) configuration, that is an isomer of (54).

8.4.5 C₆-C₁ Compounds. Benzoic Acids.

Benzoic acid and its variously oxygenated derivatives (Fig. 8.31) are very common in nature. In higher plants they are usually found as glycosides, that is conjugated with an aldose (usually D-glucose) through phenolic hydroxyls, or as esters, that is with their carboxylic group esterified with either alcohols, such as methanol, benzyl alcohol, quinic acid, sugars, etc. or phenols, as in the shikimate derived depsides, such as digallic and trigallic acids.

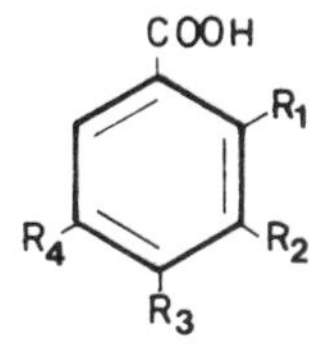

R_1	R_2	R_3	R_4	
H	H	H	H	BENZOIC ACID
OH	H	H	H	SALICYLIC ACID
H	H	OH	H	P—HYDROXYBENZOIC ACID
OH	H	H	OH	GENTISIC ACID
H	OH	OH	H	PROTOCATECHUIC ACID
H	OCH$_3$	OH	H	VANILLIC ACID
H	OH	OH	OH	GALLIC ACID
H	OCH$_3$	OH	OCH$_3$	SYRINGIC ACID

Fig. 8.31 Principal natural benzoic acids.

The benzoic acids in plants are formed in three main ways:

(a) Directly from 5-dehydroshikimic acid (see Fig. 7.3), e.g. gallic acid.

(b) From the corresponding cinnamic acids, via β-oxidation and a retro-Claisen reaction, as indicated in Fig. 8.29. This route is followed for salicylic, proto-catechuic, p-hydroxybenzoic, vanillic, syringic and gentisic acids. (Gentisic acid can also be formed by the polyketide route).

(c) On hydroxylation of another benzoic acid. For instance, on giving [^{14}C] benzoic acid to *Gaultheria procumbens* (Ericaceae) and to *Primula acaulis* (Primulaceae), labelled salicylic, protocatechuic p-hydroxybenzoic and gentisic acids were obtained.

As far as the catabolism of the benzoic acids is concerned, it is known that protocatechuic acid can be degraded to β-ketoadipic acid in fungi (Fig. 8.32). Up to now no analogous process has been found in higher plants.

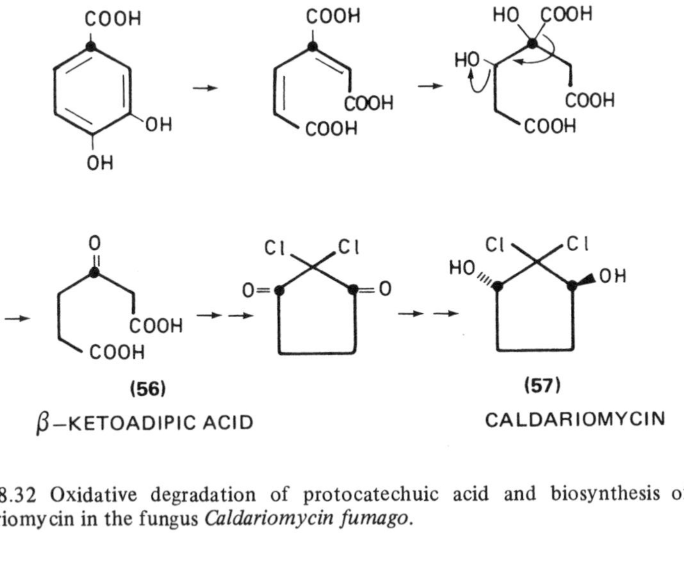

Fig. 8.32 Oxidative degradation of protocatechuic acid and biosynthesis of caldariomycin in the fungus *Caldariomycin fumago*.

The corresponding aldehydes and benzyl alcohols are found in nature together with the benzoic acids. It is generally believed that such compounds arise from a progressive reduction of the carboxylic group, after its usual conversion into the CoA thiolester. Some experimental data supporting this hypothesis have been obtained by studying the biosynthesis of salicin in *Salix purpurea*. The incor-poration of [^{14}C]phenylalanine and of [^{14}C]cinnamic acid into the salicylic alcohol obtained from the hydrolysis of salicyn were far larger than the incorporations of [^{14}C] acetic acid and of [^{14}C] glucose. These results are in agreement with the biosyntheric scheme of Fig. 8.33 and exclude alternative polyketide pathways for the synthesis of salicin.

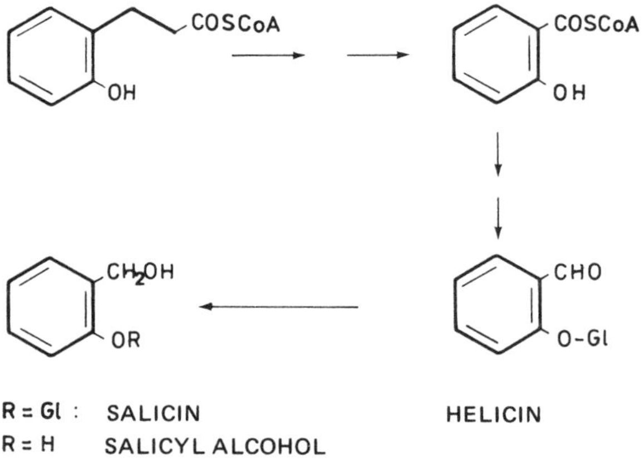

R = Gl : SALICIN   HELICIN
R = H   SALICYL ALCOHOL

Fig. 8.33 Biosynthesis of helicin and salicin in *Salix purpurea*.

It has been recently shown that in microorganisms producing both 6-methylsalicylic acid and salicylic acid, for instance *Mycobacterium fortuitum*, that the latter is exclusively biosynthesised via a shikimic acid − phenylalanine − cinnamic acid route whilst the former originates from the normal acetic acid − polyketide pathway.

8.4.6 C_6 Compounds. Simple Polyphenols

Simple polyphenols, such as hydroquinone, resorcinol, and pyrocatechol, are rarely found in higher plants. The most diverse phenol is arbutin and its methyl ether (Fig. 8.34). These polyphenols derive from the non-oxidative decarboxylation of the corresponding benzoic acids, for instance gentisic acid or its glucoside. This hypothesis is chemically reasonable, as well as having the support of some incorporation experiments. By contrast antiarol (58), isolated from *Antiaria toxicaria* (Moraceae), should be derived from an oxidative decarboxylation of gallic or syringic acids (Fig. 8.31). The simple decarboxylation of gallic acid to pyrogallol is likely to produce purpurogallin found in plant galls (see Fig. 7.6).

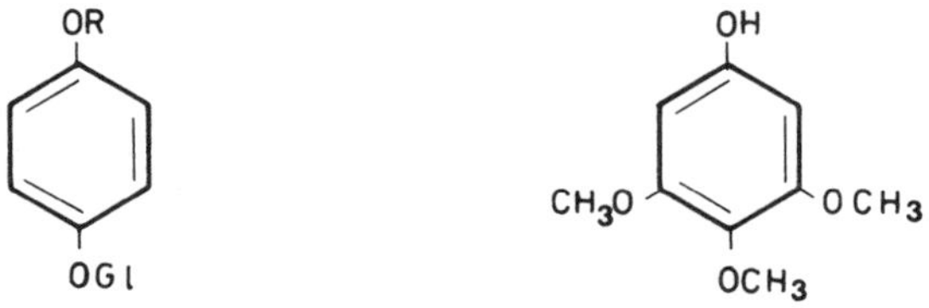

R_1 = H : ARBUTIN   (58)
R_1 = CH$_3$: ARBUTIN METHYLETHER   ANTIAROL

Fig. 8.34 Principal C_6 compounds of higher plants.

8.4.7 C_6-C_2-$(C_2)_n$ Compounds. 2-Pyrones and Stilbenes.

The lengthening of the side chain of cinnamic acids, by addition of one or more C_2 units, is widespread in the higher plants. Some compounds with important chemical and biological properties are formed. The most important and largest group includes the flavonoids and isoflavanoids, which will be described in the following sections, whilst the biosyntheses of other C_6-C_3-$(C_2)_n$ compounds and of their derivatives are described here (e.g. Fig. 8.35).

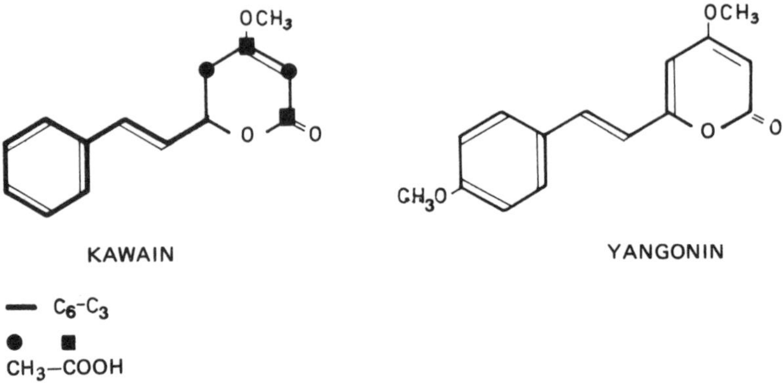

KAWAIN   YANGONIN

— C_6-C_3
● ■
CH_3—$COOH$

Fig. 8.35 Constituents of *Piper methysticum*: examples of C_6-C_3-$(C_2)_2$ types.

The mechanism for addition of the two-carton units to the side chain of cinnamic acid or of other phenylpropanoids, such as phenylpyruvic (1), hydracrylic and benzoylacetic acids (Fig. 8.31), are similar to those acting in polyketide biosynthesis (Chapter 4). The starter unit is activated by converting the C_6-C_3 acid into the corresponding acyl-CoA ester, the C_2 units are assembled on the starter (via malonyl-SCoA), and the resulting polyketide chain evolves into more stable forms. The products thus obtained are typical mixed biosynthesis metabolites; one portion of the molecule arises from shikimic acid, that is from glucose, whilst the other one comes from acetic acid. The two portions are not always distinguishable by simple structural analysis. Paracotoin for instance, a 6-aryl-2-pyrone present in paracoto bark (*Aniba pseudocoto*, Lauraceae family), can derive from either path (a) or (b) of Fig. 8.36. In these routes the respective starters are either a C_6-C_3 unit or a C_6-C_1 acid. The actual biosynthesis of paracotoin and of other, natural 6-aryl-2 pyrones has not yet been clarified. A similar uncertainty also exists for some benzophenones such as protocotoin (see Fig. 4.28). Such compounds probably originate from the corresponding benzoic acid.

It is generally believed that stilbenes, such as pinosylvin and the dihydroisocoumarins, for instance hydrangenol, are formed in plants according to the biosynthetic schemes illustrated in Fig. 8.37. Scheme (a) has been demonstrated experimentally for pinosylvin through feeding *Pinus resinosa* boughs with appropriately labelled precursors.

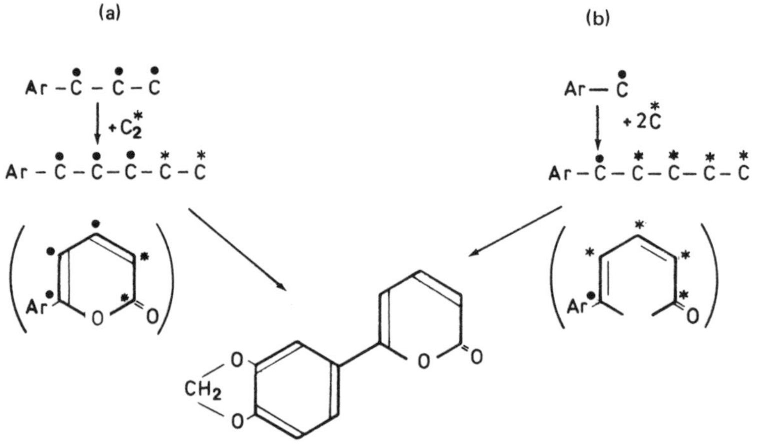

PARACOTOIN

Fig. 8.36 Possible biosynthetic routes to 6-aryl-2-pyrones.

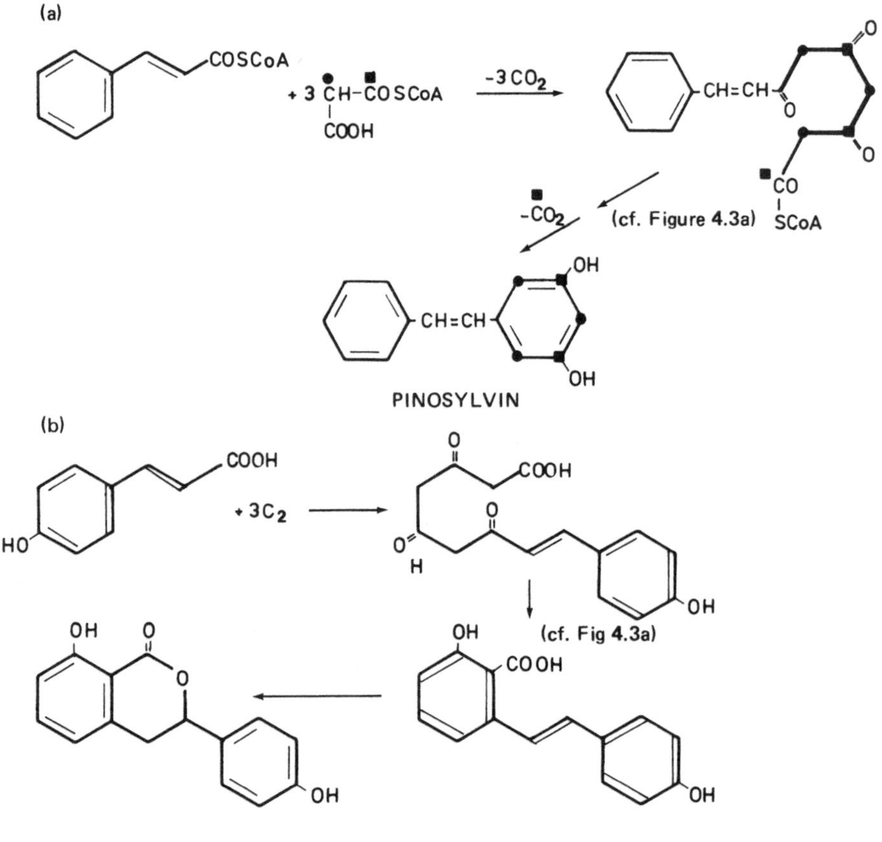

PINOSYLVIN

HYDRANGENOL   Fig. 8.37 Biosynthesis of C_6–C_3–$(C_2)_3$ compounds.

8.5 FLAVONOIDS

The *flavonoids* are polyphenolic compounds possessing 15 carbon atoms; two benzene rings are joined by a linear three carbon chain. This skeleton, shown in formula (59) (Fig. 8.38), can also be represented as the C_6-C_3-C_6 system. The flavonoids are thus 1,3-diarylpropanes. Isoflavonoids are 1,2-diarylpropanes, whilst the neoflavonoids are 1,1-diarylpropanes.

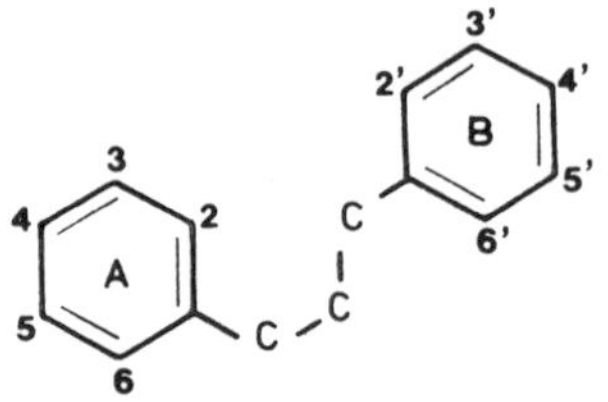

(59)

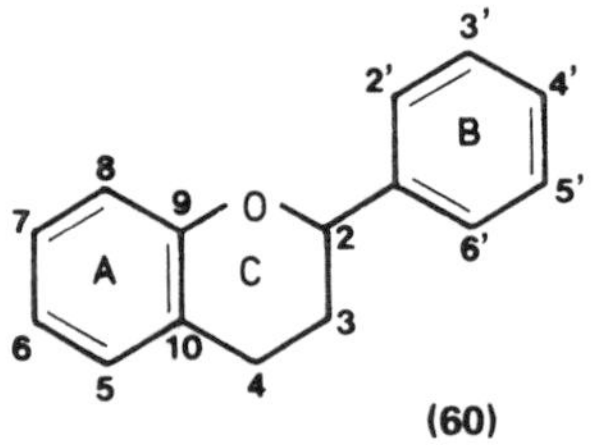

FLAVONES: e.g   AURONES : e.g

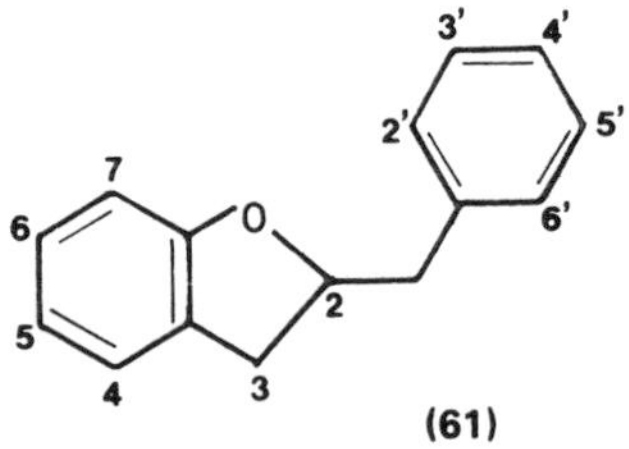

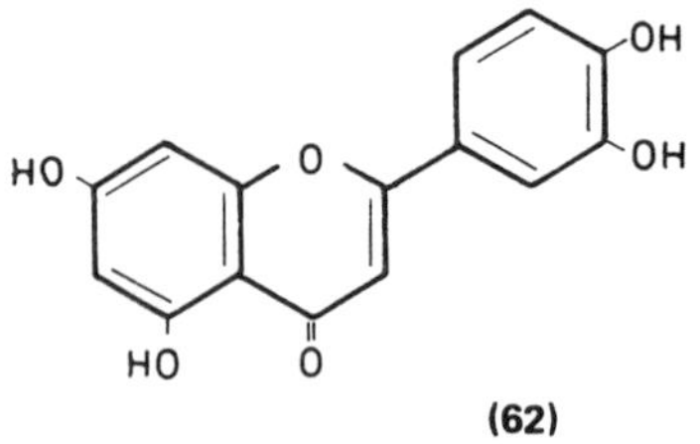

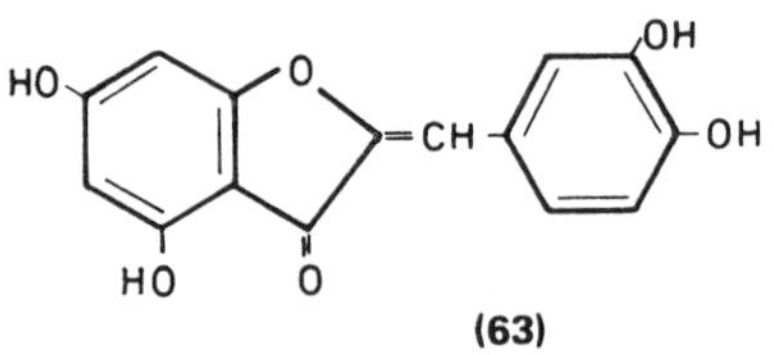

(60)   (61)

(62)   (63)

LUTEOLIN   AUREUSIDIN

CHALCONES (as glucosides) e.g

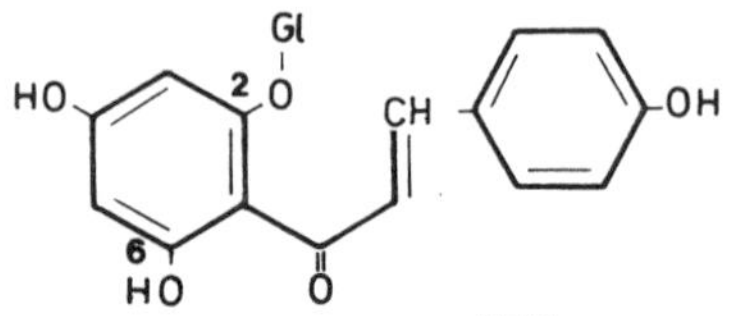

(64)

ISOSALIPURPOSIDE

Fig. 8.38 Typical flavonoids.

The term 'flavonoid', assigned to this large class of natural substances derives from the most common group of compounds, the flavones, such as (62); an oxygen bridge between the *ortho*-position of ring A and the benzylic carbon atom adjacent to ring B forms a new 4-pyrone type ring. Such heterocycles, at different oxidation levels, are present in most plants. The flavane structure (60) corresponds to the lowest oxidation level of the ring C, and is taken as the parent structure for the rational nomenclature of this group of compounds.

An oxygen bridge involving the central carbon atom of the C_3-chain occurs in a rather limited number of cases, where the resulting heterocycle is of the furan type, the C_{15}-skeleton of 2-benzyl-coumarane (61). The aurones, such as (63), belong to this structural group. The oxygen bridge is absent in chalcones, (64), which always exist in nature as glycosides.

Beside the carbon atom link, the flavonoids also have typical oxygenation patterns in their benzene rings. The substituents can be -OH, $-OCH_3$, $-O-CH_2-O-$, or -O-glycosides.

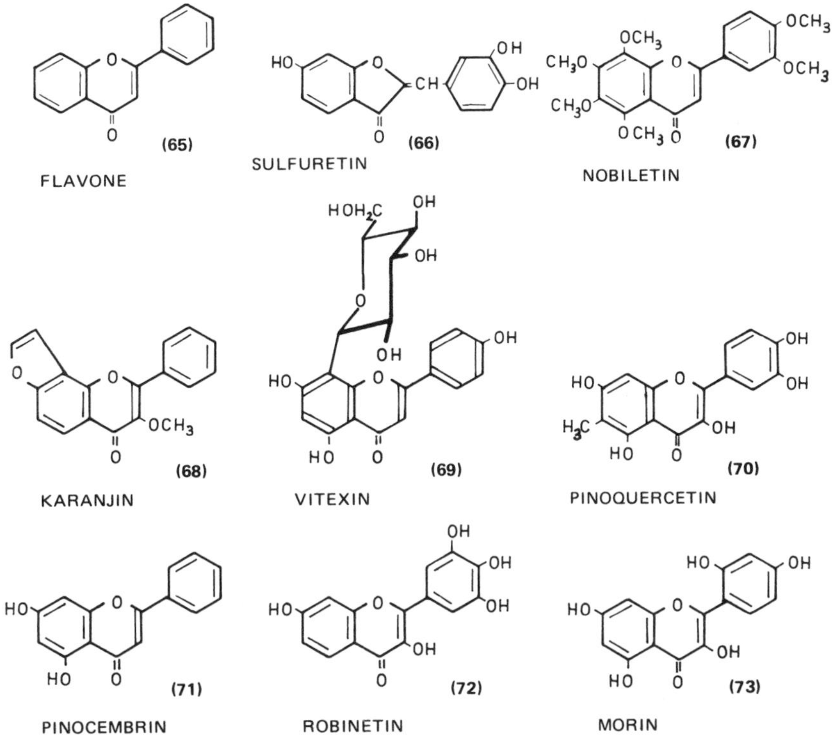

Fig. 8.39 Substitution patterns of the A and B rings in some less common flavones.

Further structure features are:

(a) Ring A generally has three alternate oxygens, at positions $2'$, $4'$ and $6'$ in the open formula (59). Particular flavonoids are given in Fig. 8.39. Compounds having more or less oxygens in their A rings are very seldon encountered (Fig. 8.39). Harborne considers the 6- and 8-hydroxylation patterns in flavones, such as in nobiletin, as a sign of a more advanced evolution in the biosynthesis of flavonoids. The nucleus A can be occasionally alkylated with prenyl or prenyl-derived units, e.g. (68), or with glycosides, as in the C-glucoside (69), or with methyl groups, for instance (70).

(b) Ring B has, in most cases, a *para*-oxygen function, or two oxygens, *para*- and *meta*- with respect to the central C_3-chain. Compounds bearing three oxygens (one *para* and two *meta*) are less frequent. Compounds with non-oxygenated B rings, or with one *ortho*-oxygen substituent, are very rarely found. The more common oxygenation patterns summarised as A (phloroglucinol) — B (catechol or phenol) can be easily explained in biogenetic terms. In the early biosynthetic stages the flavonoids move along the general path: C_6-C_3 + $n(C_2)$ → C_6-C_3-C_{2n}, where n=3 (Fig. 8.41). The resulting C_{15}-skeleton already possesses certain oxygen substituents, mostly as hydroxyl groups in the required positions connected with the formation of ring A (polyketide pathway) and with the shikimate origin of ring B (phenylalanine → cinnamic acid).

After various enzymic transformations the three central carbon atoms of the 1,3-diarylpropane skeleton can bear various functional groups, such as hydroxyls, double bonds, carbonyls, etc. The most important structural types are shown in Fig. 8.40; the compounds are listed in increasing oxidation level of the central

The flavonoids are usually present in the cells of flower tissues. The wide range of colours and shades in flowers are principally due to the presence of flavonoids, especially the anthocyanins (Fig. 8.50) which are the most important vegetable pigments after the chlorophylls and carotenes.

The origin of all the carbon atoms in the flavonoid skeleton has been experimentally checked (Fig. 8.41). There are still some doubts as to the actual structure of the phenylpropane unit used by plants as a starter for building up the triacetic chain and then ring A. Until recently it was assumed that a cinnamic acid could start the process of polyketide condensation (path (a), Fig. 8.41). It has been recently hypothesised, however, that the active C_6-C_3 unit can also be a phenylpyruic or benzoylacetic acid unit ((b) and (c), Fig. 8.41). Most chemists now believe that the cinnamic acids (*p*-coumaric, and more rarely caffeic, ferulic and sinapic acids (See Fig. 8.6)) *are* obligatory intermediates in the biosynthesis of most flavonoids. The other phenylpropane acids are implied only in secondary biosynthetic routes, which either subsequently join the principal one or are involved in the biosynthesis of particular flavonoids. These opinions rely on many incorporation experiments, some of the most meaningful being summarised in Fig. 8.42. Such experiments demonstrate that most flavonoids, including the anthocyanidins and flavanols, derive from (2*S*)-flavonoid intermediates.

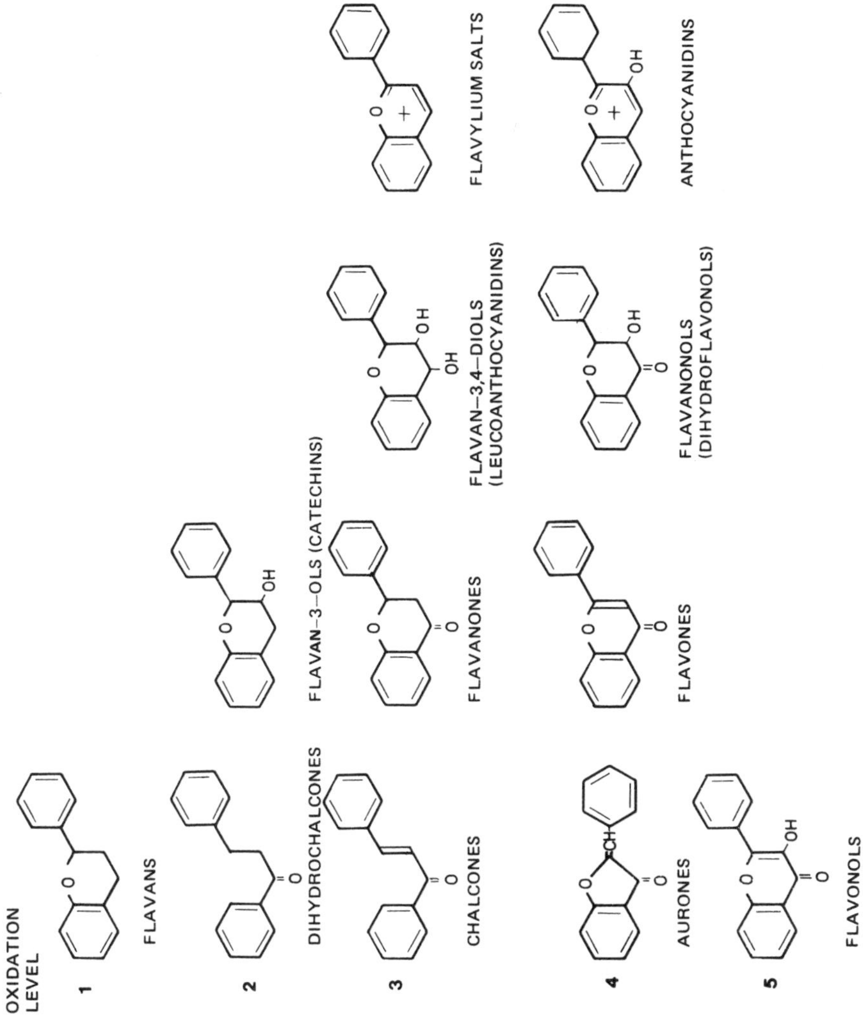

Fig. 8.40 Principal structural groups of natural flavonoids. The oxidation levels concern the three central carbon atoms: [+10 or −2H] = +1 oxidation unit.

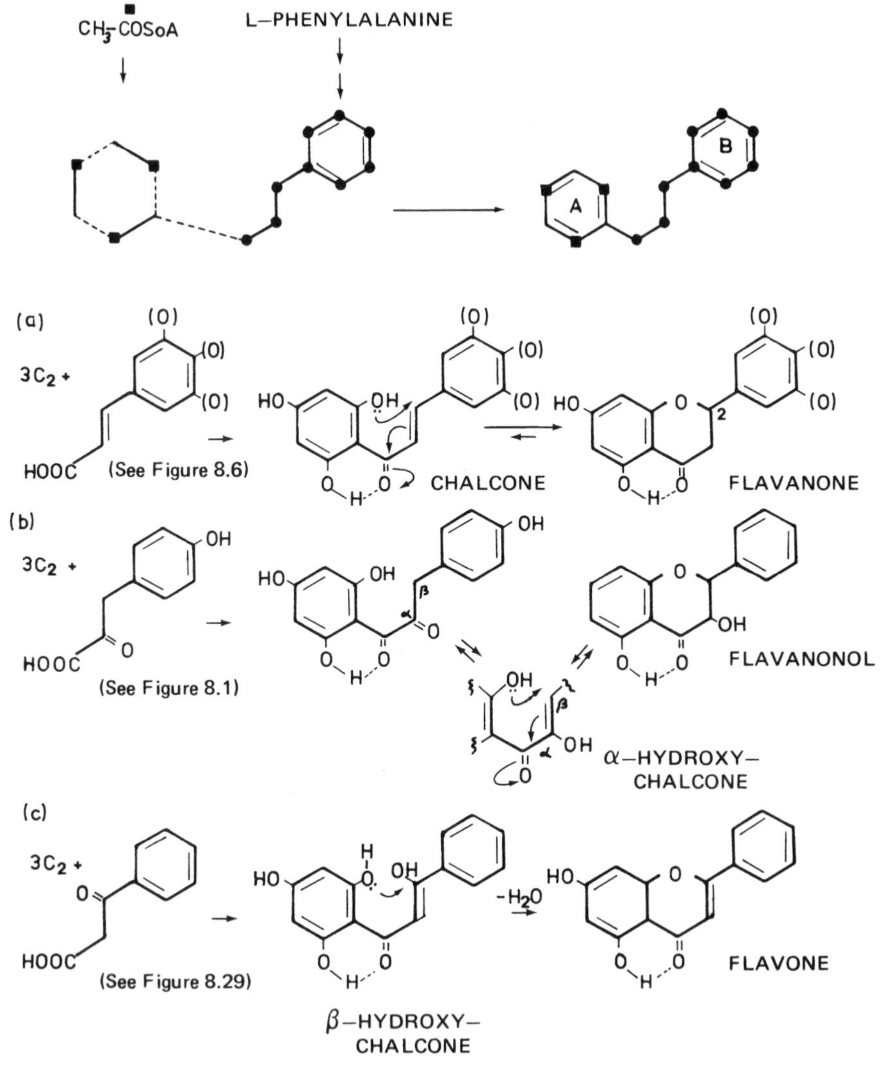

Fig. 8.41 Origin of the carbon atoms of flavonoids and possible C_6–C_3 intermediates.

There is good evidence for the *in vitro* and *in vivo* existence of an equilibrium between flavanones and the corresponding chalcones (scheme (a). Fig. 8.41). The results summarised in Fig. 8.42 can also be explained by a conversion of chalcones to flavonoids without passing through the flavanone.

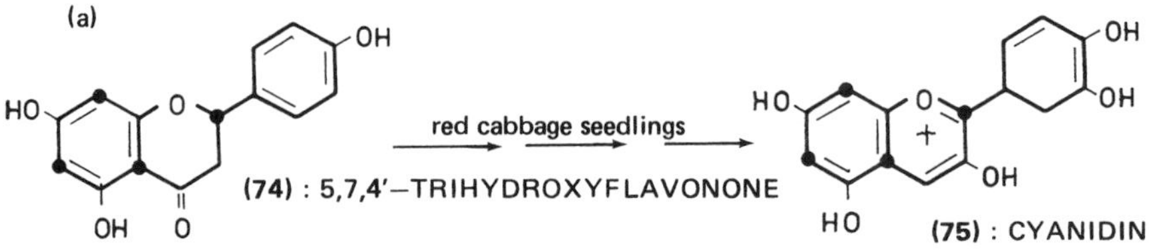

(a)

(74) : 5,7,4′-TRIHYDROXYFLAVONONE

red cabbage seedlings

(75) : CYANIDIN

[L.PATSCHKE, W.BARZ and H.GRISEBACH, *Z.Naturforsch,* [B] **19**, 1110 (1964)]

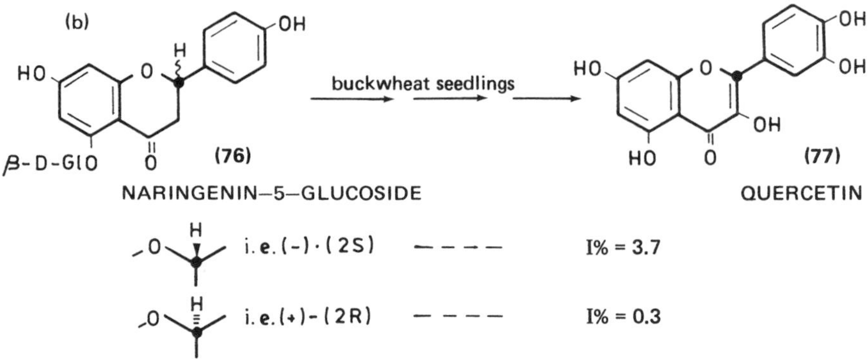

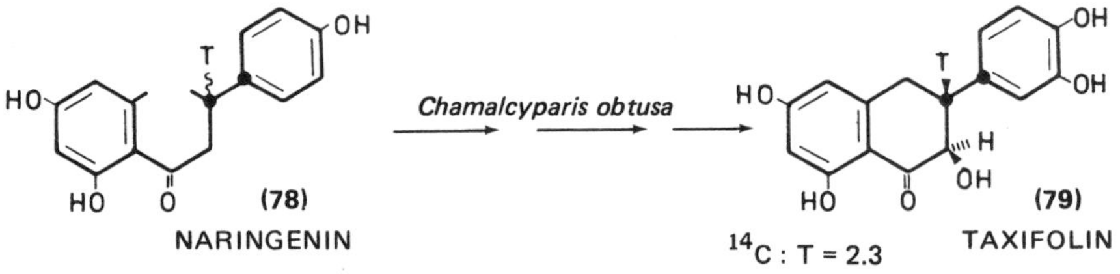

(b)

β-D-GlO

(76)

NARINGENIN—5—GLUCOSIDE

buckwheat seedlings

(77)

QUERCETIN

i.e.(-)·(2S)   — — — —   I% = 3.7

i.e.(+)-(2R)   — — — —   I% = 0.3

(78)

NARINGENIN

Chamalcyparis obtusa

(79)

TAXIFOLIN

$^{14}C : T = 2.3$

[H.GRISEBACH and S. KELLNER, *Z. Naturforsch,* [B], **446, 20,** (1965)]

Fig. 8.42 Incorporation of specifically-labelled naringenin into various flavonoids ($\bullet = {}^{14}C$).

The interconversion between chalcones and flavanones is catalysed *in vivo* by an enzyme isolated from various plant sources and which has been partially characterised. One feature of this enzymic reaction is the stereospecificity, apparent in the (S) chirality of carbon 2 of the flavanone derivative. It is not accidental that all the flavanones found in nature have the (2S)-configuration and are levorotatory. With chalcones having at least two free hydroxyl groups at positions 2 and 6 the equilibrium in an aqueous solution (acid or neutral) is completely and rapidly shifted to the flavanone form (Fig. 8.41). Both tautomeric forms are stable and isolable if only one free hydroxyl is present in these positions. The stabilisation energy of the strong hydogen bond between the carbonyl group and the *ortho*-phenolic hydroxyl group greatly influences the position of equilibrium and the interconversion rate. When only one hydroxyl is available, either for the cyclisation or for hydrogen bonding, the system tends to remain in the open form.

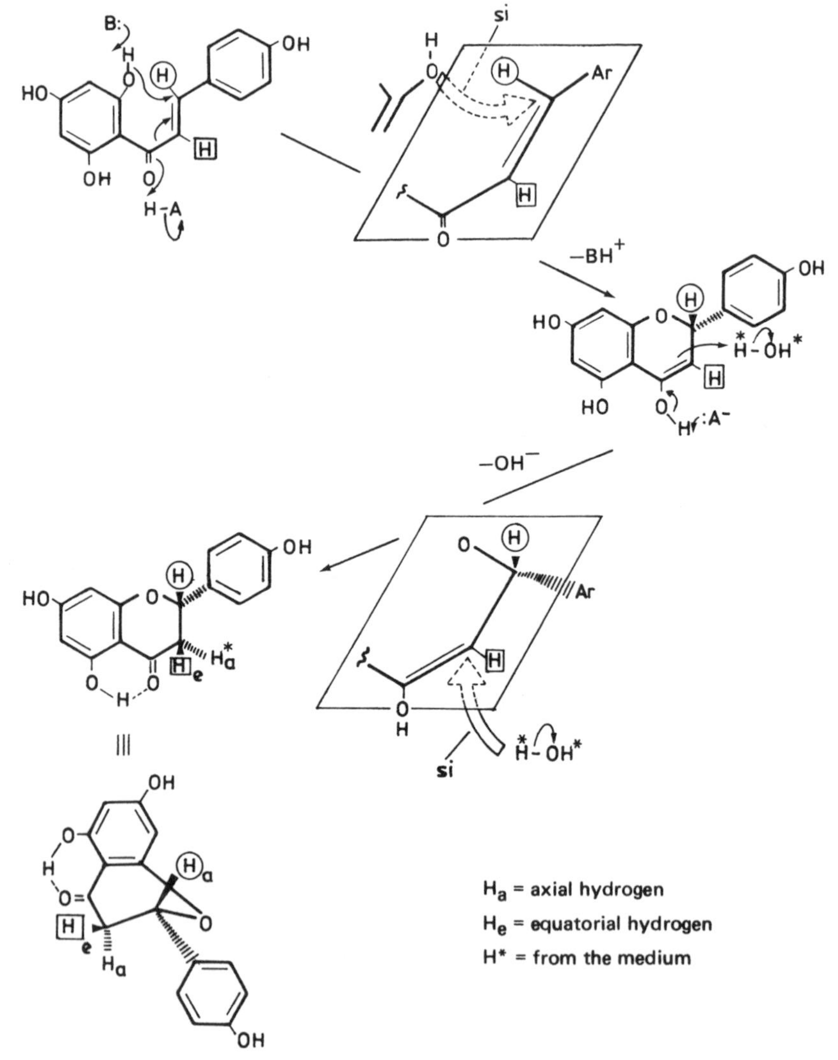

H_a = axial hydrogen
H_e = equatorial hydrogen
H^* = from the medium

Fig. 8.43 Stereochemistry of the enzymic conversion of chalcones to flavanones.

The dihydrochalcones are also rather rare in nature (Fig. 8.44); they are presumably formed by reduction of the double bond of the corresponding chalcones, although a direct polyketide synthesis starting from dihydrocinnamic acids cannot be ruled out *a priori*. Phlorizin is a dihydrochalcone present in apple tree root bark. It is a well known compound as it induces, in experimental animals, a condition similar to diabetes, with production of glycosuria caused by an alteration of glucose reabsorption mechanism in renal tubules. Its aglycone is called phloretin.

CHALCONES (Glucosides)   DIHYDROCHALCONES

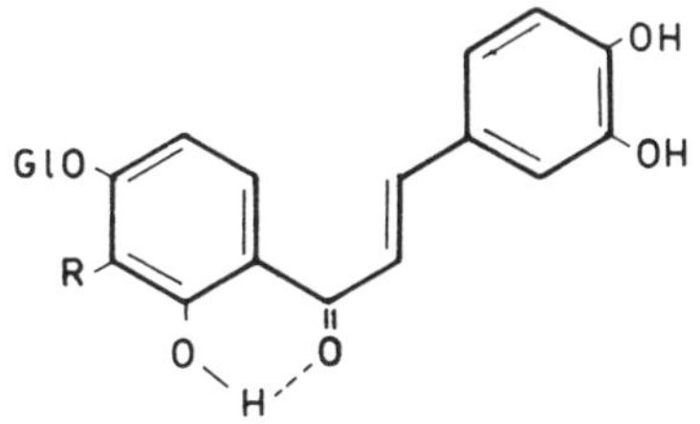

R = H   :   COREOPSIN   R = D–GLUCOSE   :   PHLORIZIN

R = OH :   MARIEN   R = L–RHAMNOSE .:   GLYCYPHYLLIN

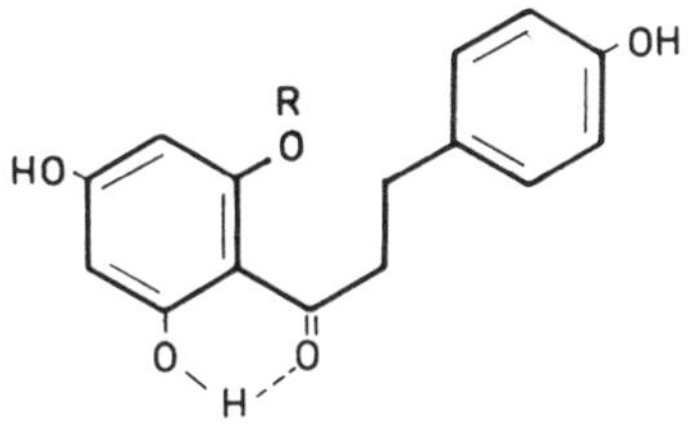

R = OCH₃ :   LANCEOLIN   R = H   :   PHLORETIN

AURONES (Glucosides)   FLAVANONES

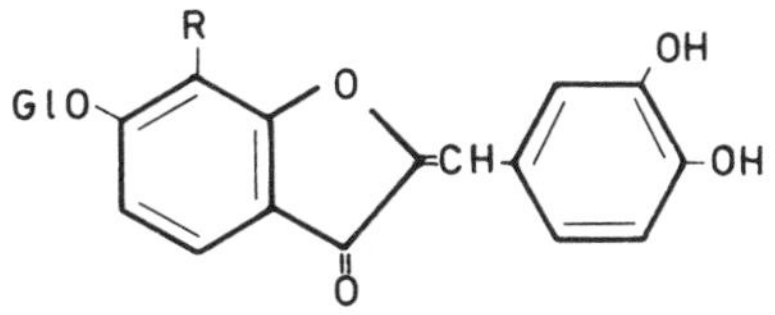

R = H : SULPHUREIN	R₁	R₂	R₃	
R = OH : MARITIMEIN	CH₃	H	H	SAKURANETIN
R = OCH₃ : LEPTOSIN	H	H	OH	ERIODICTYOL
	H	CH₃	OH	HOMOERIODICTYOL
(See also Figures 8.38 and 8.39)	H	H	OCH₃	HESPERETIN

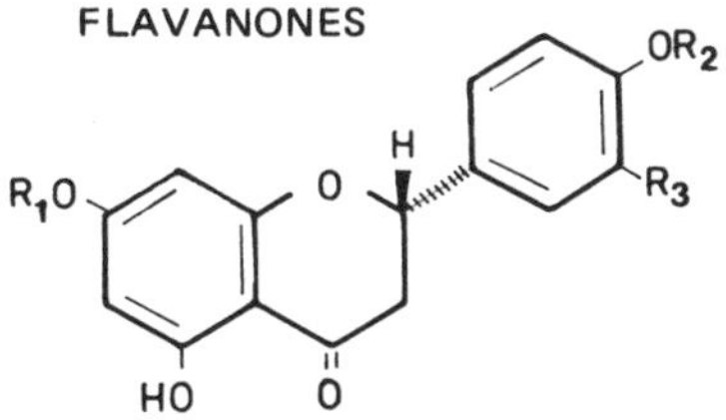

(See also Figures 8.38 and 8.42)

Fig. 8.44 Some natural chalcones, dihydrochalcones, aurones and flavanones.

If the chalcone-flavanone couple is accepted as the first biogenetic flavonoid entity, there arise problems concerning the biosynthetic paths and mechanisms whereby all, or nearly all, of the other flavonoids are produced *in vivo* from this couple. Few definite answers have been provided up to now on these steps, although the following speculations can be made. A general survey of the structures of the natural flavonoids reveals that compounds having an oxidation level of the three central carbons either equal to or higher than the chalcone-flavonone couple (Fig. 8.40) prevail in large numbers. It is logical, therefore, to assume that most flavonoids are formed through enzyme-mediated oxidation processes, which are usually selective and genetically controlled. It is also possible for flavonoid intermediates to make contact, in the plant cells, with more or less specific peroxidases and laccases, since such enzymes are widely dispersed and of a great metabolic importance in the biosynthesis of 7-hydroxycoumarins, of lignans and neolignans, of lignins and of the bis- and polyflavonoids themselves.

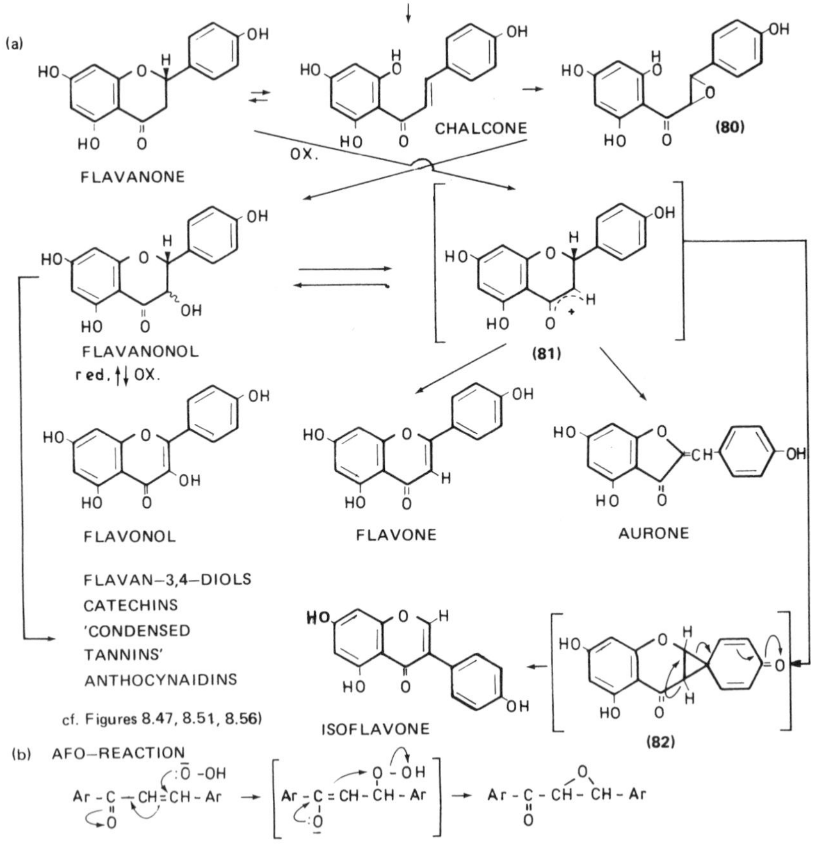

Fig. 8.45 Biogenetic relationships between flavonoids (according to Grisebach).

The first 'oxidative' hypothesis for flavonoid biosynthesis was proposed by Grisebach, who also made a detailed experimental study into the chemistry and biochemistry of flavonoids (Fig. 8.45). The main feature of Grisebach's hypothesis was the formation of an epoxide-chalcone (80), which could lead to flavonols, aurones, flavones and isoflavones, through plausible chemical mechanisms.

There is an alternative oxidation path, which involves the enolic form of the flavonone, directly affording the cation (**81**). This alternative hypothesis also gives satisfactory explanations for the biogenetic correlations amongst the various flavonoids and, especially, for the very frequent presence of an oxygen atom at position C-3 (flavanonols, flavan-3,4-diols, catechins, condensed tannins etc.). The weakest point of the epoxide hypothesis is that the intermediate epoxides e.g. (**80**), have never been found in nature. The *in vitro* epoxidation of α,β-unsaturated ketones is extremely slow with peracids and takes place in reasonable amounts only under alkaline conditions with hydrogen peroxide (AFO-reaction, scheme (b), Fig. 8.45), which are conditions totally different from those occurring *in vivo*.

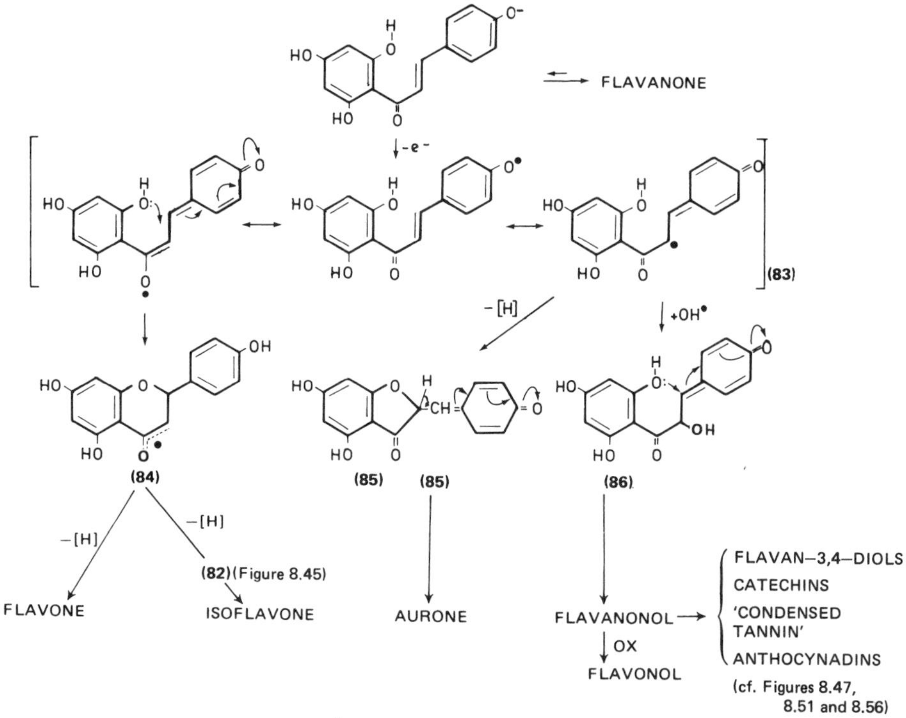

Fig. 8.46 Biogenetic hypothesis of Pelter.

The Pelter (Fig. 8.46) and Roux (Fig. 8.47) hypotheses are now widely accepted: they are both based on the phenolic oxidation of 4-hydroxychalcones.

The Pelter hypothesis is supported by a large number of *in vitro* chemical analogies, and the Roux theory by the large natural distribution of α-hydroxy-chalcones, which could also derive from *p*-hydroxyphenylpyruvic acid, as indicated in Fig. 8.41(b). A fourth hypothesis has also been suggested (Fig. 8.48).

It postulates a simultaneous phenol oxidation (by a laccase?) of the two aromatic rings of the chalcone intermediate, followed by an intramolecular radical coupling, according to a scheme very common in plants. Intermediates such as (96) and (97) (oxonium salts) should be formed, which lead to the four fundamental flavonoid structures (aurone, flavone, flavanonol and isoflavonone) and having the next oxidation level of the chalcone-flavanone couple, the oxidations being under enzymic control.

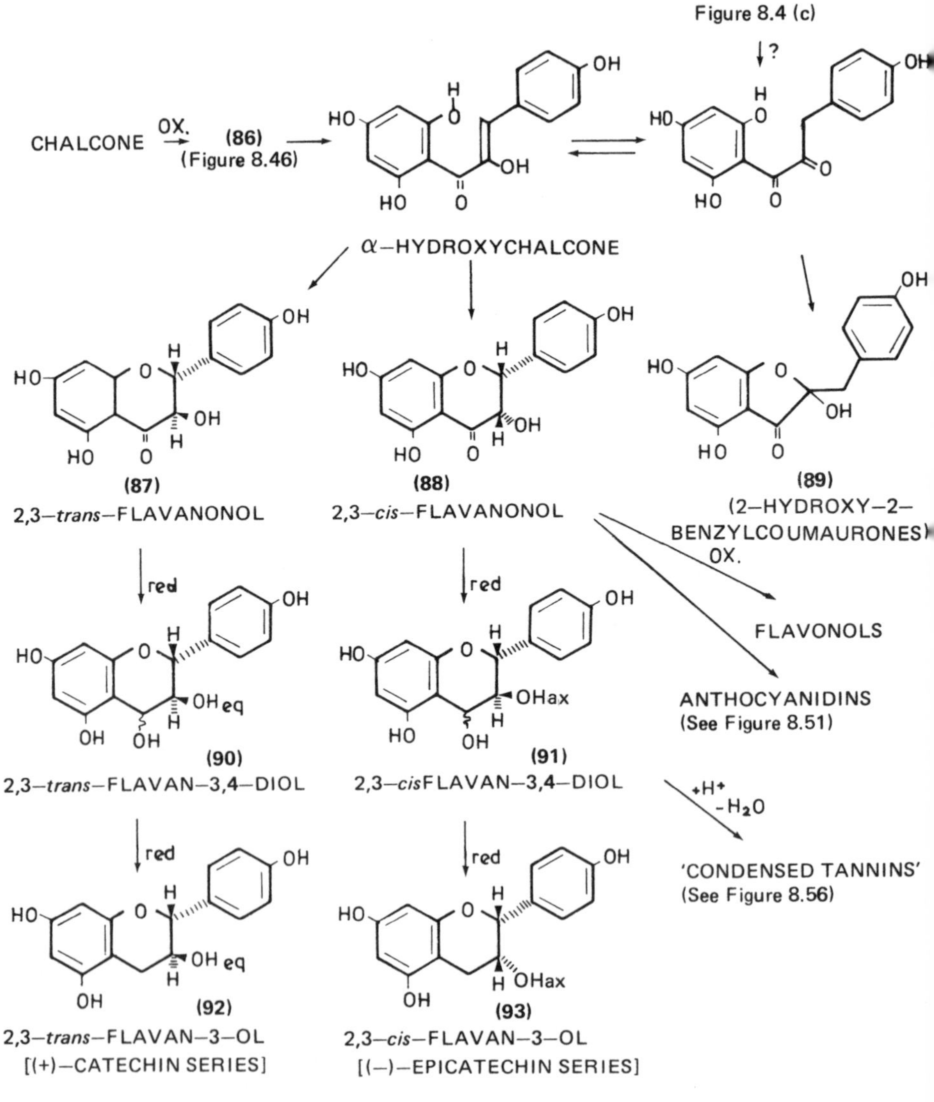

Figure 8.4 (c)

Fig. 8.47 Biogenetic hypothesis of Pelter.

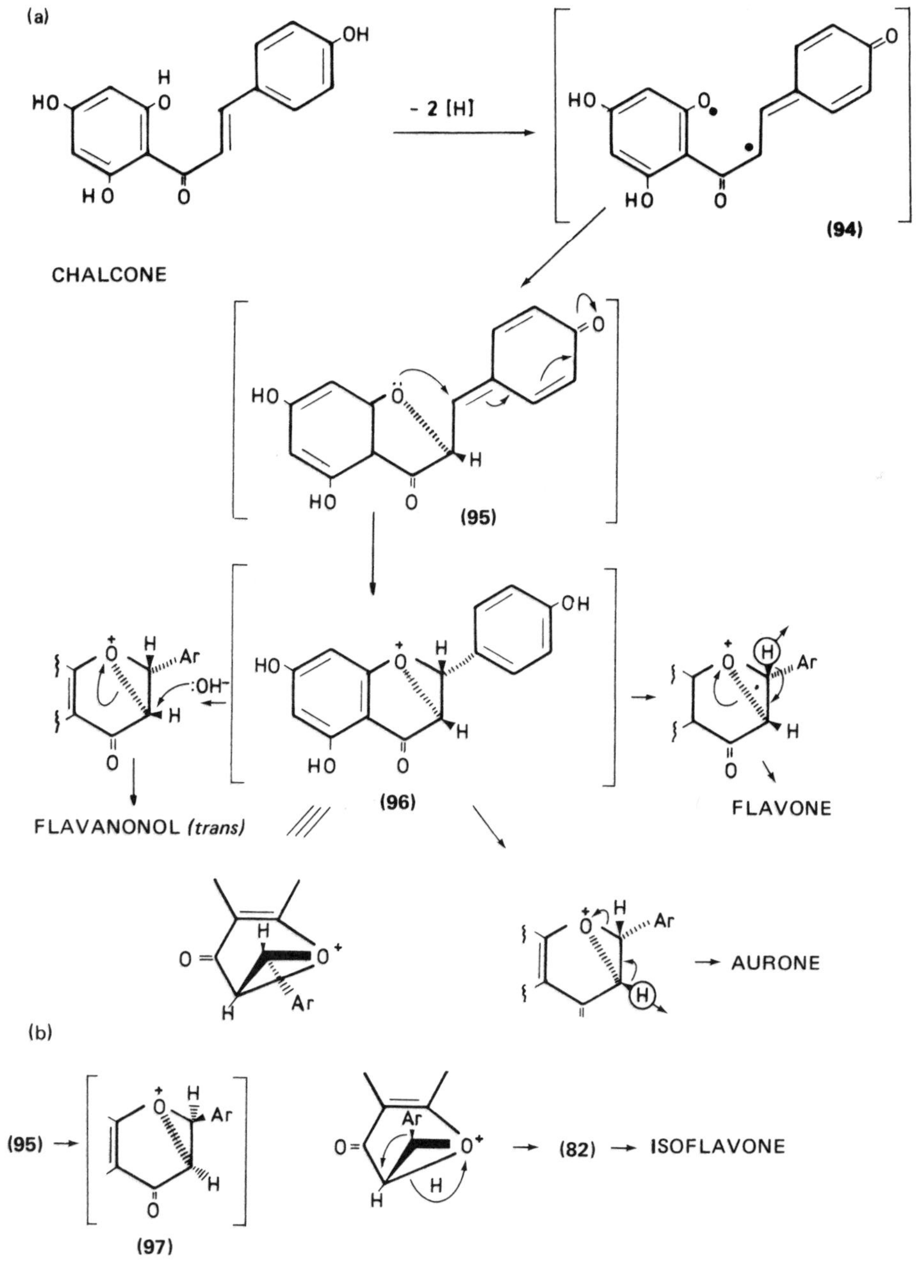

Fig. 8.48 Possible rearrangements leading to various flavanoid derivatives.

Apart from slight mechanistic differences, both the schemes of Figs. 8.46 and 8.48 can explain the formation of flavonoids having an oxygen atom in the 4' (or 2') position although they cannot account for the biosynthesis of compounds having ring B non-oxygenated. Such systems are infrequently encountered and for these many authors propose scheme (c), Fig. 8.41, e.g. (68). The various biogenetic speculations can be completed with the scheme illustrated in Fig. 8.49, including some routes which are initially reductive and which lead towards flavonoid structures in low oxidation states.

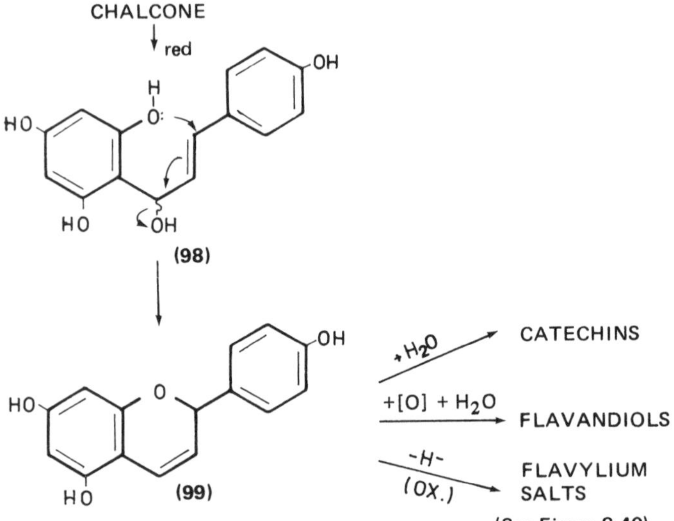

Fig. 8.49 Possible metabolic routes, initially involving reduction, leading to further flavanoid types.

Aurones. The 2-benzylidene-coumaran-3-ones, better known as aurones (Fig. 8.44), are often associated in plants with chalcones having the same oxygenation pattern as in the pairs of compounds suphurein-coreopsin (in *Cosmos sulphureus*) maritimein and marein (in *Coreopsis gigantea* and *maritima*), and leptosin and lanceolin (*Coreopsis lanceolate*) (Fig. 8.44). This fact undoubtedly has biogenetic implications and supports the hypothesis of a direct oxidative derivation of aurones from the corresponding chalcones (Figs. 8.45–8.48).

Flavones and flavanonols. Both flavones and flavanonols (the latter are also called dihydroflavonols) are widely distributed in higher plants and only a few are illustrated in Fig. 8.50. These two structures do not seem to be interconvertible *in vivo*; they should arise from two independent oxidation pathways, diverging from a common intermediate (the chalcone-flavanone couple?) (Figs. 8.45–8.48).

The flavones are very likely secondary products in plants with respect to the principal pathway, which passes instead through flavonols and leads to compounds of biological importance, such as the anthocyanin pigments.

Flavonols and anthocyanins. The flavonols (Fig. 8.50) are in an oxidation-reduction equilibrium with the corresponding flavanonols (see Fig. 8.45). Their chemical and structural analogy with the ascorbic − dehydroascorbic acid system (see Fig. 3.19) suggests that, in plants, the couple flavanonol − flavonol can also have a physiological role similar to that of Vitamin C. The flavonols are pigments contributing to the yellow and ivory shades of flowers.

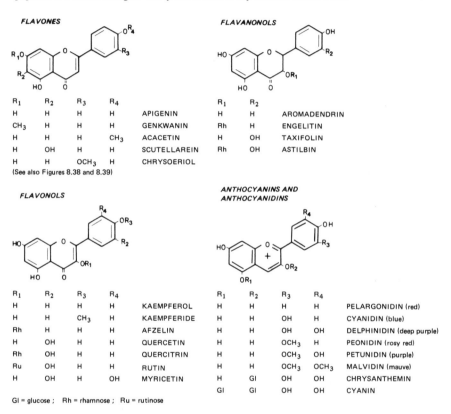

Fig. 8.50 Some natural flavones, flavonols. flavanonols and anthocyanidins.

The anthocyanins are glycosidic compounds, and are responsible for the more apparent colours of flowers (blue, purple, and red). The more common aglycones (anthocyanidins) are listed in Fig. 8.50; they can be considered as 2-phenylbenzopyrylium (flavylium) salts and are characterised by the presence of a 3-hydroxyl group. The anthocyanins carry their sugar residues (glucose, rhamnose, glactose, xylose, gentiobiose) only at position 3 if they are mono-glycosides and at positions 3 and 5 if they are diglycosides. The various shades of the colours found in nature depend on the number, type and position of the substituents in the A and B rings as well as the pH, presence of metal ions, etc.

An increase in the number of free hydroxyl groups in ring B makes the anthocyanin more intensely blue, whilst an increase in the number of methoxyl groups
shifts the colour towards the red (see Fig. 8.50). The colour is also influenced by
the medium surrounding the pigment: an anthocyanin which is red in acid, turns
to blue when the pH increases and, finally, it decomposes in strongly alkaline
solution. The colour of anthocyanins can also be modified by the association of
the pigment with metal ions, especially iron and aluminium, and with tannins.
    The biosynthesis of anthocyanins has presumably played a fundamental role
in the co-evolution of flowers and insects, as it has favoured entomophilous
pollination. It is well known that bees are attracted by yellow and blue flowers.
Diurnal butterflies prefer to rest on brightly coloured flowers, whilst nocturnal
ones almost exclusively visit white flowers, which are more apparent at evening
time, when there is poor light. Certain flowers can direct pollinating agents, as
they have strips or bands of different colours in their petals pointing to areas
where contact between the guest and the stamen anthers or the pistil stigma is
easier. Similarly, evolutional development in plant pigmentation caused by
flavonoids, may have resulted by more subtle selective processes, relying on the
preference of certain colours by the pollinating insects!
    The synthesis of the various flower pigments depends on precise informa-
tion enclosed in the genome, and this has been amply demonstrated by genetic
studies. In many cases it has been demonstrated that the presence of a substance A (such as malvidin) in a species of variety A, instead of a substance B
(very similar to, such as pelargonodin, Fig. 8.50) present in a second species or
variety B very similar to A, was directly related to slight chromosomal modifications, that is with mutations. Such mutations involve those genes controlling the
synthesis of the enzymes responsible for specific biosynthetic steps, for instance
the enzyme acting on a particular cinnamic acid e.g. p-coumaric rather than
sinapic acid in the above-mentioned example, as a starter for the polyketide
synthesis, or of the enzymes catalysing the specific hydroxylation of ring B of the
flavonoid skeleton. Many associations of the type: genes (usually denoted by an
alphabetical letter) — enzyme-biosynthetic step have been determined.
    The mechanism of formation of ring C in the anthocyanidins, that is the
formation of the 3-hydroxyflavylium salts, has still many obscure points. Haslam's
hypothesis is the most widely accepted one. This strictly correlates the biosynthesis of anthocyanins with that of the catechins (flavan-3-ols, Fig. 8.53). The
formation of anthocyanins instead of catechins, for instance during the ripening
of fruits, should be related to the minor reducing capacity of the organism.
Polyhydroxyflavanes (Flavandiols and *catechins).* The flavan-3,4-diols, widely
spread in nature (Fig. 8.53), are produced by reduction of the corresponding
flavanonols (see Fig. 8.47). They are not likely to be intermediates in the
synthesis of anthocyanidins, nor, perhaps, of catechins, despite the fact that
they form anthocyanidins on heating in a strongly acidic medium in the presence
of oxygen. (Fig. 8.52). The name of leucoanthocyanidins, previously given to

such compounds, is due to such *in vitro* behaviour. Their non-involvement in the biosynthesis of anthocyanidins is in agreement with Haslam's hypothesis (Fig. 8.51). Furthermore it has been observed how rarely in nature that the flavan-3,4-diols and the anthocyanins existing in the same plant bear the same substitution patterns in rings A and B.

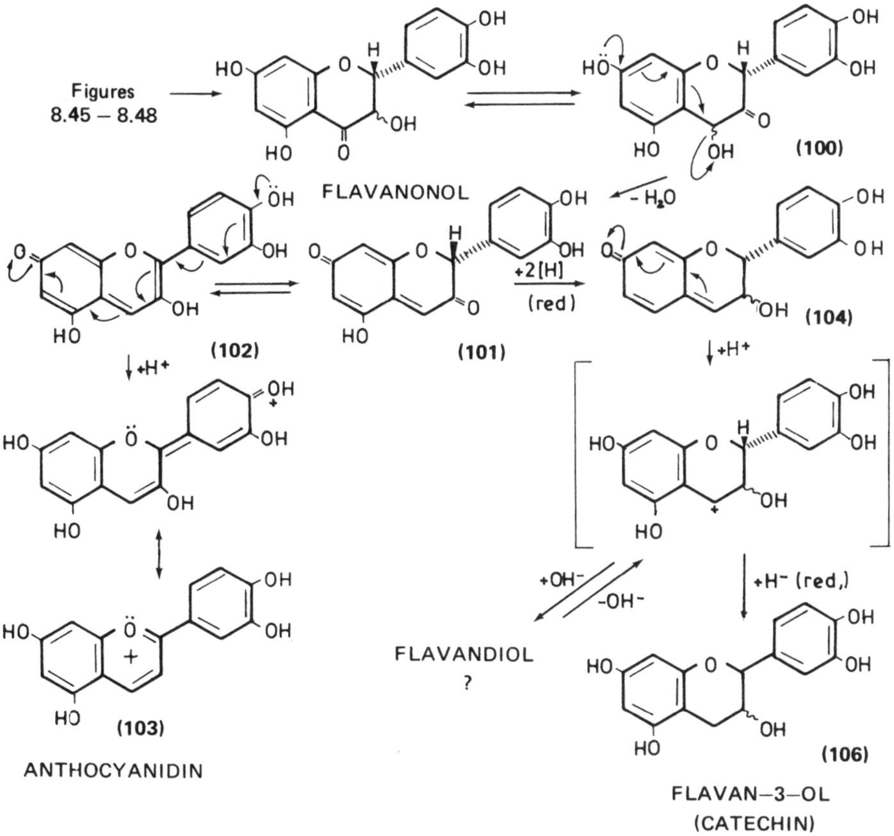

Figures 8.45 – 8.48 →

FLAVANONOL

(100)

(102)

(101) (red)

(104)

+ 2 [H]

− H₂O

+H⁺

+H⁺

(103)

ANTHOCYANIDIN

FLAVANDIOL
?

+OH⁻ / −OH⁻

+H⁻ (red₁)

(106)

FLAVAN−3−OL
(CATECHIN)

Fig. 8.51 Haslam's hypothesis on anthocyanidin and catechin biosynthesis.

The terms (+)-catechin and (−)-epicatechin are applied generally to flavan-3-ols with respective 2R,3S and 2R,3R configurations as well as to the parent compounds (Fig. 8.53). (+)Epicatechins (2S, 3S) are rare in nature, whilst no (2S, 3R) (−)-catechins have so far been isolated.

(+)-Catechin is the most common flavon-3-ol in nature and can be easily obtained in large amounts from the concentrated extract of *Uncaria gambir*. Dry tea leaves contain about 30% of flavan-3-ols: 60% of which is (−)-epigallocatechin-3-gallate, and 15% is (−)-epicatechin-3-gallate and (−)-epicatechin. In many fruits only one flavan-3-ol is exclusively present. Only speculations on the biosynthesis of the catechins exist and some schemes are shown in Figs. 8.47 and 8.51.

Another characteristic property of the polyhydroxyflavanes, of great practical importance, is their tendency to form polymers by enzymic or acid treatment. These polymers differ from the monomers in that they show tannin-like properties.

Polyflavonoids. It is relatively easy to find in nature compounds formed from two or more flavonoid units. Such dimers, oligomers and polymers can be separated into two categories, according to process of combination of the monomer units, either oxidative (dehydropolymerisation) or non-oxidative (acid catalysis).

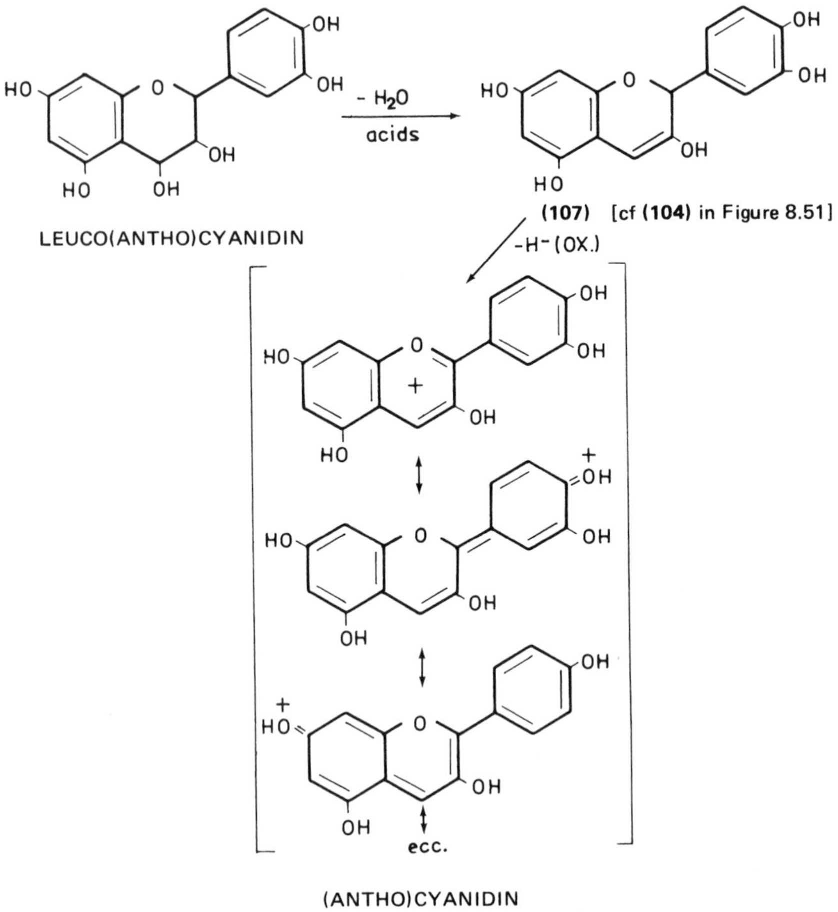

Fig. 8.52 *In vitro* conversion of leucoanthocyanidins into anthocyanidins. Some authors prefer to call leucoanthocyanidins 'leucoanthocyanidin hydrates' and compounds of the type (**107**) leucoanthocyanidins.

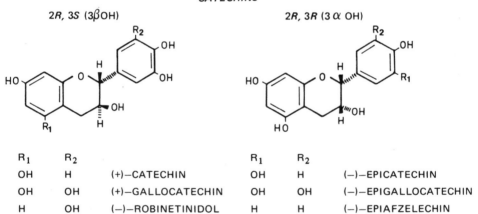

Fig. 8.53 Some natural flavandiols and catechins.

The first group includes the bis-flavones for instance (108) and (109), often present in Gymnospermae, as well as the bisflavanols, such as dehydroepigallo-catechin (110), isolated from the aqueous extract of black tea (Fig. 8.54). Black tea is a fermentation product of green tea in which the plant cells are broken and the various flavan-3-ols and oxidases are liberated, hence producing dimers, oligomers, etc. The 3'-8" and 5-5" bonds in bis-flavones and 6'-2" bonds in bis-flavonols indicate a phenol oxidation process.

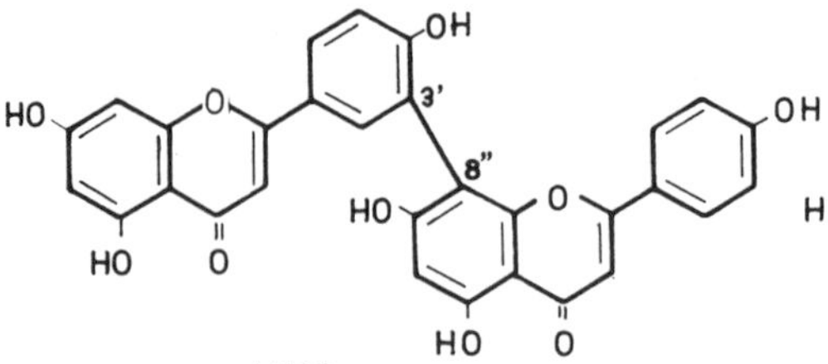

(108)
AMENTOFLAVONE

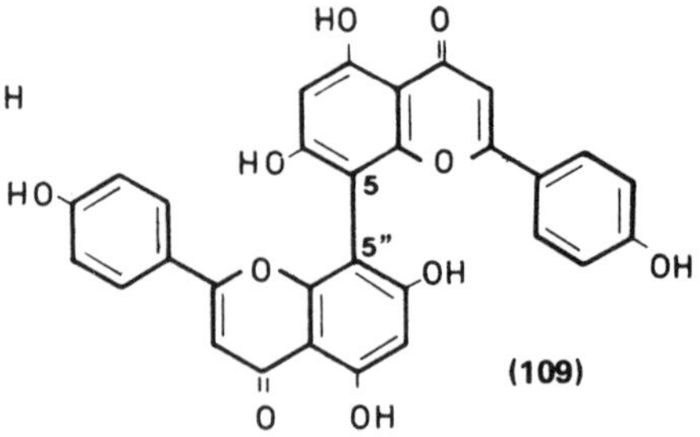

(109)
CUPRESSOFLAVONE

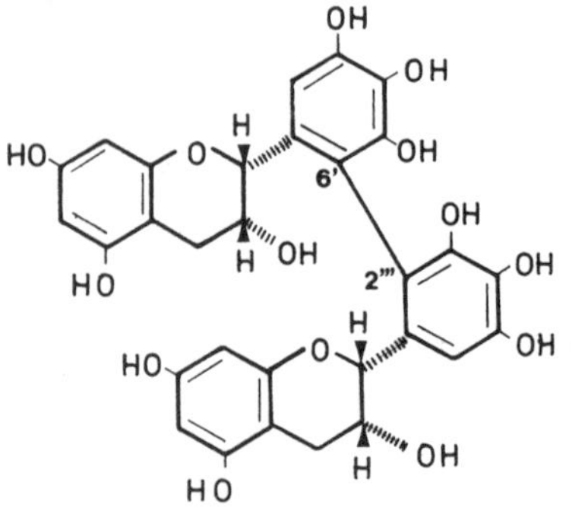

(110)
DEHYDRO−DIEPIGALLOCATECHIN

Fig. 8.54 Bis-flavonoids.

The flavanols of the type (110) represent only a small amount of the oxidised polyphenols contained in black tea. The bulk is constituted by two different fractions: the *theaflavins* and the *thearubigins.* Some typical constituents of the theaflavin fraction are shown in Fig. 8.55. The mechanism of formation of these should be similar to that postualted for the synthesis of purpurogallin in plant cells. Theaflavin should be formed via condensation of two flavanol molecules through rings B and which previously have been oxidised to *o*-quinones. Theaflavin is an orange-red, crystalline compound, obtained *in vitro* by Takimo and Imogawa in 1963, via an oxidative coupling of (−)epigallocatechin and (−)epicatechin. It is remarkable how condensation products are only known between two flavan-3-ol molecules, the former containing the pyrogallol nucleus, and the latter the pyrocatechin nucleus.

The thearubigin fraction is essentially composed of a mixture of oligomers such as (111), likely produced by acid-catalysed condensation of polyhydroxyflavanes (Fig. 8.56). According to their structural features, they belong to the family of the so-called proanthocyanidins.

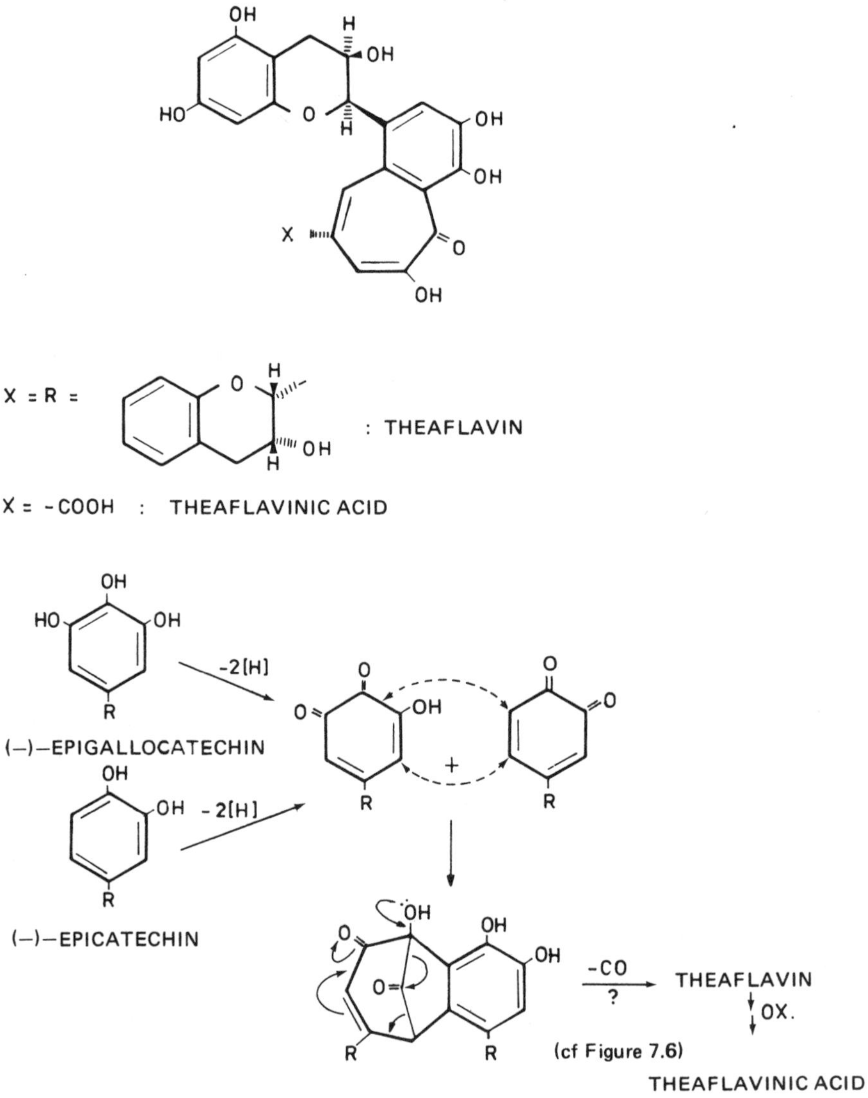

Fig. 8.55 Tropolone constituents of black tea and a possible mechanism of formation.

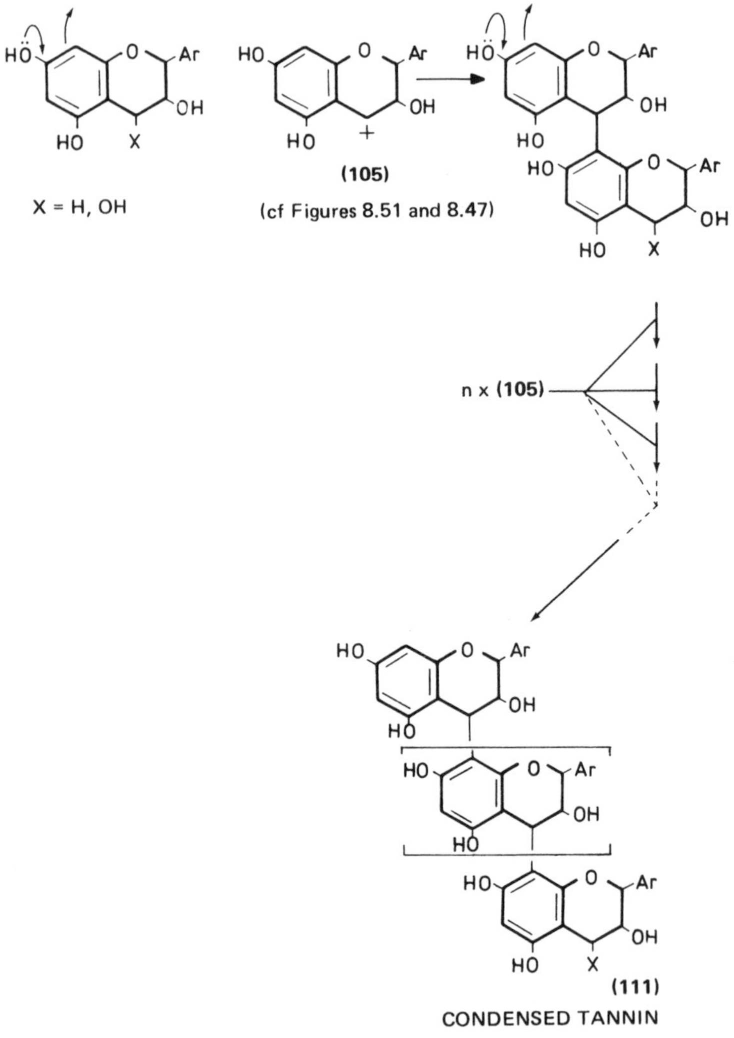

Fig. 8.56 Scheme for the formation of condensed (oligomeric) tannins.

The *proanthocyanidins* are colourless substances, present in most tissues of woody plants, and especially in fruit skins. Two features characterise these substances. First they act as precipitating agents towards proteins. Secondly, they are easily degraded in acid giving anthocyanidins (mostly cyanidin); by prolonged acid treatment, or by more vigorous hydrolysis conditions (increased mineral acidity and temperature), some complex and insoluble polymers are formed, called phlobaphenes.

The above properties explain why the proanthocyanidins are generally included with both the vegetable tannins (non-hydrolysable or condensed tannins) and, for structural and biogenetic reasons, with the polyflavonoids (Fig. 8.57). Unlike most phenolic products of plant origin, the proanthocyanidins are not found in plants as glycosides but only in their free form.

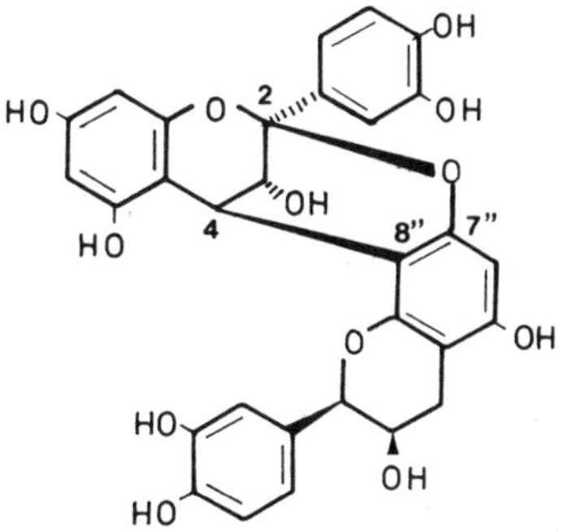

(112)
PROANTHOCYANIDIN – A2
[2 × (−)–EPICATECHIN]

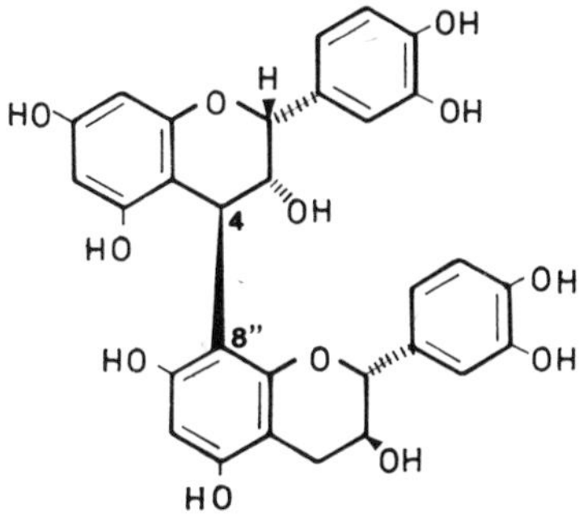

(113)
PROANTHOCYANIDIN – B1
[(+)–CATECHIN + (−)–EPECATECHIN]

Fig. 8.57 Dimeric proanthocyanidins of the A and B series.

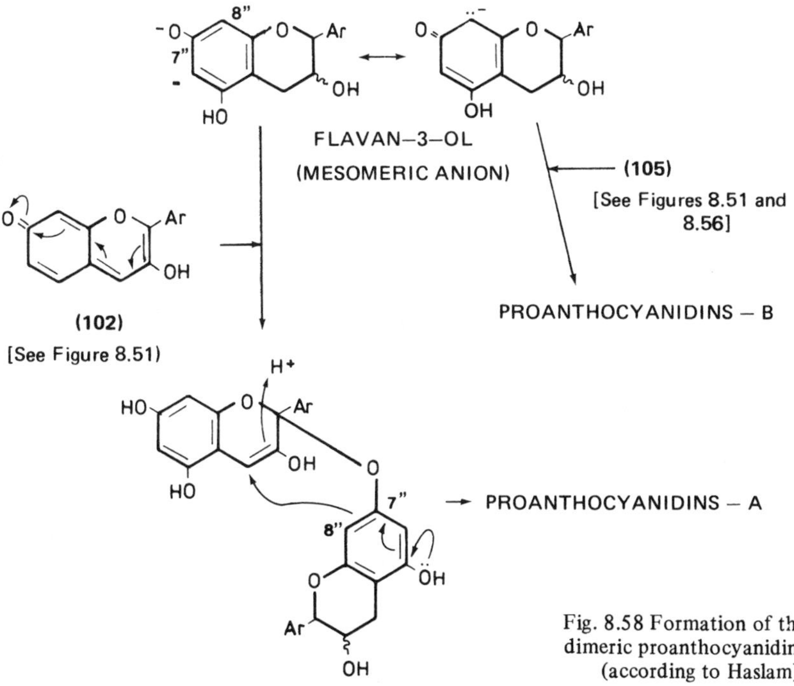

FLAVAN–3–OL
(MESOMERIC ANION)

(105)
[See Figures 8.51 and 8.56]

(102)
[See Figure 8.51)

PROANTHOCYANIDINS – B

H+

PROANTHOCYANIDINS – A

Fig. 8.58 Formation of the dimeric proanthocyanidins (according to Haslam).

The dimeric proanthocyanidins have only been carefully investigated in recent years; in many cases they have been isolated with sufficient purity in very small quantities and so their structures could only be completely elucidated with the aid of modern spectroscopic techniques. These proanthocyanidins are soluble in organic solvents and can be isolated by conventional extraction and chromatographic methods. After analysing more than sixty plant species, it was found in every case that the two monomeric units were derived from either (+)-catechin, or (−)-epicatechin, or from both. The proanthocyanidins-A are compounds in which the flavane units are joined by two bonds, such as (112), whilst the proanthocyanidins-B result from the union of two flavane units through only one C–C bond, such as (113).

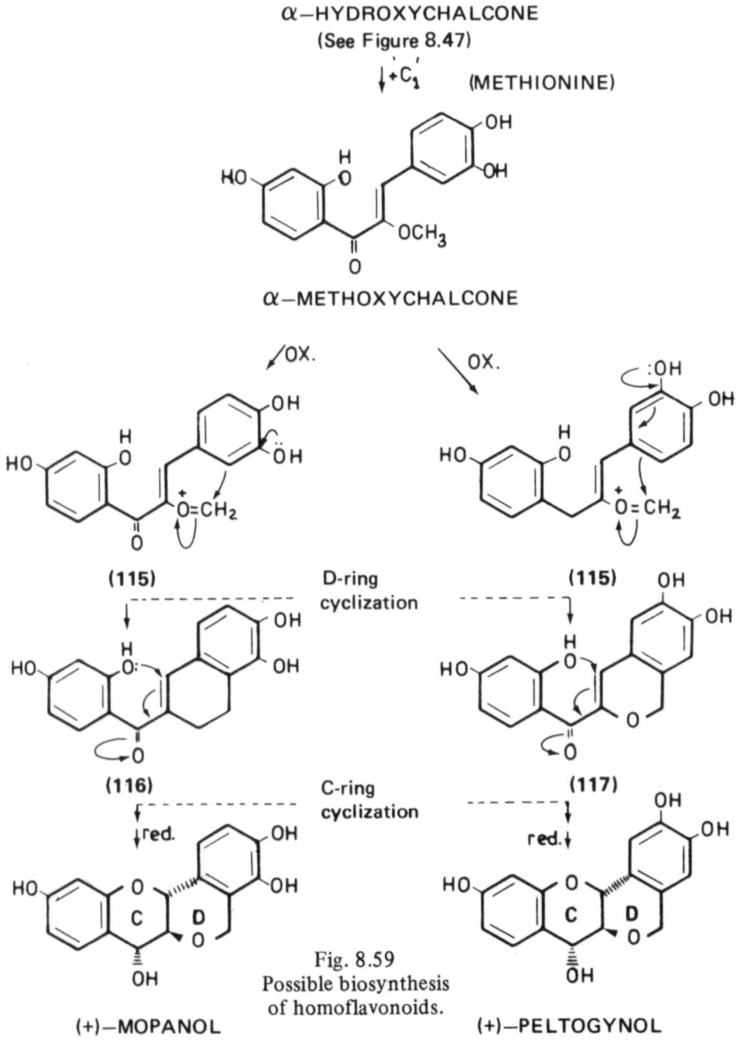

Fig. 8.59
Possible biosynthesis
of homoflavonoids.

Haslam has recently proposed an attractive hypothesis for the biosynthesis of both the A and B type proanthocyanidins (Fig. 8.58). The first should derive from an interaction between a flavan-3-ol and the quinomethide (102) (the biological equivalent of cyanidin), and the second ones from the interaction of a flavan-3-ol and the cation (105), postulated as an intermediate in the reduction process from flavanonols to catechins.

Mopanols and peltogynols. These are homoflavonoids in which the extra carbon atom present forms a further dihydropyran ring (ring D), condensed to the B and C rings (Fig. 8.59). Mopanols and peltogynols differ in the positions of the ring B oxygen atoms relative to the *extra* carbon. Roux proposed the scheme in Fig. 8.59 for the biosynthesis of (+)-mopanol and of (+)-peltogynol: the intermediates are an α-hydroxychalcone and an α-methoxychalcone (114), resulting from the former through a methylation of its enolic hydroxyl group. The oxidation of a methoxy group to $-O-CH_2OH$ and the successive nucleophilic substitution of the hydroxyl group (see 115) are likely *in vivo* process and analogies to the formation of a methylenedioxy system is also seen with the rotenoids (isoflavonoids, Fig. 8.63).

ISOFLAVONES

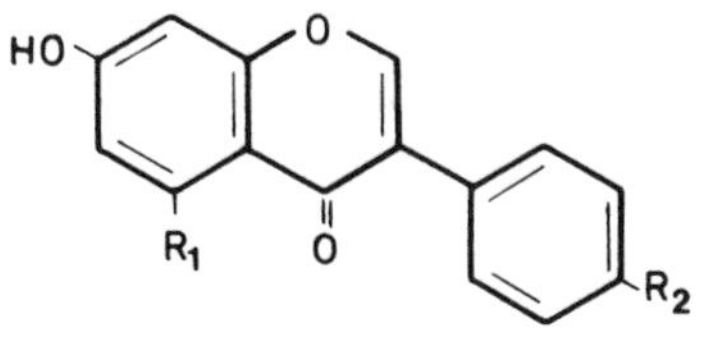

ISOFLAVANONES

R_1	R_2	
H	OH	DAIDZEIN
H	OCH_3	FORMONONETIN
OH	OH	GENISTEIN
OH	OCH_3	BIOCHANIN–A

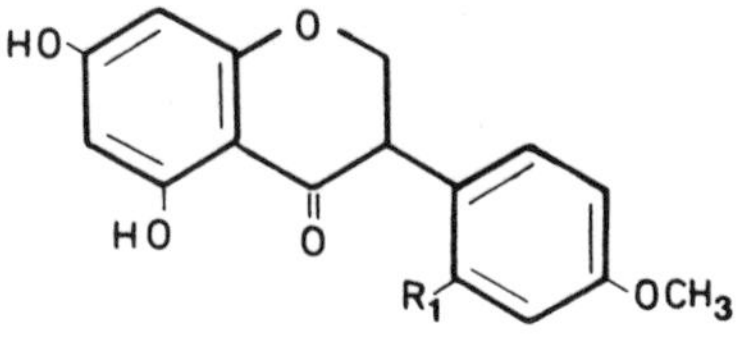

R_1	
OH	FERREIRIN
OCH_3	HOMOFERREIRIN

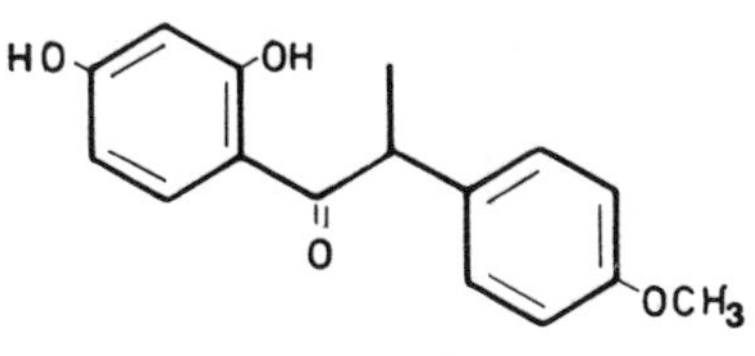

ANGOLENSIN

[See also (16) in Figure 8.12]

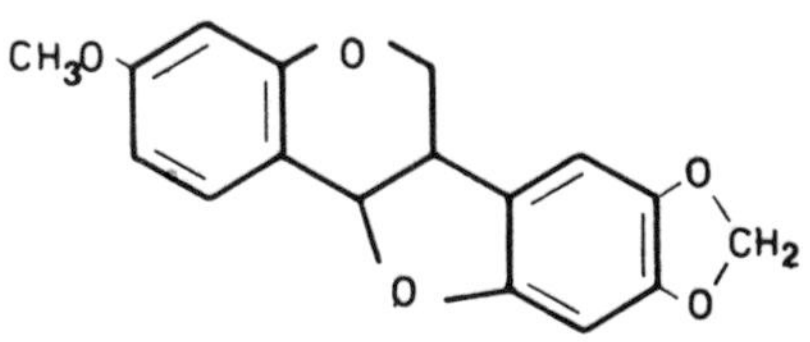

PTEROCARPIN

Fig. 8.60 Some natural isoflavonoids.

8.6 ISOFLAVONOIDS

The isoflavonoids differ from the normal flavonoids in the position of the aromatic ring B along the central propane chain. The isoflavones are particularly common in species of the Leguminosae family and are all colourless. They all show a weak estrogen activity, due to the presence of a phenolic stilbene part structure. Coumestrol (16) is thirty times more active than genistein (but 10^{-5} times less active than diethylstilbestrol) and is contained in very high quantities in certain clovers. The fertility of cattle is influenced by an abundant clover-based diet and serious agronomic problems can be created. Usually pregnant cows and sheep should not be fed with clover. The presence of isoflavones in food does not appear to influence the human female menstrual cycle.

The isoflavones are produced by the 1,2-transposition of the aromatic ring B of flavonoid intermediates (Fig. 8.61). The structure of the intermediate undergoing the 1,2-transposition and the mechanism of the shift has not been pinpointed. Two possible hypotheses are indicated in Figs. 8.45 and 8.48.

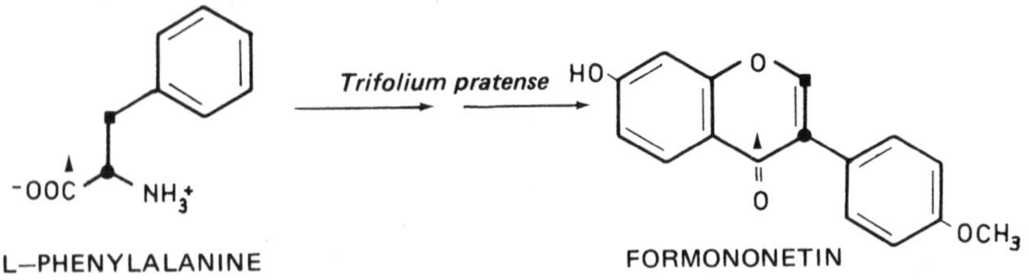

Trifolium pratense

L-PHENYLALANINE　　　　　　　　　　　**FORMONONETIN**

Fig. 8.61 Incorporation of specifically-labelled phenylalanine into formononetin.

DAIDZEIN
(See Figure 8.60)

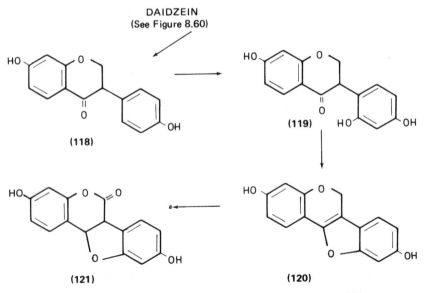

(118)

(119)

(121)

(120)

Fig. 8.62 Biosynthesis of pterocarpenes, e.g. (120), coumestanes, e.g. (121).

A hydroxyl group in the 2'-position is frequently present in isoflavonoids (Fig. 8.60). This peculiarity could be explained either by the assistance of the C-4 carbonyl to hydroxylation of ring B, which is adjacent to this carbonyl group in the isoflavone skeleton, or by the utilisation of *trans-o*-coumaric acids (interconvertible with the corresponding coumarins) for the synthesis of chalcones. The 2'-hydroxyl group allows the organisms to further elaborate the isoflavonoids to new structures, such as the coumarano-chromones or *pterocarpanes*, e.g. (120), the coumarano-couramins or *coumestanes*, e.g. (121). Of the latter type coumestrol has already been mentioned. The pterocarpanes are sometimes produced as phytoalexins, that is antifungal agents produced by plants under attack by viral or fungal agents.

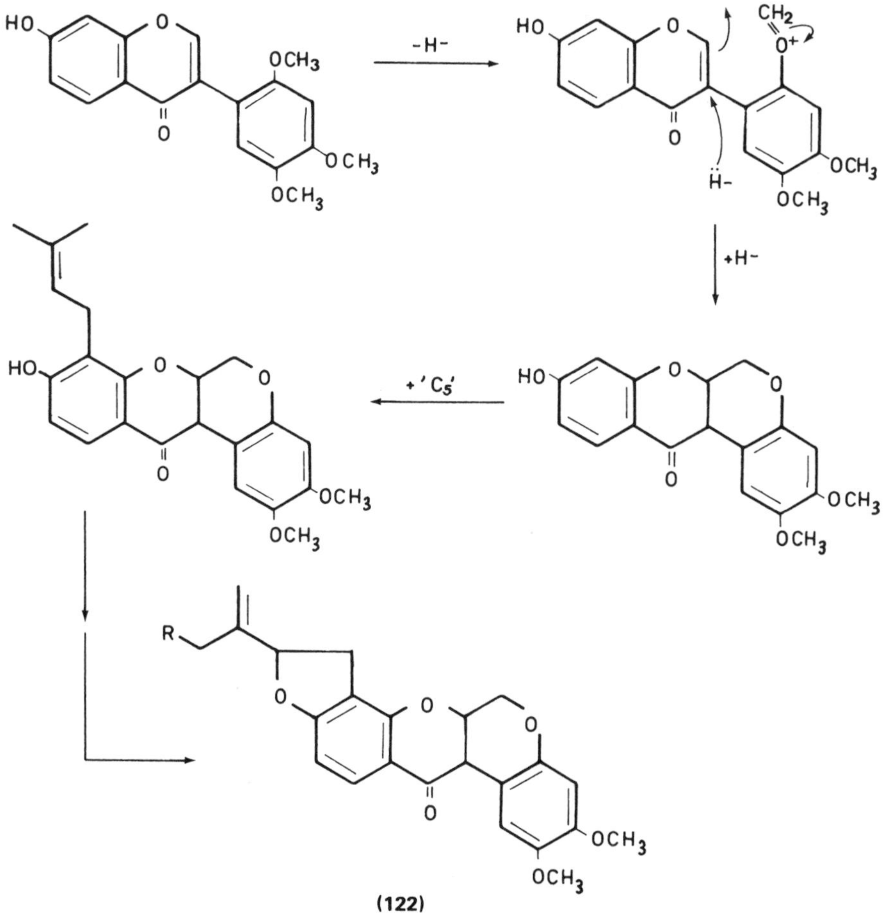

(122)

R = H ; ROTENONE

R = OH ; AMORPHIGENIN

Fig. 8.63 Biosynthesis of a typical rotenoid.

Two C_{16} isoflavonoid groups should be mentioned: the *rotenoids*, such as (122), in which an *extra* carbon atom at C-2 acts as a methylene bridge to the 2'-oxygen, producing a new ring, and the *homoisoflavanoids*, such as (123), characterised as $C_6\text{-}C_4\text{-}C_6$. The biosynthesis of the rotenoids has been widely investigated through incorporation experiments and the most recent results, obtained with *Amorpha fruticosa* (Leguminosae) plants, are in very good agreement with that schematised (Fig. 8.63) (see also the biosynthesis of the mopanol and peltogynols, Fig. 8.59).

The homoisoflavonoids have all been isolated from plants of the Liliaceae family (genus *Eucomis* and *Scilla*): there are about ten known, but their biosynthesis is still not resolved.

(a)

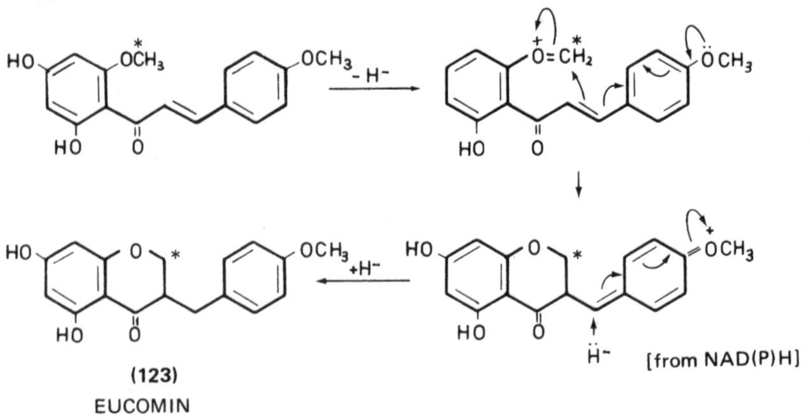

(123)
EUCOMIN

* = from methionine

Fig. 8.64 Possible biosynthesis of homoisoflavonoids.

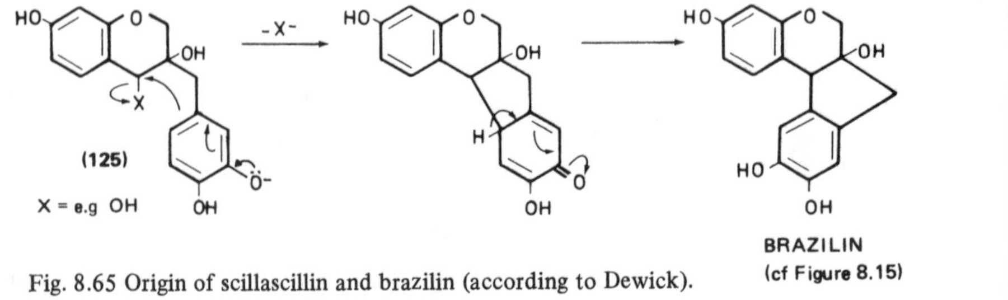

(a)

(124)   SCILLASCILLIN

(b)

(125)

X = e.g OH

BRAZILIN
(cf Figure 8.15)

Fig. 8.65 Origin of scillascillin and brazilin (according to Dewick).

Tamm and Dewick proposed a scheme (Fig. 8.64), which is in agreement with some incorporation results of labelled precursors in eucomine; however, it cannot be considered fully proven.

If confirmed, compounds of type (123) will be called '3-benzylchroman-4-ones', since they are produced from chalcone intermediates without the 1,2-transposition of the aromatic ring B, which is implicit in the homoisoflavone name. According to Dewick the 3-benzylchroman-4-ones could lead to the structures of scillascillin and brazilin (Fig. 8.65). An alternative route for the biosynthesis of brazilin is shown in Fig. 8.15 (neoflavonoids).

SOURCE MATERIALS AND SUGGESTED READING

[1] H. Grisebach and W. Ollis, (1961), Biogenetic Relationships between Coumarins, Flavonoids, Isoflavonoids and Rotenoids, *Experientia,* **17**, 4.

[2] T. A. Geissman (Ed.), (1962), *The Chemistry of Flavonoid Compounds,* Pergamon Press.

[3] J. B. Harborne (Ed.), *Biochemistry of Phenolic Compounds,* Academic Press.

[4] E. Bayer, (1966), Complex Formation and Flower Colors, *Angew. Chem. Int. Ed. Engl.,* **5**, 791.

[5] G. Billek (Ed.), (1966), *Biosynthesis of Aromatic Compounds,* Pergamon Press.

[6] G. Bilek and A. Schimpl, Biosynthesis of Plant Stilbenes in ref. 5.

[7] S. A. Brown, (1966), Lignins, *Ann. Rev. Plant Physiol.,* **17**, 223.

[8] S. A. Brown, Biosynthesis of Coumarins in ref. 5.

[9] H. Grisebach, W. Barz, K. Halbrock, S. Kellner and L. Patschke, Recent Investigations on the Biosynthesis of Flavonoids in ref. 5.

[10] A. B. Turner, (1966), Quinone Methides in Nature in *Progress in the Chemistry of Organic Natural Products,* ref. 1 of Chap. 1, **24**.

[11] M. H. Zenk, Biosynthesis of C_6-C_1 Compounds in ref. 5.

[12] T. A. Geissman, The Biosynthesis of Phenolic Plant Products in *Biogenesis of Natural Compounds,* ref. 24 of Chap. 1.

[13] J. B. Harborne, (1967) *Comparative Biochemistry of Flavonoids,* Academic Press.

[14] F. F. Nord and W. J. Schubert, The Biogenesis of Lignins in *Biogenesis of Natural Compounds,* ref. 24 of Chap. 1.

[15] M. H. Benn, (1968), On the Biogenesis of Neoflavonoids: A New Hypothesis, *Experientia,* **24**, 9.

[16] K. Freudenberg and A. C. Neish, (1968), *Constitution and Biosynthesis of Lignin,* Springer-Verlag.

[17] W. D. Ollis and O. R. Gottlieb, (1968), Biogenetic Relations Involving the Neoflavanoids and Their Congeners, *J. Chem. Soc., Chem. Comm.,* 1396.

[18] S. Yoshida, (1969), Biosynthesis and Conversion of Aromatic Amino Acids in Plants, *Ann. Rev. Plant Physiol.,* **20**, 41.

[19] A. Stoessl, (1970), Antifungal Compounds Produced by Higher Plants in *Recent Advances in Phytochemistry,* ref. 57 of Chap. 1, **3**.

[20] E. Wong, (1970), Structural and Biogenetic Relationships of Isoflavonoids in *Progress in the Chemistry of Organic Natural Products,* ref. 1 of Chap. 1, **28**.

[21] A. Pelter, J. Bradshaw and R. F. Warren, (1971), Oxidation Experiments with Flavonoids, *Phytochemistry,* **10**, 835.

[22] B. J. Deverall, Phytoalexins in *Phytochemical Ecology,* ref. 69 of Chap. 1.

[23] O. R. Gottlieb, (1972), Chemosystematic of the Lauraceae, *Phytochemistry,* **11**, 1537.

[24] J. B. Harborne, (1972-1977), Biosynthesis of Phenolic Compounds in *Biosynthesis,* ref. 30 of Chap. 1, **1-5**.

[25] L. Jurd, (1972), Recent Progress in the Chemistry of Flavilium Salts in *Recent Advances in Phytochemistry,* ref. 57 of Chap. 1, **5**.

[26] G. W. Sanderson, (1972), The Chemistry of Tea and Tea Manufacturing in *Recent Advances in Phyrochemistry,* ref. 57 of Chap. 1, **5**.

[27] E. L. Camm and G. H. N. Towers, (1973), Phenylalanine Ammonia-Lyase, *Phytochemistry,* **42**, 961.

[28] L. L. Creasy and M. Zucker, (1974), Phenylalanine Ammonia-Lyase and Phenolic Metabolism in *Recent Advances in Phytochemistry,* ref. 57 of Chap. 1, **8**.

[29] H. Grisebach and K. Hahlbroch, (1974), Enzymology and Regulation of Flavonoid and Lignin Biosynthesis in Plants and Plant Cell Suspension Cultures in *Recent Advances in Phytochemistry,* ref. 57 of Chap. 1, **8**.

[30] J. B. Harborne, Flavonoids as Systematic Markers in the Angiosperms in *Chemistry in Botanical Classification,* ref. 76 of Chap. 1.

[31] E. Haslam, *The Shikimate Pathway,* Butterworths, 1974.

[32] H. Nimz, (1974), Beech Lignin – Proposal of a Constitutional Scheme, *Angew. Chem. Int. Ed. Engl.,* **13**, 313.

[33] D. G. Roux and D. Ferreira, (1974), Hydroxychalcones as Intermediates in Flavonoid Biogenesis: The Significance of Recent Chemical Analogies, *Phytochemistry,* **13**, 2039.

[34] H. A. Stafford, (1974), Possible Multienzyme Complexes Regulating the Formation of C_6-C_3 Phenolic Compounds and Lignins in Higher Plants in *Recent Advances in Phytochemistry,* ref. 57 of Chap. 1, **8**.

[35] T. Swain, Flavonoids as Evolutionary Markers in Primitive Tracheophytes in *Chemistry in Botanical Classification,* ref. 76 of Chap. 1.

[36] von H. Wagner, (1974), Flavonoid-Glycoside in *Progress in the Chemistry of Organic Natural Products,* ref. 1 of Chap. 1, **31**.

[37] L. Farkas, M. Gabor and F. Kallay (Eds.), (1975), *Topics in Flavonoid Chemistry and Biochemistry,* Elsevier.

[38] J. B. Harborne, (1975), Flavonoid Sulphates: A New Class of Sulphur Compounds in Higher Plants, *Phytochemistry*, **14**, 1147.

[39] J. B. Harborne, T. J. Mabry and H. Mabry (Eds.), (1975), *The Flavonoids*, Chapman and Hall.

[40] J. R. L. Walker, (1975), *The Biology of Plant Phenolics*, Studies in Biology No. 54, Arnold.

[41] T. Swain, (1976), Nature and Properties of Flavonoids in *Chemistry and Biochemistry of Plant Pigments*, ref. 35 of Chap. 1, **1**.

[42] E. Wong, (1976), Biosynthesis of Flavonoids in *Chemistry and Biochemistry of Plant Pigments*, ref. 39 of Chap. 1, **1**.

[43] N. Amrhein and M. H. Zenk, (1977), Metabolism of Phenylpropanoid Compounds, *Physiol. Veg.*, **15**, 251.

[44] E. L. Camm and G. H. N. Towers, (1977), Phenylalanine Ammonia-Lyase in *Progress in Phytochemistry*, ref. 56 of Chap. 1, **4**.

[45] D. Gross, (1977), Phytoalexine und verwandte Pflanzenstoffe in *Progress in the Chemistry of Organic Natural Products*, ref. 1 of Chap. 1, **34**.

[46] J. B. Harborne, (1977), Flavonoid Sulphates in *Progress in Phytochemistry*, ref. 56 of Chap. 1, **4**.

[47] E. Haslam, (1977), Symmetry and Promiscuity in Procyanidin Biochemistry, *Phytochemistry*, **16**, 1625.

[48] H. Grisebach and J. Ebel, (1978), Phytoalexins, Chemical Defense Substances of Higher Plants?, *Angew. Chem. Int. Ed. Engl.*, **17**, 635.

[49] A. I. Gray and P. G. Waterman, (1978), Coumarins in the Rutaceae, *Phytochemistry*, **17**, 845.

[50] O. R. Gottlieb, (1978), Neolignans in *Progress in the Chemistry of Organic Natural Products*, ref. 1 of Chap. 1, **35**.

[51] Von K. Herrman, (1978), Hydroxyzimtsauren un Hydroxybenzoesauren enthaltende Naturstoffe in Pflanzen, in *Progress in the Chemistry of Organic Natural Products*, ref. 1 of Chap. 1, **35**.

[52] R. D. H. Murray, (1978), Naturally Occurring Plants Coumarins, in *Progress in the Chemistry of Organic Natural Products*, ref. 1 of Chap. 1, **35**.

[53] G. H. N. Towers and Chi-Kit Wat, (1979), Phenylpropanoid Metabolism, *Planta Medica*, **37**, 97.

[54] C.A.X.G.F. Sicherer and A. Sicherer-Roetman, (1980), The *R,S* Nomenclature of Pterocarpan Stereochemistry, *Phytochemistry*, **19**, 485.

Appendix 1
Reaction Mechanisms

An essential feature of any organic reaction is the breaking of one or more reagent bonds with the formation of new bonds. A comparison between the structures of the reagent and product indicates which bonds have disappeared and which ones have been formed, but does not provide an answer to how and why the bonds have disappeared and formed. In certain cases bonds are broken and then reformed during a reaction. Such changes are not immediately apparent by considering only the structures of products and reagents.

A study of the mechanisms of reactions, that is on their dynamic aspects, has to be made if one wants to get a deeper knowledge of the process. The reaction mechanism is the co-ordinated movement of nuclei and electrons transforming a system into another, chemically different, one. In order to understand the mechanism of an organic reaction stereochemical, kinetic, and thermodynamic parameters are employed, as well as use of analogy to related processes.

Organic reactions can be classified into two types:
(a) reduction-oxidation (redox) reactions, and
(b) non-redox processes.

Reduction-oxidation reactions are characterised by a net electron flow from the substrate (S) to a reagent (R) (substrate oxidation) or from the reagent to the substrate (substrate reduction). They can be considered as the result of two consequent half-reactions:

$$S_{red} \rightleftharpoons S_{ox} + m.e^- \qquad (1)$$

$$n.e^- + R_{ox} \rightleftharpoons R_{red} \qquad (2)$$

On multiplying (2) by $\frac{m}{n}$, and adding up side by side:

$$S_{red} + \frac{m}{n}R_{ox} \rightleftharpoons S_{ox} + \frac{m}{n}R_{red} \qquad (3)$$

The charge balance in (1), (2) and (3) (S and R can be electrically charged species) is obtained on adding an appropriate number of protons (H^+) to the left hand side or the right hand of such equations. Some examples of reduction-oxidation reactions are shown in Fig. A1.1.

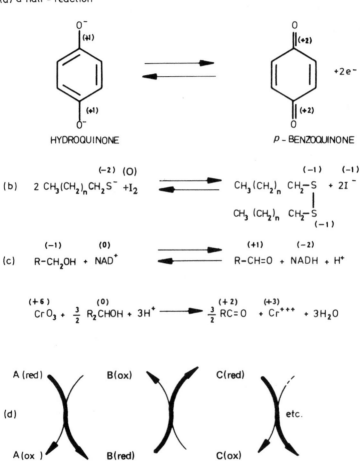

Fig. A1.1 Reduction-oxidation processes.
(Oxidation numbers are given in parentheses).

Examples of the oxidation numbers of carbon in its commonest forms are:

$$CH_4 (-4); H_3C-CH_3 (-3); HO-CH_3 (-2); CH_2=CH_2 (-2);$$

$$O=CH_2 (0); HCO_2H (+2).$$

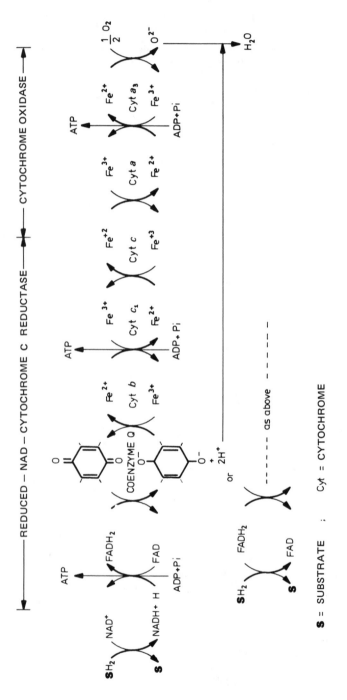

Fig. A1.2 Mitochondrial respiratory chain. Bold arrows indicate electron flow.

Often the electron migration occurs with transfer of hydrogen nuclei from, or to, the substrate (see reactions (c) and (d) of Fig. A1.1). Such transfers can occur either *separately* from the electron transfer, or *simultaneously*. In this latter instance one electron and one proton, or two electrons and one proton transfer together, respectively as a hydrogen atom or a hydride ion (that is: $H^{\bullet}$ instead of $e^- + H^+$, and H^- instead of $2e^- + H^+$). The need to distinguish between the various modes through which electrons and hydrogen nuclei are transferred might appear irrelevant if the overall reaction is considered. Mechanistically, however, the two processes are unrelated; Hamilton bases his classification of biological reduction-oxidation reactions on these differences.

In reduction-oxidation reactions the oxidation numbers of one or more substrate and reagent atoms change. In complex organic molecules the changes usually concern a specific function, such as:

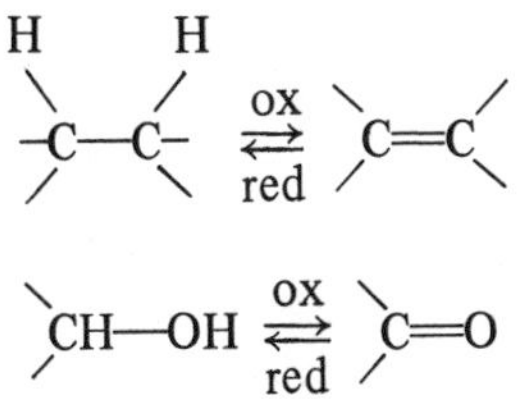

$$\begin{array}{cc} H & H \\ \backslash \;/ \\ -C{-}C{-} \\ /\;\;\backslash \end{array} \underset{\text{red}}{\overset{\text{ox}}{\rightleftharpoons}} \begin{array}{c} \backslash \;\;\;/ \\ C{=}C \\ /\;\;\;\backslash \end{array}$$

$$\begin{array}{c} \backslash \\ CH{-}OH \\ / \end{array} \underset{\text{red}}{\overset{\text{ox}}{\rightleftharpoons}} \begin{array}{c} \backslash \\ C{=}O \\ / \end{array}$$

A reduction-oxidation reaction of the type, $A_{red} + B_{ox} \rightleftharpoons A_{ox} + B_{red}$, can be connected *in vivo* with another reaction by which B_{red} is again transformed

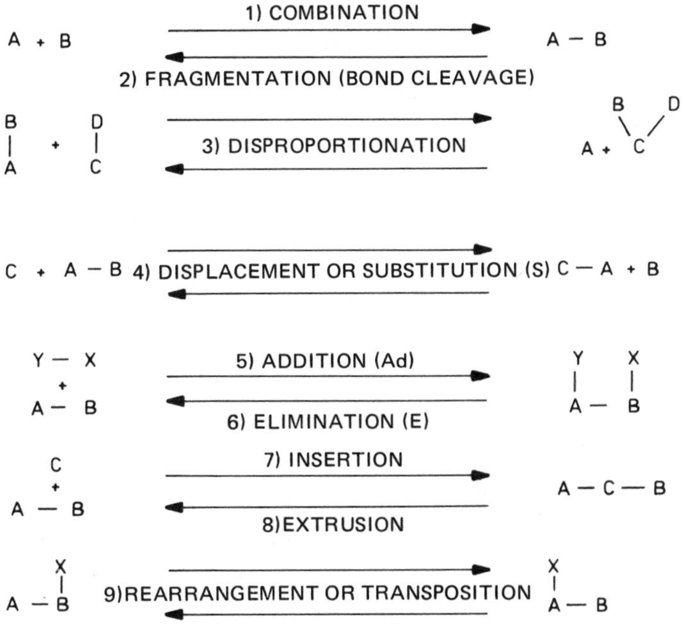

Fig. A1.3 Reactions with transfer of atoms or groups.

into its oxidised form: $B_{red} + C_{ox} \rightleftharpoons B_{ox} + C_{red}$; similarly C_{red} might be transformed into .C_{ox} by a third reaction and so on. A reactions series of such a kind is called an electron transfer system (or chain) and the substances involved are called carriers (electrons, atoms, or ions) (Fig. A1.1(e)). Reduction-oxidation chains are very frequent in nature and the mitochondrial respiratory chain (Fig. A1.2) is a classical example. The non-redox reactions are those which cannot be related to equations of the type (3). Examples include esterification, saponification, aldol and crotonic condensations, the Claisen and Mannich reactions and the addition of water to double bonds. Every organic reaction implies a variation in the number, type and distribution of the covalent bonds in the reagent. Such changes can be formally classified into various reaction types involving atom or group transfers, as indicated in Fig. A1.3.

A reaction is called a *one step* reaction if there are no distinguishable minima between reagents and products in its energy profile (free energy versus reaction coordinate; Fig. A1.4). Intermediate compounds are not formed. Multistep reactions possess a certain number of energy minima in their energy diagrams, corresponding to intermediates.

Such intermediates are termed 'transient' when their average life is extremely short and their concentration so low that they are not easily detected except by the use of sensitive spectroscopic techniques (ultraviolet, electron spin resonance, nuclear magnetic resonance, etc.). Sometimes a transient intermediate can be 'trapped', that is, blocked by reacting it with an appropriate reagent added to the reaction medium.

Each 'step' of a reaction is limited by two energy minima and shows one maximum only. At this maximum the substrate has bonds partially broken and partially formed. This point on the reaction coordinate is conventionally indicated by the symbol ‡ and by the terms *transition state* or *activated complex*. These two terms refer rigorously to two different concepts: the former (t.s.) refers to the energy, and the second one (a.c.) to the structure. In practice they are often used synonymously. In this text the term 'transition state' has always been used, to avoid confusion with the term *enzyme – substrate complex,* which has a very different meaning.

The reactions shown in Fig. A1.4 exemplify the two above quoted cases: the one step reaction is represented by a nucleophilic bimolecular substitution (S_N2, OH^- as a nucleophilic agent, second order kinetics); the two-step reaction is represented by a semimolecular nucleophilic substitution (S_N1, Cl^- as a nucleophilic agent: first order kinetics for the slower step, which determines the rate of the overall process).

An important aspect in the study of a reaction mechanism concerns the structure of the transition state and of the intermediates. These features may be used to divide the organic reactions into radical, pericyclic and ionic reaction types.

Radical reactions are recognised as those possessing amongst the reagents,

(a) (S_N2)

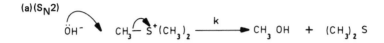

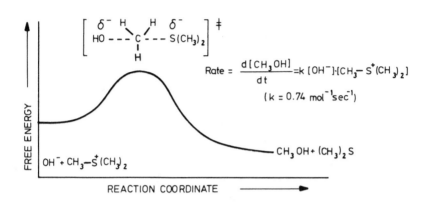

$$\text{Rate} = \frac{d[CH_3OH]}{dt} = k\,[OH^-]\cdot[CH_3-S^+(CH_3)_2]$$

$$(k = 0.74 \; mol^{-1}sec^{-1})$$

(b) (S_N1)

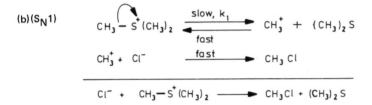

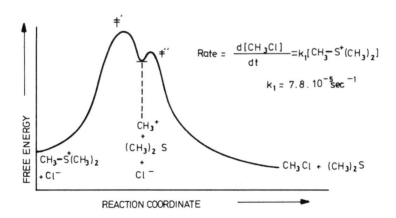

$$\text{Rate} = \frac{d[CH_3Cl]}{dt} = k_1[CH_3-S^+(CH_3)_2]$$

$$k_1 = 7.8 \cdot 10^{-5} sec^{-1}$$

Fig. A1.4 Reactions involving *a*, one step and *b*, two steps.

intermediates and products, at least one chemical species bearing an odd number of electrons, that is a radical. Radicals have an unpaired electron and have a permanent magnetic moment, so their presence and structure can be revealed through e.s.r. spectroscopy. Radical properties are also exhibited by those even-electron molecular species bearing two incompletely filled orbitals, each orbital possessing one electron. The paramagnetic oxygen molecule, in its ground state and the excited anthracene molecule are examples of such biradicals. The transition state of a radical process also possesses radical character, as some unpaired electrons are either formed or transferred.

The radical reactions encountered *in vivo* almost always follow electron transfer reduction-oxidation processes; one electron migrates each time in these reactions. A common example in secondary metabolism involves the coupling of phenoxy radicals.

The radical reactions, *in vitro*, can be initiated by the homolytic fission of a simple bond, that is through the breaking of a molecule into two fragments, each keeping one of the two bound electrons (reaction 2 of Fig. A1.3, with A and B having an odd electron number). The so-formed fragments can provoke the homolytic cleavage of other simple or multiple bonds, giving rise to S_H substitutions (reaction 4, Fig. A1.3, C and B as radicals), Ad_H additions (reaction 5 of Fig. A1.3, without Y and with X having an unpaired electron).The index *H,* at the low right of the symbols S and Ad indicates a homolytic process (similarly N means nucleophilic, and E, electrophilic). A typical feature of radical reactions is their chain propogation: there is an initial phase (e.g. $A-A \rightarrow 2A^{\cdot}$), a propagation step (e.g. $A^{\cdot} + B - C \rightarrow AB + C^{\cdot}$; $C^{\cdot} + A-A \rightarrow AC + A^{\cdot}$. etc) and a termination step (e.g. $2A^{\cdot} \rightarrow A-A$, $2C^{\cdot} \rightarrow C-C$; $A^{\cdot} + C^{\cdot} \rightarrow A-C$). The initial homolytic cleavages often require very strong reaction conditions, such as heating up to high temperatures, or irradiating with u.v. light, and are practically absent in living organisms, except when they are exposed to strong electromagnetic radiations (X-rays, γ-rays, etc.)

Pericyclic reactions are characterised by transition states in which all the σ and π bonds implied in the reaction form a continuous band enclosed on itself. Bonds breaking tend to alternate with bonds forming (Fig. A1.5). The pericyclic transition states are usually isopolar, that is they possess neither radical or ionic character. One feature of such reactions is their stereospecificity. [According to Kosower three types of transition state can exist:
(a) polar, when there is a separation or dispersion of electrical charge;
(b) radical, when a transfer of unpaired electrons is involved; and
(c) isopolar, when they show neither features of (a) or (b).]

Pericyclic reactions do not appear to be very frequent *in vivo*: the most familiar instance is the Claisen transposition of chorismic acid to prehenic acid. The occurrence of pericyclic processes cannot be ruled out, however, for many enzyme-catalysed processes, where the reaction mechanisms are still not clear.

In *ionic reactions* the bonding electrons (or the lone pair electrons of O,N etc.) are never separated, but migrate together either breaking or forming bonds. Bond fissions where the whole electron pair transfer into one atom, are called *heterolytic* processes (e.g. reaction 2, Fig. A1.3 with A a positive ion and B a negative ion).

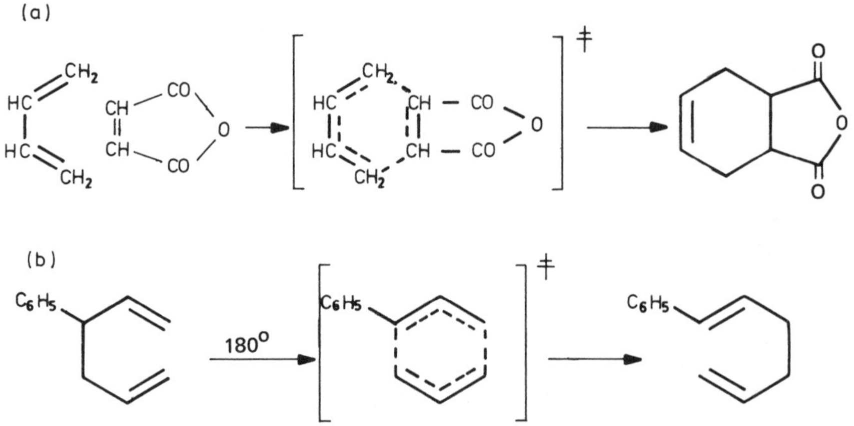

Fig. A1.5 Pericyclic reactions: *a*, Diels-Alder addition, *b*, Cope rearrangement.

The transition states of ionic reaction are always polar, that is, distinguished by a more or less marked separation and dispersion of electrical charges, (e.g. the transition states for the reactions of Fig. A1.4). For this reason ionic reactions are much more sensitive than other processes to the solvating peoperties of the medium (dielectric and polarisability properties). The solvent modifies the relative free energies of the reagents, the transition state and the products, and can markedly influence both the reaction rate and the equilibrium between reagents and products. Most biological reactions are ionic, and belong to one of the simple reactions listed in Fig. A1.6, or to a combination of them, taking due account of the action of enzymes.

When electrons migrate (indicated by the arrows in the formulae) the molecular and atomic orbitals of the system are reorganised. The reactions in Fig. A1.6 can be called *simple*, as they are single step reactions and do not involve more than four substrate atoms. Many examples of such reactions are quoted in Chapters 2-6; in most cases they are mechanistic interpretations rather than demonstrated processes. These include dissociations of the alcohol pyrophosphate esters (Fig. 2.34 and 2.40(a)), the deprotonation of methyl and methylene groups before nucleophilic additions in the aldol and Claisen condensations (Fig. 2.40(d)), electrophilic additions and rearrangements (so frequent in terpene biosynthesis), bimolecular eliminations (such as the deamination of L-phenylalanine, see Fig. 2.6), etc.

(a) DISSOCIATION (left to right) AND ASSOCIATION (left to right)

(b) DEPROTONATION (PROTONATION) BY A BASE (BY AN ACID)

(i)

(ii)

(c) BIMOLECULAR NUCLEOPHILIC SUBSTITUTION (Symbol: S_N2)

(d) BIMOLECULAR ELIMINATION (Symbol: E2)

(e) ELECTROPHILIC ADDITION (Symbol Ad_E) AND RETRO – REACTION

(a)   (of Appendix III)

(b)

* called also β – ELIMINATION (Symbol βE)

(f) NUCLEOPHILIC ADDITION (Ad_N) AND RETRO – REACTION

(g) MIGRATION TO ELECTRON – DEFICIENT ATOM

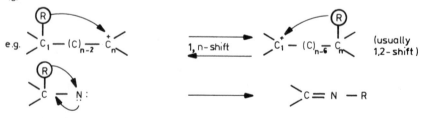

e.g.   1, n-shift   (usually 1,2- shift)

See Note bottom next page

Fig. A1.6 Fundamental heterolytic one-step reactions. It is to be noticed that the commonest multistep reactions result from a sequence of the above reactions, e.g. S_N1 consists of a dissociation followed by an association (see Figure A1.4b); E1 of a dissociation followed by β-elimination; E1cb of a deprotonation followed by a nucleophilic retroaddition, etc.

(a) ALLYLIC BIMOLECULAR SUBSTITUTION (symbol: S$_N$2')

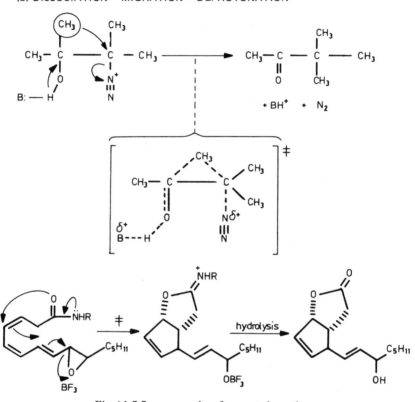

(b) DISSOCIATION — MIGRATION — DEPROTONATION

Fig. A1.7 Some examples of concerted reactions.

Note to Fig. A1.6 (opposite)

Charges in brackets are formal and reflect the displacements of electron doublets: actual charges can be obtained considering the intrinsic charge of each atom or group involved in the reaction.

X, Y = nucleophilic atom or group, e.g. X = Cl, Br, NR$_3^+$, SR$_2^+$, OH, OCOR etc;

Y = H$_2$O, OH$^-$, RCOO$^-$, Cl$^-$, Br$^-$, $\geq$C$^-$ (carbanion) etc.

B: = base (deprotonating reagent), e.g. OH$^-$, R$_3$N, H$_2$O

B(+)–H = acid (protonating reagent), e.g. H$_3$O$^+$, H$_2$SO$_4$

E$^+$ = electrophilic atom or group, e.g., Br$^+$, H$^+$, $\geq$C$^+$ (carbocation)

Z = O, OH$^+$, NR, NHR$^+$, NR$_2^+$; C$\underset{}{\overset{}{<}}$

R = H, alkyl, aryl.

Rearrangements involving large numbers of substrate atoms very often occur in nature (from 5 to 10, and sometimes more, atoms may transfer: see Figs. 5.40, 5.51 and 5.63): they can be interpreted as resulting from either successive simple reactions, separated by discrete transient intermediates, or from a concerted reaction (see Fig. 5.52).

Concerted reactions are single step reactions with a transition state having an uninterrupted chain of bonding (partial bonding) and generally extending over a large part of the molecule. According to this definition all the simple reactions depicted in Fig. A1.6 (apart from 1) are concerted, as well as the more complex ones outlined in A1.7. Often the part of the molecule involved in the concerted process consists of a conjugated system (chromophore), which contain a mobile set of electron pairs within delocalised molecular orbitals: non-conjugated double bonds may also participate provided they exist in the correct stereochemical relationship for interaction to occur (Fig. 5.52(1)).

The choice between a concerted mechanism or a multistep one is one of the most difficult problems to resolve whilst studying *in vivo* reactions.

When tackling such problems, it should be remembered that metabolic processes are usually biocatalysed and that they generally consist of a succession of various reactions, each of them taking place in the active site of a given enzyme. The free intermediates of a process must be characterised: these transfer from the site of the enzyme at which they are formed to the site of another enzyme which will transform them. Such intermediates help to define individual enzymatic reactions: a mechanism for each such enzymatic reaction can be proposed (see Fig. A1.8(a)). The initial stage (enzyme-substrate complex formation) and the final one (enzyme-product complex dissociation) are typical of all enzyme-catalysed processes. Accepting these steps two further question may be posed:

(a) Is the mechanism of the chemical transformation at the enzyme site a single reaction (concerted mechanism) or a multistep one?

(b) If it is a multi-step mechanism, what are the intermediates?

The answer to questions (a) and (b) are very difficult to determine. Even information about the stereochemistry of a reaction is not very useful in order to distinguish between a single step and a multi-step mechanism. The stereoelectronic requirements typical for a concerted reaction in solution can be completely modified by the type of 'solvation' experienced by the substrate at the enzymic active site on the other hand, even a carbocation or a carbanion at the active site may have a very high configurational stability, so that they experience complete reaction stereospecificity and thus *simulate* concerted mechanisms. Two processes are schematised in Fig. A1.8 (cases (b) and (b)), also known *in vivo*. Process (b) refers to enzymes producing a remarkably stable reaction intermediate B, characterised by a relatively large dissociation constant from the enzyme: B can competitively leave the active site, with respect to its further transformation into the compound C (see the enzyme aconitase, Fig. 3.23). In (c) the final product (B) of an enzymatic reaction is directly transferred from its site of formation to that of another enzyme (without exchange with the medium) and there it acts as a substrate of a second reaction (e.g. fatty acid synthetase, see Fig. 3.28).

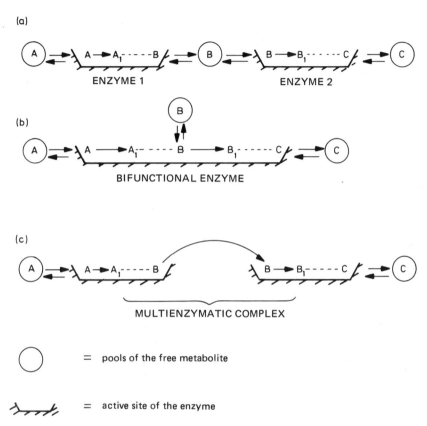

Fig. A1.8 Some biological processes.

SOURCE MATERIALS AND SUGGESTED READING

[1] K. B. Wiberg, (1963), Oxidation-Reduction Mechanism in Organic Chemistry in *Survey of Progress in Chemistry*, (A. F. Scott, Ed.), 1, Academic Press.

[2] D. V. Banthorpe, (1964), Mechanisms of Reactions of Aliphatic Compounds in *Rodd's Chemistry of Carbon Compounds*, 2nd ed. (S. Coffey, Ed.), IA, Elsevier.

[3] D. H. Hey and W. A. Waters, (1964), Free Radicals and Homolytic Reactions in *Rodd's Chemistry of Carbon Compounds*, 2nd ed. (S. Coffey, Ed.), IA, Elsevier.

[4] R. Breslow, (1965), *Organic Reaction Mechanisms,* Benjamin.

[5] D. J. Cram, (1965), *Fundamentals of Carbanion Chemistry,* Academic Press.

[6] C. J. M. Stirling, (1965), *Radicals in Organic Chemistry,* Oldbourne Press.

[7] W. A. Pryor, (1966), Free Radicals, McGraw-Hill.

[8] J. M. Tedder, A. Nechvatal, A. W. Murray and J. Carnduff, (1966-1970), *Basic Organic Chemistry,* Parts 1-3, Wiley.

[9] E. R. Altwicker, (1967), The Chemistry of Stable Phenoxy Radicals, *Chem. Rev.,* **67**, 475.

[10] D. Bethell and V. Gold, (1967), *Carbonium Ions,* Academic Press.

[11] D. J. Cram, (1968), Carbanions in *Survey of Progress in Chemistry,* (A. F. Scott, Ed.), 4, Academic Press.

[12] J. March, (1968), *Advanced Organic Chemistry: Reactions, Mechanism and Structure,* McGraw-Hill.

[13] G. A. Olah and P. Von R. Schleyer (Eds.), (1968-1977), *Carbonium Ions,* 1-5, Wiley-Interscience.

[14] H. Taube, (1968), Mechanisms of Oxidation-Reduction Reactions, *J. Chem. Educ.,* **45**, 452.

[15] B. S. Thyagarajan (Ed.), (1968-1971), *Mechanisms of Molecular Migrations,* 1-4, Wiley-Interscience.

[16] D. Bethell, (1969), Structure and Mechanism in Carbene Chemistry in *Advances in Physical Organic Chemistry,* (V. Gold, Ed.), 7, Academic Press.

[17] T. L. Gilchrist and C. W. Rees, (1969), *Carbenes, Nitrenes and Arynes,* Plenum Press.

[18] W. P. Jenks, (1969), *Catalysis in Chemistry and Enzymology,* McGraw-Hill.

[19] D. V. Banthorpe, (1970), π-Complexes as Reaction Intermediate, *Chem. Rev.,* **70**, 295.

[20] F. G. Bordwell, (1970), Are Nucleophilic Bimolecular Concerted Reactions Involving Four or More Bonds a Myth?, *Accounts Chem. Res.,* **3**, 281.

[21] M. P. Goodstein, (1970), Interpretation of Oxidation-Reduction, *J. Chem. Educ.,* **47**, 452.

[22] R. Huisgen, (1970), Kinetic Evidence for Reactive Intermediates, *Angew. Chem. Int. Ed. Engl.* **9**, 751.

[23] N. S. Isaacs, (1970), Intermediates in Organic Synthesis, *Chem. in Britain,* **6**, 206.

[24] R. W. Alder, R. Baker and J. M. Brown, (1971), *Mechanism in Organic Chemistry,* Wiley-Interscience.

[25] M. C. Caserio, (1971), Reaction Mechanisms in Organic Chemistry: Concerted Reactions, *J. Chem. Educ.,* **48**, 782.

[26] F. G. Bordwell, (1972), How Common are Base-Initiated, Concerted 1,2-Eliminations?, *Accounts Chem. Res.,* **5**, 374.

[27] C. A. Buehler, (1972) Carbenes in Insertion and Additions Reactions,

J. Chem. Educ., 49, 239.

[28] R. A. Jackson, (1972), *Mechanism: An Introduction to the Study of Organic Reactions,* Oxford University Press.

[29] P. Sykes, (1972), *The Search for Organic Reaction Pathways,* Longman.

[30] A. H. Andrist, (1973), Concertedness: A Function of Dinamics or the Nature of the Potential Energy Surface?, *J. Org. Chem.*, 38, 1772.

[31] R. P. Bell, (1973), *The Proton in Chemistry,* 2nd ed., Chapman and Hall.

[32] S. P. McManus (Ed.), (1973), *Organic Reactive Intermediates,* Academic Press.

[33] G. A. Olah, (1973), Carbocations and Electrophilic Reactions, *Angew. Chem. Int. Ed. Engl.*, 12, 173.

[34] R. A. Sneen, (1973), Organic Ion Pairs as Intermediates in Nucleophilic Substitution and Elimination Reactions, *Accounts Chem. Res.*, 6, 46.

[35] G. B. Gill and M. R. Willis, (1974), *Pericyclic Reactions,* Chapman and Hall.

[36] R. L. Huang, S. H. Goh, and S. H. Ong, (1974), *The Chemistry of Free Radicals,* E. Arnold.

[37] G. Klopman, (1974), *Chemical Reactivity and Reaction Path,* Wiley.

[38] J. P. Lowe, (1974), Is This a Concerted Reaction?, *J. Chem. Educ.*, 51, 785.

[39] D. C. Nonhebel and J. C. Walton, (1974), *Free Radical Chemistry,* Cambridge University Press.

[40] E. F. Caldin and V. Gold, (1975), *Accounts Chem. Res.*, 8, 354. *Res.*, 8, 354.

[41] P. Sykes, (1975), *A. Guidebook to Mechanism in Organic Chemistry,* 4th Ed., Longman.

[42] M. A. Wilson, (1975), Classification of the Electrophilic Addition Reaction of Olefins and Acetylenes, *J. Chem. Educ.*, 52, 495.

[43] T. C. Bruice, (1976), Some Pertinent Aspects of Mechanism as Determined with Small Molecules, *Ann. Rev. Biochem.*, 45, 331.

[44] J. M. Harris and C. C. Wamser, (1976), *Fundamentals of Organic Reaction Mechanisms,* Wiley.

[45] A. P. Marchand and R. E. Lehr (Eds.), (1977), *Pericyclic Reactions,* 2 vols. Academic Press.

[46] K. Deuchert and S. Hünig, (1978), Multistage Organic Redox System — A General Structural Principle, *Angew. Chem. Int. Ed. Engl.*, 17, 875,

[47] W. B. Jensen, (1978), The Lewis Acid-Base Definitions: A Status Report, *Chem. Rev.* 78, 1.

[48] D. C. Roberts, (1978), A Systematic Approach to the Classification and Nomenclature of Reaction Mechanisms, *J. Org. Chem.*, 43, 1473.

Principles of Stereochemistry

The structure of an organic molecule is determined by the spatial relationships existing between the component atoms. Two primary relationships exist, the *constitutional* arrangement and the *geometrical* one. The constitution of a molecule is determined by how each atom is bonded to each other. The bonding of atoms occurs by sharing of electrons between them a covalent bond is formed. Modern chemical theory describes chemical bonds in terms of quantum mechanics, in particular, on the concepts of atomic and molecular orbitals.

An atomic orbital (denoted by symbols 1s, 2s, $2p_x$, $2p_y$, $2p_z$, etc) is a mathematical function of the spatial co-ordinates describing each atomic electron. This function (wave function) describes the manner in which the electron is distributed around the nucleus and is associated with an energy value for the electron-nucleus system. The wave function thus gives the density of the electron in the space surrounding the nucleus, treating the electron as a diffuse cloud of negative charge.

Molecular orbitals originate from the interaction between two or more atomic orbitals, each belonging to a different nucleus, and they describe the behaviour of an electron which either moves around the nuclei such as to bind them together (bonding orbitals, such as the σ or π molecular orbitals), or act to keep the nuclei apart (antibonding orbitals, such as the σ^* or π^* molecular orbitals). The energy of an atomic (or molecular) orbital is that lost or gained by the atomic system(s) when an electron moves into that particular orbital from a position of zero-interaction with the nucleus or nuclei. (Zero interaction occurs, for example, when the electron and the nuclei are separated at infinite distance apart).

When a molecular orbital is formed from two atomic orbitals of the same energy, e.g. two $2p_z$ carbon orbitals, it can have an energy smaller, equal or larger than the original two atomic orbitals; these states correspond to a *bonding* orbital, a *non-bonding* orbital or an *antibonding* orbital respectively.

The *constitution* of a molecule refers to its composition in terms of which atoms are combined together to form the molecule. The constitution of a molecule, although describing the composition of it, does not give any information on the three-dimensional manner in which the atoms are held. Molecules comprised of the same atoms but having different constitutions e.g. (3) and (4) are called *constitutional isomers*. Some molecules have the same constitution but different dispositions of the atoms; their steric structure is therefore different and they are called *stereoisomers* (e.g. 5 and 4) (see Fig. A2.1).

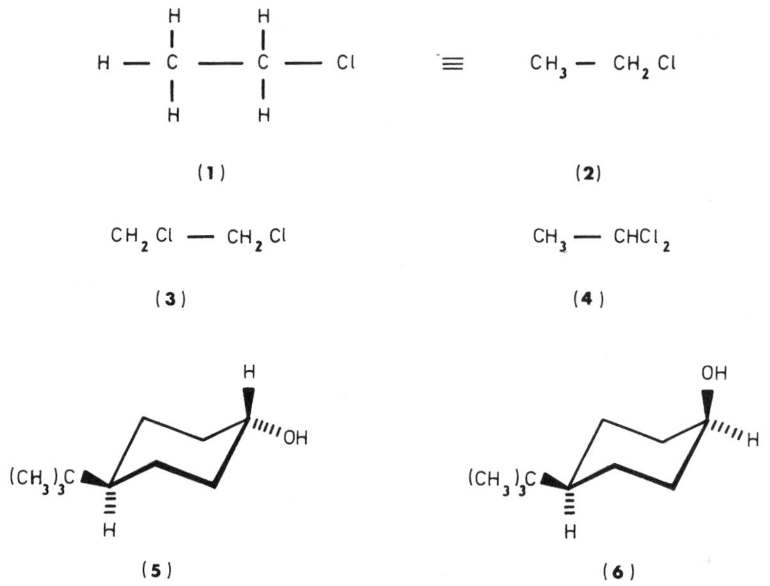

Fig. A2.1 Constitutional formulae, (1) and (2); constitutional isomers, (3) and (4), and stereoisomers, (5) and (6).

The various geometrical relationships between atoms of the same molecule depend on:
(a) the nature of the covalent bonds binding the various atoms;
(b) the intramolecular interactions between atoms which are not directly bonded to each other;
(c) the influence of intermolecular interactions between different molecules (solvation, etc.).

The distances between the nuclei of two bonded atoms is determined by their nature and by the type of bond between them: this distance, *the bond length,* is similar even in different molecules and approximates to the sum of the covalent radii of the bonded atoms (Table A2.1). If two atoms, A and C are covalently attached to a third atom, B, the direction of the bond lengths A-B and B-C subtends a *bond angle,* which depends both on the character of the atoms involved and the type of molecular orbitals existing (see Fig. A2.2).

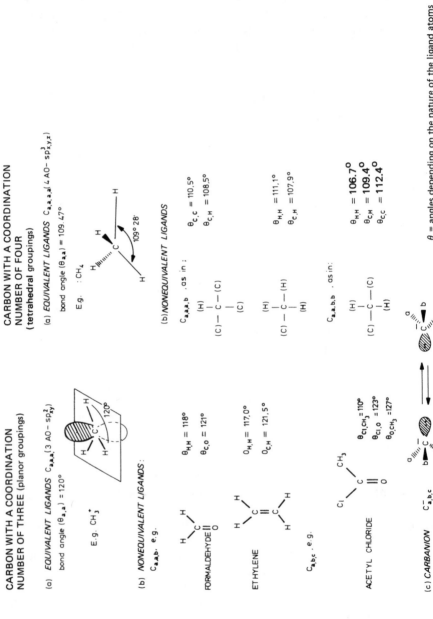

Fig. A2.2 Some typical bond angles in organic compounds.

Table A2.1
Covalent radii (Å) for principal atoms in organic molecules

	Single Bond	Double bond	Triple bond
Hydrogen	0.30	–	–
Carbon	0.77	0.665	0.60
Nitrogen	0.70	0.60	0.55
Oxygen	0.66	0.55	–
Phosphorus	1.10	1.00	–
Sulphur	1.04	0.94	–

The various types of non-bonding interactions, (b) and (c) above, have been categorised (Table A2.2).

Table A2.2
Non-covalent interactions between atoms and groups of atoms

Type of interaction	Potential energy[a]	Order of energy of the interaction kJ mole^{-1}	kcal mole^{-1}
(a) ion-ion	$E \propto -\dfrac{Z_1 Z_2}{\epsilon R}$	40–400	10–100
(b) ion induced dipole	$E \propto -\dfrac{Z_1^2 \alpha_2}{\epsilon R^4}$	4–40	1–10
(c) dipole-dipole	$E \propto -\dfrac{\mu_1^2 \mu_2^2}{\epsilon R^6 kT}$	0–4	0–1
(d) dipole-induced dipole	$E \propto -\dfrac{\mu_1^2 \alpha_2}{\epsilon^2 R^6}$	0–4	0–1
(e) dispersion (London forces)	$E \propto -\dfrac{\alpha_1 \alpha_2}{R^6}$	4–40	1–10
(f) steric repulsion	$E \propto +\dfrac{1}{R^{12}}$	0–40	0–10

[a]Z = electric charge; μ = dipole moment; α = polarisability; ϵ = dielectric constant; T = absolute temperature; k = Boltzmann constant; R = distance between the centres of the interacting systems. Subscripts 1 and 2 indicate the different groups; + represents a repulsive and − an attractive force.

If the interactions (c), (d), (e) and (f) of Table A2.2 are added a function is obtained:

$$E(R) = \frac{A}{R^6} + \frac{B}{R^{12}}$$

where A and B have values depending on the nature of the two interacting atoms. E(R) is the potential energy for the system of 'Van der Waals' forces. A minimum exists at a certain distance of R, usually indicated R_e, when the attractive forces equal the repulsive forces; at this point the distance corresponds to the sum of the two atoms *Van der Waals' radii.* The Van der Waals' radius defines the limits of steric hindrance of an atom and its contact distance to other atoms not directly bonded to it. Some values of Van der Waals' radii are given in Table A2.3; the relative size of atoms in sets of molecular models are related to these values.

Table A2.3
Van der Waals radii (Å) for principal atoms in organic molecules

Hydrogen	1.2	Sulphur	1.85
Nitrogen	1.5	Carbon (from methyl groups)	2.0
Oxygen	1.4	Carbon (from benzene)	1.85
Phosphorus	1.9		

Non-bonding forces are generally rather weak and have little effect on covalent bond lengths and bond angles. They become important between atoms separated by more than two bonds. Rotations are possible about single bonds and non-bonded interactions restrict the number of orientations possible within a molecule. The bond rotations and the restrictions thereon have important steric consequences including the existence of different rotational forms of the same molecules, which are called conformers.

CONFORMATIONAL ANALYSIS

Consider two directly bonded atoms as part of a complex molecule. The atoms will seek to reside at an equilibrium distance, r_e, from each other. The equilibrium position can be disturbed by external forces but will be opposed by repulsion forces as they are pushed closer together and the bonding forces if they are pulled apart. The energy needed to hold a system out of the equilibrium position is stored as potential energy (E_r). For very small displacements this energy can be approximated as:

$$E_r = \frac{1}{2} k_r (\Delta r)^2$$

where Δr is the change in bond length (< 0.1Å for the expression to be valid) and k_r is the linear deformation force constant for a particular bond. The k_r values are approximately constant for similar bond types (see Table A2.4). The energies involved are very high even for small deformations.

A similar situation exists for bond angle deformations. In this case:

$$E_\theta = \tfrac{1}{2} k_\theta \, (\Delta\theta)^2$$

where E_θ is the bond bending energy (angle strain, or Baeyer strain), $\Delta\theta$ is the deviation from the equilibrium angle and k_θ is the bond bending force constant. This expression holds for deformation values of $<20°$. Relatively small amounts of energy are needed to deform the bond angle.

Table A2.4
Force constants for bond distortions

Bond	k_r	(kJ mole^{-1} Å^{-2})	Angle	$k_\theta 10^3$	(kJ mole^{-1} degree^{-2})
C–H	2895	(692)*	H–C–H	29	(7.0)*
C–C	2711	(648)	C–C–C	73	(17.5)
C=O	7280	(1740)	C–C–H	50	(12.0)
C=C	5774	(1380)	C=C–H	62	(14.9)

*Values in parentheses are values in k/cal.

Rotations about a bond, A–B, are described by a relationship between the potential *torsional strain* energy and the rotation angle, ϕ. The angle ϕ is delineated as the projection of two bonds A–X and B–Y (X and Y being arbitrarily chosen amongst substituents of atoms A and B) on a plane cperpendicular to the A–B bond (Fig. A2.3). This dihedral angle expresses the reciprocal rotations of the two groups of atoms at both ends of the A–B bond. For the simplest case of a system which, after a whole revolution reproduces the starting state n times (n > 1):

$$E_\phi = \tfrac{1}{2} V_\phi \, (1 + \cos n\Delta\phi)$$

where V_ϕ is a system specific constant (see Table A2.5), n is the periodicity and $\phi = 0\text{-}360°$. V_ϕ represents the potential energy barrier separating two minima, that is two equilibrium rotamers or 'conformers'. The conformation of a molecule refers to that particular orientation adopted by it at a particular moment in time from amongst every particular orientation which can be attained by the molecule by rotation of each single bond. Because of rotational energy barriers a molecule will tend to adopt preferred conformations. The *conformation* of a molecule should be distinguished from its *configuration*. The *configuration* of a molecule is defined as the relative spatial binding of atoms to one another.

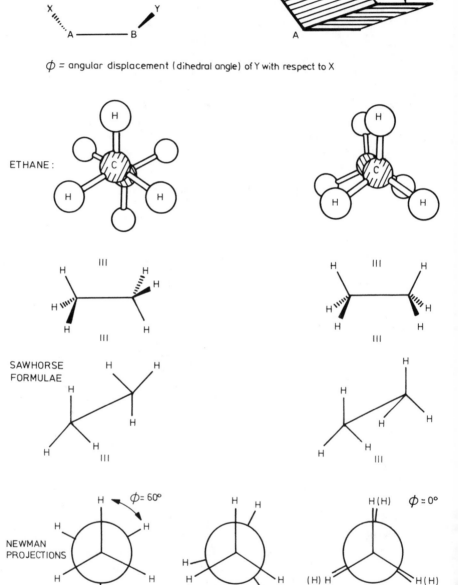

ϕ = angular displacement (dihedral angle) of Y with respect to X

Fig. A2.3 Conventions for representing structures, in different conformations, of ethane and analogues.

There are various reasons for rotational energy barriers. In the case of ethane the potential energy maxima correspond to the eclipsed conformations but this is *not* a steric effect: the distance between the eclipsed hydrogen atoms (2.3Å) is slightly shorter than the sum of their Van der Waals' radii and the estimated repulsion interaction energy between the three hydrogen couples is less than 1.5 kJ mole^{-1} which is much less than the observed value (Table A2.5). It is assumed that the torsional barrier is mainly due to electronic repulsions between the pairs of σ C–H bonds.

Table A2.5
Rotational energy barriers around single bonds

Compound	V_ϕ (kJ mole^{-1})
$CH_3\text{–}CH_3$	12.6
$CH_2\text{–}CH_2\text{–}CH_3$	13.8
$CH_3\text{–}CH(CH_3)_2$	16.2
$CH_3\text{–}C(CH_3)_3$	20.1
$CH_3\text{–}CH_2Cl$	15.0
$CH_3\text{–}OH$	4.5
$CH_3\text{–}OCH_3$	11.3
$CH_3\text{–}NO_2$	0.025
$C_6H_5\text{–}CF_3$	~0

From diagram (a) of Fig. A2.4 we deduce that the ethane molecules are practically all in the skew conformation at room temperature. Microwave spectroscopy shows that they actually rotate ± 120° at a very high frequency (~10^{10} sec^{-1}) around the C–C bond.

Substituents in the ethane molecule cause profound differences. For n-butane (two methyl substituents) the energy diagram is as shown in Fig. A2.4(b). Thus butane has three equilibrium conformations (energy minima). Two of these have the same energy: these are not identical (the two *syn*-clinal enantiomers; see below for definitions) and one is much more stable (the *anti*-periplanar conformer). There are also three energy barriers: two of the same energy, corresponding to the *anti*-clinal conformation and one, the highest energy barrier, to the *syn*-periplanar conformation. In the case of butane at room temperature the energy barriers are not too high to be completely overcome and every molecule can pass through to the different equilibrium populations of conformers.

Further stereochemical terms to be noted are:

Identical objects. Two objects are *identical* or *equal* when they are ideally superimposable to each other, through elementary translation and rotation operations. (Symbol of the identity relation: ≡).

452   **Appendix 2**

Equivalence. Two objects are *equivalent* when indistinguishable under the stated observation conditions.

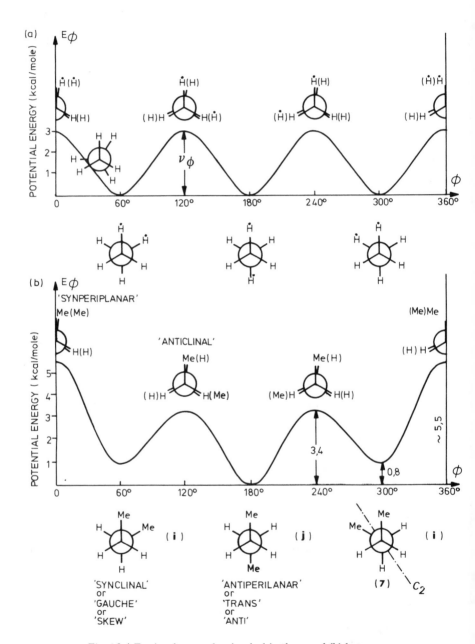

Fig. A2.4 Torsional energy barriers in (a) ethane and (b) butane.

Enantiomers (or *optical antipodes*). This term refers to the topological properties by which two objects, particularly two stereoisomeric molecular structures, exist as mirror images of each other and cannot be superimposed (symbol of the enantiomeric relation: $\frac{\circ}{\circ}$).

Diastereomer (or *diastereoisomer*). Indicates every form of stereoisomer different from enantiomers (symbol of the diastereoisomeric relation: $+\!+$). Two stereoisomers cannot simultaneously be enantiomers and diastereomers.

Chirality. An object is chiral if it can exist in two enantiomeric forms. For this to be possible, it is necessary for it to have no reflection symmetry.

Objects possessing such topological properties are more properly called *dissymmetric*; all other objects which have reflection symmetry and which are superimposable on their mirror image are called *non-dissymmetric*. The analysis of symmetry elements (mirror planes, σ; rotational axes of the order $n > 1$. C_n; alternating rotational axes of order n, S_n; symmetry centres, i) allows one to decide with certainty whether or not the object is dissymmetric (that is chiral). It is dissymmetric if it does not possess any of the above symmetry elements, in which case it also becomes *asymmetric*, or if it *only* has C_n axes (that is, it covers its original image 2,3,4 . . . n times during a whole revolution around the axis itself.) [See (7) having a C_n axis perpendicular to the bond between the two central carbon atoms passing through its middle point.]

The terms dissymmetric and non-dissymmetric refer to objects taken separately; the identity, equivalence, enantiomeric and diastereisomeric relationships imply a comparision between two or more objects. To a chemist the objects are groups of atoms or molecules.

As mentioned, molecules can change their shape with time. Any stereochemical analysis of an organic compound should therefore specify, whenever necessary, if the analysis is conducted on

(a) instant forms (conformers) of the same molecule or of different molecules;

(b) compounds comprised of a set of freely interchanging conformers, in which case their average conformation can be considered.

The latter situation is usually encountered by the bio-organic chemist and it is, therefore, appropriate to extend the different concepts (identity, enantiomerism, etc.) to compounds having free internal rotations.

(1) The superimposability (identity) of two non-rigid systems should be checked by making the appropriate spatial translations and rotations, including internal rotations about bonds possessing rotational freedom.

(2) The enantiomeric relationship implies that *every* conformation freely adopted by a compound has a corresponding *enantiomeric* conformation.

(3) The diastereomeric relationship implies that there will be no superimposability or enantiomerism between any of the conformations of the two stereoisomers.

(4) A compound is dissymmetric only if *all* of its conformers, which are possible under the observation conditions, are dissymmetric.

Returning to the case of butane, it is possible to treat the interconversion of conformers thermodynamically. At equilibrium, at the absolute temperature, T (see Fig. A2.4):

$$(\text{skew conformer, } i) \underset{k_{j,i}}{\overset{k_{i,j}}{\rightleftharpoons}} (\text{anti-conformer, } j)$$

$$\frac{k_{ij}}{k_{ji}} = K_{eq} = \frac{[j]}{[i]} = \exp\left(-\Delta G^{\circ}_{i,j}/RT\right)$$

where R = gas constant = 1.987 cal mole^{-1} $^{\circ}$K^{-1} and T = absolute temperature in $^{\circ}$K. Thus

$$\Delta G^{\circ}_{i,j} = \Delta H^{\circ}_{i,j} - T\Delta S^{\circ}_{i,j}$$

$\Delta H^{\circ}i,j$ is the potential energy difference between the two conformers, that is $E_j - E_i = 0.8$ kcal mole^{-1}. As far as the entropy term is concerned there should not be any significant differences between the two conformers i and j due to vibrations or solvation variations, etc; the respective energy states only differ in their number of degenerate energy levels: i has two geometric forms (enantiomers) of equal probability, whilst j has only one form. Thus:

$$\Delta S^{\circ}_{i,j} = -\text{Rln } 2$$

and we obtain:

$$\Delta G^{\circ}_{i,j} = (E_j - E_i) + \text{RTln } 2$$

At 25°C (298°K)

$$\Delta G^{\circ}_{i,j} = -800 + (1.987 \times 298 \times 0.693) = -390 \text{ cal mole}^{-1}$$

from which $K_{eq} = 1.93$. At 25°C this means that 66% of the butane molecules exist in the *anti*-periplanar form and 17% are in each of the two *syn*-clinal forms (see Table A2.6).

The *rate* of conversion of the i conformers into the j conformers and *vice-versa* can be estimated using the Eyring equation:

$$R = \frac{k.T}{h} \exp\left(-\Delta G^{\ddagger}/RT\right)$$

where k is the rate, k is the Boltzmann constant $(1.38 \times 10^{-33}$ J.$^{\circ}$K$^{-1})$, h is Planck's constant $(6.624 \cdot 10^{-34}$ J.s) and $\Delta G^{\ddagger}$ is the free energy of activation.

Since $\Delta G^{\ddagger} = \Delta H^{\ddagger} - T\Delta S^{\ddagger}$,

$$k = \frac{k.T}{h} \exp(-\Delta H^{\ddagger}/RT) \exp(\Delta S^{\ddagger}/R).$$

where the activation enthalpy $\Delta H^{\ddagger}$ is equal to the potential energy barrier and $\Delta S^{\ddagger}$ can be calculated on the same basis as used in calculating the equilibrium constant (i.e. comparing the energy degeneracy of the starting and transition states).

Thus: $k_{i,j} = 7.69 \times 10^{10} \, s^{-1}$ at $25°C$

where $\Delta H^{\ddagger} = 3400$ cal mole^{-1} and $\Delta S^{\ddagger} = R \ln 2$

Table A2.6

Relations between free energy differences at standard conditions* at $25°C$ ($\Delta G°_{298 °K}$), equilibrium constants and percentages of the more stable isomers in equilibria of the type $i \rightleftharpoons j$.

$\Delta G°_{(298 °K)}$ cal mole^{-1}	K_{eq}	% of the more stable isomer
0	1.00	50
119	1.22	55
240	1.50	60
367	1.86	65
502	2.33	70
651	3.00	75
821	4.00	80
1028	5.67	85
1302	9.00	90
1744	19.00	95
2722	99.00	99
4092	999.00	99.9
5456	9999	99.99

*standard conditions for i and j = 1M concentrations.

The major differences between the rotational potential energies in butane and ethane are possibly attributable to the different contributions of the Van der Waals' repulsion forces. In the butane *syn*-periplanar conformation the two methyl groups are strained against each other at a distance (2.6 Å) much shorter than the equilibrium distance (which is the sum of their Van der Waals' radii of 4Å). Analogous situations, although less dramatic, exist in the *anti*-clinal or *syn*-clinal conformers. It is possible to compute the conformational energies of molecules more complex than *n*-butane.

The conformational energy is obviously the excess of potential energy of a given conformation with respect to the more stable conformation of that molecule. One can calculate the potential energies (E) of all the possible conformations with respect to an *ideal* conformation, characterised by 'normal' bond lengths and angles (r and θ) (see Table A2.1 and Fig. A2.2), by torsional angles, $\phi = 0$, and by the absence of non-bonding forces. The total E will result from the following expression:

$$E = \sum_i (E_r)_i + \sum_j (E_\theta)_j + \sum_k (E_\phi)_k + \sum_l (E_s)_l + W$$

$\sum_i (E_r)_i$ is the sum of the linear distortion energies of all the bonds; analogously

$\sum_j (E_{re})_j$ represents the sum of the 'Baeyer strain' energies (see above) and $\sum_k (E_\phi)_k$

the sum of the 'Pitzer strain' energies (see Fig. A2.4); $\sum_l (E_s)$ is the sum of the

'steric strain' energies, that is of all the interaction of atoms non-covalently bonded to each other (see Table A2.3), excluding those already included, such as Pitzer strain energies. (For a hydrocarbon molecule the 'steric strain' energies mainly involve Van der Waals' forces.) The factor W includes every other energy type, for instance hydrogen bonds, electronic interactions between chromophores, etc.

Once the various E values have been defined, the minima can be determined: the relative conformational energies are obtained with respect to this value.

The above procedure is realisable only in particularly favourable situations, for instance with saturated hydrocarbons. Where more complex molecules are involved, estimated value E can be used to evaluate the difference in conformational energy between two conformers, not very different from each other, (we can call these α and β), if the geometrical variations between the two structures are precisely known. In these cases, the expression $(E_\beta - E_\alpha)$ is comprised of a few parameters, whose numerical values may be calculated with a fair degree of accuracy. The numerous terms corresponding to equal distances or interactions in the two conformations in fact cancel each other and it is therefore unnecessary to calculate them.

If the geometry-energy relationships between atoms, described above, are accounted for, organic compounds can be classified according to their intramolecular rotational 'freedom'. Four types of molecules can be distinguished, and each type features a particular *torsional stereoisomerism.*

(1) Molecules in which there is complete rotation around all the bonds at room temperature, at a high frequency (*free rotation*). The isolation of different conformers depends on the height of the energy barrier to bond rotation. The limit for the 'freezing' of different conformations, that is to give

different conformers having a reasonable life-time, requires energy barriers of *ca.* 20 kcal mole^{-1} at room temperature. The limit for freezing out conformations is dramatically lowered as the temperature is decreased, e.g. is only *ca.* 13 kcal mole^{-1} at $-70°C$. Butane represents one example in this class of compound.

(2) Molecules with a single C–C bond in which two rotational barriers exist (>20 kcal mole^{-1} at room temperature) to give *atropoisomers*. Certain diphenyl derivatives bearing large substituents in the *ortho*-positions belong to this class. The *ortho*-substituents introduce large Van der Waals' repulsive forces. (The energy maxima occur in the conformation in which the phenyl rings try to become co-planar). Since rotation is restricted the two atropoisomers *can* be dissymmetrical (Fig. A2.5) and therefore become enantiomeric to each other. The condition for chirality is that neither ring has a plane of symmetry perpendicular to the ring itself [a $\neq$ b and c $\neq$ d in (8)].

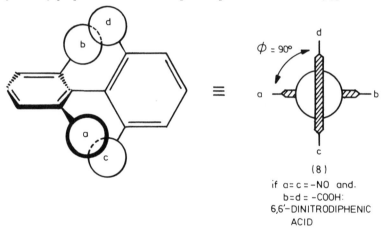

$\phi = 90°$

(8)

if a= c = –NO and.
b= d = –COOH:
6,6'–DINITRODIPHENIC
ACID

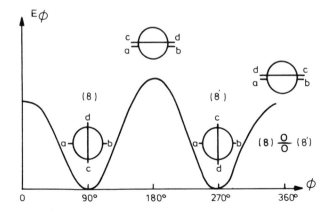

Fig. A2.5 Atropoisomerism: biphenyl derivatives.

(3) Molecules with double bonds: the rotation is prevented by higher torsional barriers. These reach 60 kcal mole^{-1} for a $\supset$C$=$C$\subset$ bond. These maxima correspond to conformations having the two carbon atoms and their substituents on perpendicular planes: the conformations corresponding to minimal energies are characterised by carbon atoms and substituents lying on the same plane. (This constitutes *geometrical* or *cis-trans isomerism,* see below.)

(4) Cyclic molecules, in which the rotations around certain simple bonds belonging to the cyclic skeleton, are limited by the strong angular, torsional, and steric strains associated with ring deformations. A familiar example is the cyclohexane ring, as it frequently appears in natural substances (terpenes, steroids, etc.).

The different conformations (stable and transitional ones) of cyclohexane and their places along the potential energy diagram are indicated in Fig. A2.6. The most stable conformation of the saturated cyclohexane ring is the *chair* form. It is rather rigid, and non-dissymmetric.

For cyclohexane at 25°C there are calculated to be about 10,000 molecules existing as chair conformers for every twist conformer. A cyclohexane molecule in the chair conformation passing through various twist conformations, goes back to its identical, initial conformation (*degenerate interconversion*) at a rate of 10^5 conversions per second at room temperature.

The different stabilities of the cyclohexane conformers are due to energy contributions of various types. For instance all the C-C systems are staggered in the chair conformation [compare (9a)] (with a minimum of torsional energy), whilst they are skewed in the twist conformation; the two hydrogen atoms of the lateral carbon atoms are eclipsed in the *boat* form (Figs. A2.3 and A2.4(a)). A strong steric strain also exists between the two hydrogens at the two ends of the boat, facing in towards each other (the interatomic distance is 1.83 Å, compared to the sum of their Van der Waals' radii of 2.4 Å).

The actual shape of the cyclohexane chair is probably silightly flatter than that calculated from molecular models having perfectly tetrahedral, sp^3-carbon atoms. Flattening slightly increases the C-C-C angle from the tetrahedral value of 109.5° to approximately 111.5°, (Fig. A2.2), and pushes apart the 1,3-diaxial hydrogens from the central axis (C_3): their unfavourable steric interaction is thus removed. If the cyclohexane ring carries substituents, their reciprocal interactions, as well as their interactions with the axial hydrogens (mostly Van der Waals' interactions) can radically modify the energy profile of Fig. A2.6. Two possible chair diastereomeric forms are possible in cyclohexane [(9) and (10), with H = CH$_3$], having different potential energies: (10) is more stable than (9) by 1.7 kcal mole^{-1} and, therefore, is by far the more abundant, see Table A2.6. Usually a substituent tends to occupy the equatorial position and the stability relative to the one with the axial substituent largely depends on the dimensions of the substituent and the length of its bond with the ring carbon (see Table A2.7).

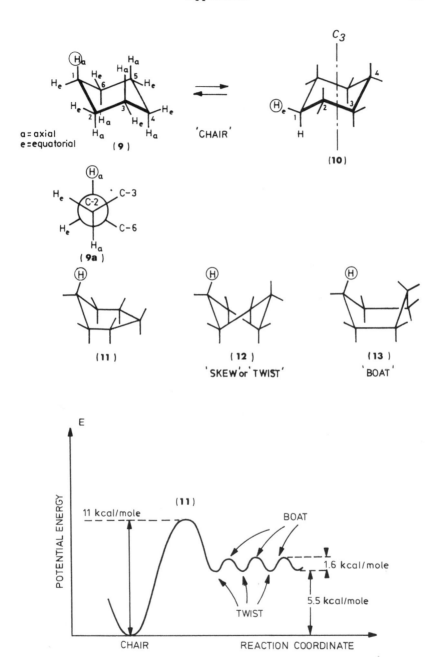

Fig. A2.6 Conformations of cyclohexane.

Table A2.7
Free energy differences between monosubstituted cyclohexanes ($C_6H_{11}X$);
$$\Delta G^\circ = G^\circ(X_{axial}) - G^\circ(X_{equatorial})$$

ΔG° (kcal mole^{-1})	Substituent (X)
1.7	$-CH_3$
1.8	$-C_2H_9$
2.1	$-CH(CH_3)_2$
~5	$-C(CH_3)_3$
3.1	$-C_6H_5$
0.7	$-OH$
0.4	$-Cl$

In Fig. A2.7 the stable conformations of the system formed by two condensed cyclohexane rings (decalin) are shown. The *trans*-fused isomer corresponds to a unique rigid conformation, whilst the *cis*-isomer can adopt two enantiomeric and interchangeable conformers. Analogous considerations apply to polyclic saturated systems (perhydrophenanthrenes, steroids, etc.).

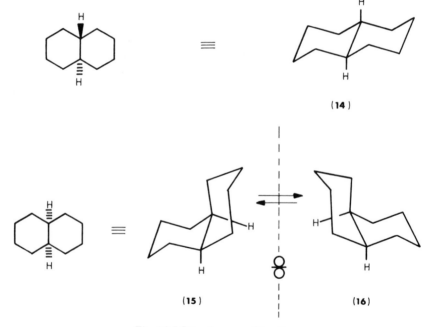

Fig. A2.7 Stereoisomers of decalin.

GEOMETRICAL STEREOISOMERISM

As already mentioned, the rotation around olefinic double bonds is hindered by very high torsional barriers. The *two* carbon atoms united by the double bond, and the other *four* atoms directly bonded to them are arranged in the same plane (σ symmetry plane: see Fig. A2.8). The planar arrangement of the groups a, b, a$'$, b$'$ is an electronic consequence of the shape of the molecular π orbital, which results from the interaction of two $2p_z$ atomic orbitals, parallely oriented, see (17). This geometrical property of olefinic systems causes a stereoisomerism known as *geometrical* or *cis-trans* stereoisomerism A $C_{a,b} = C_{a',b'}$ system (where a is different from b and a$'$ different from b$'$) can exist in two forms, (18) and (19). They have different potential energies, the non-bonding interactions (that is steric interactions) between the groups bonded to the two central carbons being different in the two cases (see the diagram 'potential energy versus torsional strain' of Fig. A2.8). The interconversion between the two diastereomers is separated by a high energy barrier, thus requiring high temperatures, photochemical excitations, or breaking and reforming of the double bond (for instance by attachment and detachment of a radical species to one of the olefinic carbon atoms). For these reasons the two diastereoisomers, commonly called *geometrical isomers,* are two different chemical compounds. The necessity of denoting them individually and non-equivocally, using the IUPAC rules of rational chemical nomenclature to define their constitutions, led to the use of the *cis* and *trans* terms. These prefixes indicate the relative positions of two ligands, each chosen between the two couples on the double bond ends [that is either a or b, and a$'$ or b$'$): *cis* means that both the chosen substituents are on the same side with respect to the longitudinal axis of the double bond, (20) for instance, referring to the olefin hydrogens]; *trans* indicates the opposite situation, compare (21). The limits of such a symbolism are readily apparent as it becomes necessary to define each time the two substituents which are in the *cis-* or *trans-*positions. A preferable nomenclature is to assign the configuration of a geometrical isomer with the (Z) and (E) symbols; this notation unequivocally defines (without further specifications) the configuration of the compound in question, once the general conventions described below are adopted.

The conventions for using (Z) and (E) are:

(1) Given the system $X_{a,b} = Y_{a',b'}$ (being X = Y = C for olefins), the priorities of each ligand pair a, b and a$'$, b$'$ are derived according to the Cahn-Ingold-Prelog sequence rules (see p. 000).

(2) If the two ligands of highest priority are on the same side with respect to the X=Y bond, the isomer is Z (from the German, *Zusammen* = together); when they are opposite the isomer is E (from the German *entgegen* = opposite). The structures (20)–(23) in Fig. A2.8 illustrate this nomenclature system.

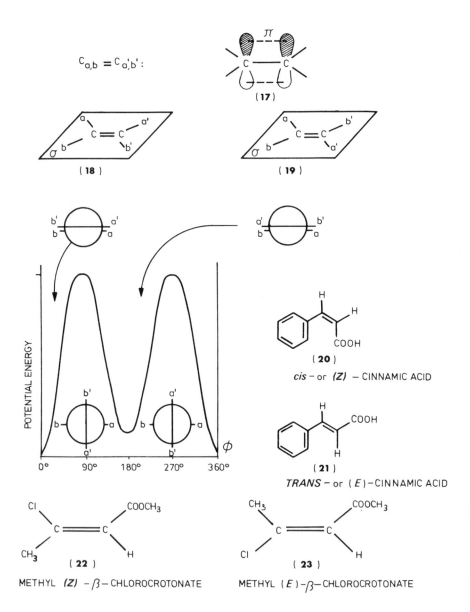

Fig. A2.8 Geometrical stereoisomers.

NON-TORSIONAL STEREOCHEMISTRY. COMPOUNDS CONTAINING
ASYMMETRIC CARBON ATOMS

The ligands of a tetracoordinated carbon atom are at the apices of a tetrahedron
and form bond angles with the central carbon of about 109.5°.
Four *different* points fixed to the apices of an ideal tetrahedron form an
asymmetric set from a topological point of view: subsequently two *enantiomeric*
forms are possible for every $C_{a,b,c,d}$ type carbon, where a, b, c, and d are either
different atoms, or different groups of atoms (see Fig. A2.9).
The stereoisomerism connected with asymmetric carbon atoms is distinguished
from the torsional one by the type of barrier opposing the stereoisomer interconversion. For asymmetric carbon atoms the configuration inversion, that is the
transformation of $C_{a,b,c,d}$ into its mirror image $C'_{a,b,c,d}$ is only possible through
the breaking and the reforming of at least one of the bonds to that carbon atom.
A fundamental stereochemical principle states: 'The chirality of an object
cannot be specified, without comparing it to another standard object, which is
itself dissymmetrical'. Two enantiomeric objects, O and O', possess both the
same qualities, having identical spatial relationships among the elements constituting them; if a second dissymmetric object is taken, in one of its enantiomeric
forms, either P or P' (let us take the P form) and associated separately to O and
O', there will be obtained two qualitatively *different* objects (O-P and O'-P).
These will then be diastereomers, because one cannot be superimposed on the
other one, either as such, or as its mirror image [see, for instance, the formulae
(26) and (27), Fig. A2.9].
A dissymmetrical object is necessary for comparing, recognising, and eventually
specifying chirality: it can be a hand, the movement of a watch hand etc. *Circularly
polarised* light is the dissymmetric agent commonly used for evaluating the
chirality of molecules. The topological property of helicity is associated with
this type of rotation: its direction of propagation is as either a right or left-handed
screw.
When interacting with chiral molecules the propagation rate of the polarised
light beam alters, as well as its degree of absorption: these variations differ
according to the direction of polarisation (left or right) and the specific chirality
of the molecules impinged upon by such a beam. The more frequently observed
and measured phenomenon is the rotation of the *polarisation plane* of a light
beam obtained by combining two beams *in phase,* circularly polarised, one
opposite to the other (*optical rotation,* O.R., see Fig. A2.10).
Other chiroptical phenomena (Cotton effects) useful for specifying the
molecular chirality are (a) optical rotation variations with the wavelength of the
impinging polarised beam (*optical rotatory dispersion,* O.R.D.) and (b) differences
between the molar extinction coefficients of the two circularly polarised beams
$(\epsilon_L - \epsilon_R)$ as a function of the wavelength (*circular dichoism.* C.D.).

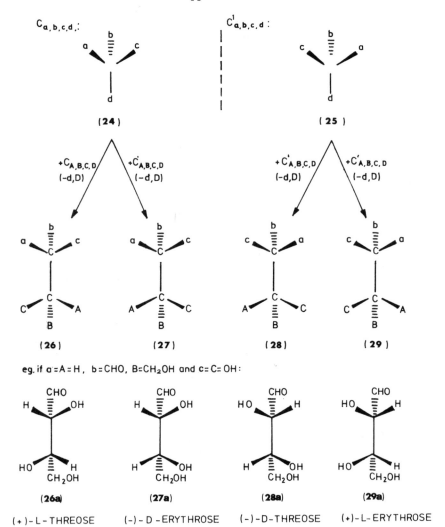

eg. if a=A=H , b=CHO, B=CH₂OH and c=C=OH :

(26a)   (27a)   (28a)   (29a)

(+)-L-THREOSE   (-)-D-ERYTHROSE   (-)-D-THREOSE   (+)-L-ERYTHROSE

RELATIONSHIPS BETWEEN STEREOISOMERS (see text for symbols)

Fig. A2.9 Stereochemistry of compounds containing one or two asymmetric carbon atoms.

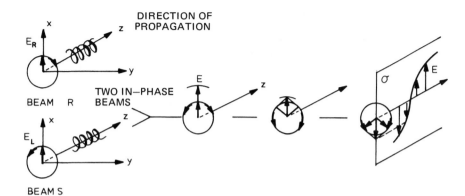

[CIRCULARLY POLARIZED LIGHT] [PLANE–POLARIZED LIGHT]

E, E$_R$, E$_L$: ELECTRIC FIELD VECTORS σ : PLANE OF POLARIZATION

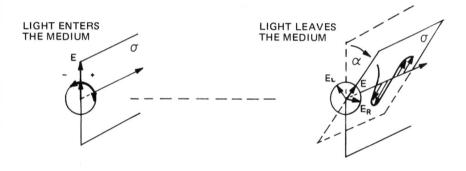

$$\alpha = \frac{180 \cdot}{\lambda}(n_L - n_R)$$

α = angle of rotation in degrees per centimeter
λ = wavelength of monochromatic light
n_L and n_R = refractive indices of the two circularly
   polarized beams

For solutions:

$$[\alpha]_\lambda^T = \frac{\alpha_\lambda^T \times 100}{l \times c}$$

$[\alpha]_\lambda^T$ = specific rotations (in degrees) at wavelength
   λ and temperature T (remember that $[\alpha]_D = [\alpha]_{589}$)

$$[\phi]_\lambda^T = \frac{M}{100}[\alpha]_\lambda^T$$

l = path length (in dm.)
c = concentration (in g/100ml of solution)
M = molecular weight

Fig. A2.10 Chiroptual effects in dissymmetric compounds.

Racemate and Enantiomeric Purity

A homogeneous substance comprised of equimolar amounts of two enantiomers is called a *racemate*. A mixture of equimolar amounts of enantiomeric molecules present as separated solid phases is called a *racemic mixture*.

For a non-equimolar mixture of enantiomers the enantiomeric purity of the *more* abundant enantiomer is given by the expression the *enantiomeric excess*. This ratio is

$$\frac{\left(\begin{array}{l}\text{moles of the more abundant}\\ \text{enantiomer}\end{array}\right) - \left(\begin{array}{l}\text{moles of the less abundant}\\ \text{enantiomer}\end{array}\right)}{\text{total moles of both enantiomers}}$$

The percent enantiomeric excess may be determined by optical rotation measurements:

$$\text{optical purity} - \frac{\text{specific rotation of the mixture}}{\text{specific rotation of pure enantiomer}} \times 100$$

A racemate has zero enantiomeric and optical purity.

The number of stereoisomers possible for a compound with n-asymmetric centres is given by the formula 2^n [the number of racemic compounds being $2^{(n-1)}$]. If there is restriction to complete rotation about the bonds of the n-asymmetric carbon atoms, the number of possible stereoisomers may be larger or smaller than 2^n. The first possibility takes place when, besides stereoisomerism due to the tetrahedral asymmetric centres, torsional stereoisomers are also present; 1,2-disubstituted cyclohexanes, with $A \neq B$ are examples (Fig. A2.11(a)). In this case both of the chair equilibrium conformers are possible stereoisomers, provided the ambient conditions prevent conformational inversions (for instance, low temperature). A reduction in the possible number of stereoisomers can however be found in rigid systems (Fig. A2.11(b).

If the two central carbon atoms in (26)–(29) are constitutionally equal, that is a = A, b = B, and c = C, the number of possible stereoisomers is 3 instead of 4 (as for tartaric acid, see Fig. A2.12). Generally, the constitutional symmetry in molecules with n (even) asymmetric carbon atoms limits the number of possible stereoisomers to $2^{n-1} + 2^{\frac{1}{2}(n-2)}$. The three isomeric forms of tartaric acid are represented by a pair of enantiomers, (40) and (41), corresponding to (26) and (28), and by a non-dissymmetric structure, called the *meso*-form (42) corresponding to (27) ≡ (29). All the *meso*-form compounds show the properties of non-dissymmetric objects (optical inactivity).

There is further possibility of a constitutionally symmetric molecule with n (even) number of asymmetric carbon atoms and a central carbon atom $C_{A,A,b,c}$, where A,A represent two chiral portions of the molecule of equal constitutions.

The two portions can be sterically different when compared in isolation, that is diastereiomeric or enantiomeric. In the latter case, the substituted carbon atom is called *pseudo-asymmetric* $[C_{(+a), (-a), b, c}]$. The possible number of stereoisimers are 2^n as a whole: $2^{n/2}$ as *meso* forms, optically inactive. On observing the pentaric acids (Fig. A2.12) one can see, in the two *meso*-forms, (45) and (46), how the two constitutionally equal groups bonded to the pseudo-asymmetric carbon (marked with a star) are related as two enantiomers. The central carbon of the dissymmetric forms is, instead, a $C_{a,a,b,c}$ carbon (non-asymmetric). Heptaric acids can also have stereo-isomeric forms with the central carbon atom bonded to two diastereomeric substituents ($C_{a,a',b,c}$: asymmetric).

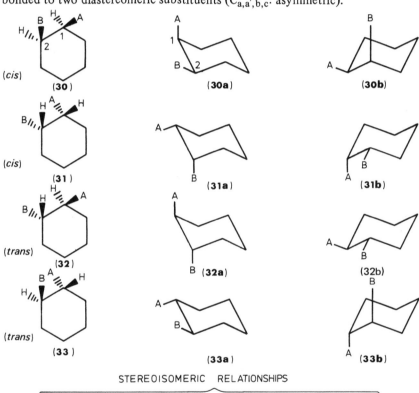

STEREOISOMERIC RELATIONSHIPS

If conformational interconversions are exceedingly rapid	If conformational interconversions are blocked
(30) $\frac{Q}{Q}$ (31); (32) $\frac{Q}{Q}$ (33); (30)⊬(32)etc.	(30a)⊬(30b); (30a)⊬(31a) etc.
if A = B : (30) ≡ (31), *meso*–form	(30a) θ (31b); (30b)θ(31a);
	(32a) θ (33b); (32b)θ(33a);
	if A=B (30a)≡(31a); (30b) ≡ (31b)
	(30a)θ(30); (31a)θ(31b)
Fig. A2.11 (contd. overleaf)	(32a) θ(33b); (32b) θ(33b)

Fig. A2.11
(contd.)

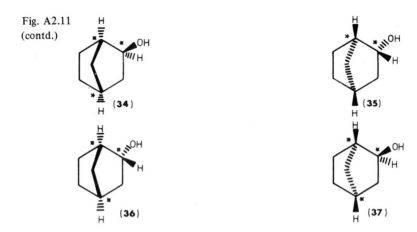

*ASYMMETRIC CARBON ATOM

(34) ⧧ (35) ; (36) ⧧ (37) ; (34) ⊬ (36) etc.

Fig. A2.11 Examples of stereoisomerism for cases of restricted rotation involving asymmetric carbon atoms.

The principal conventions and notations used for describing the stereochemistry of compounds with asymmetric atoms, especially to indicate the absolute configuration of their asymmetry, are as follows:

Formulae

Besides the usual steric formulae [for examples see (38a) and (39a)], *Fischer projections* are often used to depict acyclic compounds having free rotation around the bonds between the various asymmetric carbon atoms. The relation between these formulae and the actual steric situation of the molecule is illustrated by the correspondence between the formulae (38a-d) and (39a-d) of Fig. A2.12. Fischer formulae are extremely simple and clear and are particularly useful for representing molecules of polyhydroxylated compounds correlated to the carbohydrates (see the formulae of tartaric and pentaric acids and of glucose in Fig. A2.12). Fischer formulae are very useful for distinguishing the *threo*-isomers from *erythro*-ones. The *threo/erythro*-notations apply to systems such as $R-C_{a,b}-C_{a,c}-R'$ (being R and R' alkyl or aryl radicals equal or different to each other), and take their names from the two tetraldoses, threose (26a) and (28a), and erythrose (27a) and (29a).

The Fischer projection is obtained by drawing along the vertical line R, R' (or b and c) and the two central carbon atoms; the isomers having the two equal substituents (a) on the same side is called *erythro*; in the opposite case, *threo*. If the sawhorse formulae, and the Newman projections, relative to the two central carbon atoms (Fig. A2.2) are used, the eclipsed form with R and R' (or b and c) one in front of the other must be considered. The formulae of the ephedrines (Fig. A2.12) are typical examples.

D – (+) – GLYCERALDEHYDE [or (R) – (+) – GLYCERALDEHYDE,]

CHO   OH

C

HOH₂C   H

(38a) ≡

CHO

H — C — OH

CH₂OH

(38b) ≡

CHO

H —— OH

CH₂OH

(38c) ≡

CHO

H – C – OH

CH₂OH

(38d)

FISCHER PROJECTION FORMULAE

L –(–) – GLYCERALDEHYDE [or ( ) –(–) – GLYCERALDEHYDE]

CHO   H

C

HOH₂C   OH

(39a) ≡

CHO

HO — C — H

CH₂OH

(39b) ≡

CHO

HO —— H

CH₂OH

(39c) ≡

CHO

HO – C – H

CH₂OH

(39d)

FISCHER PROJECTION FORMULAE

TARTARIC ACIDS :

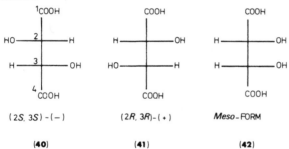

$$
\begin{array}{c}
^1COOH \\
HO-\overset{2}{\underset{}{|}}-H \\
H-\overset{3}{\underset{}{|}}-OH \\
_4COOH
\end{array}
$$

(2S, 3S) - (–)

(40)

$$
\begin{array}{c}
COOH \\
H-|-OH \\
HO-|-H \\
COOH
\end{array}
$$

(2R, 3R) - (+)

(41)

$$
\begin{array}{c}
COOH \\
H-|-OH \\
H-|-OH \\
COOH
\end{array}
$$

Meso - FORM

(42)

PENTARIC ACIDS :

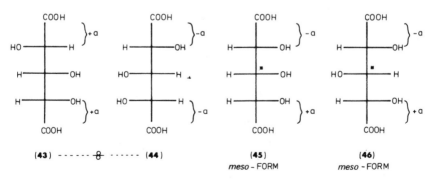

(43) - - - - - - 8 - - - - - - **(44)**

(45)
meso - FORM

(46)
meso - FORM

Fig. A2.12   (contd. overleaf)

Fig. A2.12 (contd.)

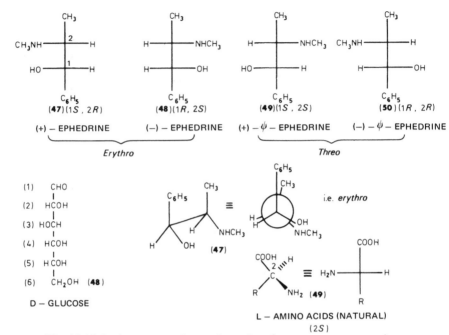

* PSEUDOASYMMETRIC CARBON ATOM : (*r*) in **(45)** e (*s*) in **(46)**

Fig. A2.12 Steric representations and notations for some common organic compounds containing asymmetric carbon atoms.

Symbols

The property of substances to rotate the plane of polarised light towards the right ($[\alpha]_D^{25} > 0$) or towards the left ($[\alpha]_D^{25} < 0$) is respectively indicated with the prefixes (+) (or *d-*, dextrorotatory) and (−) (or *l*, laevorotatory). Steric structures or molecules *cannot* be deduced immediately and unequivocally by their prefixes. (The specific rotation can change its absolute value and its sign simply by changing the solvent, although the configurations of the asymmetric centres do not change!)

In the past the symbols D and L have been used to specify the configurations of asymmetric carbon atoms, CH.R.R′.X, where X is a heteroatom. The use of these symbols is ruled by the following conventions:

(a) The Fischer projection must be oriented so that carbon atom 1 of the main chain, (or the more oxidised carbon when ambiguity arises) is placed at the top of the vertical line.

(b) The D configuration should be assigned to the carbon when the X group is viewed to the *right* hand side and the L configuration when the X group is viewed to the *left* hand side.

According to these rules, glyceraldehyde, with the absolute configuration indicated in (**38a**), is D (the symbol refers to the only asymmetric centre C-2) and its enantiomer (**39a**) is obviously L; natural alanine (**49**, R = CH₃) is L, as are all the main natural amino-acids (the carbon atoms bearing the amino and carboxylic groups are referred to); natural glucose (**48**) is D: the symbol refers in this case, as well as in all the monosaccharides, to carbon atom 5, that is in the position next to the last one according to the normal numbering of the carbo-hydrate chain.

The limits of the the D, L symbolism are clearly shown when applied to compounds in which X is not a heteroatom, or to systems such as C.R.R′.R″.X. In these cases a further convention must be introduced: the *main chain* of the carbon atoms of the type $C_{a,b,c,d}$. R. S. Cahn, C. K. Ingold and V. Prelog over-came these limitations by preparing a new system for classifying asymmetric centres in 1956: this convention is rigorous and simple and it has since been adopted among the 'Recommendations of the IUPAC Commission for Nomen-clature in Organic Chemistry, Section E: Fundamentals of Stereochemistry'.

The Cahn-Ingold-Prelog system is known as the 'Sequence Rule Procedure' and correlates the chirality of a structure (particularly of an asymmetric carbon $C_{a,b,c,d}$) to a symbol, usually *R*, or *S*; depending on the nature of the four ligands a, b, c, and d, and their distribution around the apices of an ideal tetrahedron. The ligands are assigned priorities. The conventional operations are now described (see Fig. A2.13).

1. The ligands of a tetrahedral atom $X_{a,b,c,d}$ are disposed according to a *preferential* order (also called *precedence* or *priority order*), based on to a precise sequence rule. If the preferred ligand is indicated by a, the second one by b, the following one by c, and the last one by d, the order is schematised by the notation a>b>c>d.

2. The tetrahedron is oriented so that the three ligands of the highest priority, a, b, and c, are arranged in space towards an observer: the last ligand, d, must be placed behind the plane formed by the first three, with respect to the observer.

3. The observer traces a line between the three ligands, starting from the pre-ferred one, a, and ending at the third one, c, after passing through the intermediate one, b. If a clockwise line is traced, the chirality of the tetrahedral asymmetric group $X_{a,b,c,d}$ is *R* (from the Latin, *rectus*, right); in the opposite case *S* (from the Latin, *sinister*, left).

The *sequence rule* should now be explained. It is comprised of a certain number of sub-rules, which must be applied in a rigorous order until a conclusive decision is achieved:

(1) The priority order of the ligands is determined by the atomic number of the atoms directly bonded to the chiral centre: *the larger atomic number atom* is the preferred one. In 1-bromo-1-chlorethane the preference order is Br>Cl>CH₃>H [see formula (**50**), Fig. A2.13].

(2) If two or more atoms examined have the same atomic number, the substituents on these should be examined and compared, to decide their preferential order. Some simple examples will be given (Fig. A2.13): the preferential order decision generally concerns a choice only between *two* ligands; the evaluation criteria and the procedure used can also be used for uncertain situations concerning three or four ligands.

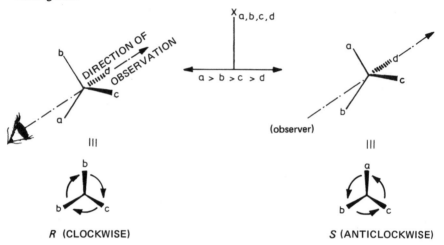

R (CLOCKWISE)   *S* (ANTICLOCKWISE)

EXAMPLES IN APPLYING THE SEQUENCE RULE:
(* indicates the tetrahedral asymmetric centre whose ligands must be arranged in order of decreasing priority)

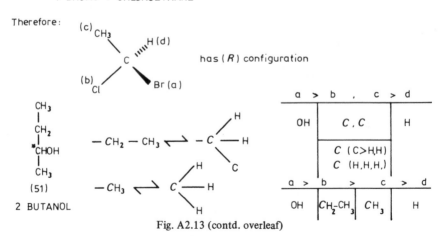

Fig. A2.13 (contd. overleaf)

Fig. A2.13
(contd.)

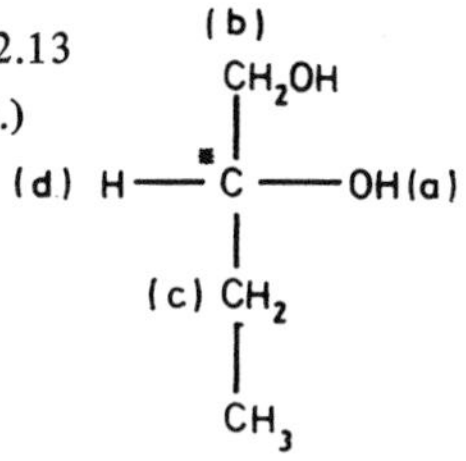

(b)
CH₂OH

(d) H —— C —— OH (a)

(c) CH₂

CH₃

(52)

1,2–BUTANEDIOL

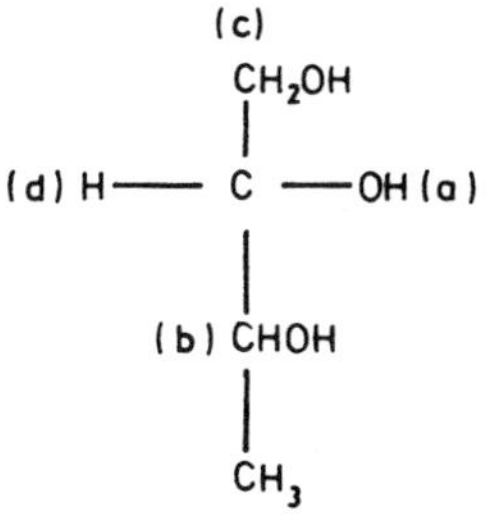

(c)
CH₂OH

(d) H —— C —— OH (a)

(b) CHOH

CH₃

(53)

1,2,3–BUTANETRIOL

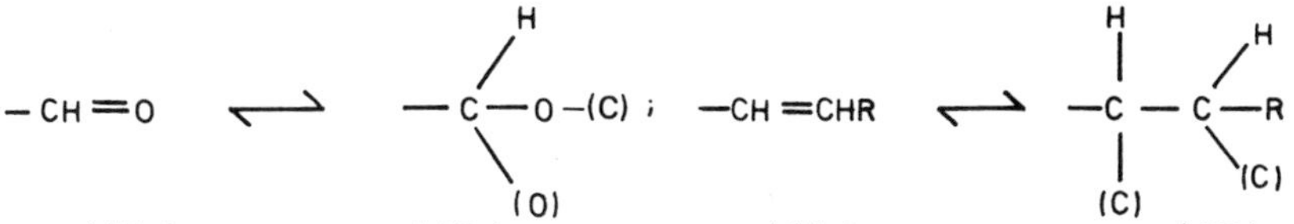

$$-CH=O \quad \rightleftharpoons \quad -\overset{H}{\underset{(O)}{C}}-O-(C) \quad ; \quad -CH=CHR \quad \rightleftharpoons \quad -\overset{H}{\underset{(C)}{C}}-\overset{H}{\underset{(C)}{C}}-R$$

(54)   (55)   (56)   (57)

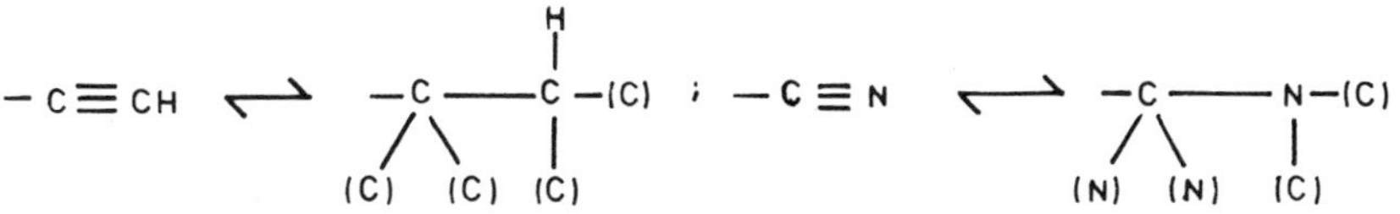

$$-C\equiv CH \quad \rightleftharpoons \quad -\underset{(C)}{C}-\overset{H}{\underset{(C)\ (C)}{C}}-(C) \quad ; \quad -C\equiv N \quad \rightleftharpoons \quad -\underset{(N)\ (N)}{C}-\underset{(C)}{N}-(C)$$

(58)   (59)   (60)   (61)

Therefore :

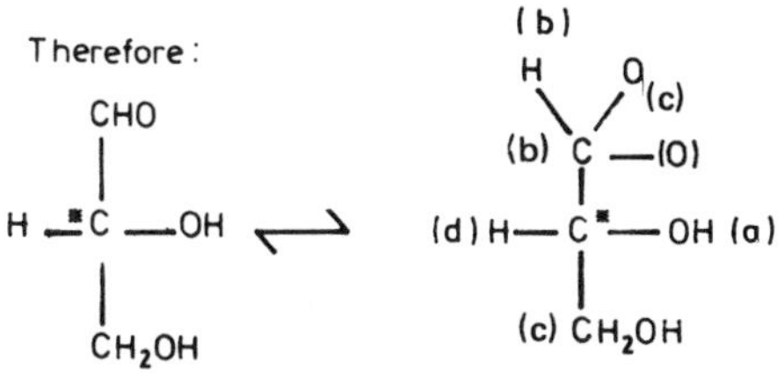

CHO

H — C —— OH

CH₂OH

(64)

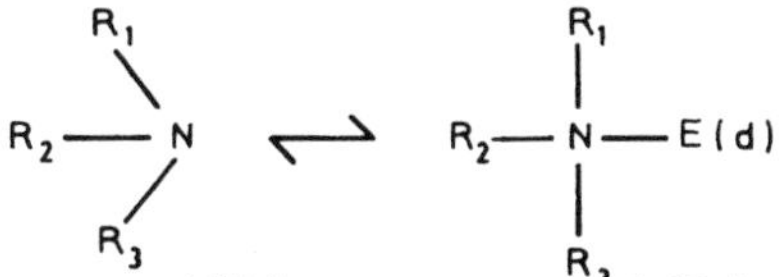

(b)
$$\overset{H}{\underset{(b)}{C}}\overset{O}{\underset{}{}}(c)$$
(d) H — C —— OH (a)

(c) CH₂OH

because
C(O,O > H) > C(O > H,H)

$$R_2-\overset{R_1}{\underset{R_3}{N}} \quad \rightleftharpoons \quad R_2-\overset{R_1}{\underset{R_3}{N}}-E(d)$$

(62)   (63)

cf (38a) and (39a) of Figure A2.12

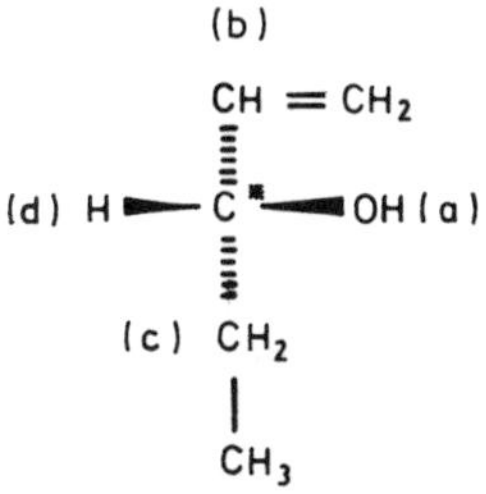

(b)
CH = CH₂

(d) H ◄ C ►► OH (a)

(c) CH₂

CH₃

(65)

(3R) – PENT – 1 – EN – 3 – OL

because — CH = CH₂ $\rightleftharpoons$ C(C,C,H)

—CH₂— CH₃ $\rightleftharpoons$ C(C,H,H)

and C (CCH) > C(C,H,H)

Fig. A2.13 (contd. overleaf)

Fig. A2.13 (b)
(contd.)

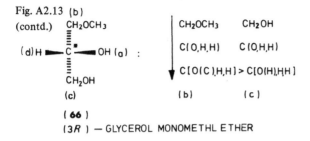

(66)

(3R) — GLYCEROL MONOMETHL ETHER

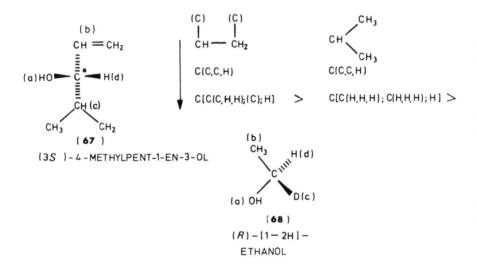

Fig. A2.13 The R-S nomenclature for asymmetric tetrahedral centres and use of the sequence rule.

In 2-butanol (51) OH>methyl, ethyl>H; the priority choice between the methyl and the ethyl falls on the ethyl; the methyl carbon has three hydrogen substituents, C(H,H,H), whilst the carbon of interest in the ethyl group is bonded to two hydrogens and one carbon, C(C>H,H); the preferred ligand of C(C>H,H) is C, which is to be preferred to the preferred ligand of C(H,H,H), which is H.

The same arguments allow an assignment to the ligand priority for 1,2-butanediol (52) for which: OH>C(O>H,H)>C(C>H,H)>H; and for the C-2 asymmetric carbon of 1,2,3-butanediol (53); it can be deduced that OH>C(O>C>H)≫C(O>H,H)>H, after comparing the C and C substituents in the second priority stage, that is C>H; the first priority stage was in fact occupied by atoms of the same atomic number in both groups (that is by one O).

(3) When the comparison directly involves an atom with a co-ordination number smaller than four, for instance, carbon atoms with double bonds (sp²) or with

triple bonds (sp), or heteroatoms such as N, O, S, etc., the ligand number must be taken to 4: (a) by *replication* for multiple bonds [duplications, such as **(54)** → **(55)**, and **(56)** → **(57)**; triplications, such as **(58)** → **(59)** and **(60)** → **(61)**] and (b) by addition of *ghost atoms* (E), with zero atomic number, for heteroatoms [see **(62)** → **(63)**]. These ghost ligands usually correspond to isolated electronic pairs, and are naturally to be put in the last place of the preference sequence (see Table A2.8). The introduction of the 'ghost atom' concept allows application of the *R,S* nomenclature to chiral centres, such as $X_{a,b,c}$, having a pyramidal geometry (X=C⁻, N, P, S, etc); these centres thus become $X_{a,b,c,d}$, where d = E. The above priority rules allow assignments to be made to the two glyceraldehydes **(38a)** and **(39a)** (see also **64**) and to pent-1-en-3-ol **(65)**.

(4) When an examination of the second level of substituent fails (that is the substituents are separated from the chiral centre X by two bonds), the third level of substituents is considered, and so on. Figure A2.13 gives examples of glycerine monomethyl ether **(66)** and 4-methylpent-1-en-3-ol **(67)**.

The principal groups present in organic molecules are listed in Table A2.8 in order of increasing priority.

Table A2.8

List of the principal atoms and groups in organic compounds, in increasing preference order according to the sequence rule.

1. ghost atom	15. cyclohexyl	29. methoxy
2. protium	16. *tert.* butyl	30. phenoxy
3. deuterium	17. isopropenyl	31. glycosyloxy
4. tritium	18. acetylenyl	32. formyloxy
5. methyl	19. phenyl	33. acetoxy
6. ethyl	20. formyl	34. benzoyloxy
7. *n*-propyl	21. acetyl	35. fluoro
8. *n*-butyl	22. benzoyl	36. mercapto (HS–)
9. isobutyl	23. carboxy	37. ethylthio (CH₃S–)
10. allyl	24. amino	38. sulpho (–SO₃H)
11. benzyl	25. methylamino	39. chloro
12. isopropyl	26. acetylamino	40. bromo
13. vinyl	27. dimethylamino	41. iodo
14. *sec*-butyl	28. hydroxy	

These former operations concern 'constitutional' differences amongst ligands. The ligands are also differentiated by steric and isotopic differences.

(5) *A larger mass isotope is preferred to a smaller mass.* This sub-rule is important in biosynthetic studies, as specifically-labelled tritium and deuterium compounds are often employed (T>D>H; see Fig. A2.13). For the sub-rules concerning steric differences between ligands, one should consult the original review articles on

the nomenclature (see the bibliography). Mention is only made here of the systems $X_{(+a), (-a), b, c}$ where (+a) and (−a) are two enantiomeric groups (when considered as isolated units). (R)-a precedes (S)-a and the prefixes (r) and (s) are used for the pseudo-asymmetric X, instead of (R) and (S).

By adding prefixes to a compound name, it is possible to specify the unambiguous stereochemistry of the compound itself. For compounds containing asymmetric carbon atoms, the prefixes (+) and (−) should be used to indicate optical activity, and prefixes such as $2R$, $3S$, $5R$ to indicate the absolute configurations of C-2, C-3, and C-5 respectively. A racemate is usually indicated by the prefix (±), to be preferred to the old notation d, l. If a compound is composed of two enantiomers of known configuration, let us say $2R$, $3S$, $5R$ for one of them, its stereochemistry is conventionally summarised by the symbol $(2RS, 3SR, 5RS)$.

The configurations discussed herein and the ones shown in the formulae are *absolute* ones; they correspond to the *actual* disposition of the atoms in the molecule considered. This raises the question of how to assign absolute configurations from experimental data. If they are geometrical isomers, such as in (E)- and (Z)olefins, either measurements of dipole moments or n.m.r. spectral information can be employed. The problem is more difficult when dealing with diastereoisomers having asymmetric centres and when making assignments to individual enantiomers. For these either *X-ray crystallography*, or *chemical correlation* methods can be employed.

The X-ray diffraction method normally gives structural parameters (internuclear distances, internuclear angles and dihedral angles) necessary for a three-dimensional representation of the molecule. This allows one to choose one pair of enantiomers among the *a priori* possibilities [for instance (26a) and (28a), instead of (27a) and (29a)], but does not establish whether the examined substance has one enantiomeric form or the other one. In other words, once the configuration of an asymmetric centre has been *arbitrarily* fixed [for instance C-5, in (48)], those of the other centres (C-2, C-3, C-4) are determined. The arbitrarily fixed configuration and all the others necessarily dependent on it, are called *relative*. Ordinary non-dissymmetric X-ray crystallography *only* determines the relative configurations of the asymmetric centres of a compound. If appropriate techniques are employed (as in the method of J. M. Bijovet), the absolute configurations are obtained. The first example of absolute configuration determination by the X-ray method took place in 1951, when Bijovet showed (+)-tartaric acid corresponded to formula (41) and thus resolved the dilemma between the two enantiomeric forms (40) and (41).

The *correlation* between two dissymmetric groups is the transformation of one into the other one, or of both into a third one, through a sequence of chemical reactions; once the configuration of one is known, that of the other one is unequivocally defined.

It may happen, with an asymmetric carbon atom, that no chemical reaction breaks a bond between the central atom C and its ligands. If so, the retention of the original configuration of that atom is assured. It can also be said that the initial compound, the final one, and all of the sequential intermediates have the same configuration (*but not necessarily the same configuration symbols, R or S,* see Fig. A2.14(a)). When, during a reaction a breaking and reforming of one σ bond to the asymmetric carbon occurs, the reaction must be necessarily stereospecific and of known stereospecificity (that is with either complete *inversion or retention,* see Fig. A2.14(b), (c)). The configuration of an asymmetric centre, obtained through chemical correlations, will be absolute or relative according to the available knowledge on the absolute configuration of the centre to which it is correlated.

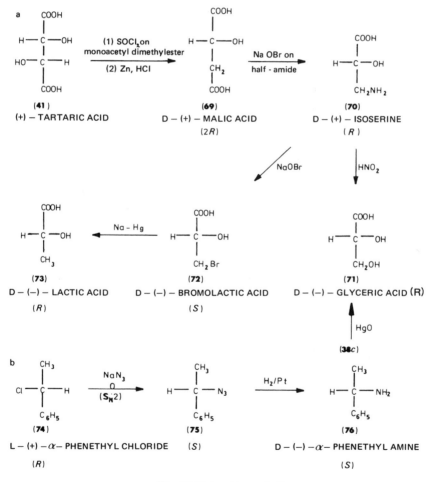

Fig. A2.14 (contd. overleaf)

Fig. A2.14 (contd.)

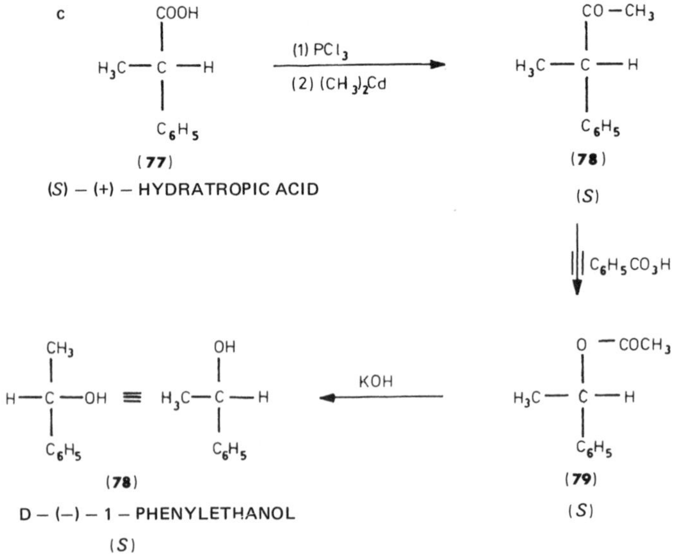

Fig. A2.14 Examples of the chemical correlation of chiral centres, (a) without affecting the asymmetric centre; (b) with inversion of the asymmetric centre and (c) with retention of configuration.

CRYPTOSTEREOCHEMISTRY

The preceding treatment concerned the goemetrical differences between stereo-isomeric molecules (enantiomers and diastereomers).

The spatial relations between portions of the same molecule, composed of single atoms or of groups of atoms bound to each other, should also be examined. These molecular portions will be generally indicated as 'groups' (G). Groups to be considered are those appearing to be equal (homomorphic) or enantiomeric (enantiomorphic) when compared in isolation from the rest of the molecule.

Two, or more, homomorphic (or two enantiomorphic) groups can have different physico-chemical properties, if they are in different environments.

The differences between the environments of two, or more, groups are constitutional when such groups are connected to the rest of the molecule through different atoms, or through some atoms with different ligands [see the hydrogens of nitrobenzene (83), in Fig. A2.15]; a purely steric difference might also exist, if the examined groups are bonded to the same atom [see (80), (81), (84), (86), (87)] or to the atoms of the same element with the same ligands [see the hydrogens of (85)]. When the environmental differences are constitutional, the physico-chemical differences of the groups are usually significant and readily apparent. It is a different matter when the environments are are only sterically different. These groups (stereoheterotopic groups) are difficult to differentiate, and cannot always be distinguished by common research methods, such as synthetic achiral reagents, and low-resolution n.m.r. spectroscopy. They can be distinguished by either using particular techniques, (e.g. n.m.r. spectroscopy) or by interacting with a chiral and stereospecific reagent (such as an enzyme) using a substrate molecule having stereoheterotopic groups specifically-labelled with an isotope. The geometrical non-equivalence of the environments surrounding the reactive centres of such molecules causes distinguishing chemical effects (selectivity, specificity, etc.). These effects have been found in all the cases so far examined of enzymatic reactions where stereoheterotopic groups have been implied.

The knowledge of the stereospecificity of an enzyme can have a fundamental importance in clarifying the mechanism of the enzymatic reaction and the structure of the enzyme-substrate complex (Chapter 2); that is why crypto-stereochemical aspects are so important in studying biocatalysed reactions.

The possible spatial relationship between groups of the same molecule are summarised in Fig. A2.15. Next should be examined: (a) how to recognise the type of stereotopic relationship when the environments have the same con-stitution, and (b) what are the actual consequences of each relationship.

There exist at least two independent criteria for evaluating the type of stereo-topic relationship; one is based on symmetry operations, whilst the other consists in separately substituting each of the related groups (G) of the examined set (molecule) with a 'test' group (G') and comparing the structures obtained. The second criterion is of simpler application, (the 'substitution criterion') and will be used for our analyses (Fig. A2.15).

Case 1. The structures containing G' (it does not matter whether chiral or achiral) instead of the various G's are equal (that is superimposable; the identity criterion). The groups G are called *homotopic,* (or equivalent) and are experi-mentally indistinguishable. For instance, if a G' group replaces either methyl groups of (83), the same molecule is obtained in both cases.

Case 2. The structures containing G' (achiral) are enantiomers (if G' is chiral, they are diastereoisomers). The G groups are *enantiotopic* and can be distinguished only by interaction with chiral agents (enzymes, or n.m.r. spectroscopy in chiral solvents; see the discussion on the fundamental principles of stereochemistry). Enantiotopic groups *cannot* be present in *linear* or *chiral* molecules.

Case 3. The structures containing G' (chiral or achiral) are *diastereomers.* The G groups are called *diastereotopic,* and can be distinguished by reagents with sufficient chemical specificity (enzymes) or with special techniques (for example, based on their non-magnetic equivalence, as in n.m.r. spectroscopy). Diastereo-topic groups can exist in chiral molecule, but not in linear ones.

The *stereoheterotopic* groups (enantiotopic or diastereotopic) most fre-quently encountered in organic molecules of biological interest correspond to pairs of equal or homomorphic substituents on sp^3 carbons, such as $C_{a,b,G,G}$. These carbon atoms are *prochiral* centres, according to current definitions (Fig. A2.16). A centre is prochiral if it becomes dissymmetric by substituting one of

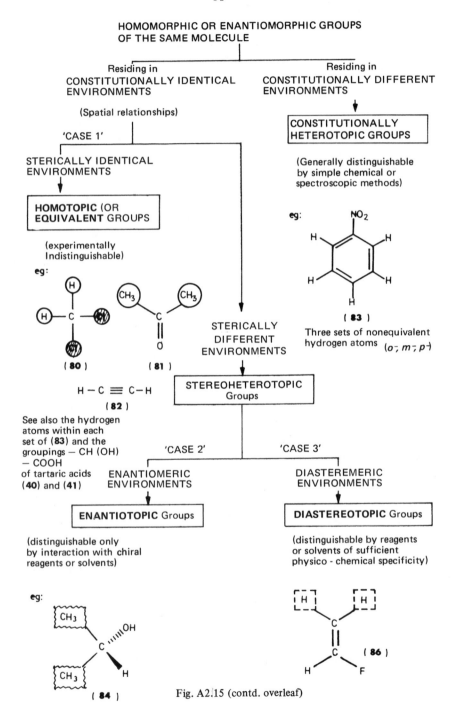

Fig. A2.15 (contd. overleaf)

Fig. A2.15 (contd.)

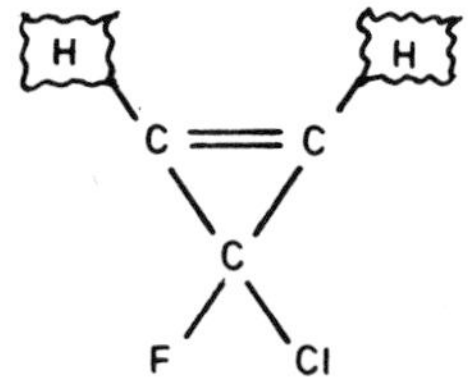

C = C
F   Cl

(85)

See also the groupings
—CH (OH) — COOH of the
meso forms of tartaric
acid (42) and of pentaric
acid (45) and (46)

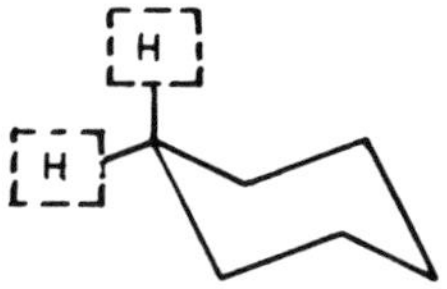

(87)

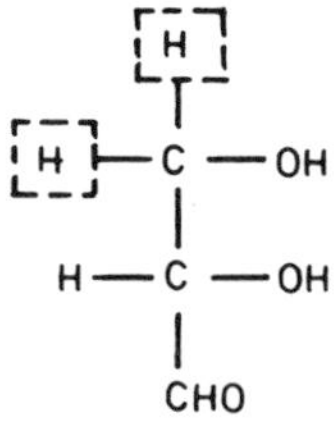

H — C — OH
H — C — OH
CHO

L – GLYCERALDEHYDE
(Fischer projection)

See also the groupings —CH(OH) — COOH
of pentaric acid (43) and (44)

Fig. A2.15 Spatial relationships between groups within molecules.

its ligands with a new ligand or by adding a new ligand to its co-ordination sphere (such as for an sp^2-carbon $\rightarrow$ sp^3 carbon). Consequently every atom bearing two enantiotopic groups is prochiral and *vice-versa* [whilst an atom bonded to two diastereotopic groups can also be non-prochiral, see (86) and (87)].

The term 'prochiral carbon' also has a wider significance than the *meso*-carbon notation used by earlier chemists. A prochiral carbon can, in fact, be of the type $C_{(+a)\ (-a),G,G}$, where (+a) and (−a) are two enantiomorphic ligands. In this latter case, on substituting G with G' the carbon becomes pseudoasymmetric, e.g. (90).

The following conventional method is adopted for naming each stereohetero-topic ligand of a carbon:

(a) One ligand (G) is ideally substituted with a test group (G'), which immediately follows G in increasing priority order, according to the Cahn-Ingold-Prelog rules; it should also occupy a position immediately between G, and the ligand of higher priority next to G; if, originally, there was a sequence a>G>b, G' has to be inserted to give a>G'>G>b. Usually G' is a heavier isotope of G, directly bonded to the prochiral centre) H is substituted with D, CH$_3$ with ^{13}CH$_3$ etc.). After such an operation the prochiral carbon becomes chiral.

(b) The new chirality is then specified, operating as previously described. If an R (or S) asymmetric centre is obtained, the ligand G, substituted with the 'test' group G', is called *pro-R* (or *pro-S*), It is usually indicated in formulae by writing the letter R (or S) in the lower right-hand corner of its symbol, e.g. H$_R$ or H$_S$. The other G ligand is automatically *pro-S* (or *pro-R*) (see formulae, Fig. A2.16).

If the prochiral atom has become pseudo-asymmetric the small letters *r* and *s* are employed.

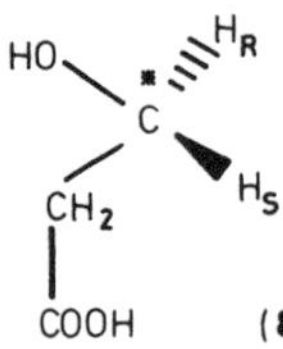

Fig. A2.16 Some examples of prochiral carbon atoms (*) are the relative rotations between enantiotopic and diasterotopic groups.

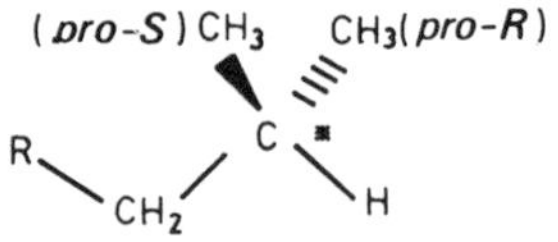

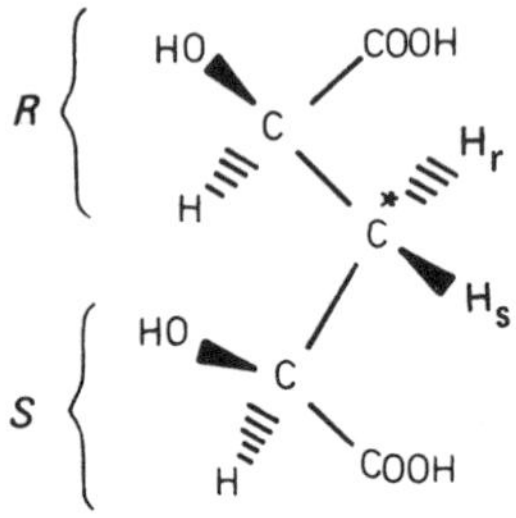

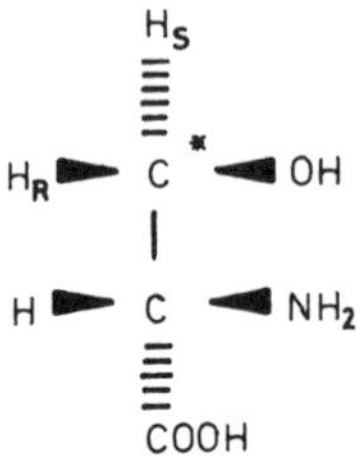

The capacity of an enzyme to recognise stereoheterotopic groups, especially in systems of the type $C_{a,G,G}$ is explained by the above-mentioned theoretical considerations. The active site of·the enzyme is a typically chiral structure; where it associates with a substrate containing prochiral carbons, it makes the enantiotopic ligands become diastereotopic and so, intrinsically, chemically non-equivalent; it also increases the differences between ligands of the substrate which are already diastereotopic.

These ideas help to clarify observations previously held in much dispute. One such dispute originated on observing how the enzyme-controlled formation of α-ketoglutaric acid, obtained from pyruvic acid and labelled $^{13}CO_2$, had an 'asymmetric' label concentrated in the carboxylic group carbon adjacent to the ketonic group. At first these results did not seem to agree with the hypotheses of an intermediate formation of oxaloacetic acid (pyruvic acid + CO_2, see p. 000) and of its conversion into α-ketoglutaric acid *via* citric acid; such an intermediate was symmetrical with respect to the carbon atom ($C_{a,b,G,G}$, type), and carried the label in one of the two homomorphic substituents G (see the Krebs' cycle, p. 000 and Fig. 3.22). A. G. Ogston showed how there was no *a priori* disagreement between such experimental results and the Krebs' cycle reaction. He suggested the mechanistic model shown in Fig. A2.17 as a plausible explanation for the discrimination operated by the aconitase enzyme (see Fig. 3.23) between the two acetic chains (today called enantiotopic chains) of the citric acid. Such a model, which can be extended to all $X_{a,b,G,G}$ type systems, known as *three-point substrate-enzyme binding,* is widely accepted. However, the reader should be warned from considering Ogston's model as a *universal* explanation for the discriminating capacity of enzymes towards stereoheterotopic groups in terms of actual reaction mechanisms.

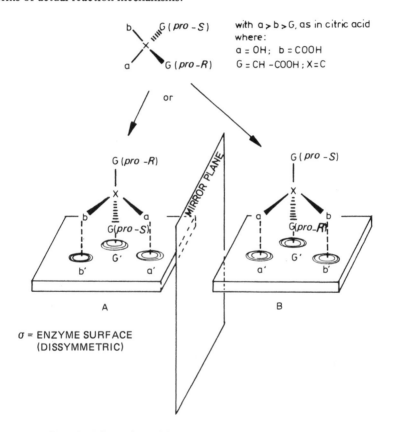

Fig. A2.17 Ogston's model ('three-point substrate enzyme binding').

Ogston's hypothesis is a possible one and agrees with the general stereochemical principles so far exposed. The three-point binding model between enzyme and substrate is attractive, as it visualises and enable one to understand the general phenomenon of *enzymatic cryptostereospecificity.* It is a mistake, however, to consider it as an actual description of biocatalysis. For the moment a clear-cut experimental proof, *for* or *against* Ogston's hypothesis is missing.

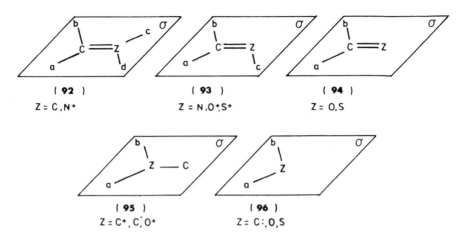

(**92**)　　　　　　　(**93**)　　　　　　　(**94**)

Z = C,N⁺　　　　Z = N,O⁺,S⁺　　　Z = O,S

(**95**)　　　　　　　(**96**)

Z = C⁺, C⁻, O⁺　　　　Z = C:, O, S

Approach of a reagent G′ to the two faces of (**94**) gives rise to the following transition states.

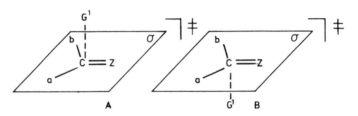

A　　　　　　　　　　　G′   B

FACES OF   σ   PLANE

(if a C_2 axis is in the plane) | (if no C_2 axis is in the plane)

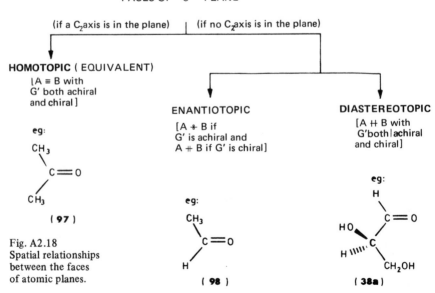

HOMOTOPIC (EQUIVALENT)

[A ≡ B with
G′ both achiral
and chiral]

eg:

CH₃

C＝O

CH₃

(**97**)

Fig. A2.18
Spatial relationships
between the faces
of atomic planes.

ENANTIOTOPIC

[A ≠ B if
G′ is achiral and
A ≠ B if G′ is chiral]

eg:

CH₃

C＝O

H

(**98**)

DIASTEREOTOPIC

[A ≠ B with
G′ both achiral
and chiral]

eg:

H

C＝O

HO

C

H

CH₂OH

(**38a**)

What has been said about spatial relations among homomorphic (or enantio-morphic) groups of a molecule can be extended to the faces of planar systems (see Fig. A2.18). Let us imagine a reagent G' attacking one such system from either face (for the sake of simplification let us imagine such an attack to be directed towards one of the atoms contained in the plane, for instance the carbonyl group carbon, and perpendicularly to it). The two transition states (A and B of Fig. A2.18), according to whether G' is chiral or not, can be identical to each other [for instance starting from (97)], or enantiomeric to each other [starting from (98), G' achiral] or diastereomeric to each other [starting from (98), with G' chiral and from (38a)]. These relations between transition states will be reflected by the composition of the reaction product mixture, which will be constituted of only one product when the transition states are equal; of two possible enantiomers formed in equimolar amounts (racemate) when the transition states are enantiomeric (and so of equal energy), or two stereoisomers with different abundances (stereoselective or stereospecific reaction) when the transition states are diastereomeric (i.e. when the energy levels are different).

One should be careful when applying these principles to enzymatic reactions; in these cases the G' group interacting with the substrate is always chiral, since it is the addition complex between the enzyme active site which is typically chiral, and the reagent, either chiral or not (H^+, H^-, carbanion, etc.), which is incorporated into the product. The biocatalysed reactions $C(sp^2) \longrightarrow C(sp^3)$ will therefore be characterised by a transition state chosen between a pair of diastereomers, either when $C(sp^2)$ has got enantiotopic faces, or when it has diastereotopic faces.

The stereospecificity always found in such transformations catalysed by enzymes can thus be explained; the product is one of the possible enantiomers if the substrate has enantiotopic faces, and Y is achiral (see Fig. A2.19), otherwise it is one of the possible diastereomers.

The analogous relationships between two equal groups bonded to a prochiral tetrahedral atom ($X_{a,b,G,G}$, particularly where X is an sp^3 carbon) and the two stereoheterotopic faces of an atom X, situated on the same plane of its ligands, are also reflected in the nomenclature system proposed by Hanson and based, in both cases, on use of the Cahn-Ingold-Prelog sequence rule. The *pro-R/pro-S* notation has already been mentioned. The method for specifying the stereo-heterotopic faces of a 'planar' atom X can be summarised (see Fig. A2.19) as:

(1) One face of the plane is oriented towards the observer.
(2) If X is dico-ordinated [such as for the O, H, and S atoms in (93) and (96)] a third 'ghost' ligand, σ, having mass number of zero is added in the plane, in an opposite direction to the two ligands already present.
(3) The priority order of the three X ligands is established according to the sequence rule: a>b>c or a>b>σ. Sometimes it is necessary to duplicate the ligand united to X by the double bond, to establish the order of ligand priority.

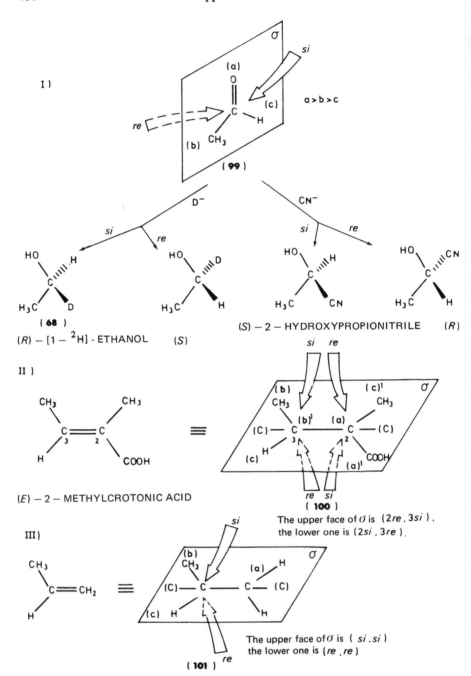

Fig. A2.19 Examples of carbon atoms lying in prochiral Sp² planes and the rotations for the stereoheterotopic faces.

(4) If the observer traces a line, going from the preferred ligand to the lower priority one, clockwise on the plane (or anticlockwise), the face looking towards him is *re* (or *si*).

The faces of planar atoms which cannot be tested as described above (being mono-co-ordinate atoms, such as the oxygen atom of the carbonyl group or bearing two equal ligands, such as carbon in the methylene group of the type $R_1R_2C=CH_2$) are assigned a notation belonging to the coinciding face of the $X_{a,b,c}$ atom, which is adjacent to them. The two vinyl group faces will therefore simply be *re,re* and *si,si* (see **101**) (see Fig. A2.19).

STEREOCHEMISTRY OF CONCERTED REACTIONS

In certain circumstances the stereochemistry of a product is determined by that of the starting material. Several factors determine the final stereochemical outcome of a reaction, including substrate stereostructure, the electronic-reaction mechanism (single or multiple step), the action of solvent and other solutes. Consider, initially, a concerted reaction, i.e. passing through a single step. For such processes we can state:

(1) 'Every transformation passing through only one transition state is stereospecific'. The transition state geometry unequivocally determines the stereochemistry of the single product.

(2) 'If an electronic reaction mechanism *a priori* allows more than one stereoisomeric transition state, one of them always has the lowest free energy and is therefore the most readily attainable'. Such a transition state leads to the more abundant (stereoselective) product or the unique (stereospecific) product. When very large free energy differences ($\Delta\Delta G^{\ddagger}$) exist between the various transition states affording a product and its stereoisomers a stereospecific process is observed. In a concerted reaction the geometry of the most stable transition state is deduced solely from the intrinsic structure of the substrate (or of the reagents) only if all solvation effects are neglected. Simplified rules have been deduced which predict the most favoured spatial orientations for the transition state, the so-called electronic requirements (see axiom 2 above) and the steric course of concerted reactions (see axiom 1 above).

These simplified rules generalise the behaviour of electron pairs in bimolecular substitution reactions (as from the work of C. K. Ingold and others) and correlate the reagent and product molecular orbitals (as from the symmetry rules of R. B. Woodward and R. Hoffmann and the stereoselection rules of K. Fukui and H. Fujimoto). Although treatment of these molecular orbital symmetry rules is beyond the scope of this text a short discussion on the 'S_N2 rule' is valuable in connection with common biosynthetic reactions. The stereochemistry of such reactions (Figs. A2.20–A2.22) is summarised by:

(1) Every concerted heterolytic reaction results from an electron pair movement along a chain of atoms; such movements are controlled: (a) the σ or π electron pairs from bonding orbitals only leave from one of the two atoms sharing them and can be associated with a new atom; a new σ or π bond is therefore formed; (b) when the electron pair belong to only one atom (an 'n' lone pair) they do not leave it; the n lone pair of one atom is shared with another atom forming a σ or π bond.

(2) When an atom exchanges an electron pair (with or without a ligand) the incoming pair approaches from the opposite side of the outgoing one, preferably in the same general direction. An example is the typical S_N2 reaction. When a doubly bonded, π electron pair attacks as a nucleophile, the movement is from above or below the plane of the sp^2 hybridised atoms and their substituents.

(3) When an atom exchanges a ligand, the overall number of electrons attached to the centre being unchanged, the oriented electron pair responsible for the bond being broken and reformed does not appear to change its initial direction (but see also the S_E2 mechanism).

According to Ingold these rather formal rules follow from physico-chemical effects: the repulsive energy between the two electron pairs associated with the same atom in a transition state serves to keep them as far from one another as possible (see Fig. A2.20, reaction 1). As mentioned earlier, a concerted reaction will pass through the intrinsically most stable transition state *only* if the solvent does not alter the intrinsic stability order of all the other possible transition states. Some quaternary ammonium salts undergo *syn*-elimination in alkaline media possibly *via* cyclic transition states requiring the participation of the solvent. The interpretation of enzymatic mechanisms must account for such facts, particularly since such reactions occur in active sites which can possess very powerful solvating properties. When the stereochemistry of an enzymatic reaction does not fit a concerted process, a multistep mechanism is usually hypothesized. Some examples are shown in Fig. A2.23.

1) ELEMENTARY NUCLEOPHILIC SUBSTITUTION (S$_N$2) : **INERSION**

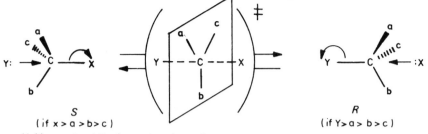

S
(if x > a > b > c)
X, Y = nucleophiles (neutral or charged)

R
(if Y > a > b > c)

2) ELECTROPHILIC SUBSTITUTION (S$_E$2) : **RETENTION**

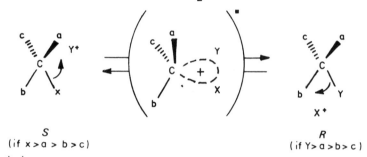

S
(if x > a > b > c)

R
(if Y > a > b > c)

X$^+$, Y$^+$ ≡ electrophiles (electron deficient, e.g. carbocations)

*According to bonding theory, both a pentacovalent transition state (T.S.) and a true intermediate (e.g. non-classical carbonium ion, see Appendix 3) are allowed.

3) CONJUGATE SUBSTITUTIONS :

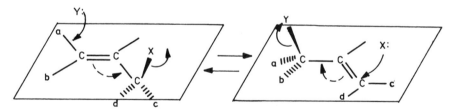

X, Y = nucleophiles
ALLYLIC NUCLEPHILIC DISPLACEMENT (S$_N$2) *SYN MECHANISM*

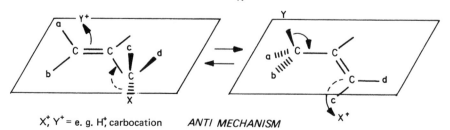

X$^+$, Y$^+$ = e. g. H$^+$, carbocation   *ANTI MECHANISM*

Fig. A2.20 (contd. overleaf)

Fig. A2.20 (contd.)

FOR THE GENERAL EQUATION

$$Y + (C = C)_k \!-\!\!- C - X \;\;\rightleftharpoons\;\; Y - C - (C = C)_k + X$$

k	X, Y = NUCLEOPHILES or RADICAL		X, Y = ELECTROPHILES	
	π –ELECTRONS IN T.S.	MECHANISM	π –ELECTRONS IN T.S.	MECHANISM
0	4 or 3	*ANTI* (INVERSION)	2	*SYN* (RETENTION)
1	6 or 5	*SYN*	4	*ANTI*
2	8 or 7	*ANTI*	6	*SYN*
3	10 or 9	*SYN*	8	*ANTI*
---	- - - -	- - - -	- - - -	- - - -

Fig. A2.20 Stereochemistry in concerted displacement reactions.

1) BIMOLECULAR ELIMINATION (E2) AND ADDITION : *ANTI MECHANISM*

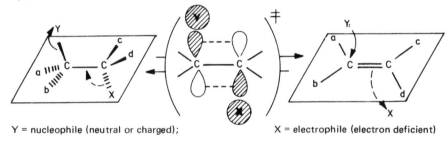

Y = nucleophile (neutral or charged);   X = electrophile (electron deficient)

2) CONJUGATE ELIMINATIONS AND ADDITIONS

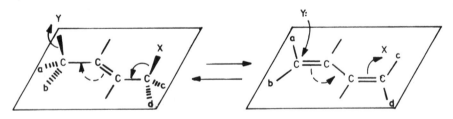

SYN MECHANISM

FOR THE GENERAL EQUATION

$$Y + (C = C)_k + X \;\;\rightleftharpoons\;\; Y - C - (C = C)_{k-1} - C - X$$

k	CHAIN LENGTH	π ELECTRONS IN T.S.	MECHANISM
1	2	2	*ANTI*
2	4	4	*SYN*
3	6	6	*ANTI*
---	- - - -	- - - -	- - - -

Fig. A2.21 Stereochemistry in concerted elimination omit additions.

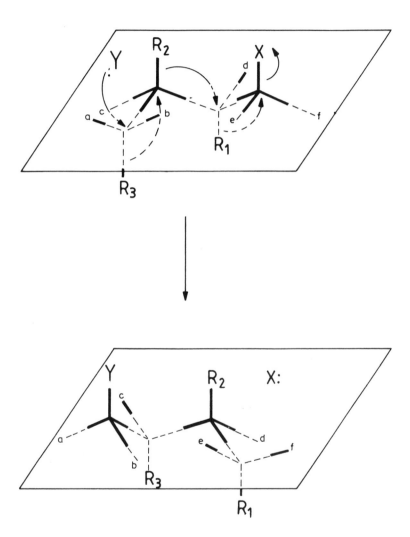

Fig. A2.22 Stereochemistry of concerted rearrangements (sequence of 1,2-shifts).

1) 1,2 – *SYN* – ELIMINATION AND ADDITION

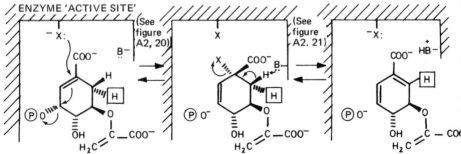

(5 - DEHYDROQUINATE
DEHYDRATASE;
of Figure 7.2)

$- H_2O$

$+ H_2$

5 - DEHYDROQUINATE   5 - DEHYDROSHIKIMATE

ENZYME 'ACTIVE SITE'

2) 1,4 – *ANTI* ELIMINATION AND ADDITION

(CHORISMATE
SYNTHETASE;
of figure 7.2)

$- P_i$

$+ P_i$

3 – ENOYLPYRUVYL –   CHORISMATE
SHIKIMATE – 5 – PHOSPHATE

ENZYME 'ACTIVE SITE'

(See figure A2, 20)   (See figure A2. 21)

Fig. A2.23 Enzymatic reactions with proposed mechanisms to explain the observed stereochemical consequences.

SOURCE MATERIALS AND SUGGESTED READING

[1] S. Cahn, C. K. Ingold and V. Prelog, (1956), The Specification of Asymmetric Configuration in Organic Chemistry, *Experientia,* **12**, 81.

[2] W. Klyne and V. Prelog, (1960), Description of Steric Relationships Across Single Bonds, *Experentia,* **16**, 521.

[3] E. L. Eliel, (1962), *Stereochemistry of Carbon Compounds,* McGraw-Hill.

[4] R. S. Cahn, (1964), An Introduction to the Sequence Rule, *J. Chem. Educ.,* **41**, 116.

[5] G. M. Campbell, (1964), Stereochemistry of Carbon Compounds in *Rodd's Chemistry of Carbon Compounds,* 2nd ed. (S. Coffey, Ed.), I/A, Elsevier.

[6] E. L. Eliel, N. L. Allinger, S. J. Angyal, and G. A. Morrison, (1965), *Conformational Analysis,* Wiley Interscience.

[7] G. Hallas, (1965), *Organic Stereochemistry,* McGraw-Hill, London.

[8] M. Hanak, (1965), *Conformation Theory,* Academic Press.

[9] S. F. Manson, (1965), Optical Activity and Molecular Dissymmetry, *Chem. in Britain,* **1**, 245.

[10] R. S. Cahn, C. K. Ingold and V. Prelog, (1966), Specification of Molecular Chirality, *Angew. Chem. Int. Ed. Engl.,* **5**, 385.

[11] K. R. Hanson, (1966), Applications of the Sequence Rule I. Naming the Paired Ligands g,g at a Tetrahedral Atom $X_{g,g,i,j}$. II. Naming the Two Faces of a Trigonal Atom $Y_{g,h,i}$, *J. Amer. Chem. Soc.,* **88**, 2731.

[12] K. Mislow, (1966), *Introduction to Stereochemistry,* Benjamin.

[13] N. L. Allinger and E. L. Eliel (Eds.), (1967-1978), Topics in Stereochemistry, **1-10**, Wiley.

[14] K. Mislow and M. Raban, (1967), *Stereochemical Relationships of Groups in Molecules* in ref. 13, **1**.

[15] J. E. Blackwood, C. L. Gladys, K. L. Loening, A. E. Petrarca, and J. E. Rush, (1968), Unabiguous Specification of Stereoisomerism about a Double Bond, *J. Amer. Chem. Soc.,* **90**, 509.

[16] J. H. Brewster, (1968), Helix Models of Optical Activity in ref. 13, Vol. 2.

[17] J. L. Carlos, (1968), Molecular Symmetry and Optical Inactivity, *J. Chem. Educ.,* **45**, 248.

[18] M. van Gorkon and G. E. Hall (1968), Equivalence of Nuclei in High-resolution Nuclear Magnetic Resonance Spectroscopy, *Quart. Rev.,* **22**, 14.

[19] R. Hoffman and R. B. Woodward, (1968), The Conservation of Orbital Symmetry, *Accounts Chem. Res.,* **1**, 17.

[20] S. I. Miller, (1968), Stereoselection in the Elementary Steps of Organic Reactions in *Advances in Physical Organic Chemistry,* (V. Gold, Ed.), Academic Press, **6**.

[21] V. Prelog, (1968), Problems in Chemical Topology, *Chem. in Britain,* **4**, 382.

[22] R. Bentley, *Molecular Asymmetry in Biology,* **1**, 1969; **2**, 1970, Academic Press.

494 Appendix 2

[23] E. L. Eliel, (1969), *Elements of Stereochemistry*, Wiley.
[24] R. C. Fahey, (1969), The Stereochemistry of Electrophilic Additions to Olefins and Acetylenes, in ref. 13, 3.
[25] D. F. Mowery, (1969), Criteria for Optical Activity in Organic Molecules, *J. Chem. Educ.*, 46, 269.
[26] D. Arigoni and E. L. Eliel, (1970), Chirality Due to the Presence of Hydrogen Isotopes at Noncyclic Positions in ref. 13, 4.
[27] J. W. Cornforth, (1970), The Chiral Methyl Group – Its Biochemical Significance, *Chem. in Britain*, 6, 431.
[28] H. G. Floss, (1970), Stereochemistry of Enzymatic Reactions at Prochiral Centers, *Naturwiss.* 57, 435.
[29] K. Fukui, (1970), Theory of Orientation and Stereoselection in *Topics in Current Chemistry*, Springer-Verlag, 15.
[30] A. J. Gordon, (1970), A Survey of Atomic and Molecular Models, *J. Chem. Educ.*, 47, 30.
[31] C. R. Petersen, (1970), Some Reflections on the Use and Abuse of Molecular Models, *J. Chem. Educ.*, 47, 24.
[32] R. B. Woodward and R. Hoffman, (1970), The Conservation of Orbital Symmetry, Verlag Chemie and Academic Press, Weinheim.
[33] K. Fukui, (1971), Recognition of Stereochemical Paths by Orbital Interaction, *Accounts Chem. Res.*, 4, 57.
[34] E. L. Eliel, (1971), Recent Advances in Stereochemical Nomenclature, *J. Chem. Educ.*, 48, 163.
[35] R. G. Pearson, (1971), Symmetry Rules for Chemical Reactions, *Accounts Chem. Res.*, 4, 152.
[36] D. W. Slocum, D. Sugarman and S. P. Tucker, (1971), The Two Faces of D and L Nomenclature, *J. Chem. Ed.*, 48, 597.
[37] J. F. Stoddart, (1971), *Stereochemistry of Carbohydrates*, Wiley-Interscience.
[38] J. L. Abernethy, (1972), The Concept of Dissymmetric Worlds, *J. Chem. Educ.*, 49, 455.
[39] W. L. Alworth, (1972), *Stereochemistry and Its Application in Biochemistry*, Wiley-Interscience.
[40] D. H. R. Barton and O. Hassel, (1972), Fundamental Contributions to Conformational Analysis, in ref. 13, 6.
[41] O. Cori, (1972), Complementary Rules to Define R or S Configuration, *J. Chem. Educ.*, 49, 461.
[42] J. D. Donaldson, S. D. Ross, (1972), *Symmetry and Stereochemistry*, Wiley.
[43] R. J. Gillespie, (1972), *Molecular Geometry*, Van Nostrand-Reinhold.
[44] C. L. Perrin, (1972), The Woodward-Hoffman Rules – An Elementary Approach, *Chem. in Britain*, 8, 163.
[45] J. Sicher, (1972), The *syn* and *anti* Steric Cours in Bimolecular Olefin Formin Eliminations, *Angew. Chem. Int. Ed. Engl.*, 11, 200.

[46] S. Wolfe, (1972), The Gauche Effect. – Some Stereochemical Consequences of Adjacent Electron Pairs and Polar Bonds, *Accounts Chem. Res.,* **5**, 102.

[47] A. H. Andrist, (1973), Concertedness: A Function of Dynamics or the Nature of the Potential Energy Surface?, *J. Org. Chem.,* **38**, 1772.

[48] D. Whittaker, (1973), *Stereochemistry and mechanism,* Oxford University Press.

[49] A. Ault, (1974), Test for Chemical Shift and Magnetic Equivalence in NMR, *J. Chem. Educ.,* **51**, 729.

[50] R. Bucourt, (1974), The Torsion Angle Concept in Conformational Analysis, in ref. 13, 8.

[51] M. K. Kaloustian, (1974), The Electrostatic Dimension in Conformational Analysis, *J. Chem. Educ.,* **51**, 777.

[52] W. Klyne and J. Buckingham, (1974), *Atlas of Stereochemistry – Absolute Configurations of Organic Molecules,* Chapman and Hall.

[53] W. K. Li and T. C. W. Mak, (1974), Bond Angle Relationship in Some AX_nY_m Molecules, *J. Chem. Educ.,* **51**, 571.

[54] J. P. Lowe, (1974), Is This a Concerted Reaction?, *J. Chem. Educ.,* **51**, 785.

[55] D. H. Wertz and N. L. Allinger, (1974), The Gauche Hydrogen Interaction as the Basis of Conformational Analysis, *Tetrahedron,* **30**, 1579.

[56] J. Weyer, (1974), A Hundred Years of Stereochemistry – The Principal Developments Phases in Retrospect, *Angew. Chem. Int. Ed. Engl.,* **13**, 591.

[57] E. L. Eliel, (1975), Conformational Analysis – The Last 25 Years, *J. Chem. Ed.,* **52**, 762.

[58] F. D. Gunstone, (1975), *Guidebook to Stereochemistry,* Longman.

[59] K. R. Hanson, (1975), Reactions at Prochiral Centers, *J. Biol. Chem.,* **250**, 8309.

[60] W. B. Jennings, (1975), Chemical Shifts Nonequivalence in Prochiral Groups, *Chem. Rev.,* **75**, 307.

[61] S. A. Kaloustian and M. K. Kaloustian, (1975), Determining Homotopic, Enantiotopic, and Diastereotopic Faces in Organic Molecules, *J. Chem. Educ.,* **52**, 56.

[62] P. M. Scopes, (1975), Application of the Chiroptical Techniques to the Study of Natural Products in *Progress in the Chemistry of Organic Natural Products,* ref. 1 of Chap. 1, **32**.

[63] R. Bentley, (1976), The Use of Biochemical Methods for Determination of Configuration, in *Application of Biochemical Systems in Organic Chemistry,* ref. 98 of Chap. 2, **1**.

[64] Fundamental Stereochemistry – IUPAC 1974 Recommendations for Section E, (1976), in *Pure and Applied Chemistry – Official Journal of IUPAC,* **45**, Pergamon Press.

[65] K. R. Hanson, (1976), Concepts and Perspectives in Enzyme Stereochemistry, *Ann. Rev. Biochem.,* **45**, 307.

[66] J. B. Jones, (1976), Stereochemical Considerations and Terminologies of

Biochemical Importance in *Application of Biochemical Systems in Organic Chemistry*, ref. 98 of Chap. 2, 1.

[67] J. Pearce and E. Glynn, (1976), *Stereochemistry – An Introductory Programme with Models*, Wiley.

[68] R. G. Pearson, (1976), *Symmetry Rules for Chemical Reactions*, Wiley.

[69] V. Prelog, (1976), Chiralty in Chemistry, *Science*, **193**, 17; *J. Mol. Catal.*, **1**, 159.

[70] F. H. Clarke, (1977), New Skeletal-Space-Filling Models, *J. Chem. Educ.*, **54**, 230.

[71] M. Gielen, (1977), From the Concept of Relative Configuration to the Definition of *erythro* and *threo*, *J. Chem. Educ.*, **54**, 673.

[72] H. B. Kagan (Ed.), (1977), Stereochemistry. Fundamentals and Methods, a multivolume work, Georg Thieme, Stuttgart, **1**, (Spectrometric Methods); **2** (Dipole Moments, CD, ORD); **3** (Determination of Configurations by Chemical Methods); **4** (Absolute Configurations. One Asymmetric Carbon Atom). Further vols. in preparation.

[73] N. S. Zefirov, (1977), The Problem of Conformational Effects, *Tetrahedron*, **33**, 3193.

[74] R. Bentley, (1978), Ogston and the Development of Prochirality Theory, *Nature*, **276**, 673.

[75] J. Reisse, R. Ottinger, P. Bickart and K. Mislow, (1978), Intrinsic Asymmetry and Diastereotopism, *J. Amer. Chem. Soc.*, **100**, 911.

[76] R. A. Dietzel, (1979), Determination of Chiral Molecule Configuration Using the ±1,2,5 Rule, *J. Chem. Educ.*, **56**, 451.

[77] E. Juaristi, (1979), The Attractive and Repulsive Gauche Effects, *J. Chem. Educ.*, **56**, 438.

[78] H. Kagan, (1979), *Organic Stereochemistry*, Halsted Press – Wiley.

[79] C. A. Kingsbury, (1979), Conformations of Substituted Ethanes, *J. Chem. Educ.*, **56**, 431.

[80] B. Testa, (1979), *Principles of Organic Stereochemistry*, M. Dekker.

[81] E. L. Eliel, (1980), Stereochemical Non-Equivalence of Ligands and Faces (Heterotopicity), *J. Chem. Educ.*, **57**, 52.

Non-Classical Carbocations

σ-Bridged non-classical carbocations are familiar species to the organic chemist. A non-classical cation was originally proposed by C. L. Wilson as a possible intermediate (1) in the rearrangement of camphene hydrochloride (2) to iso-bornyl chloride (3) (see also Fig. 5.18: 44-45). N. Meerwein had noticed the same rearrangement earlier but had hypothesised the formation of the classical carbocation (5) as a reaction intermediate. Wilson examined the mechanism and kinetics for the rearrangement and conluded that there was a fast equilibrium between the two classical isomeric carbocations (4) and (5), or, alternatively, a unique intermediate (1).

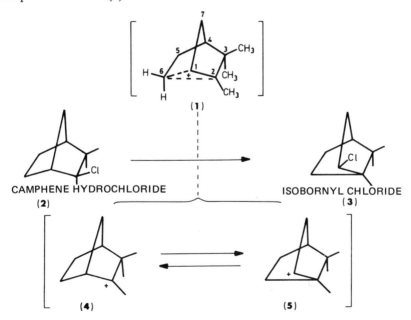

Fig. A3.1 Possible intermediate carbocations, classical (4) and (5) and non-classical (1).

Species similar to (1), stabilised by the mesomeric effect, were hypothesised in many other reactions in the following years. Apart from the methyl cation, every known aliphatic, alicyclic and bicyclic cation has been described in non-classical terms (See Fig. A3.2). The concept of the non-classical cation became fashionable largely owing to the works of Roberts, Winstein, Bartlett, and Sargent. In the 1960's H. C. Brown, questioned the basis for such assumptions since many of the results attributed to the intermediacy of non-classical ('delocalised') carbocations can be explained just as well by postulating a rapidly equilibrating set of classical carbocations. The reasons for his stand have been presented in a review by him (see bibliography). That non-classical carbocations can exist has been demonstrated for certain systems [e.g. the norbornyl system (12)] by G. Olah. He found that in certain, highly acidic media these species could be observed spectroscopically. However, such evidence does not necessarily prove the nominal existence of such species whenever reactions proceed through rearrangement of a carbocation species into another one.

The actual situation can be summarised as follows: most experiments, which were considered in the past to give evidence for the formation of non-classical cations, can also be interpreted in terms of traditional stabilisation factors of classical carbocations (inductive effects, conjugation, hyperconjugation and relief of steric strain) and the fast equilibration between isomeric carbocation species.

In rearrangements involving carbocationic intermediates, there are three possible *a priori* cases: (A) formation of an essentially static classical cation, which is subsequently transformed into products without rearrangement; (B) a cation rapidly reaching a thermodynamic equilibrium among the isomeric structures in the time interval between its formation and conversion into the products; (C) formation of a non-classical carbocation (σ bridged).

Cases A, B and C are exemplified by reactions in Fig. A3.2.

The concept of the non-classical carbocation was enthusiastically received by organic chemists in the 1950–60s' and biogenetic studies were considerably affected by it. Non-classical carbocations have been proposed as intermediates in many biosynthetic processes. These include the hypotheses on the origin of the various triterpenic structures described in Chapter 5. Sometimes non-classical carbocations were advocated simply because they were fashionable, although there was no actual need for them in order to explain the reaction mechanism. (A general mechanism for the biocatalysed reaction is given at the end of this Appendix: in many instances it represents a plausible alternative hypothesis to that based on non-classical carbocations).

Some definitions for various types of positively charged carbon species are appropriate:

Carbonium ions are species with one or more carbon atoms bearing a positive electrical charge: the charge can be whole or partial and it determines the physico-chemical behaviour of such a species. If the electrical charge is formed

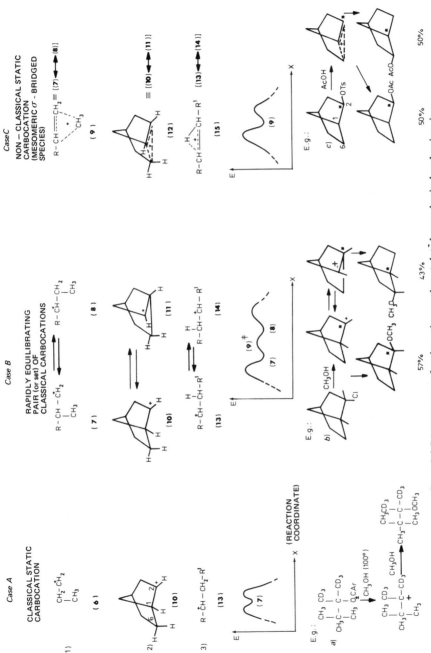

Fig. A3.2 Diverse types of carbocations and postulated non-classical carbonium ions.

and it eventually disappears during a reaction, the carbonium ions are called incipient, unless they correspond to a minimum of the potential energy versus a reaction coordinate (Fig. A1.4b). The words *carbocations*, or *carbonium ions* are more generally used when referring to the latter, i.e. to discrete reaction intermediates.

Carbocations can be of three types:

(1) Delocalised carbocations, where the charge is delocalised by interaction with a neighbouring π bond, e.g. the allyl cation, $(CH_2\text{----}CH\text{----}CH_2)^+$.

(2) Carbocations with a charge localised on an sp^2 hydridised carbon (Fig. A3.2). Such classical carbocations were more properly called *carbenium ions* by Olah (protonated carbenes), as the carbon atom only shares six electrons. The less common positively charged sp-hydridised carbon atoms can be included in this class, such as the oxocarbenium ion, $-\overset{+}{C}=O$ (resonance stabilised as $-C\equiv\overset{+}{O}$) and the vinylic cation $-\overset{+}{C}=C$.

(3) Penta- or tetra-coordinated carbocations, the non-classical carbocations. These consist of one carbon atom bound to four or five ligands through three simple, independent σ-bonds, each overlapping with a ligand, and a three-centre two-electron bond, simultaneously involving the fourth and fifth ligands (CL_5^+) or the fourth ligand and one of the previous ones (CL_4^+). Penta- and tetra-coordinated carbocations include the *non-classical-carbocations* (σ-bridged ions) treated in this Appendix.

There are various definitions for non-classical carbocations (σ-bridged). Bartlett's definition is extremely simple and concise: a fundamental state having delocalised σ-bond electrons. Some conventional representations usually employed for non-classical carbocations are given (Fig. A3.3).

In this volume Winstein's original notation is employed: mesomerism between cononical forms is stressed: each corresponds to the possible formation and collapse of a non-classical cation('σ-route' via **10** and **11** and 'π-route' via **16**.)

In order to explain how a non-classical carbocation can be formed by the interaction between a carbenium ion, or an incipient carbocation, and a single bond in the 2, 3 position, that is via the so-called 'σ route' (reaction c, Fig. A3.2), it is convenient to consider some other known examples of *neighbouring group participation*.

Substituents in an organic molecule can influence its reactivity in three distinct ways: (a) they alter the polarisation of the reaction centre by electronic interactions along the molecular skeleton (inductive or conjugative effects); (b) they increase or reduce steric access to the reaction site; (c) they stabilise (or destabilise) a transition state, or an intermediate through transient formation of a full or partial bond with the reaction centre. This latter substituent effect is called *intramolecular participation* and takes place in a manner analogens to intermolecular nucleophilic or electrophilic catalysis. If the consequent increase in reaction rate involves the rate-determining step (the slowest step) the phenomenon is also indicated by the term *anchimeric assistance*.

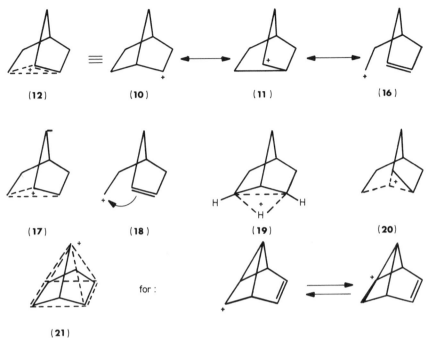

(21)

Fig. A3.3 Some of the more common notations employed to describe non-classical cations: of Winstein **(12)**; Olah **(17)** and **(20)**; Dewar **(18)**; Klopman **(19)**. Formula **(21)** represents an extreme representation for a carbocation of the non-classical type.

Solvolysis of ω-methoxybutyl and ω-methoxypentyl-*p*- bromobenzene-sulphonate **(22)**, n = 4 and 5, respectively, represents two typical examples where anchimeric assistance occurs. As indicated in Fig. A3.4 the methoxy group of these compounds stabilises the carbocation resulting from the hetero-lysis of the $-C-OBs$ bond, thus forming oxonium ion intermediates, **(24)** and **(25)**. The compounds **(22)**, n = 4, and, to a minor extent **(22)**, n = 5, require a smaller activation energy for solvolysis than their lower or higher homologues: The oxonium-bridged ring of the latter is either too large or too small, and the intramolecular participation of the electron donating group is unfavoured. The variations of the activation energies are shown in Table A3.1.

Other nucleophilic groups capable of anchimeric assistance to an S_N reaction are: $-OH$, $-NH_2$, $-SR$, $-Br$, $-COO^-$, COOR, $-CONHR$, $-COR$. In every case the electron pair stabilising the positive reaction centre is a lone pair of a heteroatom (which becomes an 'onium' ion). Intramolecular attack of this type may also afford neutral cyclic compounds, such as in reaction (b), of Fig. A3.4.

Of the results presented in Table A3.1 those of **(22)**, n = 2 should be noted, i.e. in which the neighbouring group is very close to the reacting centre. No catalytic enhancement is noted. This does not necessarily mean that anchimeric

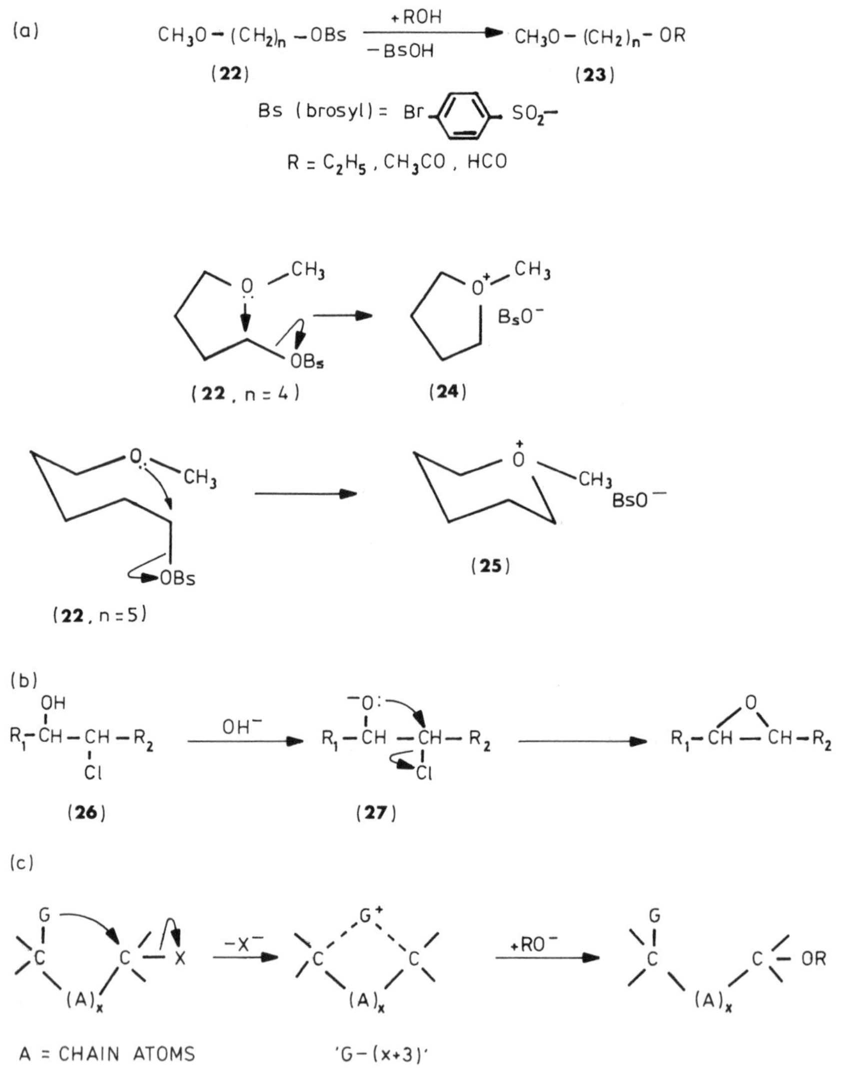

(a)

$$CH_3O-(CH_2)_n-OBs \xrightarrow[\;-BsOH\;]{+ROH} CH_3O-(CH_2)_n-OR$$

(22)   (23)

Bs (brosyl) = Br ⟨⟩ SO_2-

$R = C_2H_5 , CH_3CO , HCO$

(22, n = 4)   (24)

(22, n = 5)   (25)

(b)

$$R_1-\overset{\underset{|}{OH}}{CH}-\overset{\underset{|}{Cl}}{CH}-R_2 \xrightarrow{OH^-} R_1-\overset{-O:}{CH}-\overset{\underset{|}{Cl}}{CH}-R_2 \longrightarrow R_1-\overset{O}{CH}-CH-R_2$$

(26)   (27)

(c)

A = CHAIN ATOMS   'G-(x+3)'

Fig. A3.4 Some examples of intramolecular participation.

assistance does not occur but that its effect might be masked by secondary electron effects such as inducive effects of the methoxyl group. The activation energy $(\Delta G^{\ddagger})$ of a reaction is made up of two parts, the enthalpy term $\Delta H^{\ddagger}$ and an entropy term $(-T\Delta S^{\ddagger})$, where $\Delta G^{\ddagger} = \Delta H^{\ddagger} - T\Delta S^{\ddagger}$ and the reaction rate is much faster as $\Delta G^{\ddagger}$ is smaller. In reactions forming rings the transition state $\Delta H^{\ddagger}$ is the more positive term due to interactions between the non-bonded atoms and bond distortions, and the smaller the energy of the new intramolecular bond; $\Delta S^{\ddagger}$ is more negative as the degree of freedom of motion of the system decreases. These effects can be summarised for different ring sizes, as follows:

	number of ring atoms		
	3	5–6	10
$\Delta H^{\ddagger}$	large, positive	small, positive	large, negative
$\Delta S^{\ddagger}$	small, negative	medium, negative	large, negative

Thus 5 or 6-membered carbon rings are favoured by enthalpy but three-membered rings are favoured by entropy. The intramolecular participation can be indicated with the symbol 'G--n', where G is the nucleophilic group acting as a 'bridge', and n is the number of atoms of the resulting ring. The reactions of Fig. A3.4 can be classified as MeO-5 (22)-(24); MeO-6, (22)-(25); O⁻ --3, (26)-(27).

Table A3.1

Relative rates of solvolysis of $MeO(CH_2)_n$ OBs (22)

Compound	*Relative Rate*		
n	EtOH(75°C)	acetic acid (25°C)	formic acid (75°C)
[$CH_3(CH_2)_3$ OBs]	1.00	1.00	1.00
2	.25	.28	.10
3	.67	.63	.33
4	20.4	657.0	461.0
5	2.8	123.0	32.6
6	1.19	1.16	1.13

An important effect of intramolecular participation is the resulting stereo-specificity of a reaction. For example nucleophilic substitutions (for instance solvolysis), can take place with configuration retention at the sp^3 carbon under-going substitution. In such cases the overall reaction consists in a double bimolecular nucleophilic substitution with inversion (Fig. A3.5(a)). The stereochemical implications of neighbouring group participation in a substitution process are shown in Fig. A3.5.

When *d,l-erythro*-3-bromobutan-2-ol (28a and 28b) is treated with concen-trated hydrobromic acid, *meso*-2,3-dibromobutane (30) is obtained, whilst *d,l-threo*-3-bromobutane-2-ol (31a + 31b) gives *d,l*,-dibromobutane (33a and 33b), also obtained from *threo*-3-bromobutan-3-ol in an optically active state (31a or 31b). These data agree with the intermediate formation of bromonium ions (29) and (32).

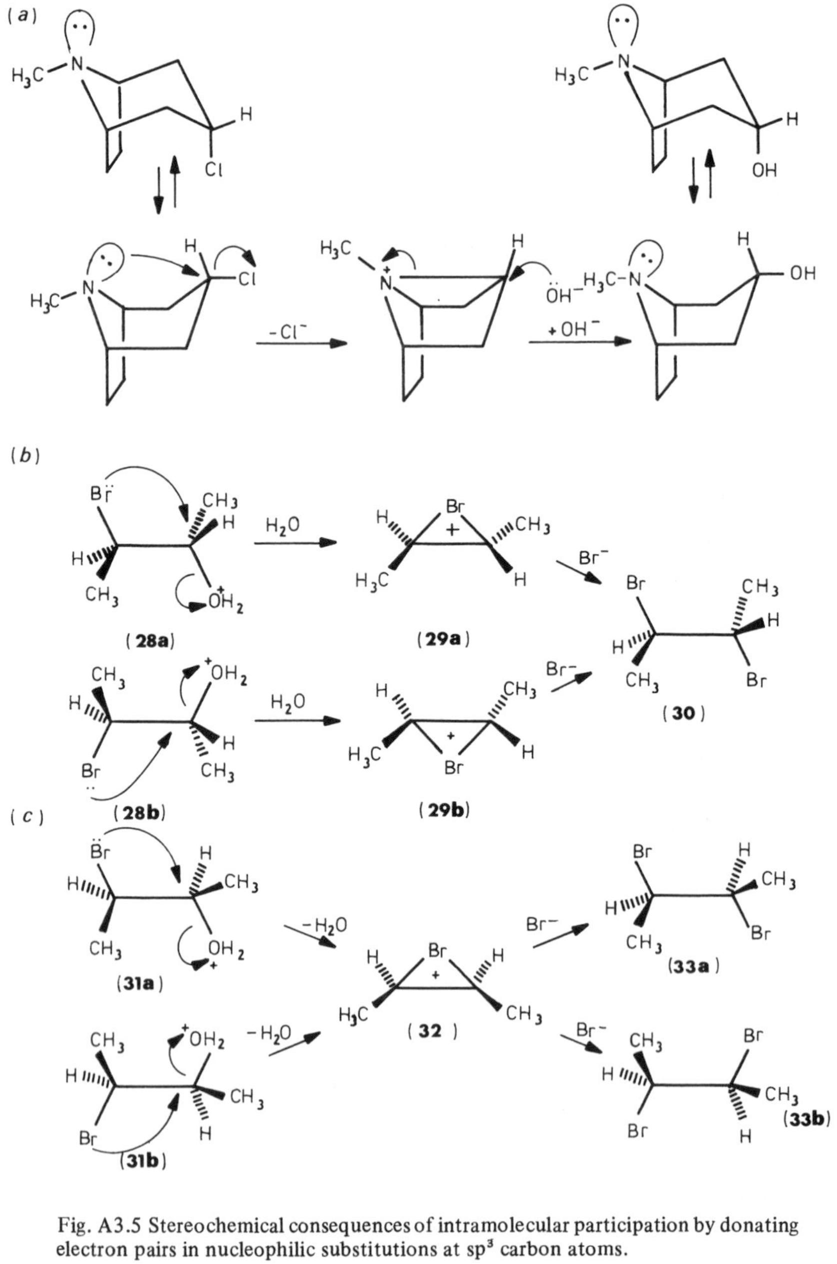

Fig. A3.5 Stereochemical consequences of intramolecular participation by donating electron pairs in nucleophilic substitutions at sp³ carbon atoms.

The neighbouring group can also be an unsaturated carbon atom with available π electrons. The formation of phenonium ions, e.g. (35), is a well-known example, since they are intermediates in reactions implying an incipient carbocation in the β position of the side chain of alkylbenzene derivatives (Fig. A3.6). The electron pair forming the bridge originally occupies a six centre (delocalised) π molecular orbital. The higher stability of the phenonium cation with respect to a cation with a localised positive charge on C–β, derives from the delocalisation of such a charge. Spiro-structures obtained by freezing the phenomium ion have been isolated, for instance eliminating a proton affords spirodienones (37) and (38).

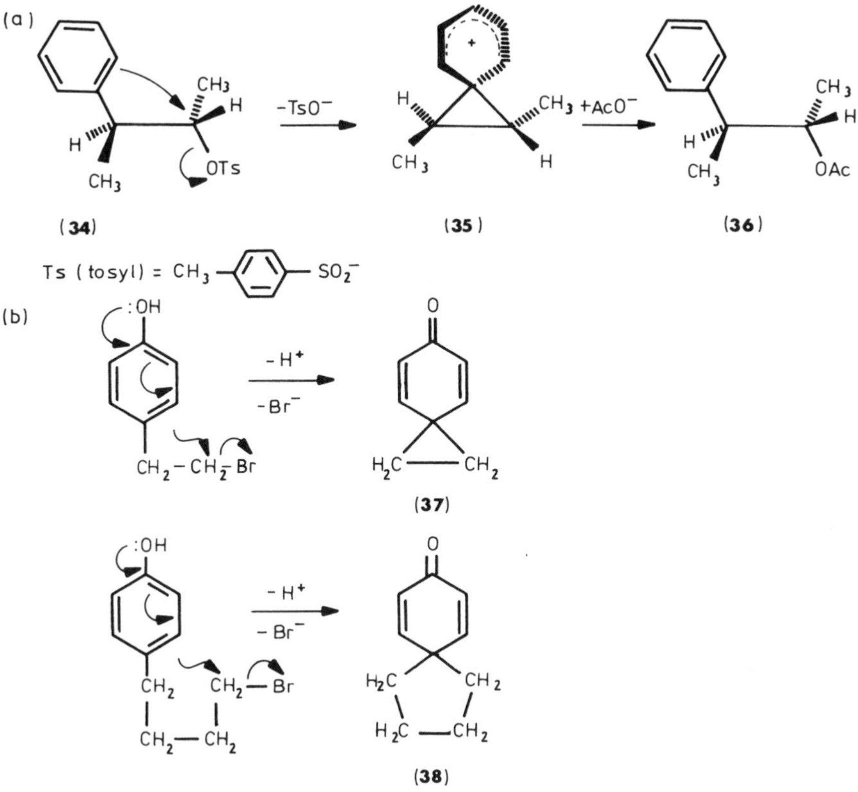

Fig. A3.6 Examples of intramolecular participation involving delocalised π electrons.

The π bond of an isolated olefinic double bond may interact with an incipient carbocation, provided the relative position of the three carbon atoms allows overlapping of the three p_z atomic orbitals (the three orbitals being the two originally associated with the olefinic bond, and the one from the carbonium centre) (see Fig. A3.7).

In the allylic cation (39) the p_z atomic orbitals overlap exclusively in a π type way (these are delocalised π electrons). In the homoallylic case (40) there is a small degree of σ overlap; the σ overlapping is more extensive in the 7-norbornenyl cation (41) and in the 2-norbornyl one (42). In these latter cases the electrons delocalised on three carbon atoms are essentially σ, and so (41) and (42) are 'σ-bridged non-classical cations' according to Bartlett's definition.

Reactions (c) and (d) (Fig. A3.7) are examples of non-classical cation formation, through the 'π route' [according to the symbol used for intramolecular participation, (41) should be indicated as π-3,4 and (42) as π-5,6]. Their formation is clearly noticeable by the Dewar notation [(18) Fig. A3.3].

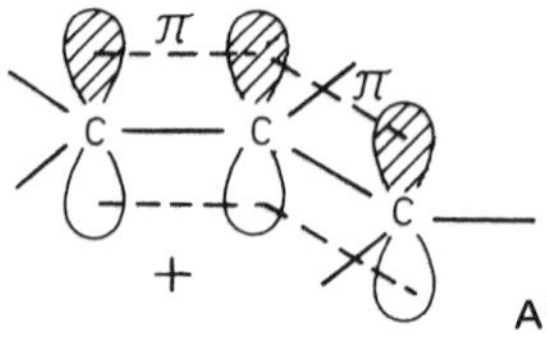

ALLYL CATION

(39)

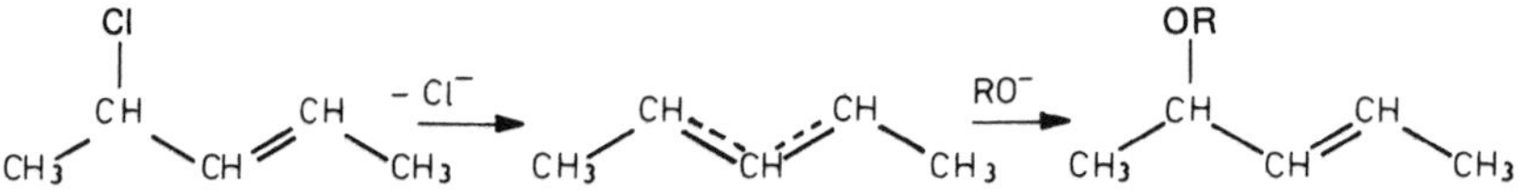

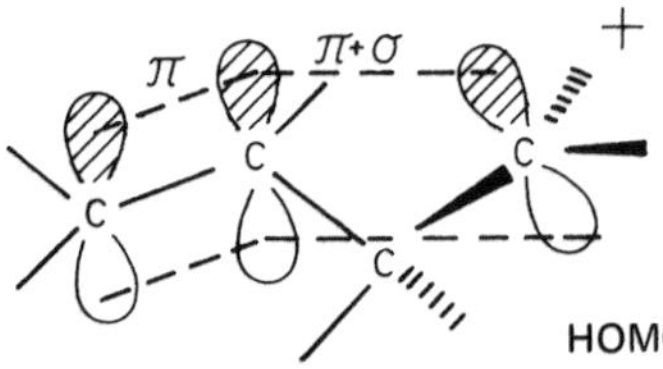

HOMOALLYLIC CATION

(40)

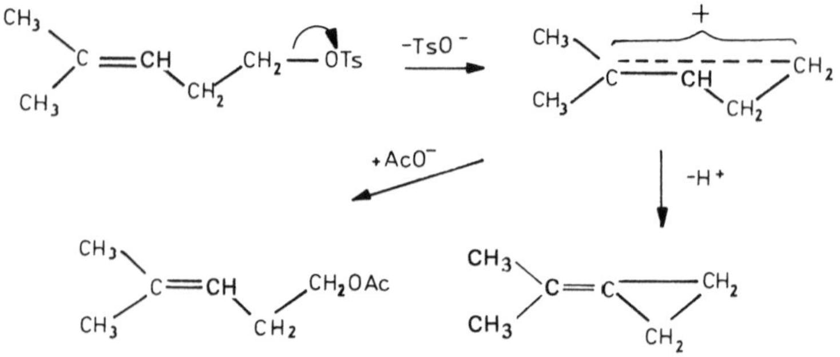

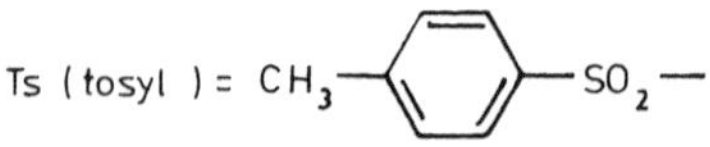

Fig. A3.7 (contd. overleaf)

Fig. A3.7 (contd.)

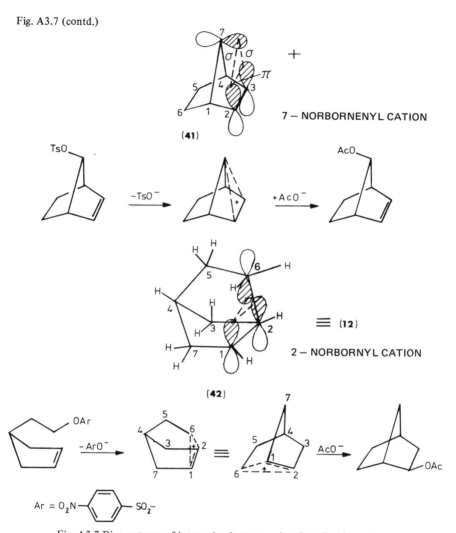

7 – NORBORNENYL CATION

2 – NORBORNYL CATION

$Ar = O_2N$—⟨benzene⟩—SO_2^-

Fig. A3.7 Diverse types of interaction between π bonds and carbocations.

If it is accepted that a σ electron pair adjacent to a nucleophilic substitution centre (particularly a single C–C or C–H bond) may also act as an intramolecular nucleophilic agent, similar non-classical carbocations can originate either through the already described π-route, or a σ-route (Fig. A3.8).

For the 2-norbornyl cation (12), an example of the σ-route is given by catalysis of *exo*-norborn-2-yl-*p*-toluenesulphonate (reaction c, Fig. A3.2). The bond between C-6 and C-1 of the norbornyl structure can easily interact with the incipient carbocation at C-2. (This σ-route could be indicated with the general notation CL_3-3, where L is a ligand, united through a simple bond to the carbon atom acting as a 'bridge'). The possible formation and transformations for a σ-bridged carbocation are shown in Fig. A3.8. The stereochemical aspects of the various transformations have been particularly stressed: non-classical carbocations have been and still are hypothesised as intermediates in order to explain steric features of biosynthesis. *Syn*-additions to double bonds *in vivo* have been explained with mechanisms of the type $(44) \rightarrow (45) \rightarrow (54)$.

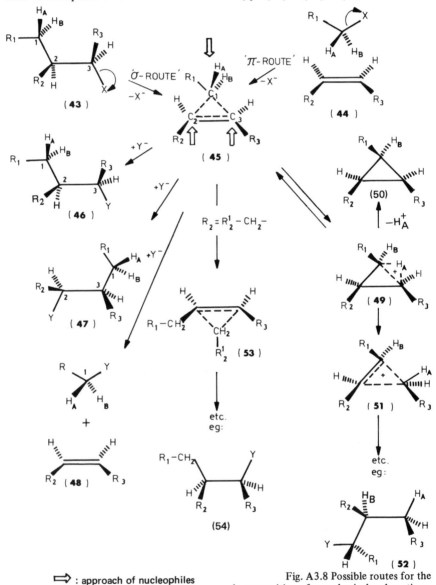

⇨ : approach of nucleophiles

Fig. A3.8 Possible routes for the decomposition of non-classical carbocations.

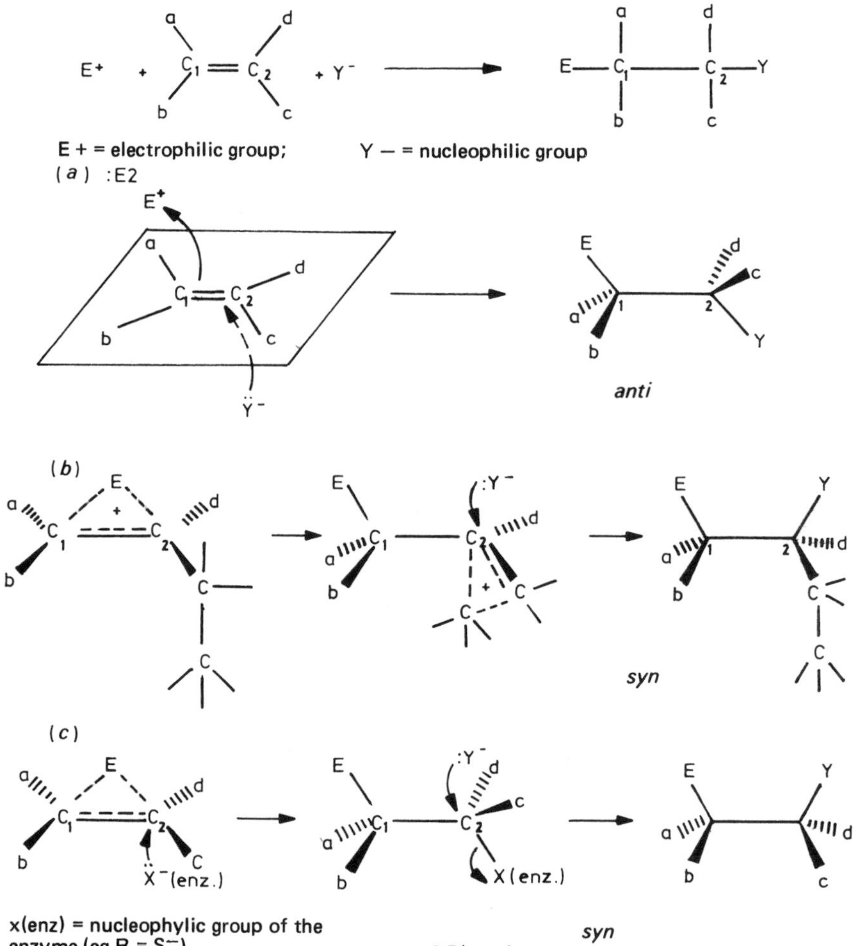

E + = electrophilic group;   Y − = nucleophilic group
(a) : E2

anti

(b)

syn

(c)

x(enz) = nucleophilic group of the
enzyme (eg R = S⁻)   syn

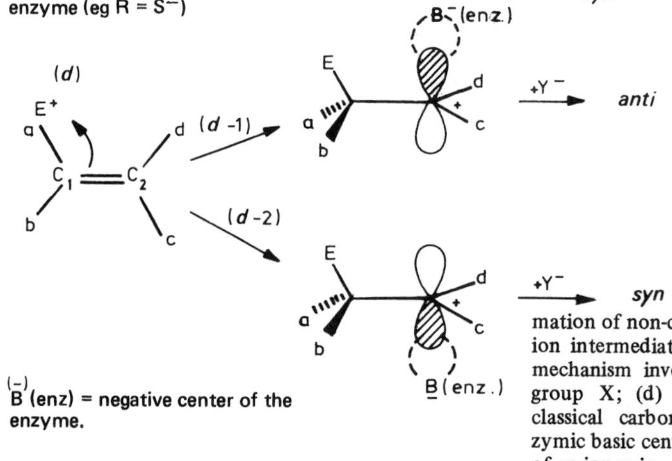

(d)

$\underset{(-)}{B}$ (enz) = negative center of the
enzyme.

anti

syn

Fig. A3.9 Possible mechanisms for the stereospecificity of enzymically-controlled 1,2-additions: (a) concerted addition; (b) intramolecular participation with formation of non-classical carbonium ion intermediates; (c) the 'in-out' mechanism involving an enzymic group X; (d) stabilisation of a classical carbonium ion by enzymic basic centre with formation of an ion pair.

It must be stressed that the nucleophilic substitutions proceeding with retention of configuration and the 1,2-*syn*-additions (possibly non-concerted) could also take place *in vivo* through an intermolecular participation of an enzyme centre to the reaction in question. Such a participation should consist in an 'in-out' intervention of a nucleophilic group X of the active site on the incipient carbocation [Cornforth's 'X group' hypothesis: via (c) of Fig. A3.9], or in the stabilisation of the configuration of a classical carbocation owing to formation of an intimate ion pair with a basic centre of the enzyme [via (d-2) of Fig. A3.9].

On considering these two other mechanistic hypotheses, the importance of the non-classical cations in enzyme-catalysed biosynthetic schemes appears to be diminished.

SOURCE MATERIALS AND SUGGESTED READING

[1] B. Capon, (1964), Neighbouring Group Participation, *Quart. Rev.*, **18**, 45.

[2] P. D. Bartlett, (1965), *Nonclassical Ions,* Benjamin.

[3] H. C. Brown, (1966), The Norbornyl Cation: Classical or Non Classical?, *Chem. in Britain*, **2**, 199.

[4] G. D. Sargent, (1966), Bridged, Non-classical Carbonium Ions, *Quart. Rev.*, **20**, 301.

[5] H. C. Brown, (1967, Feb. 13), A Symposium-in-print', *Chem. Engng. News*, **45**, 87.

[6] G. A. Olah, (1967, Mar. 27), Stable, Long-lived Carbonium Ions, *Chem. Engng. News,* **45**, 77.

[7] S. Winstein, (1969), Nonclassical Ions and Homoaromaticity, *Quart. Rev.,* **23**, 141.

[8] W. R. Dolbier, (1970), Neighbouring Group Participation, *J. Chem. Educ.,* **47**, 42.

[9] G. A. Olah, (1972), The Electron Donor Single Bond in Organic Chemistry, *Chem. in Britain*, **8**, 281.

[10] H. C. Brown, (1973), The Question of σ-Bridging in the Solvolysis of 2-Norbornyl Derivatives, *Accounts Chem. Res.*, **6**, 377.

[11] V. Buss, P. R. Schleyer and L. C. Allen, (1973), The Electronic Structure and Stereochemistry of Simple Carbonium Ions in *Topics in Stereochemistry*, ref. 13 of Appendix II, 7.

[12] G. A. Olah, (1973), Carbocations and Electrophilic Reactions, *Angew. Chem. Int. Ed. Engl.,* **12**, 173.

[13] H. C. Brown, (1975), Explorations in the Nonclassical Ion Area, *Tetrahedron*, **32**, 179.

[14] B. Capon and S. P. McManus, (1976), *Neighboring Group Participation*, Plenum Publishing Corporation.

[15] G. A. Olah, (1976), The σ-Bridged 2-Norbornyl Cation and Its Significance to Chemistry, *Accounts Chem. Res.,* **9**, 41.

[16] H. C. Brown, (1977), The Nonclassical Ion Problem, Plenum Publishing Corporation.
[17] J. B. Lambert, H. W. Mark, A. G. Holcomb and E. Stedman Magyar, (1979). Inductive Enhancement of Neighboring Group Participation, *Accounts Chem. Res.*, **12**, 317.

Isotopic Effects

If an atom in a molecule is substituted with an isotopic nucleus [for instance ^1H (protium, H) with ^2H (deuterium, D) or with ^3H (tritium, T), ^{14}N with ^{15}N, ^{12}C with ^{13}C or ^{14}C, ^{16}O with ^{17}O or ^{18}O, etc.] small changes in some chemical properties are usually observed (such as reaction rates and chemical equilibria) and in some physical properties (such as m.p., b.p., etc.) of the compound itself. D_2O has m.p. 3.82°C, and b.p. 101.42°C.

The small variations from the normal molecular and macroscopic properties caused by such isotopic substitutions are generally called *isotopic effects*.

The amount of a macroscopic isotopic effect may be estimated and will increase with:

(a) the degree of isotopic substitution (obtainable by measuring, spectroscopically, the relative abundance of the isotopic nuclides).

(b) the ratio between the mass of the substituted nuclide and its isotope. In certain cases the degree of the isotopic effect depends on the position of the isotopically substituted atom in the molecule.

Herein the 'kinetic isotopic effect' (K.I.E.) will be discussed; it is manifest as a variation of the reaction rate. Its great importance in the study of enzymatic reactions is also described. The kinetic isotope effect relative to a certain reaction, e.g. reaction (1), and to a certain substrate A is quantitatively expressed by the ratio $(k_Z/k_{Z'})$, where k_Z is the rate constant for the transformation of A, containing the nuclide in a certain position, and $k_{Z'}$ is the analogous constant for A observed when Z has been substituted by the isotope Z'; k_Z and $k_{Z'}$ are experimentally obtained by kinetic measurements.

$$A\,(+B) \xrightarrow{\;k\;} \text{products} \tag{1}$$

for a first order reaction in A involving a monomolecular reaction:

$$V = \frac{-d[A]}{dt} = k[A]$$

For a bimolecular reaction of first order in A and in B.

$$V = \frac{-d[A]}{dt} = k[A][B]$$

Two kinetic isotope effects are recognised (Fig. A4.1). The *primary* effect appears when the isotopically substituted atom is directly involved in the reaction, that is when one of its bonds is broken. The *secondary* effect occurs when the isotopically substituted atom is only indirectly implied and none of its bonds is broken; in such a case the isotopic effect must be attributed to a change in vibration of at least one of its bounds in passing from the ground state to the transition state. Secondary isotope effects are usually much weaker than primary ones and become smaller as the distance of the isotope from the reaction centre is increased.

PRIMARY KINETIC ISOTOPE EFFECT

(a)

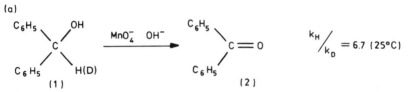

$$k_H/k_D = 6.7 \ (25°C)$$

(P. SYKES, *A Guidebook to Mechanism in Organic Chemistry* 4th ed. Longman, London, 1975, p. 46)

(b) $^{12}_{(14)}CH_3I + OH^- \longrightarrow ^{12}_{(14)}CH_3OH + I^-$ $k_{12_C}/k_{14_C} = 1{,}09 \ (25°C)$

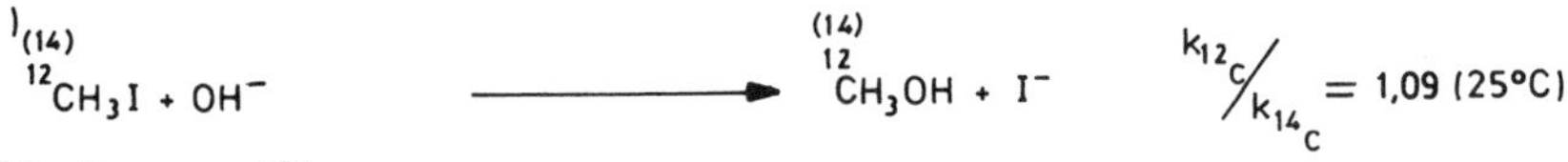

(*ibed:m* , p.47)

(c) $C_6H_5 - CH(D)O$

$C_rO_3 \nearrow C_6H_5 - COOH$ $k_H/k_D = 4.3$

$\searrow OH^-$ Cannizzaro $C_6H_5 - COOH + C_6H_5 - CH_2(D)OH$ $k_H/k_D = 1.8$

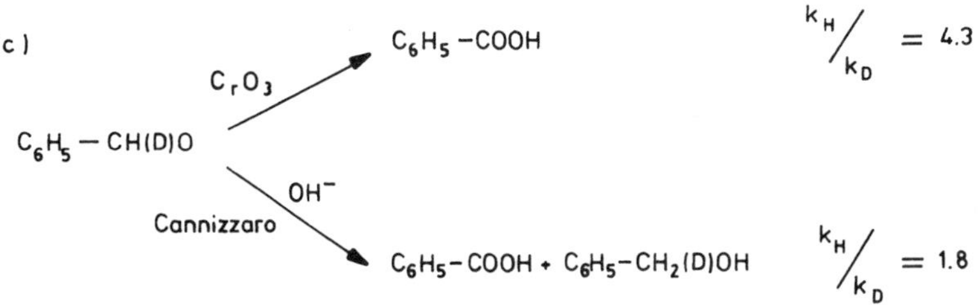

(K. B. WIBERG, *J. Am. Chem. Soc.,* **76**, 5371 (1954))

(d) $C_6H_5 - \overset{CF_3}{\underset{OH}{CH(D)}} \xrightarrow{MnO_4^-} C_6H_5 - \overset{O}{\overset{\|}{C}} - CF_3$ $k_H/k_D = 16$

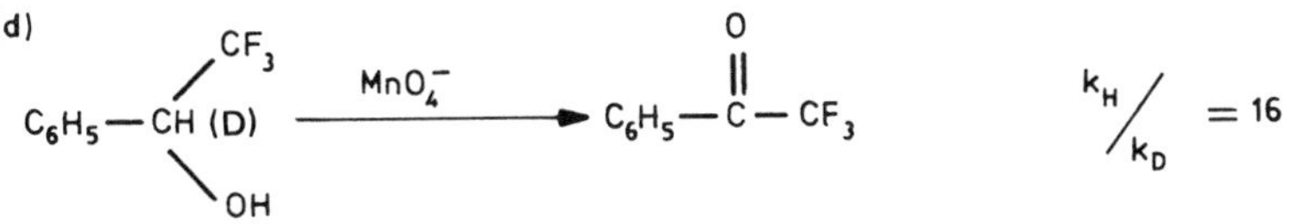

(R. STEWARD and R. VAN DER LINDEN, *Tetrahedron Letters,* 1960, 28)

Fig. A4.1 (contd. overleaf)

Fig. A4.1 (contd.)

(e) *SECONDARY KINETIC ISOTOPE EFECT*

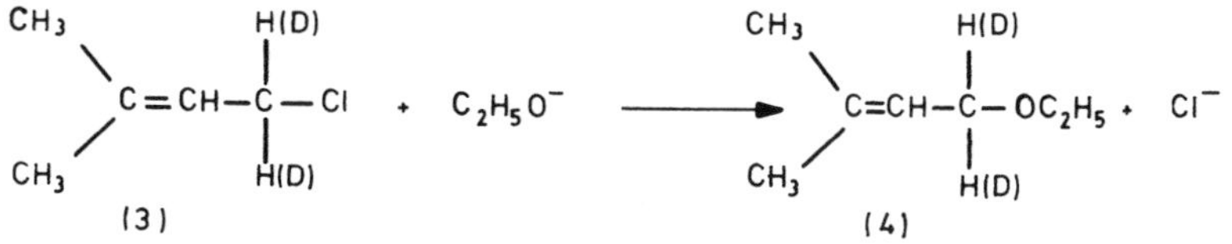

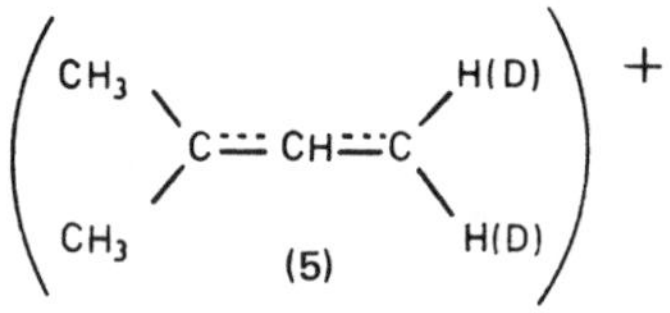

$$\underset{CH_3}{\overset{CH_3}{\diagdown}}C=CH-\underset{\underset{H(D)}{|}}{\overset{\overset{H(D)}{|}}{C}}-Cl \;+\; C_2H_5O^- \longrightarrow \underset{CH_3}{\overset{CH_3}{\diagdown}}C=CH-\underset{\underset{H(D)}{|}}{\overset{\overset{H(D)}{|}}{C}}-OC_2H_5 \;+\; Cl^-$$

(3)   (4)

$$\left(\underset{CH_3}{\overset{CH_3}{\diagdown}}C \cdots CH \cdots C\overset{H(D)}{\underset{H(D)}{\diagup}} \right) \;+\qquad k_H/k_D = 1.20$$

(5)   (for two deuteriums)

(V. BELANIC-LIPOVAĆ, S. BORČIČ, and D. E. SUNKO, *Croat. Chem. Acta.*, **37**, 61 (1965))

(f) TERTIARY ALKYL CHLORIDE SOLVOLYSES, : CUMULATIVE β-DEUTERIUM EFFECTS (80 per cent aqueous alcohol, 25°C)

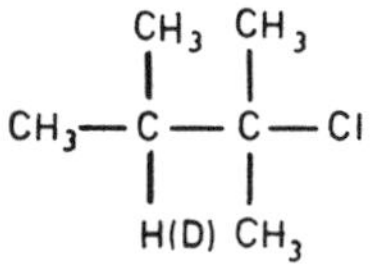

$$CH_3-\underset{\underset{H(D)}{|}}{\overset{\overset{CH_3}{|}}{C}}-\underset{\underset{CH_3}{|}}{\overset{\overset{CH_3}{|}}{C}}-Cl \quad k_H/k_D = 1.28 \quad ; \quad CH_3-CH_2(O_2)-\underset{\underset{CH_2}{|}}{\overset{\overset{CH_3}{|}}{C}}-Cl \quad k_H/k_D = 1.40$$

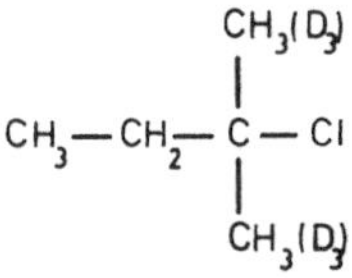

$$CH_3-CH_2-\underset{\underset{CH_3(D_3)}{|}}{\overset{\overset{CH_3(D_3)}{|}}{C}}-Cl \quad k_H/k_D = 1.77 \quad ; \quad CH_3-CH_2(D)-\underset{\underset{CH(D)}{|}}{\overset{\overset{CH_3(D_3)}{|}}{C}}-Cl \quad k_H/k_D = 2.35$$

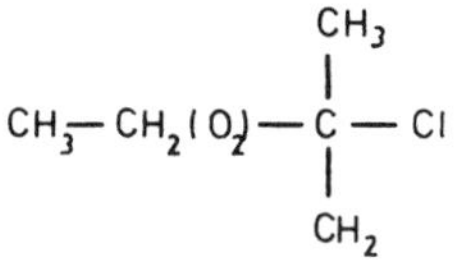

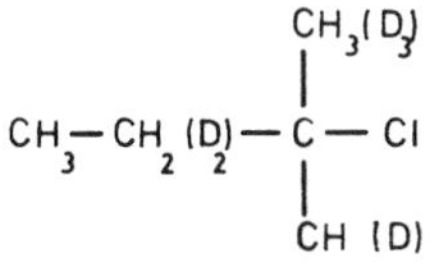

(V. J. SHINER Jr., *J. Am. Chem. Soc.*, **75**, 2925 (1953); **76**, 1603 (1954))

(g) INVERSE KINETIC ISOTOPE EFFECT

$$\underset{C_6H_5}{\overset{H}{\diagdown}}C=C\underset{H(D)}{\overset{COOH}{\diagup}} \xrightarrow[MnO_4^-]{k_1} C_6H_5-\underset{\underset{O}{|}}{\overset{\overset{H}{|}}{C}}-\underset{\underset{O}{|}}{\overset{\overset{COOH}{|}}{C}}-H(D) \xrightarrow{k_2} \begin{matrix} C_6H_5CHO \\ + \\ O=CH(D)-COOH \\ + \\ MnO_2^- \end{matrix}$$

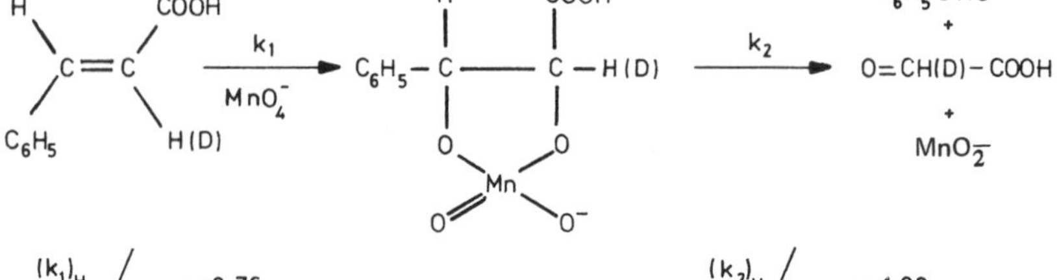

$$(k_1)_H / (k_1)_D = 0.76 \qquad (k_2)_H / (k_2)_D = 1.09$$

(D. J. LEE and J. R. BROWNRIDGE, *J. Am. Chem. Soc.*, **95**, 3033 (1973))

Fig. A4.1 Examples of kinetic isotope effects 51.

THEORY OF THE KINETIC ISOTOPE EFFECT

A rigorous theoretical treatment of the knetic isotope effect is beyond the scope of this book. Only the simplest treatment of the primary effect, known at the *semi-classical interpretation* will be given.

As a basis for our discussion let us consider reactions of type (2), where X and Y represent atoms, or atomic groups, either electrically neutral or charged, between which a proton (or a deuteron) is being transferred, or a hydrogen (or deuterium) atom, or a hydride (or deuteride) ion.

$$X\!-\!H\,(D) + Y \xrightarrow{\;kH\,(kD)\;} X + Y\!-\!H\,(D)$$

These reactions have been chosen because they are both simple and important in enzymatic processes. Such problems are frequent in nature and the preparation of specifically deuterated compounds is easy.

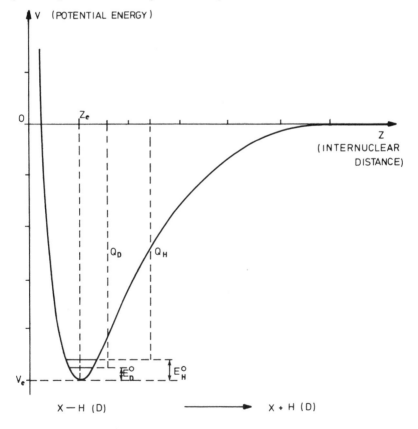

Fig. A4.2 Potential energy curve for a system of two nuclic X–H(D) showing the zero point energy (vibratin). (The symbol E refers to the energy of one mole.)

The theoretical considerations and conclusions obtained for the pair X-H; X-D can be extended to other pairs of isotopically substituted bonds, particularly to the couple X-H; X-T (see Table A4.1).

The semiclassical interpretation of the $k_Z/k_{Z'}$ ratio (in our case k_H/k_D) is based on two fundamental postulates: (a) the potential energy curve, or energy surface as a function of the internuclear distances in a system (substrate molecule, or substrate + reagent), is not modified by isotopic substitutions; (b) the bonds are assumed to act as ideal harmonic oscillators, whose vibrations can be explained by quantum mechanical principles.

Let us firstly consider a type (2) reaction consisting of the homolytic or heterolytic dissociation of the X-H(D) bond. The potential energy curve of such a bond is shown in Fig. A4.2 and it is plotted in the electronic ground state versus the internuclear X-H (or D) distance. The quantum mechanical treatment of the X-H bond (harmonic oscillator) leads to the conclusion that the energy associated with the lowest possible vibrational level overcomes the minimum of the potential energy curve by $\epsilon_H^o = \frac{1}{2} h \nu_{X-H}$ [ν is the oscillator frequency in s^{-1}, according to purely classical theory, that is from equation (3); h is Planck's constant, $6.625 \cdot 10^{-27}$ erg.s]. Analogously, for the X-D bond this is $\epsilon_D^o = \frac{1}{2} h \nu_{X-D}$. ϵ^o is called the *vibrational energy at zero point,* and is related to the nuclear masses by the classical equation:

$$\nu^2 = \frac{f}{4\pi^2\mu} \tag{3}$$

where f is the bond strength constant (identical for both X-H and X-D), and μ is the reduced mass of the system. For instance,

$$\mu_{X,H} = \frac{m_X \cdot m_H}{m_X + m_H} \tag{4}$$

m = mass of the nuclide X,
m_H = mass of the nuclide protium.
On writing (3) for the systems X-D and X-H, and dividing throughout:

$$\frac{\nu^2_{X-D}}{\nu^2_{X-H}} = \frac{\mu_{X,H}}{\mu_{X-D}} \tag{5}$$

If (4) is accounted for, (5) becomes:

$$\frac{\nu_{X-D}}{\nu_{X-H}} = \sqrt{\frac{m_H(m_X + m_D)}{m_D(m_X + m_H)}} \tag{6}$$

as $m_X + m_D \simeq m_X + m_H$ this simplifies to

$$\nu_{X-D} \simeq \sqrt{\tfrac{1}{2}} \cdot \nu_{X-H} \tag{7}$$

ν_{X-D} and ν_{X-H} have been accurately measured many times by infra-red spectro-scopy and the best average value for their ratio is 0.741. The small difference from the value $\sqrt{\tfrac{1}{2}}$ partly arises from the reduced masses being slightly higher than m_H and m_D, and partly on the anharmonicity of the oscillating system.

The difference between the energies of the two X-H and X-D bonds at the zero point can be calculated through the relation:

$$\epsilon^{\circ}_{H} - \epsilon^{\circ}_{D} = \tfrac{1}{2} h \, (\nu_{X-H} - \nu_{X-D}) = \tfrac{1}{2} h \, \nu_{X-H} \, (1 - 0.741) =$$

$$0.130 \, h \cdot \nu_{X-H} \tag{8}$$

Analogously, for the X-H and X-T bonds:

$$\epsilon^{\circ}_{H} - \epsilon^{\circ}_{T} = 0.185 \, h \cdot \nu_{X-H} \tag{9}$$

The zero point energy differences for other isotopes are much smaller than those for H:D and H:T, for two reasons: (a) the coefficient of equations such as (8) and (9) is reduced when the nuclide masses increase [e.g. 0.0286 for ^{16}O; ^{18}O, see (6)]; (b) the frequency ν also decreases when the reduced mass of the oscillating bond increases.

If we refer to molar quantities, (8) becomes:

$$E^{\circ}_{H} - E^{\circ}_{D} = .130 \, N \, h \, \nu_{X-H} = 3.72 \cdot \bar{\nu}_{X-H} \cdot 10^{-4} \, kcal/mole$$

N being Avogadros's number, and $\bar{\nu}$ the wavenumber in cm^{-1} [The relation between frequency and wavenumber is $\nu = \bar{\nu}.c$, where c is the speed of light; so $\nu(s^{-1}) = 2.9979 \, .\bar{\nu}(cm^{-1}). \, 10^{10}$].

The difference between the zero point vibrational energies $(E^{\circ}_{H} - E^{\circ}_{D})$ causes a kinetic isotope effect in breaking the X-H (D) bond, according to equation (13): this relationship is obtained from equations (11) and (12), themselves obtained by the well known exponential expression relating rate constant to the activation energy [and, using the nomenclature of Fig. A4.2, $Q = -(Ve + E)$]

$$\frac{k_H}{k_D} = \frac{\exp(-Q_H/RT)}{\exp(-Q_D/RT)} \tag{11}$$

that is, $$\frac{k_H}{k_D} = \exp. \{[(Ve + E^{\circ}_{H}) - (Ve + E^{\circ}_{D})]/RT\} \tag{12}$$

or $\quad \dfrac{k_H}{k_D} = \exp\left[(E^{\circ}_H - E^{\circ}_D)/RT\right]$ $\qquad$ (13)

Where R (gas constant) $= 1.987.10^{-3}$ kcal mole^{-10} k^{-1}; T $=$ absolute temperature. On considering equation (10) we obtain:

$$\frac{k_H}{k_D} = \frac{0.1872 \cdot \bar{\nu}_{X-H}}{T} \qquad (14)$$

expressing the isotopic effect as a function of the wave number and the temperature. The k_H/k_D values, calculated for the more common single bonds through relationship (14) are shown in Table A4.1.

Table A4.1

Theoretical maximum values for kinetic primary isotope effects for the pairs ^{1}H:^{2}H and ^{1}H: ^{3}H at 25°C*

Bond	(cm^{-1})	$E^{\circ}_H - E^{\circ}_D$ (k cal. mole^{-1})	k_H/k_D	k_H/k_T
C–H	2900	1.078	6.18	13.43
N–H	3100	1.153	7.01	16.06
O–H	3300	1.227	7.94	19.21
S–H	2500	0.930	4.80	9.38

*Only stretching vibrations are considered.

If the process (2) takes place in one step, and not in two as in the above hypothesis, the potential energy curve for the system versus the reaction coordinate X can be presented as in Fig. A4.3.

It can be deduced from this that the activation energy $Q^{\ddagger}$ for transferring the protium (or deuterium) nuclide from X to Y is smaller than that necessary for the above described two-step hydrogen transfer. $Q^{\ddagger}$ is no longer obtained as the difference between the dissociation electronic energy $(-Ve)$ and the zero point energy; in a concerted mechanism $Q^{\ddagger}$ equals the difference between the actual energy of the transition state $(V^{\ddagger} + E^{\ddagger})$ and the substrate energy in its lowest level $(Ve + E^{\circ})$. Equation (12) then becomes

$$\frac{k_H}{k_D} = \frac{\exp\left\{-\left[(V^{\ddagger} + E_H^{\ddagger}) - (Ve + E^{\circ}_H)\right]/RT\right\}}{\exp\left\{-\left[(V^{\ddagger} + E_D^{\ddagger}) - (Ve + E^{\circ}_D)\right]/RT\right\}} \qquad (15)$$

that is:

$$\frac{k_H}{k_D} = \exp\left\{\left[(E^{\circ}_H - E^{\circ}_D) - (E_H^{\ddagger} - E_D^{\ddagger})\right]/RT\right\} \qquad (16)$$

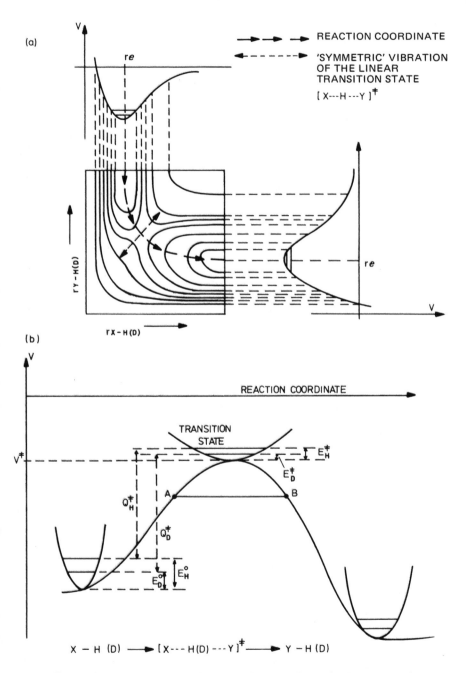

Fig. A4.3 Potential energy surface (a) and origin of the kinetic isotope effect (b) for a reaction involving transfer of hydrogen.

This means that the ratio k_H/k_D will be equal to that obtained from (14) (see Table A4.1) multiplied by a factor γ, depending on the difference in the zero point energy between the protium transferring transition state, and the deuterium transferring one; $\gamma = \exp\ [-(E_H^{\ddagger} - E_D^{\ddagger})/RT]$. If the zero point energies of the transition states are analysed in an analogous manner to the ground states of the X–H and X–D bonds we obtain $(E_H^{\ddagger} - E_D^{\ddagger}) \geqslant 0$, and so $\gamma \leqslant 1$. Therefore the k_H/k_D ratios listed in Table I should represent *maximum* possible values for type (2) reactions: lower experimental values should indicate one-step processes, with transition states characterised by partial bonds between H (or D) and X and Y. The eventuality that $(E_H^{\ddagger} - E_D^{\ddagger}) > (E^{\circ}_H - E^{\circ}_D)$ i.e. $k_H/k_D) < 1$, is extremely unlikely because, in the transition state, the transferred atom H (or D) should oscillate with an f force constant *larger* than that of the bond being broken [e.g. see (3)]. When a bond strength increases, the frequency of its associated harmonic oscillator also grows, together with the energy difference the zero point, due to the isotope substitution [see (10)]. Numerous experimental measures of primary kinetic isotope effects have been made in recent years. For most hydrogen transfer reactions, k_H/k_D ratios varying from 1 to the maximum value of Table 1 have been found. Occasionally effects *larger* than those predicted by the semi-classical treatment have been observed. Quantitative interpretations of these anomalous isotope effects can be found in specialised texts (e.g. the monograph by R. P. Bell, see Bibliography). Two important phenomena have been omitted in the above treatment, which can sometimes contribute to the primary kinetic isotope effect. These are the bond *bending vibrations* and *tunnelling*. The bending vibrations can be accounted for by a treatment substantially similar to that already described for stretching vibrations: the differences in $E^{\circ}_H - E^{\circ}_D$ are increased, and so are the theoretical maximum values for k_H/k_D and k_H/k_T (see Table A4.2).

Table A4.2

Table A4.1 values corrected for the contribution of bending vibrations contributions

Bond	$E^{\circ}_H - E^{\circ}_D$ (k cal. mole^{-1})	k_H/k_D	k_H/k_T
C–H	1.35	10	27
N–H	1.43	11	33
O–H	1.50	13	39
S–H	1.21	8	19

Tunnelling is a consequence of the Heisenberg uncertainty principle. According to this principle the position and the momentum (mass × rate) of a particle cannot be stated simultaneously in an exact way. The principle is fully satisfied by a quantum mechanical treatment of the oscillating bond: in fact, at the zero point, the system does not have energy V_e, and nuclei *still* at a distance r_e. The imprecise nature of the position Δr and of the momentum along the r axis, $\Delta(mv_r$, are related by:

$$\Delta r. \; \Delta(mVr) \simeq h/2\pi \tag{17}$$

When the mass is small, such as for H, D and T, the uncertainty in its momentum is also small and, therefore, the uncertainty in its position can reach values of the same order of magnitude as the distances covered by the nuclei during reactions leading to transition states, such as shown in Fig. A4.3. In other words, because of the uncertainty in its position a nucleus can overcome the barrier of a potential energy curve, (for instance from A to B, Fig. A4.3) passing from one slope to the other one without proceeding over the top and so an energy barrier smaller than the activation energy is required. For the above reasons the tunnel effect will be more effective for smaller nuclides [see (17)]: the rate reaction for protium will be increased with respect to the one of deuterium and tritium; and the k_H/k_D and k_H/k_T ratios will be larger than the ones obtained with the above semi-classical treatment.

PRIMARY KINETIC ISOTOPIC EFFECTS AND ENZYMATIC REACTIONS

Secondary kinetic isotopic effects are usually neglected in studying the stereo-specificity of biocatalysed reactions involving labelled isotopes: primary effects in contrast, should always be borne in mind, especially for the couples H:D and H:T. (Secondary isotope effects are helpful in helping to determine transition state structures in enzymatic reactions).

Consider the enzymatic dehydrogenation reaction of a primary alcohol (6) to aldehyde (7) (Fig. A4.4), using a D or T labelled alcohol in place of the hydrogens bonded to the alcoholic carbon. Assume the substrate molecules are accepted in the active site of the enzyme with the same facility, with or without the isotopic label, in a fast and reversible process. This implies a zero secondary isotope effect in forming the substrate-enzyme complex (such an admission is usually acceptable, although not rigorously true). Finally suppose the breaking of the C–H bond is the slowest step, thus limiting the overall rate of the enzymatic process (the rate-limiting step).

In order to forecast the primary kinetic isotopic effect on the isotopic composition of the reaction mixture one should account for certain features of the enzyme and the substrate:
(a) the stereospecificity of the enzyme; (b) the isotopic purity of the substrate; (c) the enantiomeric purity of the substrate stereospecifically labelled in the hydroxymethyl group with D or T (that is, in H_R or H_S).

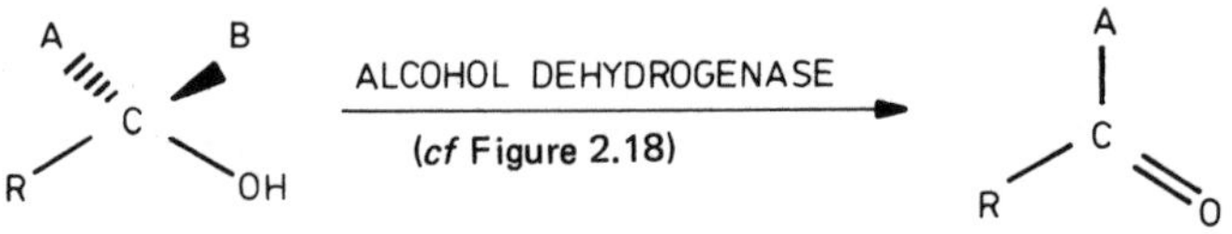

a : A, B $= {}^1H$ (A = pro-S; B = pro-R)

b : A, B $= D$

c : A $= {}^1H$; B $= D$

d : A $= D$; B $= {}^1H$

e : A $= {}^1H$; B $= T$

f : A $= T$; B $= {}^1H$

g : A, B $= {}^1H$; C $= {}^{14}C$

a : A $= {}^1H$

b : A $= D$

c : A $= T$

d : A $= {}^1H$; C $= {}^{14}C$

Fig. A4.4 Cases for isotope effects in an alcohol to aldehyde process catalysed by an enzyme.

First case: the enzyme is not stereospecific towards H_R and H_S (such a situation also exists for oxidations with achiral reagents, such as Ag^{2+}, CrO_3, etc.). The enantiomeric purity of the labelled substrate is in this case unimportant and the isotopic purity only should be considered.

A pure dideuterated alcohol [molecules D_2-alcohol (6b) = 100%] will produce the corresponding pure monodeuterated aldehyde [molecules D_1-aldehyde (7b) = 100%], although at a lower rate (k_D) than the one observed in the non-labelled alcohol (k_H). If the alcohol is purely monodeuterated [D_1-alcohols (6c) and (6d) = 100%]. The C–D bond will be broken more slowly than the C–H bond and a mixed aldehyde will be obtained, composed of the non-deuterated species (7a) and of the deuterated one (7b) this latter prevailing. At every moment of such a transformation (assuming it is irreversible) there will be:

$$\frac{\text{molecules (7b) (D}_1\text{-aldehyde)}}{\text{molecules (7a) (D}_0\text{-aldehyde)}} = \frac{k_H}{k_D} \tag{18}$$

that is:

$$\frac{\text{(D}_1\text{-aldehyde)}}{\text{(D}_1\text{-alcohol)}_{\text{reacted}} - \text{(D}_1\text{-aldehyde)}} = \frac{k_H}{k_D} \tag{19}$$

Let us write $(D_1\text{-alcohol})_{\text{reacted}}$ = total aldehyde formed = 100: then:

$$\% \, (D_1\text{-aldehyde}) = \frac{100}{1 + k_D/k_H} = \text{deuterium retention} \qquad (20)$$

On assuming $k_H/k_D = 6$.

$$(D_1\text{-aldehyde}) = 85.7\%$$

If the alcohol is a mixture of monodeuterated and non-deuterated molecules, for instance D_1-alcohol, (6c) and (6d) = 70%, and D_0-alcohol (6a) = 30%, the isotopic compositions will vary with time. The (6a) molecules will in fact react at almost double the rate of (6c) and (6d) and so the deuterium contact will be initially concentrated in the alcohol species, later on, as the conversion proceeds, the deuterium content in the aldehyde increases until the 60% theoretical value is achieved at completed conversion.

At the end of the reaction:

$$\frac{(D_1\text{-aldehyde})}{(D_0\text{-aldehyde})_{\text{ex } D_1\text{-alcohol}}} = \frac{k_H}{k_D} \qquad (18a)$$

that is

$$\frac{(D_1\text{-aldehyde})}{(D_1\text{-alcohol})_{\text{reacted}} - (D_1\text{-aldehyde})} = \frac{k_H}{k_D} \qquad (19a)$$

as in (18) and (19). As $(D_1\text{-alcohol})_{\text{reacted}} = 70\%$ (that is, of 100 molecules transferred into the aldehyde, 70 are D_1-alcohol), there is:

$$\% \, (D_1\text{-aldehyde}) = \frac{70}{1 + \dfrac{k_D}{k_H}} \qquad (20a)$$

If $k_H/k_D = 6$; D_1-aldehyde = 60%.

A situation for molecules containing only one isotopically labelled atom and diluted into a very large number of unlabelled identical molecules, is typical for tritiated substrates. In such cases the tritium retention or loss is evaluated with the help of an internal standard, usually the substrate being labelled with ^{14}C. That means that the $[^{14}C,T]$ compound is actually a mixture of (6a), (6e), (6f) and (6g) molecules: the abundance of tritiated species is measured by the $T/^{14}C$ ratio. For the above reasons, if we operate with tritiated compounds, the degree of reaction completed must be taken in due account. When a *complete conversion* has occurred, there is:

$$\frac{\text{(retained tritium)}}{\text{(lost tritium)}} = \frac{k_H}{k_T} \tag{21}$$

[compare with (18a)]
that is:

$$\frac{T/^{14}C \text{ (aldehyde)}}{(T^{14}C \text{ (initial alcohol)} - T/^{14}C \text{ (aldehyde)}} = \frac{k_H}{k_T} \tag{22}$$

During intermediate states of conversion, $T/^{14}C$ aldehyde values much lower than the final ones will be read, whilst the unreacted alcohol will show $T/^{14}C$ values much larger than the initial ratio.

Second case: the enzyme is stereospecific, and distinguishes between the stereo-heterotopic atoms H_R and H_S of (6) *(only one of these being eliminated).*
    This case differs from the former one, as the alcohol molecules have only one reactive atom, and not two (for instance H_R if dehydrogenases such as LADH and YADH are considered). With a non-stereospecific reagent, the oxidation of the hydroxymethyl group would equally involve both the hydrogen atoms, which would be eliminated at equal rates, *apart from isotopic effects.* The isotopic substituent now influences the reaction rate only if present in a reactive position (H_R in this example).
    (Actually the isotopic substituent in the non-reactive position, H_S, weakly influences the rate of the enzymatic process by a secondary kinetic isotope effect because of its proximity to the reaction centre. That is, when the $C(H)–H$ bond breaks, $k_{H(H)}$ and $k_{H(D \text{ or } T)}$ as well as a $k_{D \text{ or } T(H)}$ rates should be distinguished, where the nuclide in parentheses is geminal to the one involved in the oxidation. In this second case discussion of such effects is omitted.)
    The following situations should therefore be considered:
(a) *Enantiomerically pure substrate, with the heavy isotope in the non-reactive position of the alcohol,* (H_S).
Result: total retention of the isotope in the aldehyde.
For the initial mixture of (6a) and (6d) molecules, throughout the conversion of (6) into (7), there is:

$$\frac{(S)\text{-}(D_1\text{-alcohol})}{(D_0 - \text{alcohol})} = \frac{(D_1\text{-aldehyde})}{(D_0\text{-aldehyde})} \tag{23}$$

The same thing will take place with a $[T, {}^{14}C]$ alcohol (6f + 6g + 6a):

$$T/^{14}C \text{ (aldehyde)} = T/^{14}C \text{ (alcohol)} \tag{24}$$

(b) *Enantiomerically pure substrate with the heavy isotope present in the reactive position of the alcohol* (H_R).

The labelled molecules will react more slowly than the unlabelled ones. *Result:* a remarkable concentration of the isotope in the untransformed alcohol will be noticed, whilst the aldehyde produced will be unlabelled.

(c) *The substrate is not enantiomerically pure* (it can be racemic, that is composed of an equal number of **6c** and **6d** molecules). These situations must be treated taking into account the considerations made in (a) and in (b). Let us consider a highly labelled mono-deuterated alcohol (100% D_1-molecules), comprised of a mixture of 70% (*R*) enantiomer (**6c**), and 30% (*S*) enantiomer (**6d**). The dehydrogenation reaction will initially take place on the unlabelled molecules at the reactive centre, which are, in this example, impure (**6d**) ones.

The first aldehyde molecules will therefore be deuterated to a large extent (D_1-aldehyde $\gg$ 30%) and only at the end of the reaction will the D_1aldehyde (**2b**) = 30%. The analysis of an aldehyde isolated when the enzymatic reaction is not completed could lead to a mistake, which is attributing the enzyme the opposite stereospecificity to the actual one; this is because of the apparent deuterium retention in an alcohol considered as pure (*R*)-[D_1].

When data on enzymatic reactions involving stereospecifically labelled substrates made by synthesis are examined, it is good practice:

(1) to check the enantiomeric purity of the substrate as much as possible;

(2) check that the reaction is complete;

(3) to test both forms of the enantiomerically labelled substrate with the enzyme.

SOURCE MATERIALS AND SUGGESTED READING

[1] K. B. Wiberg, (1955), The Deuterium Isotope Effects, *Chem. Rev.,* **55**, 713.

[2] J. Bigeleisen and M. Wolfsberg, (1958), 'Theoretical and Experimental Aspects of Isotope Effects in Chemical Kinetics in *Advances in Chemical Physics,* Wiley-Interscience, **1**.

[3] L. Melander, (1960), *Isotope Effects on Reaction Rate,* Ronald Press.

[4] F. H. Westhimer, (1961), The Magnitude of the Primary Kinetic Isotope Effect for Compounds of Hydrogen and Deuterium, *Chem. Rev.,* **61**, 265.

[5] J. F. Thompson, (1963), *Biological Effects of Deuterium,* McMillan.

[6] C. J. Collins, (1964), Isotopes and Organic Reaction Mechanisms in *Advances in Physical Organic Chemistry,* (V. Gold, Ed.), **2**, Academic Press.

[7] W. H. Saunders, (1966), Kinetic Isotope Effects in *Survey of Progress in Chemistry,* (A. F. Scott, Ed.), **3**, Academic Press.

[8] H. Simon and D. Palm, (1966), Isotope Effects in Organic Chemistry and Biochemistry, *Angew. Chem. Int. Ed. Engl.,* **5**, 920.

[9] C. J. Collins and N. S. Bowman (Eds.), (1970), *Isotope Effects in Chemical Reactions,* Van Nostrand-Reinhold.

[10] V. Gold, (1970), Application of Isotope Effects, *Chem. in Britain,* **6**, 292.

[11] J. L. Hollenberg, (1970), Energy States of Molecules, *J. Chem. Educ.,* **47**, 2.

[12] J. H. Richards, (1970), Kinetic Isotope Effects in Enzymic Reactions in *The Enzymes,* ref. 40 of Chap. 2, 2.

[13] M. B. Neiman and D. Gal, (1971), *The Kinetic Isotope Method and Its Application,* Elsevier.

[14] A. R. Battersby, (1972), Some Applications of Tritium Labelling for the Exploration of Biochemical Mechanisms, *Accounts Chem. Res.,* **5**, 148.

[15] A. Fry, (1972), Isotope Effect Studies of Elimination Reactions, *Chem. Soc. Rev.,* **1**, 163.

[16] S. E. Scheppele, (1972), Kinetic Isotope Effects as Valid Measure of Structure-Reactivity Relationships. Isotope Effects and Non-classical Theory, *Chem. Rev.,* **72**, 511.

[17] M. Wolfsberg, (1972), Theoretical Evaluation of Experimentally Observed Isotope Effects, *Accounts Chem. Res.,* **5**, 225.

[18] R. P. Bell, (1973), *The Proton in Chemistry,* 2nd ed., chap. 12, Chapman and Hall.

[19] R. P. Bell, (1974), Recent Advances in the Study of Kinetic Isotope Effects, *Chem. Soc. Rev.,* **3**, 513.

[20] W. H. Saunders, (1974), Kinetic Isotope Effects in *Investigation of Rates and Mechanisms of Reactions,* (Ed. E. S. Lewis), (Techniques of Chemistry, Ed. A. Weissberger, **IV**, 3rd ed.), Wiley-Interscience.

[21] K. R. Hanson, (1975), Reactions at Prochiral Centers, *J. Biol. Chem.,* **250**, 8309.

[22] P. A. Roch (Ed.), (1976), *Isotopes and Chemical Principles,* ACS Symposium No 11, The Chemical Society.

Appendix 5

Key to Numbering and Classification of Enzymes

1. *Oxidoreductases*
    1.1 Acting on the CH–OH group of donors
        1.1.1 With NAD or NADP as acceptor
        1.1.2 With cytochrome as an acceptor
        1.1.3 With O_2 as acceptor
        1.1.99 With other acceptors

    1.2 Acting on the aldehyde or keto-group of donors
        1.2.1 With NAD or NADP as acceptor
        1.2.2 With a cytochrome as an acceptor
        1.2.3 With O_2 as acceptor
        1.2.4 With lipoate as acceptor
        1.2.99 With other acceptors

    1.3 Acting on the CH–CH group of donors
        1.3.1 With NAD or NADP as acceptor
        1.3.2 With a cytochrome as an acceptor
        1.3.3 With O_2 as acceptor
        1.3.99 With other acceptors

    1.4 Acting on the CH–NH_2 group of donors
        1.4.1 With NAD or NADP as acceptor
        1.4.3 With O_2 as acceptor

    1.5 Acting on the C–NH group of donors
        1.5.1 With NAD or NADP as acceptor
        1.5.3 With O_2 as acceptor

    1.6 Acting on reduced NAD or NADP as donor
        1.6.1 With NAD or NADP as acceptor
        1.6.2 With a cytochrome as an acceptor

REFERENCES

[1] T. E. Barman, (1969), *Enzyme Handbook,* (A catalogue of enzyme data), Springer-Verlag, I and II; (1974), Supplement 1.
[2] *Enzyme Nomenclature,* (1973), Recommendations (1972) of the Commission on Biochemical Nomenclature on the Nomenclature and Classification of Enzymes together with their Units and the Symbols of Enzyme Kinetics, Elsevier.
[3] Nomenclature of Multiple Forms of Enzyme, (1978), IUPAC-IUB Commission on Biochemical Nomenclature (CBN), Recommendations 1976, *Eur. J. Biochem.,* **82**, 1.

Appendix 6

Arrangements of Orders and Families in Spermatophyta and Alphabetical List of Orders and Families

ARRANGEMENTS OF ORDERS AND FAMILIES IN SPERMATOPHYTA*

I. Gymnospermae

Orders	Families		
1. Cycadales	1. Cycadaceae		
2. Ginkgoales	1. Ginkgoaceae		
3. Conferales	1. Pinaceae	3. Cupressaceae	5. Cephalotaxaceae
	2. Taxodiaceae	4. Podocarpaceae	6. Araucariaceae
4. Taxales	1. Taxaceae		
5. Gnetales	1. Gnetaceae	2. Ephedraceae	3. Welwitschiaceae

II. Angiospermae†
A. MONOCOTYLEDONEÆ

1. Pandanales	1. Typhaceae	2. Pandanaceae	3. Sparganiaceae
2. Helobiae	1. Potamogetonaceae	4. Scheuchzeriaceae	6. Butomaceae
	2. Najadaceae	5. Alismaceae	7. Hydrocharitaceae
	3. Aponogetonaceae		
3. Triuridales	1. Triuridaceae		
4. Glumiflorae	1. Gramineae	2. Cyperaceae	
5. Principes	1. Palmae		
6. Synanthae	1. Cyclanthaceae		
7. Spathiflorae	1. Araceae	2. Lemnaceae	
8. Farinosae	1. Flagellariaceae	6. Eriocaulaceae	10. Commelinaceae
	2. Restionaceae	7. Thurniaceae	11. Pontederiaceae
	3. Centrolepidaceae	8. Rapateaceae	12. Cyanastraceae
	4. Mayacaceae	9. Bromeliaceae	13. Philydraceae
	5. Xyridaceae		
9. Liliflorae	1. Juncaceae	4. Haemodoraceae	7. Taccaceae
	2. Stemonaceae	5. Amaryllidaceae	8. Dioscoreaceae
	3. Liliaceae	6. Velloziaceae	9. Iridaceae
10. Scitamineae	1. Musaceae	3. Cannaceae	4. Marantaceae
	2. Zingiberaceae		
11. Microspermae	1. Burmanniaceae	2. Orchidaceae	

*Reproduced with permission from T. Swain (Ed.), 'Chemical Plant Taxonomy' Academic Press, New York and London, 1963.

Orders	Families

B. DICOTYLEDONEÆ

I. *Archichlamydeae*

Orders	Families		
1. Verticillatae	1. Casuarinaceae		
2. Piperales	1. Saururaceae	3. Chloranthaceae	4. Lacistermaceae
	2. Piperaceae		
3. Salicales	1. Salicaceae		
4. Garryales	1. Garryaceae		
5. Myricales	1. Myricaceae		
6. Balanopsidales	1. Balanopsidaceae		
7. Leitneriales	1. Leitneriaceae		
8. Juglandales	1. Juglandaceae		
9. Batidales	1. Batidaceae		
10. Julianiales	1. Julianiaceae		
11. Fagales	1. Betulaceae	2. Fagaceae	
12. Urticales	1. Ulmaceae	2. Moraceae	3. Urticaceae
13. Proteales	1. Proteaceae		
14. Santalales	1. Myzodendraceae	4. Grubbiaceae	7. Loranthaceae
	2. Santalaceae	5. Olacaceae	8. Balanophoraceae
	3. Opiliaceae	6. Octoknemataceae	
15. Aristolochiales	1. Aristolochiaceae	2. Rafflesiaceae	3. Hydnoraceae
16. Polygonales	1. Polygonaceae		
17. Centrospermae	1. Chenopodiaceae	4. Cynocrambaceae	7. Portulacaceae
	2. Amarantaceae	5. Phytolaccaceae	8. Basellaceae
	3. Nyctaginaceae	6. Aizoaceae	9. Caryophyllaceae
18. Ranales	1. Nymphaeaceae	7. Berberidaceae	13. Eupomatiaceae
	2. Ceratophyllaceae	8. Menispermaceae	14. Myristicaceae
	3. Trochodendraceae	9. Magnoliaceae	15. Gomortegaceae
	4. Cercidiphyllaceae	10. Calycanthaceae	16. Monimiaceae
	5. Ranunculaceae	11. Lactoridaceae	17. Lauraceae
	6. Lardizabalaceae	12. Anonaceae	18. Hernandiaceae
19. Rhoeadales	1. Papaveraceae	3. Cruciferae	5. Resedaceae
	2. Capparidaceae	4. Tovariaceae	6. Moringaceae
20. Sarraceniales	1. Sarraceniaceae	2. Nepenthaceae	3. Droseraceae
21. Rosales	1. Podostemaceae	7. Pittosporaceae	13. Eucommiaceae
	2. Tristichaceae	8. Brunelliaceae	14. Platanaceae
	3. Hydrostachyaceae	9. Cunoniaceae	15. Crossosomataceae
	4. Crassulaceae	10. Myrothamnaceae	16. Rosaceae
	5. Cephalotaceae	11. Bruniaceae	17. Connaraceae
	6. Saxifragaceae	12. Hamamelidaceae	18. Leguminosae
22. Pandales	1. Pandaceae		
23. Geraniales	1. Geraniaceae	8. Cneoraceae	15. Vochysiaceae
	2. Oxalidaceae	9. Rutaceae	16. Tremandraceae
	3. Tropaeolaceae	10. Simarubaceae	17. Polygalaceae
	4. Linaceae	11. Burseraceae	18. Dichapetalaceae
	5. Humiriaceae	12. Meliaceae	19. Euphorbiaceae
	6. Erythroxylaceae	13. Malpighiaceae	20. Callitrichaceae
	7. Zygophyllaceae	14. Trigoniaceae	
24. Sapindales	1. Buxaceae	8. Corynocarpaceae	15. Icacinaceae
	2. Empetraceae	9. Aquifoliaceae	16. Aceraceae
	3. Coriariaceae	10. Celastraceae	17. Hippocastanaceae
	4. Limnanthaceae	11. Hippocrateaceae	18. Sapindaceae
	5. Anacardiaceae	12. Salvadoraceae	19. Sabiaceae
	6. Cyrillaceae	13. Stackhousiaceae	20. Melianthaceae
	7. Pentaphylacaceae	14. Staphyleaceae	21. Balsaminaceae

Orders	Families		
25. Rhamnales	1. Rhamnaceae	2. Vitaceae	
26. Malvales	1. Elaeocarpaceae	4. Tiliaceae	7. Sterculiaceae
	2. Chlaenaceae	5. Malvaceae	8. Scytopetalaceae
	3. Gonystilaceae	6. Bombacaceae	
27. Parietales	1. Dilleniaceae	11. Frankeniaceae	21. Turneraceae
	2. Eucryphiaceae	12. Tamaricaceae	22. Malesherbiaceae
	3. Ochnaceae	13. Fouquieriaceae	23. Passifloraceae
	4. Caryocaraceae	14. Cistaceae	24. Achariaceae
	5. Maregraviaceae	15. Bixaceae	25. Caricaceae
	6. Quiinaceae	16. Cochlospermaceae	26. Loasaceae
	7. Theaceae	17. Winteranaceae	27. Datiscaceae
	8. Guttiferae	18. Violaceae	28. Begoniaceae
	9. Dipterocarpaceae	19. Flacourtiaceae	29. Ancistrocladaceae
	10. Elatinaceae	20. Stachyuraceae	
28. Opuntiales	1. Cactaceae		
29. Myrtiflorae	1. Geissolomataceae	8. Punicaceae	14. Myrtaceae
	2. Penaeaceae	9. Lecythidaceae	15. Melastomaceae
	3. Oliniaceae	10. Rhizophoraceae	16. Onagraceae
	4. Thymelaeaceae	11. Nyssaceae	17. Haloragidaceae
	5. Elaeagnaceae	12. Alangiaceae	18. Hippuridaceae
	6. Lythraceae	13. Combretaceae	19. Cynomoriaceae
	7. Sonneratiaceae		
30. Umbelliflorae	1. Araliaceae	2. Umbelliferae	3. Cornaceae

II. *Sympetalae* (= *Metachlamydeae*)

Orders	Families		
1. Ericales	1. Clethraceae	3. Lennoaceae	5. Epacridaceae
	2. Pyrolaceae	4. Ericaceae	6. Diapensiaceae
2. Primulales	1. Theophrastaceae	2. Myrsinaceae	3. Primulaceae
3. Plumbaginales	1. Plumbaginaceae		
4. Ebenales	1. Sapotaceae	3. Symplocaceae	4. Styracaceae
	2. Ebenaceae		
5. Contortae	1. Oleaceae	3. Gentianaceae	5. Asclepiadaceae
	2. Loganiaceae	4. Apocynaceae	
6. Tubiflorae	1. Convolvulaceae	8. Solanaceae	15. Columelliaceae
	2. Polemoniaceae	9. Scrophulariaceae	16. Lentibulariaceae
	3. Hydrophyllaceae	10. Bignoniaceae	17. Globulariaceae
	4. Boraginaceae	11. Pedaliaceae	18. Acanthaceae
	5. Verbenaceae	12. Martyniaceae	19. Myoporaceae
	6. Labiatae	13. Orobanchaceae	20. Phrymaceae
	7. Nolanaceae	14. Gesneriaceae	
7. Plantaginales	1. Plantaginaceae		
8. Rubiales	1. Rubiaceae	3. Adoxaceae	5. Dipsacaceae
	2. Caprifoliaceae	4. Valerianaceae	
9. Cucurbitales	1. Cucurbitaceae		
10. Campanulatae	1. Campanulaceae	3. Brunoniaceae	5. Calyseraceae
	2. Goodeniaceae	4. Stylidiaceae	6. Compositae

ALPHABETICAL LIST OF ORDERS AND FAMILIES

I. Orders*

Aristolochiales	A 15	Juglandales	A 8	Ranales	A 18
Balanopsidales	A 6	Julianiales	A 10	Rhamnales	A 25
Batidales	A 9	Leitneriales	A 7	Rhoeadales	A 19
Campanulatae	S 10	Liliiflorae	M 9	Rosales	A 21
Centrospermae	A 17	Malvales	A 26	Rubiales	S 8
Coniferales	G 3	Microspermae	M 11	Salicales	A 3
Contortae	S 5	Myricales	A 5	Santalales	A 14
Cucurbitales	S 9	Myrtiflorae	A 29	Sapindales	A 24
Cycadales	G 1	Opuntiales	A 28	Sarraceniales	A 20
Ebenales	S 4	Pandales	A 22	Scitamineae	M 10
Ericales	S 1	Pandanales	M 1	Spathiflorae	M 7
Fagales	A 11	Parietales	A 27	Synanthae	M 6
Farinosae	M 8	Piperales	A 2	Taxales	G 6
Garryales	A 4	Plantaginales	S 7	Triuridales	M 3
Geraniales	A 23	Plumbaginales	S 3	Tubiflorae	S 6
Ginkgoales	G 2	Polygonales	A 16	Umbelliflorae	A 30
Glumiflorae	M 4	Primulales	S 2	Urticales	A 12
Gnetales	G 5	Principes	M 5	Verticillatae	A 1
Helobiae	M 2	Proteales	A 13		

II. Families†

A. GYMNOSPERMAE

Araucariaceae	3.6	Ephedraceae	5.2	Podocarpaceae	3.4
Cephalotaxaceae	3.5	Ginkgoaceae	2.1	Taxaceae	4.1
Cycadaceae	1.1	Gnetaceae	5.1	Taxodiaceae	3.2
Cupressaceae	3.3	Pinaceae	3.1	Welwitschiaceae	5.3

B. MONOCOTYLEDONEÆ

Alismataceae	2.5	Gramineae	4.1	Potamogetonaceae	2.1
Amaryllidaceae	9.5	Haemodoraceae	9.4	Rapateaceae	8.8
Aponogetonaceae	2.3	Hydrocharitaceae	2.7	Restionaceae	8.2
Araceae	7.1	Iridaceae	9.9	Scheuchzeriaceae	2.4
Bromeliaceae	8.9	Juncaceae	9.1	Sparganiaceae	1.3
Burmanniaceae	11.1	Lemnaceae	7.2	Stemonaceae	9.2
Butomaceae	2.6	Liliaceae	9.3	Taccaceae	9.7
Cannaceae	10.3	Marantaceae	10.4	Thurniaceae	8.7
Centrolepidaceae	8.3	Mayacaceae	8.4	Triuridaceae	3.1
Commelinaceae	8.10	Musaceae	10.1	Typhaceae	1.1
Cyanastraceae	8.12	Najadaceae	2.2	Velloziaceae	9.6
Cyelanthaceae	6.1	Orchidaceae	11.2	Xyridaceae	8.5
Cyperaceae	4.2	Palmae	5.1	Zingiberaceae	10.2
Dioscoreaceae	9.8	Pandanaceae	1.2		
Eriocaulaceae	8.6	Philydraceae	8.13		
Flagellariaceae	8.1	Pontederiaceae	8.11		

C. DICOTYLEDONEÆ*

Acanthaceae	S 6 18	Actinidiaceae	A 27 –	Akaniaceae	A 24 –
Aceraceae	A 24 16	Adoxaceae	S 8 3	Alangiaceae	A 29 12
Achariaceae	A 27 24	Aizoaceae	A 17 6	Amarantaceae	A 17 2

Family				Family				Family			
Anacardiaceae	A	24	5	Crossosomataceae	A	21	15	Lacistemaceae	A	2	4
Ancistrocladaceae	A	27	29	Cruciferae	A	19	3	Lactoridaceae	A	18	11
Anonaceae	A	18	12	Cucurbitaceae	S	9	1	Lardizabalaceae	A	18	6
Apocynaceae	S	5	4	Cunoniaceae	A	21	9	Lauraceae	A	18	17
Aquifoliaceae	A	24	9	Cynocrambaceae	A	17	4	Lecythidaceae	A	29	9
Araliaceae	A	30	1	Cynomoriaceae	A	29	19	Leguminosae	A	21	18
Aristolochiaceae	A	15	1	Cyrillaceae	A	24	6	Leitneriaceae	A	7	1
Asclepiadaceae	S	5	5	Datiscaceae	A	27	27	Lennoaceae	S	1	3
Balanophoraceae	A	14	8	Diapensiaceae	S	1	6	Lentibulariaceae	S	6	16
Balanopsidaceae	A	6	1	Dichapetalaceae	A	23	18	Limnanthaceae	A	24	4
Balsaminaceae	A	24	21	Dilleniaceae	A	27	1	Linaceae	A	23	4
Basellaceae	A	17	8	Dipsacaceae	S	8	5	Loasaceae	A	27	26
Batidaceae	A	9	1	Dipterocarpaceae	A	27	9	Loganiaceae	S	5	2
Begoniaceae	A	27	28	Droseraceae	A	20	3	Loranthaceae	A	14	7
Berberidaceae	A	18	7	Ebenaceae	S	4	2	Lythraceae	A	29	6
Betulaceae	A	11	1	Elaeagnaceae	A	29	5	Magnoliaceae	A	18	9
Bignoniaceae	S	6	10	Elaeocarpaceae	A	26	1	Malesherbiaceae	A	27	22
Bixaceae	A	27	15	Elatinaceae	A	27	10	Malpighiaceae	A	23	13
Bombacaceae	A	26	6	Empetraceae	A	24	2	Malvaceae	A	26	5
Boraginaceae	S	6	4	Epacridaceae	S	1	5	Marcgraviaceae	A	27	5
Brunelliaceae	A	21	8	Ericaceae	S	1	4	Martyniaceae	S	6	12
Bruniaceae	A	21	11	Erythroxylaceae	A	23	6	Melastomataceae	A	29	15
Brunoniaceae	S	10	3	Eucommiaceae	A	21	13	Meliaceae	A	23	12
Burseraceae	A	23	11	Eucryphiaceae	A	27	2	Melianthaceae	A	24	20
Buxaceae	A	24	1	Euphorbiaceae	A	23	19	Menispermaceae	A	18	8
Cactaceae	A	28	1	Eupomatiaceae	A	18	13	Monimiaceae	A	18	16
Callitrichaceae	A	23	20	Fagaceae	A	11	2	Moraceae	A	12	2
Calycanthaceae	A	18	10	Flacourtiaceae	A	27	19	Moringaceae	A	19	6
Calyceraceae	S	10	5	Fouquieriaceae	A	27	13	Myoporaceae	S	6	19
Campanulaceae	S	10	1	Frankeniaceae	A	27	11	Myricaceae	A	5	1
Capparidaceae	A	19	2	Garryaceae	A	4	1	Myristicaceae	A	18	14
Caprifoliaceae	S	8	2	Geissolomataceae	A	29	1	Myrothamnaceae	A	21	10
Caricaceae	A	27	25	Gentianaceae	S	5	3	Myrsinaceae	S	2	2
Caryocaraceae	A	27	4	Geraniaceae	A	23	1	Myrtaceae	A	29	14
Caryophyllaceae	A	17	9	Gesneriaceae	S	6	14	Myzodendraceae	A	14	1
Casuarinaceae	A	1	1	Globulariaceae	S	6	17	Nepenthaceae	A	20	2
Celastraceae	A	24	10	Gomortegaceae	A	18	15	Nolanaceae	S	6	7
Cephalotaceae	A	21	5	Gonystilaceae	A	26	3	Nyctaginaceae	A	17	3
Ceratophyllaceae	A	18	2	Goodeniaceae	S	10	2	Nymphaeaceae	A	18	1
Cercidiphyllaceae	A	18	4	Grubbiaceae	A	14	4	Nyssaceae	A	29	11
Chenopodiaceae	A	17	1	Guttiferae	A	27	8	Ochnaceae	A	27	3
Chalaenaceae	A	26	2	Haloragiaceae	A	29	17	Octoknemataceae	A	14	6
Chloranthanceae	A	2	3	Hamamelidaceae	A	21	12	Olacaceae	A	14	5
Cistaceae	A	27	14	Hernandiaceae	A	18	18	Oleaceae	S	5	1
Clethraceae	S	1	1	Himantandraceae†	A	18	–	Oliniaceae	A	29	3
Cneoraceae	A	23	8	Hippocastaneaceae	A	24	17	Onagraceae	A	29	16
Cochlospermaceae	A	27	16	Hippocrateaceae	A	24	11	Opiliaceae	A	14	3
Collumelliaceae	S	6	15	Hippuridaceae	A	29	18	Orobanchaceae	S	6	13
Combretaceae	A	29	13	Humiriaceae	A	23	5	Oxalidaceae	A	23	2
Compositae	S	10	6	Hydnoraceae	A	15	3	Pandaceae	A	22	1
Connaraceae	A	21	17	Hydrophyllaceae	S	6	3	Papaveraceae	A	19	1
Convolvulaceae	S	6	1	Hydrostachyaceae	A	21	3	Passifloraceae	A	27	23
Coriariaceae	A	24	3	Icacinaceae	A	24	15	Pedaliaceae	S	6	11
Cornaceae	A	30	3	Juglandaceae	A	8	1	Penaeaceae	A	29	2
Corynocarpaceae	A	24	8	Julianiaceae	A	10	1	Pentaphylacaceae	A	24	7
Crassulaceae	A	21	4	Labiatae	S	6	6	Phrymaceae	S	6	20

Family				Family				Family			
Phytolaccaceae	A	17	5	Rutaceae	A	23	9	Tamaricaceae	A	27	12
Piperaceae	A	2	2	Sabiaceae	A	24	19	Theaceae	A	27	7
Pittosporaceae	A	21	7	Salicaceae	A	3	1	Theophrastaceae	S	2	1
Plantaginaceae	S	7	1	Salvadoraceae	A	24	12	Thymelaeaceae	A	29	4
Platanaceae	A	21	14	Santalaceae	A	14	2	Tiliaceae	A	26	4
Plumbaginaceae	S	3	1	Sapindaceae	A	24	18	Tovariaceae	A	19	4
Podostemaceae	A	21	1	Sapotaceae	S	4	1	Tremandraceae	A	23	16
Polemoniacea	S	6	2	Sarraceniaceae	A	20	1	Trigoniaceae	A	23	14
Polygalaceae	A	23	17	Saururaceae	A	2	1	Tristichaceae	A	21	2
Polygonaceae	A	16	1	Saxifragaceae	A	21	6	Trochodendraceae	A	18	3
Portulacaceae	A	17	7	Scrophulariaceae	S	6	9	Tropaeolaceae	A	23	3
Primulaceae	S	2	3	Scytoperalaceae	A	26	8	Turneraceae	A	27	21
Proteaceae	A	13	1	Simarubaceae	A	23	10	Ulmaceae	A	12	1
Punicaceae	A	29	8	Solanaceae	S	6	8	Umbelliferae	A	30	2
Pyrolaceae	S	1	2	Sonneratiaceae	A	29	7	Urticaceae	A	12	3
Quiinaceae	A	27	6	Stachyuraceae	A	27	20	Valerianaceae	S	8	4
Rafflesiaceae	A	15	2	Staphyleaceae	A	24	14	Verbenaceae	S	6	5
Ranunculaceae	A	18	5	Sterculiaceae	A	26	7	Violaceae	A	27	18
Resedaceae	A	19	5	Stylidiaceae	S	10	4	Vitaceae	A	25	2
Rhamnaceae	A	25	1	Styracaceae	S	4	4	Vochysiaceae	A	23	15
Rhizophoraceae	A	29	10	Symplocaceae	S	4	3	Winteranaceae	A	27	17
Rosaceae	A	21	16	Stackhousiaceae	A	24	13	Zygophyllaceae	A	23	7
Rubiaceae	S	8	1								

*A = Archichlamydae (Dicotyledoneae). G = Gymnospermae. M = Monocotyledoneae. S = Sympetalae (Dicotyledonæ). Figures refer to order number (see Key).

† Figures refer to order and family number (see Key).

Key—Orders: Spathiflorae M 7 ≡ Monocotyledoneae, order 7. Families: Ochnaceae A 27 3 ≡ Archichlamydaceae (Dicotyledoneae), order 27 (Parietales), family 3.

Index